Defect Recognition and Image Processing in Semiconductors 1995

Other titles in the Series

The Institute of Physics Conference Series regularly features papers presented at important conferences and symposia highlighting new developments in physics and related fields. Recent titles include:

141: Compound Semiconductors 1994
Papers presented at the 21st International Symposium, San Diego, CA, USA
Edited by H Goronkin and U Mishra

144: Narrow Gap Semiconductors 1995
Papers presented at the 7th International Conference, Santa Fe, NM, USA
Edited by J L Reno

145: Compound Semiconductors 1995
Papers presented at the 22nd International Symposium, Cheju Island, Korea
Edited by J-C Woo and Y S Park

146: Microscopy of Semiconducting Materials 1995
Papers presented at the Institute of Physics Conference, Oxford, UK
Edited by A G Cullis and A E Staton-Bevan

Defect Recognition and Image Processing in Semiconductors 1995

Proceedings of the Sixth International Conference held in Boulder, Colorado, 3–6 December 1995

Edited by A R Mickelson

Institute of Physics Conference Series Number 149
Institute of Physics Publishing, Bristol and Philadelphia

CODEN IPHSAC 149 1–370 (1996)

British Library Cataloguing in Publication Data

A catalogue record for this book is available from the British Library.

ISBN 0 7503 0372 7

Library of Congress Cataloging-in-Publication Data are available

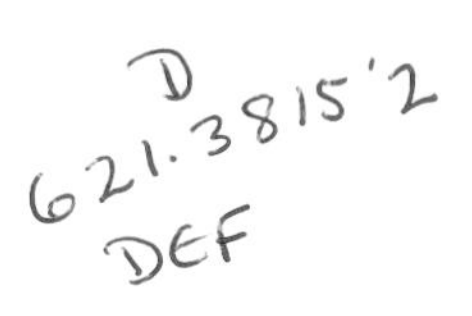

Published by Institute of Physics Publishing, wholly owned by The Institute of Physics, London
Institute of Physics Publishing, Techno House, Redcliffe Way, Bristol BS1 6NX, UK
US Editorial Office: Institute of Physics Publishing, The Public Ledger Building, Suite 1035, 150 South Independence Mall West, Philadelphia, PA 19106, USA

Printed in the UK by Galliard (Printers) Ltd, Great Yarmouth, Norfolk

Contents

Poster section

Preface

The most recent Defect Recognition and Image Processing (DRIP) Conference, held in Allenspark, Colorado from 3 to 6 December 1995, was the sixth in an international series that had its inception in Montpellier, France in 1985. The conference has been held in Monterrey, California, USA in 1987, Tokyo, Japan in 1989, Wilmslow, UK in 1991 and Santander, Spain in 1993. The DRIP Conference series has been one that has emphasized measurement techniques for defect recognition and processing of the data taken using these techniques. This distinguishes DRIP from a number of conference series that place primary emphasis on the study of specific material systems. Although I am personally somewhat new to the series, having first attended the 1993 Santander conference, my understanding is that many of the techniques of defect recognition presently in use were first described during the DRIP series.

This DRIP conference consisted of 45 orally presented papers, of which 13 were invited, and 23 poster papers. Of these 68 works, 61 appear in this proceedings. As usual at DRIP, these papers covered a large number of different defect recognition techniques. Using rather loose definitions of techniques, one could say that 9 papers treated aspects of electron microscopy including cathodoluminescence, 8 papers dealt with laser scanning tomography, 7 papers with photoluminescent microscopy, 3 with surface photovoltage mapping, 3 with deep level transient spectroscopy, 2 with atomic force microscopy and 1 each with transmission topography, surface imaging, photoemission microscopy, mobile current sensing, electrooptic sampling and positron annihilation mapping. A number of other papers used multiple techniques to get different pictures of defect structures, with 6 of these papers treating the GaAs material system, 6 silicon and 1 each lithium niobate, ZnCdSe, GaInP, InGaAs and TiN. Although there was some review material in the invited talks, the vast majority of material presented was new, representing recent research results.

The conference began with an invited session of five talks which contained an element of overview together with new results. Reviews included talks on photoluminescence, electron microscopy and laser scanning tomography. Of the two other papers, one discussed the defects caused by dry etching and the other introduced defect studies now being carried out in the ferroelectric crystal system of $LiNbO_3$, which is widely used as an electrooptic material in integrated optics. The second session was one that included eight talks which had as a common thread the gallium arsenide material system. The next session was the poster session, whose various talks spanned the topics of the conference. There was then a session that had electron beam techniques as the common thread, a session which dealt with defect studies of optical materials, a session on defect characterization, and then a session which covered a variety of techniques, whose papers dealt primarily with silicon, and finally a session on laser scanning tomography.

DRIP is now ten years old but it is still as viable as at its inception due to its ever-changing themes and participants. Each new DRIP brings a new set of topics and attracts a new set of scientists to supplement the continuing topics and ongoing participants. We expect this process to continue in the forthcoming DRIPs VII and VIII, which will take us up to the millenium.

I wish to thank all the people who participated at the conference as delegates, speakers, invited speakers, chairpeople and committee members. Special thanks must go to my administrative assistant, Julie Fredlund, who made all this possible. Without Julie, there would have been no DRIP VI. I should also like to thank Karen MacKenzie for her help at the conference and the conference personnel at Aspen Lodge. Special mention should also go to Juan Jiménez for imparting to me his wisdom gained from chairing DRIP V, wisdom which served as a guide for chairing DRIP VI.

A R Mickelson

Inst. Phys. Conf. Ser. No 149
Paper presented at DRIP 95

Electron microscopy techniques for evaluating epitaxial and bulk III-V compound semiconductors

C Frigeri

CNR-MASPEC Institute, via Chiavari 18/A, 43100 Parma, Italy

Abstract. Electron microscopy is an important technique to study interfaces and microdefects in advanced III-V compound semiconductors. The paper briefly reviews some of the TEM methods used to this purpose and shows examples of their application to the characterization of epitaxial structures such as InGaAs/GaAs and GaAs/Ge as well as processed substrates like implanted InP.

1. Introduction

Among the techniques for defect recognition the electron microscopy ones (TEM, SEM) are characterized by having a very high spatial resolution. In particular, by TEM such resolution is of some nanometers in the conventional two-beam diffraction contrast mode that extends down to the atomic level in the case of High Resolution TEM (HREM) by which it is possible to determine the atomic structure of any interface and crystal defect. HREM is also a chemically sensitive method because of the peculiar dependence of the intensity of some reflections on the scattering factor of the chemical elements (Ourmazd et al 1990). The (200) dark field (DF) mode also exploits the dependence of the (200) reflection intensity on the atomic scattering factor. The (200) DF mode has a spatial resolution of 0.5 nm (Cerva 1991) and an accuracy of 3% for AlGaAs (Bithell and Stobbs 1989). By these methods chemical mappings with subnanometer resolution, e. g. of layer composition in layered epistructures, can be obtained by using cross-sectional specimens. TEM is thus an indispensable tool for characterizing interface faults and microdefects in as-grown and processed epitaxial and bulk III-V semiconductors. After a brief survey of the (200) DF and HREM methods, the paper reports on the applications of TEM to the study of defects in three III-V semiconductor systems, namely interface faults due to intrinsic material properties in InGaAs/GaAs layers, crystal defects due to growth conditions in GaAs/Ge heterostructures and due to processing in implanted bulk InP.

2. TEM Methods

2.1. (200) dark field method

In the III-V compound semiconductors with the zincblende structure the reflections in the TEM of the type $h + k + l = 4n + 2$, $n = 0, 1, ..$, such as (200) and (420), are sensitive to the difference of the atomic scattering factors of the elements constituting the semiconductor and very useful to extract compositional information. The (200) reflection is mostly used. Its structure factor F_{200} in a III-V compound is given by (Petroff et al 1977)

$$F_{200} = 4\,(f_{III} - f_{V})$$

where f is the atomic scattering factor of either group III or V element. f is proportional to the atomic number of the element considered (Hirsch at al 1977).

For a ternary compound like $In_xGa_{1-x}As$ of composition x, F_{200} is then

$$F_{200} = 4 [x f_{In} + (1 - x) f_{Ga} - f_{As}] \approx 4 x [f_{In} - f_{Ga}]$$

since f_{Ga} and f_{As} are nearly equal. In the kinematical approximation (Petroff 1977, Cerva 1991, Kightley et al 1991) the diffracted intensity is then

$$I_{200} (In_xGa_{1-x}As) = K | F_{200} |^2 = 16 K x^2 (f_{In} - f_{Ga})^2 \qquad (1)$$

with K a constant that also includes specimen thickness, whereas for GaAs

$$I_{200} (GaAs) = K | F_{200} |^2 = 16 K (f_{Ga} - f_{As})^2 \qquad (2)$$

which is very small. For $Al_xGa_{1-x}As$ the (200) intensity is still given by eq. (1) with f_{In} replaced by f_{Al}. The (200) intensity of a ternary compound is thus much different from that of GaAs and also depends on the composition x through eq. (1). For the determination of the absolute value of x the ratio of the intensity of the ternary (eq. 1) to that of GaAs (eq. 2) is used in order to eliminate the dependence on K (Cerva 1991).

In the (200) dark field method only the (200) diffracted beam is made to pass through the objective aperture of the TEM and contribute to the image. On a positive print AlGaAs always appears brighter than GaAs for all Al compositions. The behaviour of InGaAs is more complex. An InGaAs layer looks brighter than GaAs for $x > 0.45$ and darker for $x < 0.45$ (Cerva 1991). Moreover, for $x < 0.45$, the (200) DF intensity decreases for $0 < x < 0.22$ and increases for $0.22 < x < 0.45$, having a minimum at $x = 0.22$ for which InGaAs appears the darkest. In our experiments the InGaAs layers with $x = 0.22$ exhibited a darker contrast than those with $x < 0.22$, thus confirming the calculations by Cerva (1991). The $In_xGa_{1-x}As$ layers discussed in this paper all had $x \leq 0.22$.

2.2. HREM method

A HREM image from <110> cross sectional specimens is obtained by letting the six reflections of the lowest order (two (200) beams and four (111) beams) pass through the objective aperture along with the transmitted beam. The interference between these seven beams gives the atomic lattice image of the specimen. HREM images taken in this way contain both a purely structural information on the positions of the atoms, because the intensity of the (111) reflections is only weakly dependent on the atomic scattering factor differences, as well as a compositional information through the chemically sensitive reflections of the {200} type, as explained in § 2.1.

In order to semi-quantitatively evaluate the interface roughness and compositional homogeneity of epitaxial III-V layers by the (200) DF and HREM methods the experimental TEM cross sectional images were digitized in a computer by means of a scanner. The digitized micrographs were then scanned by using EMMPDL (Electron Microscopy and Microanalysis Public Domain Library) programs, in particular the fftImage 1.25 package, to obtain the (200) DF or HREM intensity profile across the layers. A scanning rectangular box about 20 nm large in the <011> direction parallel to the interfaces and perpendicular to the growth direction was used so as to have intensity integration over a large area of the interfaces.

A detailed presentation of the various procedures to analyse crystal defects by the conventional two-beam diffraction contrast method is given in Hirsch et al (1977).

3. Results and Discussion

3.1. InGAs/GaAs system

InGaAs-based single and multi-quantum well structures are currently used to fabricate lasers and HEMTs. Either GaAs or AlGaAs can be used as confinement barrier to the InGaAs well.

The overall quality of the InGaAs layers in such structures is not as good as expected. In MBE-grown samples, the major reason for this has been considered to be indium segregation in the growth direction (Moison et al 1989, Nagle et al 1993). Little has been done by TEM to study such phenomenon in MOVPE-grown layers (Frigeri et al 1994, Höpner et al 1995).

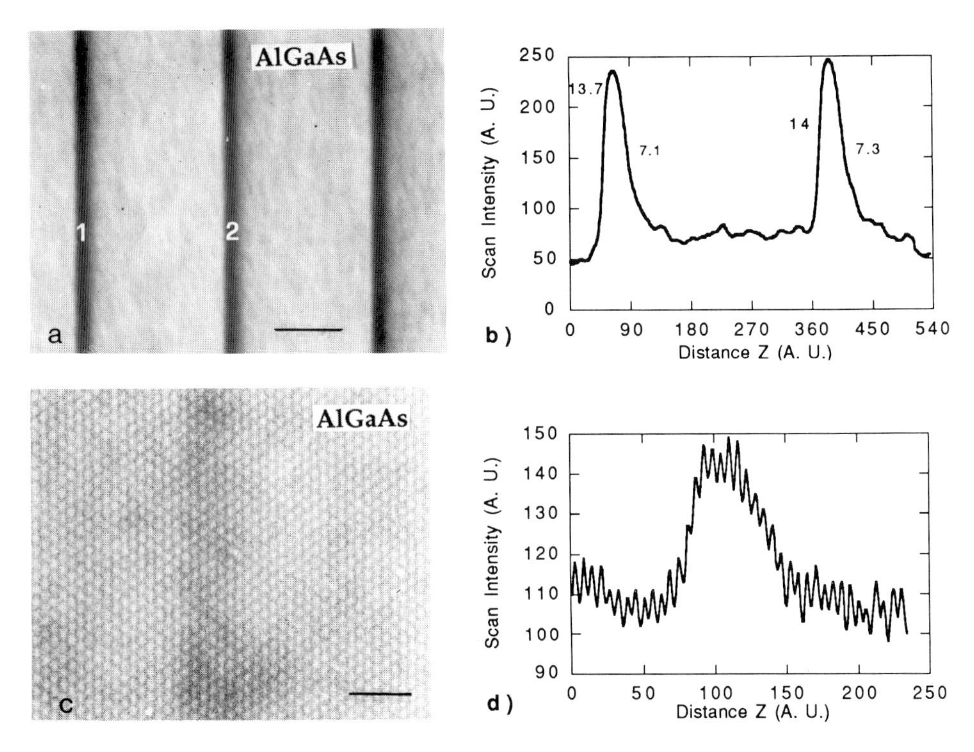

Fig. 1- a) (200) DF TEM image of a section of an InGaAs/AlGaAs MQW (bar = 10 nm) and b) intensity profile, with inverted contrast, across the InGaAs QWs 1 and 2 in a). c) HREM image and d) its intensity profile, with inverted contrast, of a well in the InGaAs/ AlGaAs MQW of a), bar = 2 nm. Growth direction z points to the right (sample top).

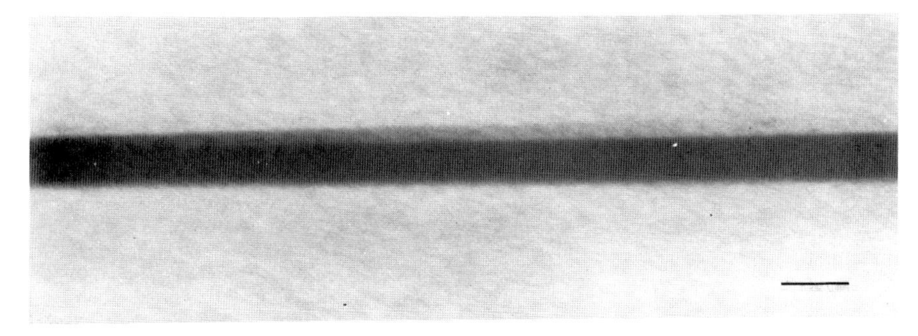

Fig. 2 - (200) DF image of an $In_{0.15}Ga_{0.85}As$ SQW grown on a vicinal substrate. Growth direction upwards. Bar = 20 nm.

Fig. 1 a) is the (200) DF TEM image of a part of a MQW grown by MOVPE and made up of ten couples of $In_{0.17}Ga_{0.83}As/Al_{0.25}Ga_{0.75}As$ layers. The InGaAs layers are dark in the

DF image. Fig. 1 a) shows that the AlGaAs-on-InGaAs interface is rougher than the InGaAs-on-AlGaAs interface. This is confirmed by the asymmetry of the (200) DF intensity profiles taken across the layers that are shown in fig. 1 b) along with the measured slopes, in arbitrary units, at each interface. The slopes are a direct indication of the concentration gradient across the interfaces, and therefore of their sharpness, as from eq. (1) it is

$$\partial I_{200} / \partial z = 32\, K\, x\, (f_{In} - f_{Ga})^2\, \partial x / \partial z$$

where z is the growth direction. The slope at the AlGaAs-on-InGaAs interface is smaller than that at the InGaAs-on-AlGaAs one indicating that the former interface is less sharp than the latter one. Similar results are obtained by HREM images of the same sample (fig. 1 c). In the intensity profile in fig. 1 d) each peak corresponds to a row of atoms perpendicular to the growth direction. By taking the interfacial width as the distance across the interface for which 20%-80% of the total contrast change between the layers occurs, from fig. 1 d) it is seen that the AlGaAs-on-InGaAs interface roughness extends over ~ 3.2 MLs corresponding to a concentration gradient of ~0.19 % nm^{-1}. Such roughness at the top interface has to be ascribed to In segregation in the growth direction and can be a measure of it. It has been shown elsewhere by TEM (Frigeri et al 1994) that In segregation into the overgrowth barrier also takes place when the barrier is GaAs, in agreement with previous results by other techniques (Moison et al 1989, Nagle et al 1993). (200) DF has shown that In segregation is more effective when the barrier is AlGaAs rather than GaAs (Frigeri et al 1994) very likely because of the greater exchange energy of the In-Al couple with respect to the In-Ga couple (Gerard 1993).

When InGaAs layers are MOVPE-grown on vicinal GaAs substrates, the main fault of the top interface of the InGaAs layer(s) is no longer the roughness associated with In segregation but rather the presence of macrosteps on the top of the InGaAs layer. An example is shown in the (200) DF image of fig. 2 that refers to a nominally 14 nm thick $In_{0.15}Ga_{0.85}As$ SQW grown on a GaAs substrate misoriented 2° off (100) towards [110]. The cap and buffer layer is GaAs. The macrosteps are characterized, and detected, by the fact that the (200) DF intensity decreases laterally along them in a nearly monotonic way passing from the value typical of the InGaAs layer underneath to that typical of the GaAs barrier/cap layer (fig. 2). This would indicate that there is a decrease of In content along the macrosteps which produces a strong compositional inhomogeneity inside the SQW layers that is expected to negatively affects their optical and electrical properties. A possible reason for the lateral compositional inhomogeneity can be sought for in the dissimilar tendency of the group III elements to attach to step sites during layer growth (Harimoto et al 1994). The formation of macrosteps for growth on vicinal substrates is due to bunching of the atomic steps associated with the 2° offcut of the substrate, whereby steps wider and higher than the atomic steps can form (Cox et al 1990).

The macrosteps have a lateral extension of ~ 180 nm and are as high as 4-8 nm (fig. 2) (Frigeri et al 1995 a). Due to this large lateral extension the macrosteps cannot be analysed as a whole by means of HREM due to the limited field of view of HREM. Because of this and due to its high sensitivity to compositional changes (200) DF TEM is the most suitable method to study such macrosteps.

It has been seen that growth on vicinal substrates also increases the density of misfit dislocations, i. e. decreases the critical thickness for their generation (Frigeri et al 1995 a).

3.2. GaAs/Ge system

The presence and type of defects in heteroepitaxies can also be due to growth conditions. This can be monitored by TEM analysis and an example is shown for the case of GaAs/Ge heterostructures MOVPE-grown under different growth conditions by changing the V/III ratio from 1.3 to 13.3. Fig. 3 a) shows that for Ga-rich flows (V/III ≤ 3.3) the defects detected by TEM are typical misfit dislocations (MDs) due to plastic relaxation of the lattice mismatched layer as its thickness has exceeded the critical thickness ($h_c = 0.28\ \mu m$) for generation of MDs (Matthews 1975). In fig. 3 a) the MDs are not at 90° to each other because the substrate is miscut by an angle of 6° off (100). On the other hand, for As-rich flows no MD was detected but only stacking faults even in layers as thick as $\sim 5 \cdot h_c$ suggesting that the use of high growth rates inhibits the formation of misfit dislocations. This can be explained by considering that the concept of critical thickness h_c and the mechanism of misfit strain relief through the formation of a grid of <110> MDs when h_c is exceeded only applies for a 2D layer-by-layer growth (Matthews 1975). When high growth rates are used the layers very likely grow by a 3D island mechanism (Ernst and Pirouz 1989) rather than by a 2D mechanism so that the model of plastic relaxation of the misfit strain when h_c is exceeded does not apply (Matthews 1975). In such a case stacking faults are generated due to stacking errors at the {111} island facets or due to island coalescence. Growth islands have indeed been observed in the samples discussed here (Frigeri et al 1993). Once the stacking faults have been generated at growth islands, the misfit strain accumulated in the layer can partially be relieved by the edge components of the partial dislocations bordering the stacking faults, although in a less efficient way than the 60° misfit dislocations, and this can contribute to increase the value of the layer thickness at which eventually misfit dislocations can be generated (Frigeri et al 1995 b).

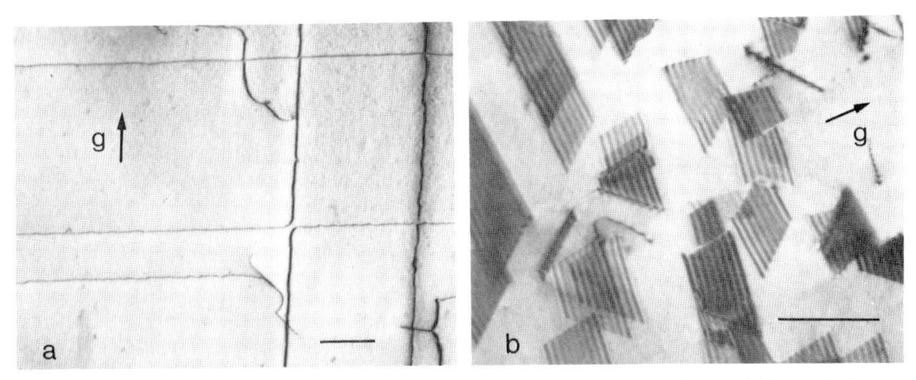

Fig. 3 - Two-beam diffraction contrast TEM image in plan view of GaAs/Ge layers grown under a) Ga-rich, b) As-rich conditions. **g** = $[\bar{2}\bar{2}0]$. Bars = 0.5 μm.

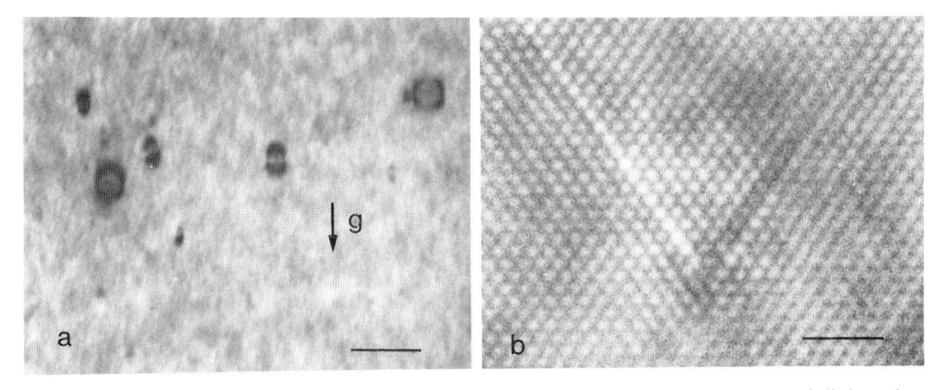

Fig. 4 - Fe-implanted InP. a) Two-beam diffraction contrast TEM image of dislocation loops. **g** = [220], bar = 0.1 μm. b) Cross-sectional HREM image of a stacking fault tetrahedron. Bar = 2 nm.

3.3. Implanted InP

The process of ion implantation and subsequent annealing also introduces defects that are typically studied by TEM. Much work is carried out nowadays on Fe-implanted InP to create a semi-insulating (buried) layer. Even if the desired electrical isolation of the implanted layers is achieved, very often the annealing treatment does not succeed in removing all crystal defects associated with the implant and annealing processes. The elimination of such defects is desirable as they are generally localized inside the implanted layer or at the implanted layer/substrate interface where they can cause device failure. Fig. 4 shows the defects still present in 200 KeV Fe-implanted InP after annealing at 650 °C for 2 hours (Frigeri et al 1995 c). The as-implanted layer was amorphous and its recovery upon annealing has produced two types of defects, both located at the original amorphous/crystalline interface, namely extrinsic dislocation loops (fig. 4 a) and intrinsic stacking fault tetrahedra (fig. 4 b). The latter ones are due to solid phase epitaxial regrowth and form when an amorphous region recrystallizes in contact with a crystalline one (De Cooman et al 1987). Their intrinsic nature has been determined by the shift of the (200) lattice plane fringes away from the tip of the defect (De Cooman et al 1987, Frigeri et al 1995 c) (fig. 4 b). The dislocation loops are end-of-range defects as they form at the original a/c interface (Jones et al 1988). They are of the 1/3<111> Frank type and very likely due to condensation of non-equilibrium interstitial P and In recoil atoms created by the implantation process. Gettering of the implanted species at such end-of-range loops is also possible (Jones et al 1988, Sadana et al 1982). It has been observed that, by increasing the annealing time from 1.5 to 2 hours, loop coarsening occurs. This indicates that further annealing in necessary for loop disappearance as this is achieved by loop shrinkage after an initial stage of loop coarsening (Meekison 1994).

Acknowledgments

The author is indebted for provision of samples to Drs A Di Paola, D M Ricthie and F Vidimari (MOVPE InGaAs), G Attolini and C Pelosi (MOVPE GaAs/Ge), A Carnera and A Gasparotto (implanted InP) and to Mr F Longo for skilful technical assistance.

References

Bithell E G and Stobbs W M 1989 *Phil. Mag. A* **60** 39
Cerva H 1991 *Appl. Surf. Sci.* **50** 19
Cox H M, Aspnes D E, Allen S J, Bastos P, Hwang D M, Mahajan S, Shahid M A and Morais P C 1990 *Appl. Phys. Lett.* **57** 611
De Cooman B C, McKernan S, Carter C B, Ralston J R, Wicks G W, Eastman L F 1987 *Phil. Mag. Lett.* **56** 85
Ernst F and Pirouz P 1988 *J. Appl. Phys.* **64** 4526
Frigeri C, Attolini G, Pelosi C and Longo F 1993 *Inst. Phys. Conf. Ser.* **135** 343
Frigeri C, Di Paola A, Gambacorti N, Richtie D M, Longo F and Della Giovanna M 1994 *Mater. Sci. Eng.* B **28** 346
Frigeri C, Di Paola A, Richtie D M, Longo F, Brinciotti A, Riva M and Vidimari F 1995 a *Proc. 9th Int. Conf. on Microscopy Semiconducting Materials,* (Bristol: Institute of Physics) in press
Frigeri C, Attolini G, Pelosi C and Longo F 1995 b *Mater. Sci. Technol.* submitted
Frigeri C, Carnera A and Gasparotto A 1995 c *Appl. Phys. A* **61** in press
Gerard J M 1993 *J. Crystal Growth* **127** 981
Hiramoto K, Tsuchiya T, Sagawa M, and Uomi K 1994 *J. Crystal Growth* **145** 133
Hirsch P, Howie A, Nicholson R B, Pashley D W and Whelan M J 1977 *Electron Microscopy of Thin Crystals* (Malabar: Krieger)
Höpner A, Seitz H, Rechenberg I, Bugge F, Procop M, Scheerschmidt K and Queisser H J 1995 *Phys. Stat. Sol (a)* **150** 427
Kightley F, Kiely C J and Goodhew P 1991 *Inst. Phys. Conf. Ser.* **117** 595
Jones K S, Prussin S and Weber E R 1988 *Appl. Phys. A* **45** 1
Nagle J, Landesman J P, Larine M, Mottet C and Bois P 1993 *J. Crystal Growth* **127** 550
Matthews J W 1975 in *Epitaxial Growth, pt. B* Matthews J W ed (New York: Academic) 560
Meekison C D 1994 *Phil. Mag. A* **69** 379
Moison J M, Guille C, Houzay F, Barthe F and Van Rompay M 1989 *Phys. Rev. B* **40** 6149
Ourmarzd A, Baumann F H, Bode M and Kim Y 1990 *Ultramicroscopy* **34** 255
Petroff P M 1977 *J. Vac. Sci. Technol.* **14** 973
Sadana D K, Washburn J and Booker G R 1982 *Phil. Mag. B* **46** 611

Inst. Phys. Conf. Ser. No 149
Paper presented at DRIP 95

Defect Studies in $LiNbO_3$

Alan R Mickelson§

Guided Wave Optics Laboratory, Department of Electrical & Computer Engineering, University of Colorado, Boulder, CO 80309-0425, USA

Abstract. A program has been initiated here at the University of Colorado's Guided Wave Optics Laboratory (GWOL) to study defects in lithium niobate ($LiNbO_3$) using a variaty of techniques. Lithium niobate is the material of choice for practical integrated optics and is growing in its economic importance. Although it is unfortunately too early to present any major results of measurements from this program, we herein will discuss some previous results of material measurements in GaAs as well as device measurements in $LiNbO_3$ to try to indicate the motivation for the research decisions we have made.

1. Introduction

Lithium niobate remains the material of choice for electrooptic devices such as GHz modulators and switches. Despite the fact that the material has been in laboratory use in devices since the late 1960s, little is known of the defect structures in this material. This is in great part due to the fact that $LiNbO_3$ modulators have in no way found the commercial application of silicon circuits, gallium arsenide lasers, or quaternary laser materials. However, lithium niobate devices are being used in progressively more applications. A barrier to more widespread application at present is cost, which is in great part not material cost but processing cost. Were this cost to drop significantly, one would expect the use of lithium devices to appear in many more applications.

What we will present here will be more a case for why we are launching a program of defect studies in lithium niobate than specific results, as we are just beginning to obtain specific results. We have just begun to analyze optical images of defects in lithium niobate, despite the fact that we actually have been using laser scanning tomography (LST)-like techniques for some time to measure loss in channel waveguide devices. We will also briefly discuss an earlier defect study in GaAs and why we think the results of this study may be applicable to the explanation of such effects as the magnitude of the loss and channel crosstalk in indiffused $LiNbO_3$ devices. We will then discuss the techniques we are presently using to address the problem of defect structure in lithium niobate. We will present no conclusion, as this work is far from concluded.

§ E-mail mickel@schof.colorado.edu

2. LST for Lithium Niobate

Figure 1 illustrates an LST image taken from an undoped lithium niobate wafer. The tomography setup is illustrated in Figure 2 and is discussed in another article in this volume (Benhaddou and Mickelson, 1996). As the maximum number of defects we could see with our tomograph's pixilation and with the 20× objective is on the order of $10^{10}\,\mathrm{cm}^{-3}$, we would estimate that we have a precipitate density on the order of several times $10^{8}\,\mathrm{cm}^{-3}$ here. This compares with silicon defect densities of $10^{7}\,\mathrm{cm}^{-3}$ or so. As is evident, lithium niobate has high precipitate density—on the order of that seen in gallium arsenide (GaAs). The defects seen in the tomograph may also have an origin that is analogous to that in GaAs. It is generally thought that the defects seen in laser scanning tomographs (LSTs) of GaAs, whether they be dislocations, phase boundaries, and/or grain boundaries, are associated with arsenic-rich regions. In lithium niobate, the comparable structures are thought to be closely related to niobium-rich regions. In GaAs, the defects that are due to grain boundaries can be seen to move under various heat treatments or even to disappear. This effect is probably due to the excess arsenic dissolving out of the grains and substituting itself into the almost stochiometric lattice, residing in good part at vacancy sites. Interestingly enough, we know of no study in lithium niobate to reveal the effects of annealing steps. Such a study could well serve as a good test of the niobium grain hypothesis.

As discussed above, many defects appear in LSTs taken on gallium arsenide, and the conventional wisdom is that these are due to arsenic-rich regions. This would tend to indicate that even a quite pure mixture of gallium and arsenide would reveal such defect structure. Contaminants could lead to still other deleterious effects. Lithium niobate has many contaminants. It is generally accepted that there can be Na, Si, S, Ca, and Pt abundances all at or greater than the 1 ppm level. Fe is one of the impurities (or dopants) that is controlled, as it is well-known that its abundance has a strong effect on the so-called optical damage threshold (or photorefractive coefficient). Yet, for integrated optical quality lithium niobate, the Fe concentration is generally only specified at being less than 1 ppm. Other dopants are not routinely monitored at all.

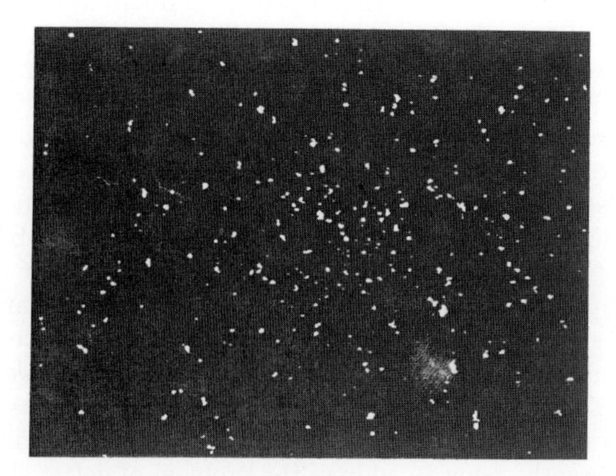

Figure 1. Laser scanning tomography (LST) image of a 500 μm × 500 μm area of a piece of as-grown $LiNbO_3$.

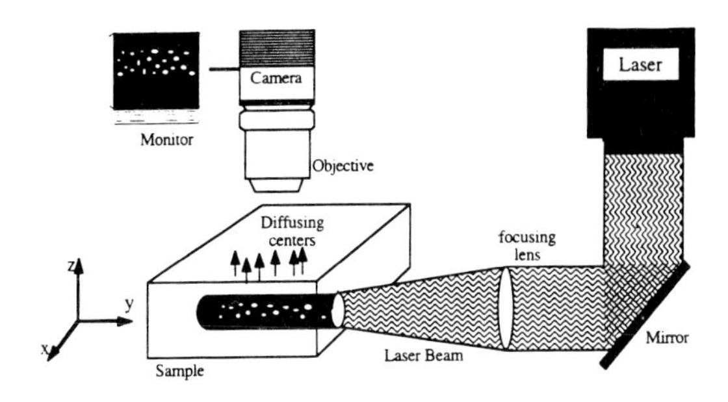

Figure 2. A setup used for generating tomographic images.

3. An Electrooptic Coefficient Mapping Study

A technique originally set up here at GWOL for mapping of potentials in MMIC circuits (Hjelme and Mickelson, 1992; Hjelme et al, 1993; and Radisic et al, 1993) and that was discussed at the previous DRIP (Biernacki et al, 1994; and Biernacki et al, 1994) and that is schematically depicted in Figure 3, has more recently been used to map the electrooptic (r) coefficient in GaAs and lithium niobate substrates that have been treated in some manner. In the initial study, the object was to study the spatial potentials in the vicinity of an ohmic contact. Figure 4 illustrates the test structure. As the electrode configuration of the structure is quite simple, the fields can be found analytically. When the sampling of the structure was performed, however, traces such as that illustrated in Figure 5 resulted. The trace in (a) looks uncalibrated but is calibrated in the same manner as the trace in (b). (More discussion is given to the sampling in an accompanying paper in this volume (Biernacki and Mickelson, 1996).) The conclusion is that the electrooptic coefficient falls off rather rapidly. Indeed, in a paper in the last DRIP, Fillard et al (1994) presented data which showed that ohmic contact formation led to complicated precipitate structures underneath the contact. These structures varied greatly from technology to technology used to make the contact, but there was also high variability within a given technology, even within a single given substrate. The presence of large precipitates wihin a substrate could certainly disrupt the local ferroelectric order and greatly suppress an electrooptic coefficient.

To try to explain the variability of the precipitate structure under a contact, we hypothesized (with only little justification, if any) that substrate defects such as stacking faults on precipitate clusters could influence the activation energies for the diffusion of the contact-forming species on the substrate surface. Now, of course, one fabricates FETs on top of epilayers of perhaps a micron or two thickness on top of the substrate. However, as was shown at the last DRIP in Baeumler et al (1994), defect structures, especially stacking faults, will grow right up through epilayers of several microns thickness. As there were no real hard numbers available, it was of course possible to come to a qualitative description of what we had observed. We would postulate that, if this were the explanation in GaAs, it could well also be the explanation in other materials which had a comparable number of defects. Certainly, then, a similar phenomenon should occur in lithium niobate. The processing of integrated optics generally requires

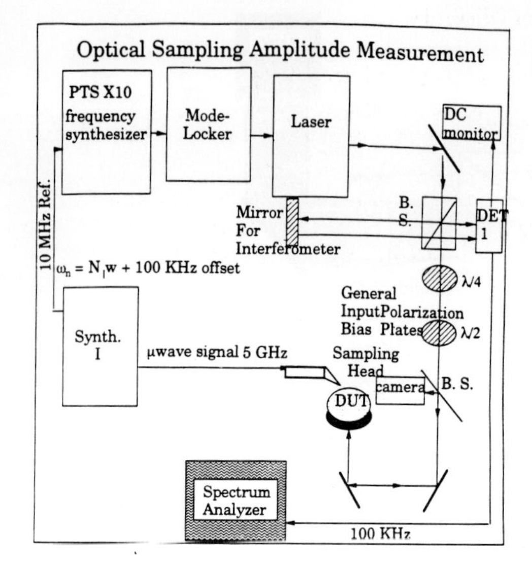

Figure 3. Schematic depiction of the setup used to map potentials in a high-speed integrated circuit.

Figure 4. A test structure for testing the potentials in the neighborhood of an ohmic contact.

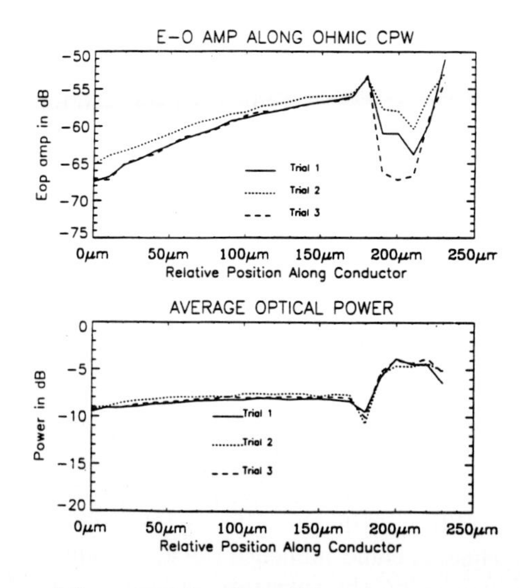

Figure 5. A trace of the potential underneath an ohmic contact illustrating a rapid drop in the signal.

several steps to take place at elevated temperature, ranging from initial post cleaning bakes through an indiffusion step, an annealing step, and then also perhaps several vacuum chamber metallizations. A slight bit of uncertainty in each of these steps due to a random distribution of substrate defects, which may or may not lie in the diffusion path, could be expected to lead to yield problems.

4. Some Device Studies in Lithium Niobate

The two really great questions that arise in integrated optics are

1. Why are the losses of planar devices more or less independent of the employed technology, so many orders of magnitude higher than those of optical fibers?; and
2. Why do integrated optical coupling structures have such large crosstalk?

After a recent study, we feel that we have found a coupling between loss and crosstalk and an inkling of how we might eventually get quantitative answers to the above questions.

Figure 6 is a depiction of an integrated optical directional coupler, and Figure 7 illustrates the results of some crosstalk as a function of wavelength measurements made of H^+ $LiNbO_3$ (that is, proton exchanged and annealed lithium niobate) devices some years ago (Januar et al, 1993). There have been a number of such studies carried out.

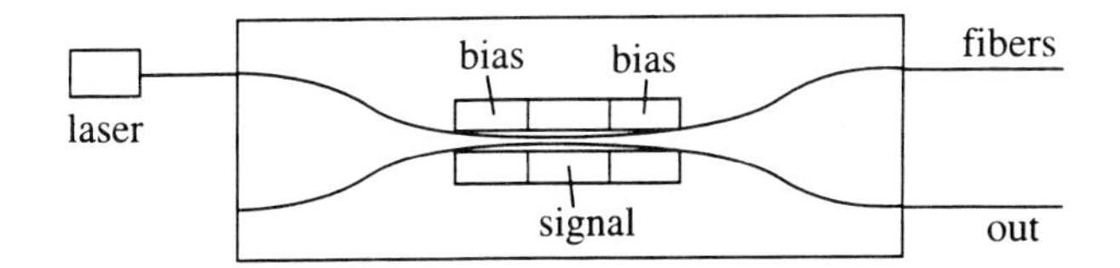

Figure 6. Schematic depiction of an integrated optical directional coupler.

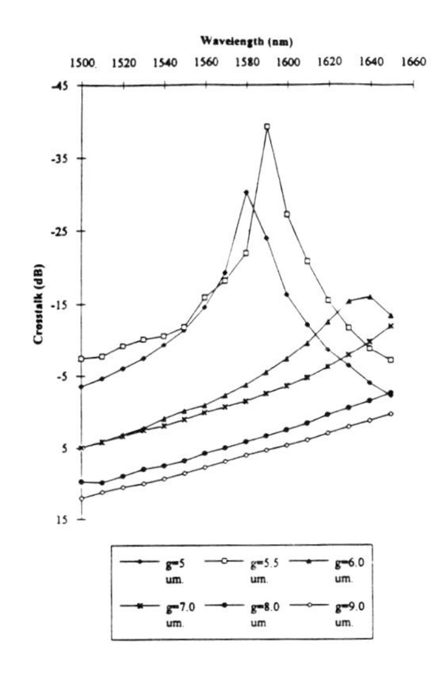

Figure 7. Measurements of the crosstalk as a function of wavelength in an integrated optical directional coupler.

We have recently analyzed the results of a number of crosstalk studies and tried a plethora of different explanations to describe the cause of the crosstalk. Our conclusions were that the crosstalk arose from rapid index fluctuations—that is to say, a Rayleigh or Mie scattering-type mechanism. Although many explanations of crosstalk over the years had invoked slow (compared to a wavelength) variations of the index over the propagation path, it is not really surprising that the fast ones would cause it. Perusal of LST outputs indicate that there is a lot of scattering at high angles. Use of the waveguide capture aperture, together with the total scattering loss magnitude, rapidly seems to give one the kind of crosstalk values one observes. The question then arises, what causes so much scattering. We feel that the answer lies in defect-mitigated diffusion. That is, planar integrated optical devices quite generally require both in-diffusions and then afterwards other processing steps that also require temperature treatments. If diffusion coefficients have variations that are sub wavelength, as is observed in LST as well as photoluminescent microscopy (PLM), even if macroscopic in the sense that they are averaged over many lattice sites, such diffusions will lead to a roughness sufficient to cause significant scattering and therefore integrated optic loss.

5. Defect Analysis Techniques at GWOL

As we have mentioned previously, we have already set up laser scanning tomography (LST) and coefficient and crosstalk characterization. We are also working on PLM, as well as planning work on phase scanning microscopy (PSM) and peak fringe scanning microscopy (PFSM). We will certainlyreport results at the next DRIP—that is, DRIP VII in 1997.

References

Baeumler M et al 1994 *Defect Recognition and Image Processing in Semiconductors and Devices* Institute of Physics Conf. Series Number 135, (Bristol, UK: Institute of Physics Publishing) 169–173

Benhaddou D and Mickelson A R 1996 *Defect Recognition and Image Processing in Semiconductors and Devices* (Bristol, UK: Institute of Physics Publishing), to appear

Biernacki P et al 1994 *Defect Recognition and Image Processing in Semiconductors and Devices* Institute of Physics Conf. Series Number 135, (Bristol, UK: Institute of Physics Publishing) p 403–406

Biernacki P et al 1994 *Defect Recognition and Image Processing in Semiconductors and Devices* Institute of Physics Conf. Series Number 135, (Bristol, UK: Institute of Physics Publishing) p 407–410

Biernacki P, Lee H and Mickelson A R 1995 Evaluation of of defect-related diffusion in semiconductors by electrooptic sampling, *IEEE Journ of Quant Elect*, to appear

Biernacki P and Mickelson A R 1996 *Proc. Defect Recognition and Image Processing in Semiconductors and Devices VI* (Bristol, UK: Institute of Physics Publishing), to appear

Fillard J P et al 1994 *Proc. Defect Recognition and Image Processing in Semiconductors and Devices* Institute of Physics Conf. Series Number 135, p 109–116

Hjelme D R and Mickelson A R 1992 Voltage Calibration of the direct electrooptic sampling technique, *IEEE Trans on Microwave Th and Tech* **40** 1941–1950

Hjelme D R, Yadlowsky M J and Mickelson A R 1993 Two-dimensional mapping of microwave potential on MMICs using electrooptic sampling, *IEEE Trans on Microwave Th and Tech* **41** 1149–1158

Januar I et al 1992 Wavelength sensitivity in directional couplers, *Journ Lightwave Tech* **10** 1202–1209

Radisic V et al 1993 Experimentally verifiable modeling of coplanar waveguide discontinuities, *IEEE Trans on Microwave Th and Tech* **41** 1524–1533

Inst. Phys. Conf. Ser. No 149
Paper presented at DRIP 95

Microscopic defect characterization of damage-introduced regions in GaAs due to dry-etching processes

Y. Mochizuki[1)], M. Mizuta[1)], T. Ishii[1)*] and A. Mochizuki[2)]

[1)] Fundamental Research Laboratories, NEC Corporation,
34 Miyukigaoka, Tsukuba, Ibaraki 305, Japan

[2)] ULSI Device Development Laboratories, NEC Corporation,
2-9-1 Seiran, Ohtsu, Shiga 512, Japan

Abstract. Effects of dry-etching processes on GaAs are studied by the ODMR technique, in which artificially provided luminescence from heterostructured samples are monitored. Based on this technique, depth resolution and near-surface sensitivity is improved and a microscopic defect characterization of the damaged regions is realized. Damage-induced defects are detected after a reactive-ion etching process. At the same time, an occurrence of hydrogen passivation of grown-in defects (Ga-interstitials) is detected. These results are found to be consistent with the photoreflectance data, with which a distinction can be made between defect-rich and hydrogenated regions depending on the depth from the wafer surface.

1. Introduction

Dry-etching techniques such as reactive-ion etching (RIE) are widely employed in modern device fabrication processes. Use of RIE is essential in realizing submicron patterns but a serious drawback has long been recognized. The process causes a degradation of electrical and/or optical properties in the near-surface region of semiconductor wafers[1]: it is often encountered that carrier concentration decreases or luminescence efficiency becomes lower. These effects are collectively known as the dry-etching damage. Since such modifications are detected in the range far exceeding the projected range of implanted plasma species, it is generally speculated that formation and diffusion of point defects are playing important roles, which take place in the near room-temperature environment during the plasma exposure.

In spite of growing importance of elucidating the identity of the dry-etching damage, little has been experimentally clarified concerning its microscopic character except for limited cases[2]. This is mostly due to a difficulty in applying structure-sensitive tools such as ESR or related techniques. Although the damage-induced change is a drastic one, it is introduced to a depth only of the order of 100 nm from the wafer surface.

In this paper, we describe our attempt of applying optically-detected magnetic resonance (ODMR) to this problem. ODMR, as compared to the conventional ESR, has an advantage that the measurement is more sensitive in the near-surface region. Our approach here is to combine this technique with a heterostructured sample in order to drastically enhance the near-surface sensitivity and/or depth resolution of spin resonance experiments. We discuss the results in conjunction with those obtained by photoreflectance (PR) measurements, in which the analysis we developed provides further insight into the nature of RIE-induced damage.

2. Depth-resolved ODMR

2.1 Detection scheme of ODMR

Generally, probed region in ODMR is defined by that of photoluminescence (PL), which is determined by photoabsorption and minority carrier diffusion. For GaAs, this is typically a few micron from the surface, being still larger than the introduction range of dry-etching damage (~100nm). A better resolution in depth, however, can be achieved if a sample contains a specific luminescent layer whose location can be defined by crystal growth techniques. By extracting this PL, emitted at different photon energy from those arising from the rest of the structure, one can perform a depth-resolved ODMR experiment.

Choice of the artificial PL process (defect-spectator process in depth-resolved ODMR) is based on the lifetime consideration. Since we are interested in the defect-induced recombination paths, which affects the monitored radiative process, recombination rates for these competing channels should be comparable. Further, the values have to be also comparable to the magnetic dipole transition rate of ESR, which is usually in the range of μs^{-1}. The example we will describe in the following uses donor-acceptor pair (DAP) transition at shallow impurities. In this recombination process, the lifetime is rather long (~0.1μs) due to inter-impurity separation.

2.2 Experimental

Figure 1 shows the sample structure. The luminescent layer was a thin (50 nm) GaAs film co-doped with Si ($1x10^{16}$ cm^{-3}) and Be ($3x10^{16}$ cm^{-3}), denoted as a GaAs:D,A film, which was then sandwiched by AlGaAs cladding layers and positioned within the near-surface region to which damage-introduction is expected. A single quantum well was also inserted at a deeper location (115 nm from the surface) in order to avoid a high level of carrier injection into the GaAs:D,A layer. Throughout the whole structure, it is only the GaAs:D,A layer which emits at the near-bandgap energy of GaAs. Therefore, by extracting PL in this energy range, the probed region of spin resonance can be limited within this layer.

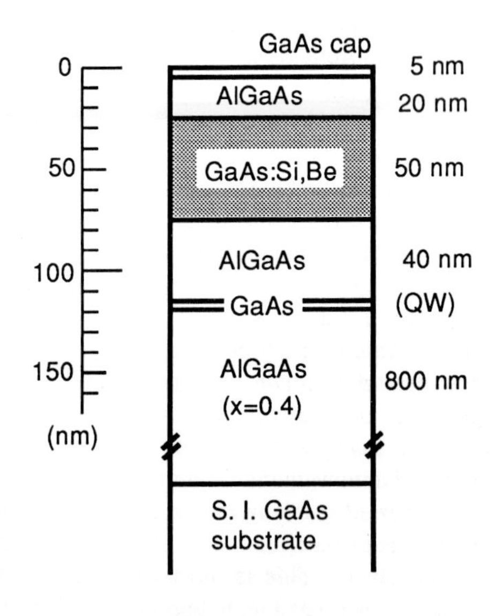

Fig.1 Layer structure of GaAs:D,A sample used in the ODMR study.

The effect of RIE was studied by first depositing a 230 nm-thick SiO_2 film on to the wafer, and then completely removing it by CHF_3/O_2 plasma. This is a dry etchant for dielectric films and have very small etching rate for GaAs. Therefore, the heterostructure of the sample is not destroyed.

Our ODMR setup consists of a 4T split-coil superconducting magnet and a 35 GHz microwave system, which produces the maximum incident power of 100 mW. The PL was excited by an Ar^+-ion laser (488 nm, ~1mW/cm^2) and was monitored by a Si photodiode. The shallow DAP-PL was selected by placing an optical filter. In the measurement, the applied microwave power was square-chopped with scanning the magnetic field and a corresponding change in the PL intensity was recorded.

2.3 Results
The PL spectrum in the near-bandgap energy region of GaAs was found to be dominated by the DAP recombination and no excitonic emission was detected under the aforementioned excitation condition. After the RIE treatment, the intensity of the DAP-PL was decreased to about 70% with no noticeable change in the spectral shape. The intensity reduction is suggestive of the introduction of additional recombination centers due to the damage.
The ODMR spectra, obtained by monitoring this DAP-PL is shown in Fig. 2. We observed a different types of defects before and after the RIE. All of them quenches the PL intensity (with $\Delta I_{PL}/I_{PL}$~10^{-4}), so the defects are competing with the shallow DAP recombination. On the contrary, when we monitored a deep level luminescence at around 0.8 eV, a PL-enhancing resonance, probably due to triplet bound excitons[3], was observed. However, for a deep PL, it is difficult to specify from which layer it is emitted. In fact, the deep triplet resonance did not show any change due to the RIE treatment. Thus, we point out the importance of the artificially provided PL process (shallow DAPs, in this case) in performing the depth-resolved ODMR experiments.

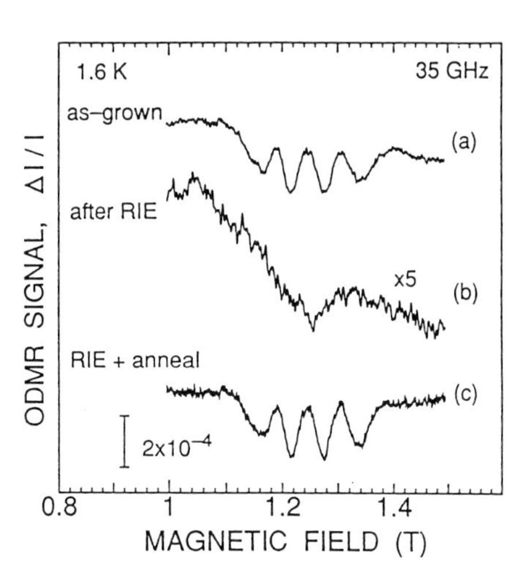

Fig.2 ODMR spectra before and after the RIE. The subsequent annealing was performed at 450°C for 30 min.

In Fig. 2, the grown-in defect observed before performing the RIE shows a four-line resonance which was found to be almost isotropic. The shape is accounted for by the combination of hyperfine splittings due to ^{69}Ga and ^{71}Ga nuclei and the defect can be assigned as Ga-interstitials (Ga_i)[4,5]. After the RIE, this resonance disappears and another broad resonance appears. The latter is speculated to be due to the damage-induced defects, which now act as recombination centers. Although the spectral shape resembles the one reported in neutron-irradiated GaAs, for which several model have been suggested, the identity is unclarified.

Actually, Ga_i defects were not destroyed by the RIE since the subsequent annealing at 450°C recovered the original signal. The disappearance after the RIE is rather attributed to the hydrogen passivation effect, which is discussed in the next section.

3. Photoreflectance: Fermi-level pinning versus dopant neutralization

Since introduction of damage-induced defects is not the only mechanism which changes the carrier concentration, we also performed PR study. The PR spectroscopy is a convenient method to evaluate the change in electrical properties of plasma-processed GaAs. The principle relies on the analysis of Franz-Keldysh oscillations (FKOs)[6] appearing on the high-energy side of the fundamental gap transition. Intervals of the energy extrema yields strength of near-surface electric field. Combined with repeated step-etching, the technique has been applied to the dry-etching-damage problem[7]. Based on our analysis method, however, it is further possible to obatain information on the mechanism which leads to the reduced carrier concentration.

A decrease in carrier concentration of a doped wafer can be accounted for by two mechanisms. One is the Fermi-level pinning or charge compensation due to damage-induced point defects. The other is the neutralization of dopant atoms via dopant-defect or dopant-impurity pair formation. We emphasize the roles of these two contributions because reductoin of carrier concentration occurs for both n-type and p-type semiconductors. Using the PR method, we have detected both of the effects playing the dominant roles, depending on the depth from the surface. Therefore, the technique provides a way to quantitatively evaluate these two factors.

Figure 3 shows the near-surface field of an n-type MOCVD GaAs ($n=1.7x10^{17}$ cm^{-3}) as a function of etched depth, which is estimated from the energy periods of FKOs in the PR spectra. After the RIE with CHF_3/O_2, the field becomes smaller than in the reference sample in the near-surface region up to 200 nm from the surface. The values, however, recover to the original ones after a subsequent annealing at 450°C for 30 min.

The surface electric field, F, deduced from PR measurements are widely used to estimate dopant concentrations, N_d, in epitaxial wafers based on the relation[7]

$$F = \sqrt{(2eV_s N_d) / \varepsilon_s} \tag{1}$$

However, this formalism assumes a uniform distribution of dopant concentration and is not applicable to the analysis of data in Fig. 3 (after RIE), which have a variation with depth

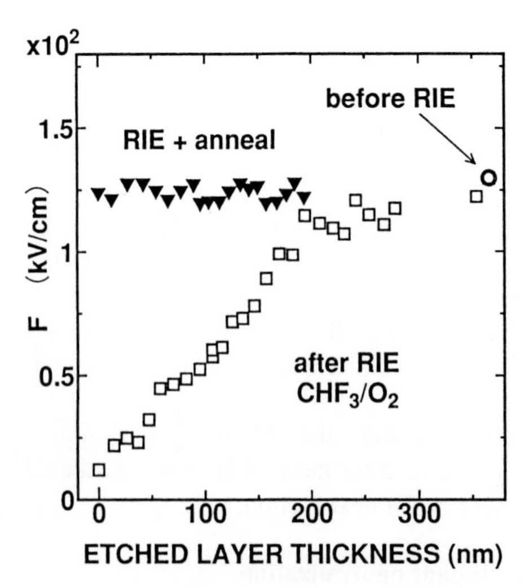

Fig.3 Profiles of surface electric field obtained by the step-etched PR method.

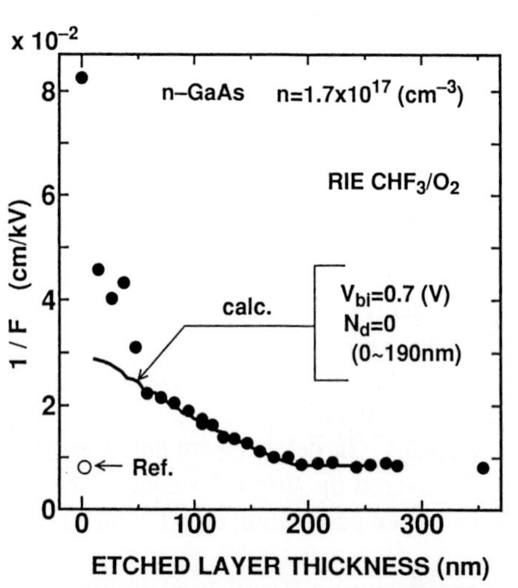

Fig. 4 Reciprocal field as a function of etched layer thickness for the sample after the RIE. The solid curve was calculated by integrating the Poisson's equation for an ideal *i*/*n*-junction.

from surface. We first simplify the problem by approximating the system with an abrupt i/n junction, where the i-layer, from the depth 0 to l, is a pure (intrinsic) semiconductor. This corresponds to an assumption in which the reduction of carrier concentration is due to a complete dopant neutralization. Upon repeated wet-etching, the surface field at each step (after removing thickness d from the original surface) is analytically given by

$$F(d) = \frac{e \bullet N_d}{\varepsilon_s} \bullet \left\{ \sqrt{(l-d)^2 + \frac{2\varepsilon_s V_{bi}}{eN_d}} - (l-d) \right\} \sim \frac{V_{bi}}{l-d} \tag{2}$$

This is a straightforward expression. If the thickness of the i-layer is large or the doping concentration, N_d, is large, the device is a parallel-plate capacitor in which the field is inversely proportional to the plate distance. The step-etching simply decreases this distance. Thus, it is convenient to plot the experimental data in reciprocal field as a function of etched depth, as in Fig. 4 . This should yield a linear relation unless the remaining i-layer thickness finally becomes so small and depletion of the n-layer becomes non-negligible. The important point is that the slope of the straight line is determined only by the inverse of surface band bending (Fermi-level difference between at the surface and in n-layer), which is roughly half the gap of GaAs, and is not affected by other parameters.

In fact., the data taken after the CHF_3/O_2-RIE show a reasonable agreement with the calculated ones when the etched-off layer thickness is larger than 50 nm. A significant disagreement, on the contrary, is noted as long as the etched depth is smaller than 50 nm. (The calculated points in Fig 4 are not based on (2) but were obtained by numerically integrating the Poisson's equation.) This means that, in the region 50~200 nm from the surface, the nature of the i-layer is similar to an intrinsic semiconductor in that the potential gradient across the layer is constant. We also simulated the deviation from the linear dependence when the residual donor concentration in the i-layer becomes larger. The calculation indicated that if it is larger than about 3×10^{15} cm^{-3}, apparent deviation should have been observed. Therefore, the most likely mechanism for the carrier reduction in the region 50~200 nm after the RIE is the dopant-neutralization or dopant deactivation. In the shallower region, on the other hand, the field variation suddenly deviates and suggests the predominance of other mechanism. Since the filed (i.e. the potential gradient) is small, Fermi-level pinning due to damage-induced defects (deep levels) is suggested.

Our results demonstrate that the effects of damage-induced Fermi-level pinning and dopant neutralization can be distinguished by the PR method. We emphasize that, using the conventional capacitance method, only position of depletion-layer edge can be measured. Thus, it is difficult to clarify the reduction mechanism of carrier concentration and to quantitatively evaluate the contributions of the aforementioned two factors.

The PR results suggest that donor atoms are neutralized during the RIE process. The most plausible cause of this is the so-called hydrogen passivation effect[8]. When hydrogen atoms are diffused into semiconductors, they form pair defects with shallow dopants (as well as with deep defects) to kill their electrical activity. We thus performed a SIMS analysis and confirmed that hydrogen atoms are indeed incorporated after the RIE. Their concentration was found to be in excess of the number of dopants, at least to the depth where dopant neutralization was observed.

4. Discussions

The results obtained by the depth-resolved ODMR method are consistent with the conclusion obtained from the PR measurements. The inserted GaAs:D,A film covers the depth of 25~75nm from the surface. As noted from the PR study, this is the region where the major contribution (to the modification of electrical property) changes from the defect-induced Fermi-level pinning to the unintentional hydrogenation (50 nm from the surface). This explains why we observed both the introduction of damage-induced defects together with the passivation of grown-in Ga_i defects in ODMR.

The most puzzling issue concerning the dry-etching damage is the mechanism of rapid defect migration in the near-room temperature environment. At least, the observed stability of Ga_i defects is not compatible with the rapid introduction of the RIE damage, and thus, they are not likely to be playing a significant role, in spite of the general speculation that they would be fast diffusers. (Generation of cation-site Frenkel pairs, however, becomes important under a heavier irradiation, as revealed by the ODMR study of AlAs/GaAs quantum-well structures[9].) Furthermore, our study clarifies that, in discussing the kinetics, it is important to pay attention to the simultaneous incorporation of hydrogen, which may make it rather difficult to purely access the bombardment-induced portion of the phenomena. The source of hydrogen can be the etchant gas itself (like CHF_3/O_2) but we also detected a similar degree of unintentional hydrogenation, both electrically and from the mass spectrometry, even when the dry-etchant does not nominally contain hydrogen (SF_6). In this case, the source might be water vapor adsorbed on the chamber walls and so on.

5. Conclusions

In conclusion, a novelty of combining semiconductor heterostructures with the ODMR technique is demonstrated and is shown to be useful in providing microscopic information on the dry-process-induced damage in GaAs. We observed the introduction of damage-related defects together with the hydrogen passivation of grown-in Ga_i defects. Furthermore, analysis of PR spectra has also led to the detection of these two contributions

Acknowledgments

The authors acknowledge Satoshi Ideshita for the MBE growth of the wafers used in the ODMR study. They also thank Toshihiro Ishii for his assistance in the PR measurements.

References

* Department of Physics, Tsukuba University.

[1] Pearton S J, Ren F, Fullowan T R, Kopf R F, Hobson W S, Abernathy C R, Katz A, Chakrabarti U K and Swaminathan V 1991 *Mat. Sci. Forum* **83-87** 1439.

[2] Benton J J, Kennedy M A, Mechel J and Kimmerling L C 1991 *Mat. Sci. Forum* **83-87** 1433.

[3] Kennedy T A and Wilsey N D 1985 *Phys. Rev B* **32** 6942, Gislason H P, Rong F and Watkins G D 1985 *Phys. Rev. B* **32** 6945

[4] Kennedy T A, Magno R and Spencer M G 1988 *Phys. Rev. B* **37** 6325.

[5] Trombetta J M, Kennedy T A, Tseng W and Gammon D 1991 *Phys. Rev. B* **43** 2458.

[6] Aspnes D E and Stunda A A 1973 *Phys. Rev. B* **7** 4605.

[7] Wada K and Nakanishi H 1994 *Mat. Sci. Forum* **143-147** 1433.

[8] Pearton S J, Corbett J W and Stavola M 1992 *Hydrogen in Crystalline Semiconductors* (Springer, Berlin).

[9] Mochizuki Y, Mizuta M and Mochizuki A 1995 *18th Int. Conf. on Defects in Semicond, Sendai* (to be published in *Mat. Sci. Forum*)

Inst. Phys. Conf. Ser. No 149
Paper presented at DRIP 95

IR-LST a powerful non-invasive tool to observe crystal defects in as-grown silicon, after device processing, and in heteroepitaxial layers

G Kissinger[1], J Vanhellemont[2], D Gräf[3], C Claeys[2], and H Richter[1]

[1]) Institute of Semiconductor Physics, P.O. Box 409, D-15204 Frankfurt (Oder), Germany
[2]) IMEC, Kapeldreef 75, B-3001 Leuven, Belgium
[3]) Wacker Siltronic GmbH, P.O. Box 1140, D-84479 Burghausen, Germany

Abstract. One of the main advantages of infrared light scattering tomography (IR-LST) is the wide range of defect densities that can be studied using this technique. As-grown defects of low density and very small size as well as oxygen precipitation related defects that appear in densities up to some 10^{10} cm^{-3} can be observed. As-grown wafers with a "stacking fault ring" were investigated in order to correlate the defects observed by IR-LST with the results of Secco etching and alcaline cleaning solution (SC1) treatment revealing flow pattern defects (FPDs) and crystal originated particles (COPs), respectively. These wafers were studied after a wet oxidation at 1100 °C for 100 min. In processed CZ silicon wafers it was possible to identify stacking faults and prismatic punching systems directly from the IR-LST image. Brewster angle illumination is a special mode to reveal defects in epitaxial layers in a non-destructive way. Misfit dislocations in the interface between a $Ge_{0.92}Si_{0.08}$ layer and a silicon substrate were studied using this mode that allows to observe very low dislocation densities.

1. Introduction

More than 15 years are gone since Ogawa and Moriya have developed the light scattering tomography (LST) technique [1, 2] that is based on the principles of ultramicroscopy [3, 4] as a new observation tool with new possibilities due to the scanning principle, electronic image processing, and infrared laser beam illumination. One of the main advantages of this technique is the wide range of defect densities that can be observed ranging from 10^5 to some 10^{10} cm^{-3} making it valuable for the whole defect spectrum that occurs after crystal growth, as well as after wafer processing. The possiblity to follow the change of defects after processing in the same sample offers interesting possibilities in the field of defect engineering and materials science.

At first GaAs was the most explored material by LST [see e.g. 5, 6] but the method has already found its place in the diagnostics of defects in silicon [see e.g. 7, 8]. In the following examples are given demonstrating defect studies by IR-LST in a variety of silicon materials.

2. Experimental

For the IR-LST studies the MILSA IRHQ-2 infrared light scattering tomograph of Ratoc was used. The system is equipped with a YAG-laser of 1060 nm wavelength for the 90°scattering mode and a laser diode of 870 nm wavelength for the Brewster angle illumination mode. A more detailled description of the used illumination modes is given elsewere [9].

Non-agitated Secco etchant [10] was used for 30 min at room temperature in order to develop the flow pattern defects (FPDs) on the as-grown silicon wafers. Crystal originated particles (COPs) were observed by a scanning surface inspection tool after a treatment of the as-grown wafers with an SC1 cleaning solution (NH_4OH : H_2O_2 : H_2O = 1 : 1 : 5) at 85 °C for 8 h.

Bevelled sections of processed wafers with bevel angles of about 6° were etched for 2 min with the Yang etchant [11].

As-grown p-type 150 mm Czochralski (CZ) silicon wafers from crystals with a "stacking fault ring" grown in a combined interstitial/vacancy rich regime were studied by IR-LST. The about 1 cm wide ring, the area between the the vacancy supersaturated center and the interstitial supersaturated edge region, was located about 35 mm away from the wafer edge.

The fabrication process of the studied processed wafers is described in detail in Ref. [12].

3. Results and discussion

3.1. As-grown silicon wafers

Due to their small size and low density, few methods are available to study the nature of the as-grown crystal defects. Well established are invasive methods like etching with non-agitated Secco etchant in order to determine FPD and Secco etch pit (SEP) densities [13], SC1 cleaning to develop COPs [14], and copper or lithium decoration techniques [15]. Using wide visible field scanning, these very low defect densities can be easily revealed by IR-LST as LST defects (LSTs). Now the question arises if all these methodes reveal the same type of defects. For wafers of crystals grown in the vacancy rich regime COPs and FPDs have been correlated and have been connected with D-defects that are known to degrade the gate oxide integrity [16- 18]. Also correlations between LSTs and FPDs were found [9, 19].

This contribution is concerned with the correlation of COPs, FPDs, and LSTs in wafers with a "stacking fault ring" allowing a statement for interstitial and vacancy rich regions as well as for the ring area. Fig. 1 shows the results of this study obtained from one wafer. The ring is located at about 30 mm away from the wafer edge. It is obvious that only in the vacancy rich region inside of the ring defects were observed in a good correspondence between all three graphs. The distribution of the FPDs and LSTs appears in a clustered manner. Most of the maxima of the LST graph also can be found in the FPD graph. However, the density of the FPDs is one order of magnitude lower than the density of the LSTs. This was found also in 150 mm and 200 mm wafers grown in the vacancy rich regime under different cooling rates [9]. In case that the LSTs become very small the deviation between LST and FPD densities becomes distinctly larger implying that IR-LST is able to detect smaller defects than Secco etching. In the ring as well as in the interstitial rich region outside the ring the LST and FPD densities are under the detection limit and the COP density is extemely low.

The nature of the D-defects is still an unsolved question. Recently, it was stated that the defects causing the gate oxide breakdown are large vacancy clusters and oxide precipitates [20]. IR-LST would be able to detect both. Oxide precipitates can be detected for radii down to 20 nm. It is expected that in case of vacancy precipitates this limit is lower because of the higher difference in the refractive index resulting in a higher intensity of the scattered light. After a short time anneal at 1000 °C for 15 min which is known to

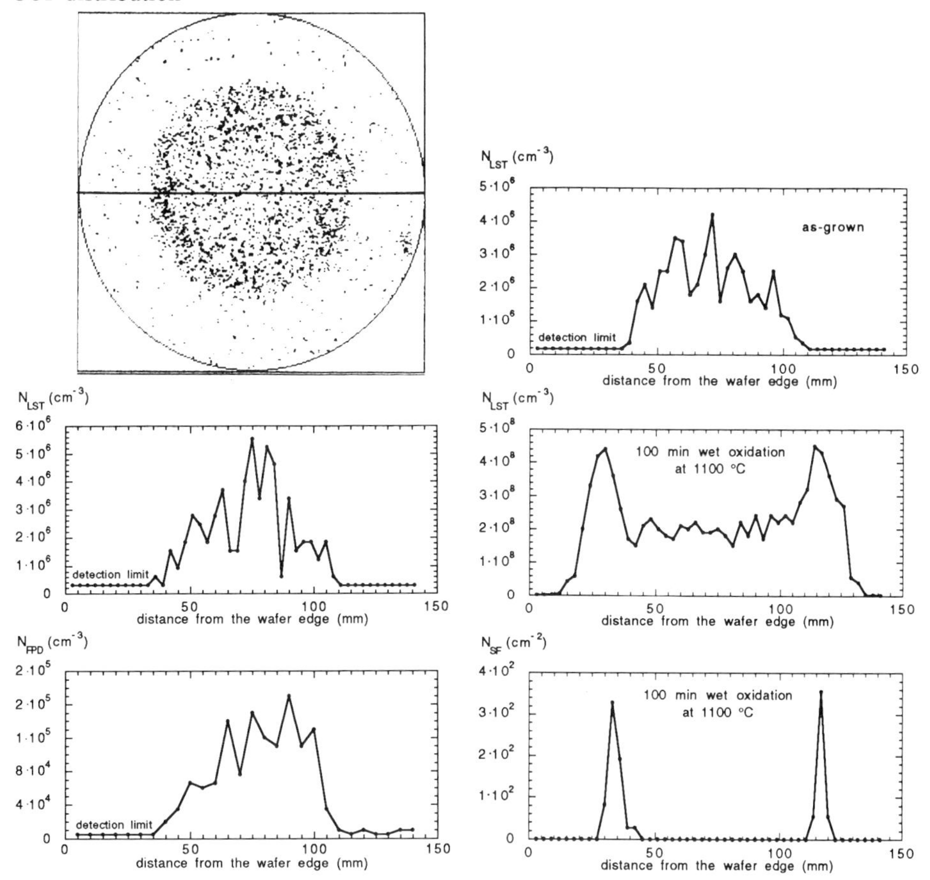

Fig.1 COP distribution, density of LST defects (N_{LST}), and density of FPDs (N_{FPD}) measured along the same line on a 150mm silicon wafer. The line is marked in the COP distribution figure that has the same scale like the density graphs.

Fig.2 Density of LST defects (N_{LST}) and density of stacking faults (N_{SF}) measured along the same line on a 150 mm silicon wafer before and after a wet oxidation at 1100 °C for 100 min.

dissolve as-grown oxide precipitates up to a certain size [21] nothing has changed in the density distribution of the LSTs. Moreover, it was observed that they grow to larger sizes during gate oxidation [22]. This supports the assumption that the D-defects are large grown-in oxide precipitates as suggested by Hourai et al. [23].

IR-LST offers the unique possibility to observe the same area of a sample after processing because this method is non-invasive. The neighbouring wafer of the same crystal as the one studied first was used for such experiments. Fig. 2 shows the change in the defect density distribution in a wafer with a ring after wet oxidation at 1100 °C for 100 min. In the ring area a remarkable increase of the defect density was observed which is clearly higher than the increase in the D-defect region inside the ring. However, the size of the defects is clearly lower in the ring area than in the D-defect region as concluded from differences in the scattering intensity of the precipitates. Stacking faults, typically formed in ring areas during oxidation, were observed by etching only in a small band around the maximum of the LST density as seen in the lowest graph of Fig.2. This indicates that a lot of very small oxide precipitate nuclei smaller than 20 nm must exist in the ring area which grow to a larger size during oxidation and become visible by IR-LST. These results are in agreement with the results in Ref. [24].

3.2. Processed silicon wafers

During device processing the defects originated in the as-grown wafers grow to larger sizes. Bulk defect densities of processed wafers determined by IR-LST correlate very well with defect densities obtained by etching with the Yang etchant [9]. Oxide precipitates release their strain resulting from their volume expansion during growth via punching of prismatic loops in <110> directions or via emission of self-interstitials that can form an extrinsic stacking fault around the precipitates [25, 26]. It was possible to identify both defect arrangements directly from the IR-LST image [27]. Fig. 3 and 4 show examples for a punching system and a stacking fault, respectively. The punching systems can be identified because of their characteristic defect arrangement where the punched loops appear as point-shaped scatterers aligned in <110> directions with a central oxide

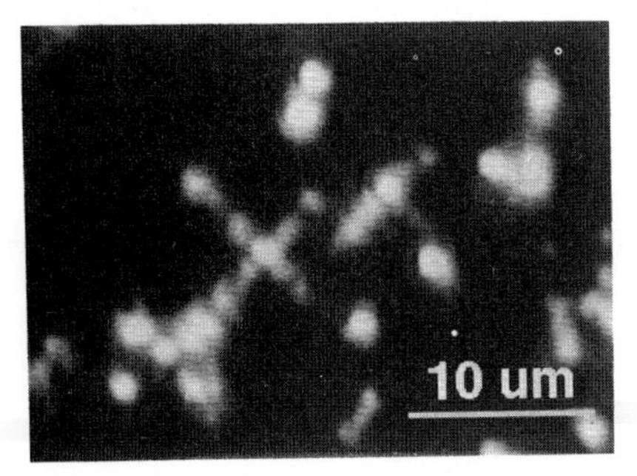

Fig. 3 Prismatic punching system observed by IR-LST in a processed (001) silicon wafer

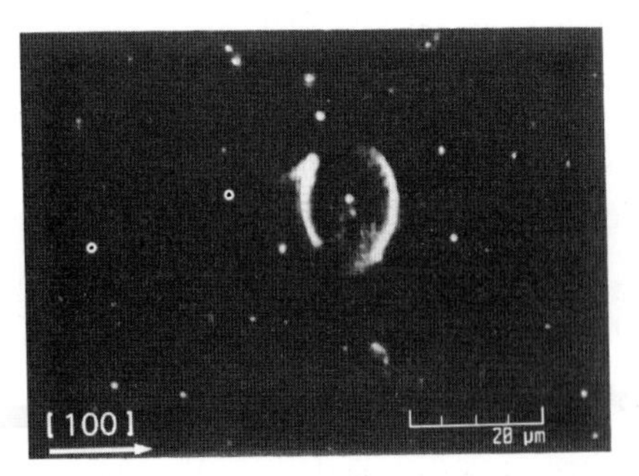

Fig. 4 Stacking fault observed by IR-LST in a processed (001) silicon wafer

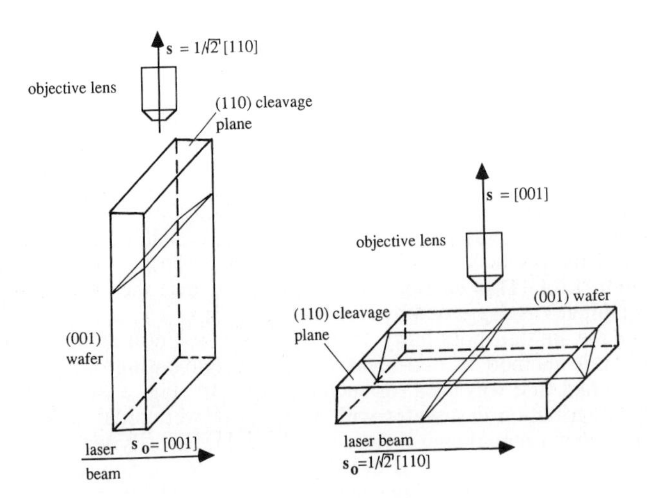

Fig. 5 Conditions to observe stacking faults in (001) silicon wafers by IR-LST under plan-view conditions (left) and for the cross-sectional view (right)

precipitate.
The IR-LST features of stacking faults are due to scattering of the Frank-type dislocation loop surrounding the stacking fault. They appear as ellipses with a central scatterer being the originating oxide precipitate. However, not all possible sets of stacking faults on {111} planes in silicon become visible. This depends on the orientation of the scattering vector **g**, being the difference between the unit vector of the observation direction **s** and the unit vector of the incident laser beam $\mathbf{s_0}$, to the normal of the stacking fault plane **n**. Fig. 5 shows all planes on which stacking faults were observed for both imaging conditions possible for cleaved (001) wafers. The stacking faults stay invisible if **g** is perpendicular to **n**. The b/a ratio of the ellipse depends on the angle between **n** and **s** and the dislocation line **l** is in perfect extinction for $\mathbf{s_0}$ being parallel to **l**.

3.3. Heteroepitaxial $Si_{(1-X)}Ge_X$ layers

Due to the lattice mismatch of 4.2 % in the Si-Ge system heteroepitaxial Ge_XSi_{1-X} layers on silicon are strained. They can relax their stress via formation of misfit dislocations in the interface between layer and substrate if the layer exceeds a critical thickness. Once nucleated the misfit dislocations are formed via gliding of their threading parts that intersect the surface. Such misfit dislocations can be observed using the Brewster angle illumination mode that is non-destructive [8]. However, only one set of misfit dislocations is visible in one image because the dislocation line has to be perpendicular to the polarization of the incident beam [9]. Fig. 6 shows as an example a 270 nm thick $Ge_{0.92}Si_{0.08}$ layer on a silicon substrate. Because of the low germanium concentration in the layer, the misfit is comparatively small. This results in a relaxation mechanism that is characterized by the appearence of misfit dislocations in bundles. Therefore, the intensity of the misfit dislocations in Fig. 6 is very different because the resolution of IR-LST is too low to resolve the single misfit dislocations if their distance is lower than 2 μm. However, the number of dislocations hidden in one line could be estimated from the scattering intensity.

The intensity maxima observable on the dislocation lines are crossing-points with misfit dislocations perpendicular to the observed ones. As observed by TEM, here the Hagen-Strunk mechanism [28] is acting by at least forming a dislocation arm moving upwards. It becomes visible as an intensity maximum.

In heteroepitaxial layers with a higher lattice mismatch the misfit dislocations are distributed more uniformly and the scattering intensity is nearly equal but low for all dislocations. Threading dislocations can be observed also as intensity maxima [9]. Fig. 7 shows in its center a short misfit dislocation with a threading dislocation on both ends. It must have been nucleated at the surface by a dislocation loop which glides on a (111) plane

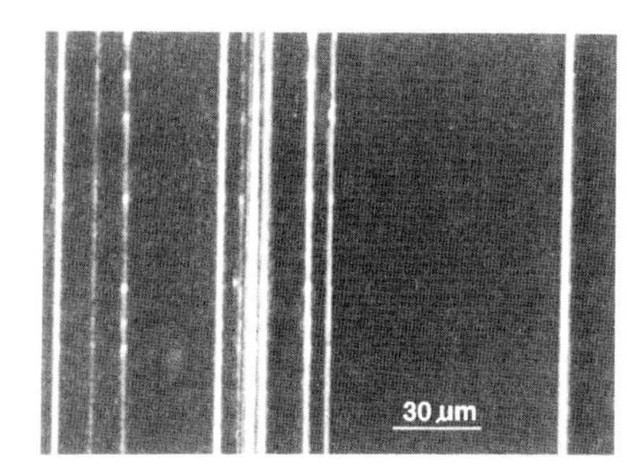

Fig. 6 Misfit dislocations observed by Brewster angle light scattering tomography in a $Ge_{0.92}Si_{0.08}$ layer on silicon

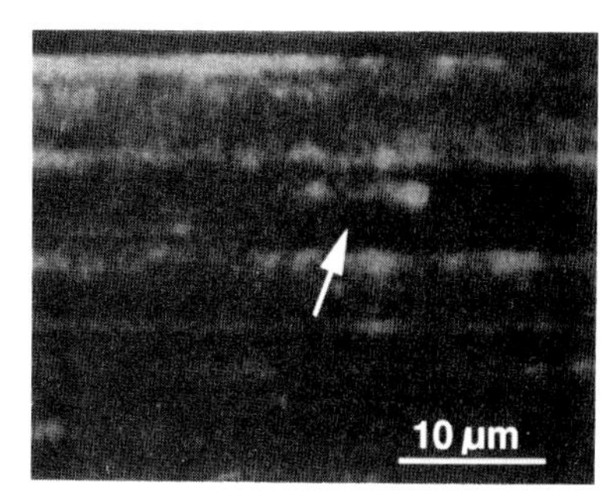

Fig.7 Misfit dislocation having two threading parts observed in a $Ge_{0.15}Si_{0.85}$ layer on silicon by Brewster angle light scattering tomography (arrow)

down to the interface and the threading dislocations start moving through the wafer on the same gliding plane simultaneously forming the misfit dislocation. This is a well known nucleation mechanism for misfit dislocations in SiGe layers [29].

4. Summary

It has been demonstrated that IR-LST can be applied successfully to study a wide variety of defects in silicon. As-grown D-defects as well as oxygen precipitation related defects can be detected in a non-invasive way. In some cases like extended stacking faults and prismatic punching systems it is possible to identify them directly on the image.

The unique possibility to observe the same area after processing offers interesting studies. In this way it was shown that in the "ring" region of as-grown wafers small oxide precipitate nuclei exist that stay invisible for IR-LST because their radius is lower than 20 nm.

Misfit dislocations in the interface between a SiGe layer and a silicon substrate can be observed in a non-destructive way using the Brewster angle illumination mode of IR-LST. So it is be possible to observe the first misfit dislocation in a sample or to study their development after different processing steps.

References

[1] Moriya K and Ogawa T 1978 *J. Cryst. Growth* **44** 53
[2] Moriya K and Ogawa T 1980 *Phil. Mag. A* **41** 191
[3] Vand V, Vedam K, and Stein R 1961 *J. Appl. Phys.* **73** 2551
[4] Tajima M and Iizuka T 1976 *Jap. J. Appl. Phys.* **15** 651
[5] Ogawa T 1989 *J. Cryst. Growth* **96** 777
[6] Gall P, Fillard J P, Castagne M, Weyher J L, and Bonnafe J 1988 *J. Appl. Phys.* **64** 5161
[7] Moriya K, Hirai K, Kashima K, and Takasu S 1989 *J. Appl. Phys.* **66** 5267
[8] Taijing L, Toyoda K, Nango N, and Ogawa T 1991 *J. Cryst. Growth* **114** 64
[9] Kissinger G, Vanhellemont J, Gräf D, Zulehner W, Claeys C, and Richter H 1995 *Electrochem. Soc. Proc. Vol.* **95-30** 156
[10] Secco d´Áragona F 1972 *J. Electrochem. Soc.* **119** 948
[11] Yang K H 1984 *J. Electrochem. Soc.* **131** 1140
[12] Kissinger G, Vanhellemont J, Simoen E, Claeys C, and Richter H 1995 *Mat. Sci. & Eng. B* in press
[13] Yamagishi H, Fusegawa I, Fujimaki N, and Katayama M 1992 *Semicond. Sci. Technol.* **7** A135
[14] Ryuta J, Morita E, Tanaka T, and Shimanuki Y 1990 *Jap. J. Appl. Phys.* **29** L1947
[15] De Kock A J R 1973 *Philips Res. Rept. Suppl. No.1*
[16] Yamagishi H, Fusegawa I, Takano K, Iino E, Fujimaki N, Ohta T, and Sakurada M 1994 *Electrochem. Soc. Proc. Vol.* **94-10** 124
[17] Brohl M, Gräf D, Wagner P, Lambert U, Gerber H A, and Piontek H, 1994 *Electrochem. Soc. Proc. Vol.* **94-2** 619
[18] Wagner P, Brohl M, Gräf D, Lambert U 1995 *Mat. Res. Soc. Proc. Vol.* **378** 17
[19] Umeno S, Sadamitsu S, Murakami H, Hourai M, Sumita S, and Shigematsu T 1993 *Jap. J. Appl. Phys.* **32** L699
[20] Park J-G, Kirk H, Cho K-C, Lee H-K, Lee C-S, and Rozgonyi G A 1994 *Electrochem. Soc. Proc. Vol.* **94-10** 370
[21] Falster R private communication
[22] Vanhellemont J, Kissinger G, Gräf D, Kenis K, Depas M, Mertens P, Lambert U, Heyns M, Claeys C, Richter H, and Wagner P 1996 this volume
[23] Hourai M, Nagashima T, Kajita E, Miki S, Shigematsu T, Okui M 1995 *J. Electrochem. Soc.* **142** 3193
[24] Marsden K, Sadamitsu S, Hourai M, Sumita S, and Shigematsu T 1995 *J. Electrochem. Soc.* **142** 996
[25] Tan T Y and Tice W K 1976 *Phil. Mag.* **34** 615
[26] Patel J R, Jackson K A, and Reiss H 1977 *J. Appl. Phys.* **48** 5279
[27] Kissinger G Vanhellemont J, Claeys C, and Richter H 1995 *J. Cryst. Growth* in press
[28] Hagen W and Strunk H 1978 *Appl. Phys.* **17** 85
[29] Matthews J W, Mader S, and Light T B 1970 *J. Appl. Phys.* **41** 3800

Inst. Phys. Conf. Ser. No 149
Paper presented at DRIP 95

Measurement and mapping of materials parameters for gallium arsenide wafers by infrared transmission topography

M. G. Mier

Solid State Electronics Directorate, Wright Laboratories; WL/ELDM, Wright-Patterson Air Force Base OH 45433-7323

D. C. Look and D. C. Walters

University Research Center, Wright State University, Dayton OH 45435

Abstract. Polished wafers of semiinsulating (SI) undoped GaAs or of doped conducting GaAs are important for manufacture of monolithic microwave integrated circuits or of junction light emitters. For SI wafers, the EL2 defect causes the SI state, and variation in EL2 density can cause on-wafer variations in device isolation and other device properties. With conducting materials, crystalline dislocations cause dark-line defects and other recombination centers that limit carrier lifetime in fabricated lasers. High free carrier concentration leads to low series resistance ohmic contacts and is very desirable in semiconductor lasers. In the process of evaluating these materials, we have found that infrared transmission measurements can provide dense data on device-pertinent materials parameters for correlation to device parameters [1]. Our custom color maps of these materials parameters keyed to color histograms of the measurement data can provide informative presentations of very large data sets for comparison to device measurements [1,2]. For example, we discuss nondestructive neutral and total EL2 density measurements in SI GaAs [1] and nondestructive dislocation density and free carrier concentration measurements in GaAs:Si [2], both by infrared topography. When test devices can be fabricated at known positions on the materials evaluation wafer (or on a nearby wafer from the same boule), sensitive comparison of materials evaluation data to measured device performance can be achieved. We show that topographic color maps allow meaningful comparisons of the materials measurements to device measurements at different spacings.

[1] See, for example, Sewell J S et. al., 1989 *J. Electron. Mat.* **18**, 191; Look D C et. al. 1989 *J. Appl. Phys.* **65**, 1375; and Mier M G et. al., 1992 *Solid State Electron.* **35**, 319
[2].Look D C et. al., 1994 *Appl. Phys. Lett.* **65**, 2188

1. Introduction

Our purpose in measuring materials parameters on gallium arsenide wafers is, first, to provide feedback at every stage of materials growth and processing that has already been done, and, second, to predict how materials will perform after subsequent processing, usually into some kind of device. We will consider two classes of gallium arsenide wafers: semiinsulating (SI) wafers [1] that are typically used as substrates for electronic devices, and doped conducting wafers such as the silicon-doped wafers [2] used in the fabrication of solar cells and edge- emitting lasers. For the SI wafers, the device-pertinent materials parameters that can be easily measured by infrared transmission are neutral and total EL2 density, which can be measured nondestructively [1], and dislocation density, which requires etching the wafer [1]. For silicon-doped gallium arsenide, EL2 density is typically very small, and the materials parameters that can be measured readily are dislocation density and free carrier concentration , both of which can be measured nondestructively [2]. With appropriate optics, all of these quantities can be measured to half-millimeter resolution or better on the wafer.

This brings us to the second topic of this conference: meaningful displays of truly large quantities of data. We have developed a unique colormapping display of these data. Our technique is to calculate a color-coded histogram of values measured, then plot the colors on a map of the wafer. We find this scheme very useful in providing a meaningful display of huge quantities of data.

2. Apparatus

Data are acquired from the apparatus shown schematically in Fig. 1. The wafer is mechanically scanned past a beam from a tungsten-halogen light source which is apertured and focused through a quarter-meter monochromator and an electromechanical chopper into a 0.5 mm square spot on the sample. Infrared light transmitted by the sample is detected by a thermoelectrically-cooled germanium diode detector operating in the low-noise zero-bias mode. The detector output is synchronously detected in a commercial lock-in amplifier, the amplitude digitized and stored in a file on the controlling computer with the on-wafer coordinates. Measurement of the 16,597 locations on a three-inch wafer typically requires about an hour, though data can be acquired faster. For 100-mm wafers we reduce the resolution to 0.6 mm and measure 19,287 locations.

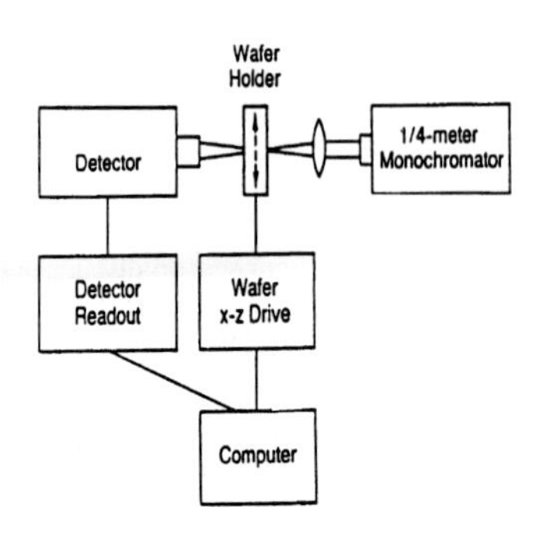

Fig. 1 Schematic of the apparatus

3. Data display

Comprehending the meaning of these datasets and comparing these data to the smaller datasets which obtain from other measurement techniques (e.g, photoluminescence and Hall effect measurements on pieces cut from the wafer) can be very difficult. We find that ranking the data into, typically, fourteen bins and assigning a color to each bin, then plotting that color at each location where the measured value corresponds to a bin range, gives an easily interpreted colormap of the measured values keyed to the value's location on the wafer. Any correlations between datasets are immediately seen by visual inspection, even when the datasets have very different resolutions on the wafer. In Fig. 2 we compare

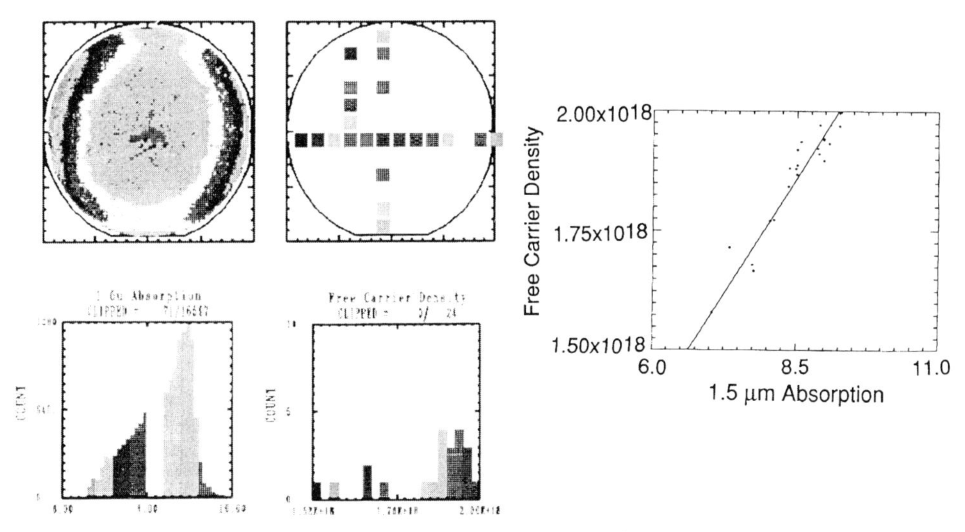

Fig. 2. 1.5 μm absorption map and histogram of values (on left). Hall-effect-measured free carrier density measurements on 6 mm square pieces and histogram of values (middle). Correlation plot of absorption vs. free carrier density (on right).

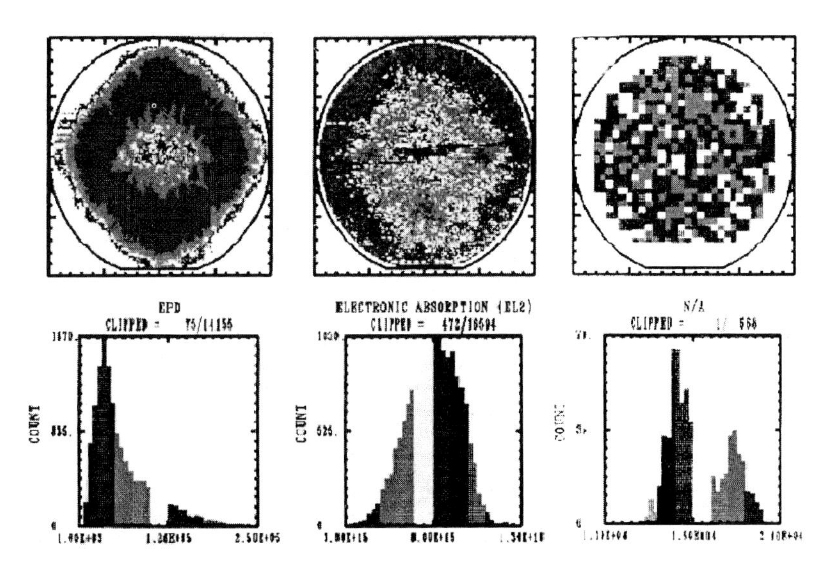

Fig. 3. From left, dislocation density, EL2 density, and ungated, unrecessed FET source-drain saturation current for a semiinsulating GaAs wafer.

absorption of a three-inch silicon-doped gallium arsenide wafer at 1.5 μm (0.5 mm resolution, 16,597 locations) to Hall-effect-measured free carrier density (24 locations). The Hall-effect samples were sized 6 mm X 6 mm, so 144 absorption values (corresponding to the area of the Hall-effect sample) were averaged for comparison to the Hall-effect-measured values. Fig. 3 shows the dislocation density, neutral EL2 density, and ungated, unrecessed source-drain saturation current (an early stage in FET fabrication) for a three-inch SI gallium arsenide wafer, an example of comparing datasets with similar (0.5 mm) on-wafer resolution.

4. Comparison of materials measurements to device measurements

Any attempt to compare a materials measurement to a device measurement requires either a nondestructive materials measurement (so that devices may be processed on the same wafer) or the assumption that the materials properties correspond on adjacent (or nearby) wafers. Further, the material property must vary outside of the usual statistical uncertainties; then a variation in device properties can be correlated with this variation in a materials property. This is complicated by the size of devices patterned on the wafer, which is usually larger than the resolution of an optical materials measurement. Finally, there must be enough corresponding locations on the wafer that we can see corresponding patterns (if they exist). With all these conditions and caveats, it is not surprising that the literature contains many publications that find little or no correlation between any materials measurement and any device measurement. Obviously there would be no correlation between totally different materials. Our observation is that, under the proper conditions, materials measurements do indeed show variations that correspond to variations in device measurements.

Our colormapping technique provides an ideal means to investigate relatively obscure correlations between materials properties and device properties. In Fig. 4 we display four *dc* measurements (source-drain resistance, gated source-drain saturation current, pinchoff voltage, and associated gate voltage) for devices patterned on a wafer from the same boule and adjacent to the materials measurement wafer in Fig. 3. Finally, in Fig. 5 we display three microwave measurements (*rf* source-drain resistance, small device S_{22}, and MIMIC device S_{22}) on this same device wafer. The colormaps of the device measurements show decreasing similarity to the materials measurement maps as more processing is done on the wafer, an effect that is easily discerned with colormaps but scarcely shows up at all in grayscale maps.

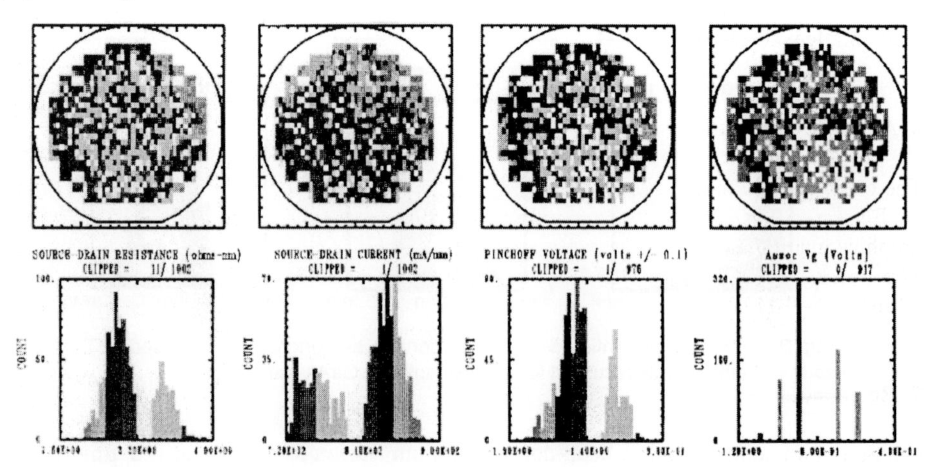

Fig. 4. Source-drain resistance, gated I_{dss}, pinchoff voltage, and associated gate voltage for a wafer from the same boule and adjacent to the wafer in Fig. 3.

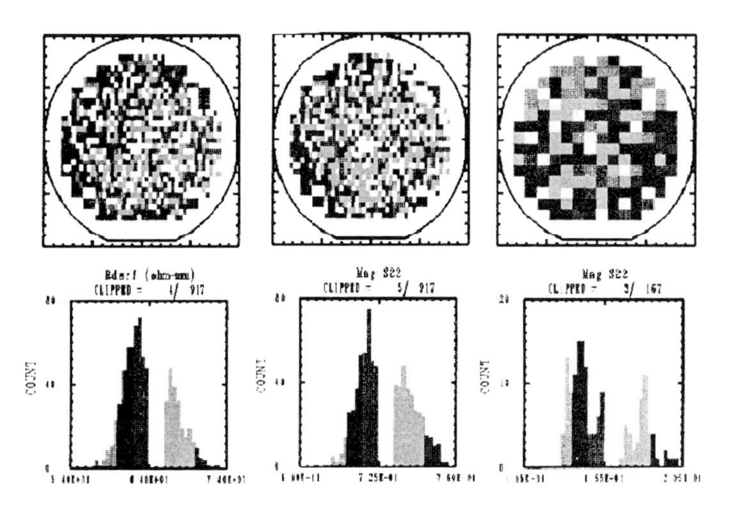

Fig. 5. *rf* source-drain resistance, small-device S_{22} magnitude, and MIMIC device S_{22} magnitude.

5. Discussion

There are several requirements for meaningful comparison of two datasets, especially if the datasets have different resolutions on-wafer. First, each dataset must have elements that vary from each other by more than the experimental uncertainty. Second, the values of measurements at corresponding locations on the wafer must be compared. If on-wafer resolutions are different, the higher resolution measurement values must be averaged over the area corresponding to each lower resolution measurement (as in Fig. 2). Both of these requirements can be met using computer programs to produce colormaps of the data. Visual inspection of the colormaps then quickly reveals any rough correlations and more detailed mathematical correlations can be carried out as desired.

6. Acknowledgments

The original Fortran-coded colormapping software using STI library graphics was written by D. Elsaesser. We thank J. Sizelove for creating the colormaps, J. Demers for help with graphics manipulations, and D. L. Beasley for constructing the apparatus and sample holders. The work of DCL and DCW was performed with AFOSR support under Contract F33615-95-C-1619.

7. References

[1] See, for example, Sewell J S et. al. 1989 *J. Electron. Mat.* **18**, 191, Look D C et.al., 1989 *J. Appl. Phys.* **65**, 1375, Look D C et. al. 1989 *J. Electron. Mat.* **18**, 487, and Mier M G et. al., 1992 *Solid State Electron.* **35**, 319

[2] D C Look et. al., 1994 *Appl. Phys. Lett.* **65**, 2188

Inst. Phys. Conf. Ser. No 149
Paper presented at DRIP 95

Device Degradation and Defects in GaAs

K. Wada, H. Fushimi, N. Watanabe, and M. Uematsu
NTT LSI Laboratories, Atsugi, Kanagawa 243-01, Japan

Abstract. This paper describes degradation behavior and factors of heterojunction bipolar transistors of the GaAs/AlGaAs system. Impurity-related degradation and point-defect-related degradation are discussed. Hydrogen is found to be a typical example of the impurity-related factor. We propose that intrinsic and extrinsic Frenkel-pair generation induced by surface-Fermi-level-pinning are the most probable mechanism for incorporating degradation factors related to point defects in heavily doped n-type and p-type layers.

1. Introduction

Device degradation is one phenomenon in which the nature and behavior of defects could be unveiled. In fact, dislocation-related problems in metal-semiconductor field effect transistors (MESFETs) [1] and semiconductor heterojunction laser diodes (LDs) [2] in GaAs-based systems evidently show how point defects interact with dislocations. It is therefore important to achieve a deeper understanding of defect physics from the degradation phenomena in order to develop electronic and/or photonic devices.

Recently, another type of degradation phenomenon has emerged in AlGaAs/GaAs heterojunction bipolar transistors (HBTs): current gain reduction during long-term operation [3]. Since HBTs are minority-carrier injection (MCI) devices, the degradation can be classified into a category including those of LDs or light emitting diodes (LEDs). Nonetheless, it is strongly expected that new aspects of defect physics will appear, since MCI basically occurs in the heavily doped base layer. This situation is significantly different from LDs and LEDs in which MCI occurs in the undoped layer.

This paper briefly introduces three types of degradation of HBTs: the base dopants are Be, C with unintentional H, and C without H. The current understanding of the degradation mechanism is described.

2. Experimental

In general, device structures are so sophisticated that it is not easy to classify factors responsible for degradation, i.e, material factors and process-induced factors. Since HBTs are no exceptions, we used simple p-n diodes instead of the HBT structures and employed wet etching instead of dry etching in order to isolate material factors.

In HBTs with Be-doped base layers, the degradation is induced by Be enhanced diffusion toward the emitter layer under MCI, resulting in a shift of the emitter/base junction into the AlGaAs layer [3]. Thus, the current gain decreases quickly because of disappearance of the heterojunction barrier against holes. In order to measure such Be anomalous diffusion, Esaki tunnel (p^+-n^+ GaAs) diodes were used [4]: They enable one to analyze Be diffusivity from the reduction in its peak current density. The epi-structures were prepared by molecular beam epitaxy (MBE) and Be acceptors had a concentration of 4 x 10^{19} cm^{-3} and Si donors 1 x 10^{19} cm^{-3}. Experimental details are given in Ref. 4.

In HBTs with C-doped base layers, the emitter/base junction shift observed in Be-doped HBTs was not detectable [5]. However, long-term operation increases leakage in the base current at low base-bias-voltage in a way that the ideality factor increases more than two. This reduces the current gain, leading to degradation. Here, we used p^+-GaAs:C / n-$Al_{0.3}Ga_{0.7}As$ heterojunction diodes, which simulate the base/emitter heterojunction of HBTs [6]. The structures with C acceptors with a concentration of 3 x 10^{19} cm^{-3} and with Si donors in 2 x 10^{18} cm^{-3} were grown by metalorganic vapor phase epitaxy (MOVPE). Experimental details are shown in Ref. 6.

In both cases, injection current densities ranged from 1 to 2 kA/cm^2 which corresponds well to the base current density of HBTs. Current injection was performed at temperatures between room temperature and 200 °C. In the Esaki diodes, the peak current density was monitored. In the heterojunction diodes, current leakage was measured at low bias voltage, here 0.7 V. Both measurements were done at room temperature.

3. Results

3.1 Be-doped HBTs.

The diffusivity of Be in the Esaki diodes can be analyzed by measuring the reduction in peak current density. Typical results of Be diffusivity obtained are shown in Fig. 1. The thermal activation energy was 1.8 eV which corresponds closely to the one obtained by thermal diffusion experiments of Be in GaAs [7]. However, it was dramatically reduced to 0.6 eV with MCI. The diffusivity enhancement was reduced with the duration of MCI. The diffusivity in Fig. 1 is the one extrapolated to the initial stage (t->0).

On the other hand, the intensity of electro-luminescence from the diodes increased with the MCI time [8]. This suggests that the diffusivity enhancement should be accompanied by reduction of recombination centers. Indeed, the rate of increase of the luminescence is almost identical to the rate of reduction of enhanced Be diffusivity. We have thus proposed a model in which the diffusivity is enhanced in terms of recombination enhanced defect reaction (REDR). In this model the reduction of diffusivity enhancement is due to injection annealing of recombination centers and the activation energy reduction of 1.2 eV should be related to the value of energy obtained from the non-radiative recombination process at recombination centers.

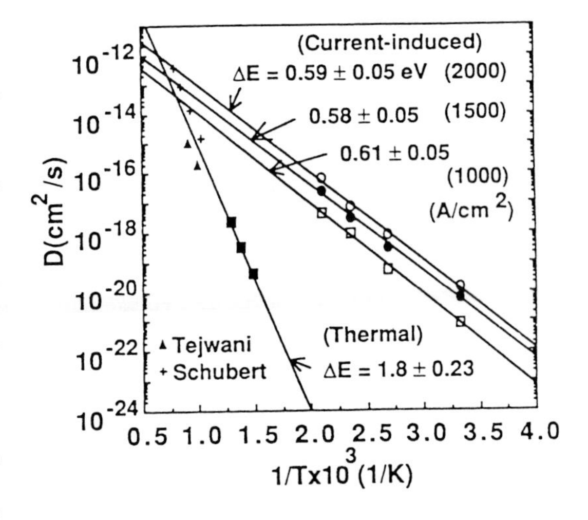

Figure 1 Recombination-enhanced Be diffusion.

Since Be is known to undergo kick-out diffusion (Be_{Ga} + I_{Ga} <--> Be_i) [9], the diffusivity enhancement may require an increase in I_{Ga} concentration. According to Hobson, et al. [10], I_{Ga} is abundant since it is generated during the growth of the n^+ layer due to surface-Fermi-level-pinning. It is likely that I_{Ga} would form complexes with impurities or I_{Ga} themselves in the n^+ layer of the Esaki diodes because of its instability. Assuming that these complexes would act as recombination centers and can release I_{Ga} through injection annealing of the centers, they diffuse into the p^+ layer to kick Be_{Ga} off into an interstitial site, which enhances Be diffusivity. This model should be experimentally confirmed.

3.2. C-doped HBTs.

In C-doped HBTs, degradation occurs with an increase in the base leakage current [6]. We found that H acts as a degradation factor in the following two ways: Isolated H donors and C-H complexes. H impurities are incorporated in the C-doped base layer at a concentration of more than 1×10^{18} cm^{-3}. The isolated H donors that are incorporated during the cooling stage of the growth start to increase the base current leakage at low bias voltage under MCI. The degradation is quite rapid. It is not yet clear how the increase in leakage current is related to hydrogen. Wel will come back it later.

C-H complexes increase the base current leakage as well. The increase rate is much slower than that by the isolated H donors. The degradation related to C-H complexes should start with the decomposition of C-H complexes. The leakage current increase was found to be accompanied by an increase in holes, i.e., reactivation of C in the base layer. The increase rate corresponds closely to the C reactivation rate on a quantitative basis. Since C was passivated by H, the reactivation means that C-H complexes decompose under MCI, resulting in isolated H [11]. As noted above, once isolated H is formed, the current leakage increases.

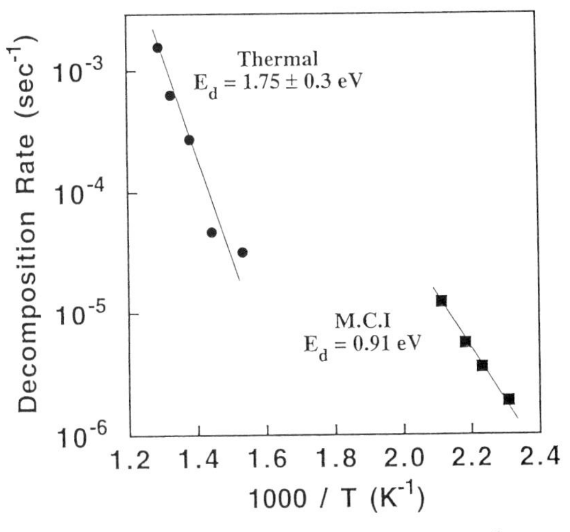

Figure 2 Decomposition rate of C-H complexes.

The activation energy of C-H decomposition is about 1.7 eV without MCI while it is about 0.9 eV with MCI, as shown in Fig. 2 [6]. Since C-H complexes do not act as recombination centers, the decomposition mechanism is not related to REDR. Breuer, Jones, Oberg and Briddon (hereafter referred to as BJOB) recently calculated the decomposition energy considering the charge state [12]. According to them, the thermal decomposition energy is 1.8 eV, while it is reduced to 0.9 eV when a C-H complex captures a single electron. Our result is in good agreement with their calculation. This indicates that the decomposition enhancement mechanism under MCI is related to a charge state effect (CSE).

The increase rate of the leakage current is proportional to the square of the injection current density, i.e., I^2. This requires two electrons for the decomposition of C-H complexes. BJOB have proposed that another electron must be consumed in changing the charge state of H^0 to H^-. This is because an isolated H is a negative U type defect as predicted so far [13] where H^0 states are unstable. If their model is right, our I^2 dependence would be the direct confirmation that isolated H has a negative U character.

3.3 C-doped HBTs without H.

Very slow degradation is observed under MCI even if most H are evacuated by thermal annealing. This degradation is also accompanied by current leakage at low bias voltage. The cause of leakage is not clear, but it is not H-related.

It has been reported that hole concentration is decreased by high temperature thermal annealing and the lattice constant is simultaneously restored [14]. Following Vegard's law, the decrease in hole was quantitatively explained by the decrease in C impurities on the As sublattices. Thus, the decrease in C impurities on the As sublattices causes both the hole decrease and the lattice constant restoration.

The decrease of C impurities on the As sublattice can be explained as follows. According to Baraff and Schulter, V_{As} should be abundant as a doubly ionized donor, when the Fermi level is located near the valence band maximum [15]. Thus, a high concentration of V_{As} donors is required in the heavily-C-doped p-type GaAs base. V_{As} donors, needed to equilibrate the p^+ GaAs:C system, can be generated in at least two ways: intrinsic and extrinsic F-pair generation on the As sublattices (As_{As} <--> V_{As} + I_{As} and C_{As} <--> V_{As} + C_i). Schottky type V_{As} generation is less probable because of Fermi-level-pinning at the sample surface [16]. In the latter case, it is obvious that C acceptors become inactive and the lattice constant is restored. In the former case, C is deactivated by forming complexes with I_{As} and is even kicked off by I_{As} into interstitial sites C_i.

It has also been reported that photoluminescence (PL) intensity is decreased by high-temperature thermal annealing [14]. The formation of recombination centers indicated by PL could also be related to the carbon deactivation. Recombination centers are often interstitial type defects.

If V_{As} by the F-pair generation has a high enough concentration, then they cause hole concentration to saturate. Figure 3 shows a simple calculation predicting hole saturation effects in the C-doped layer (p = $[C_{As}] - 2\,[V_{As}]$). In this calculation, we employed the way proposed by Walukievicz [17]. The solid circles are experimental results. The calculation indeed reproduces the saturation behavior in GaAs:C system.

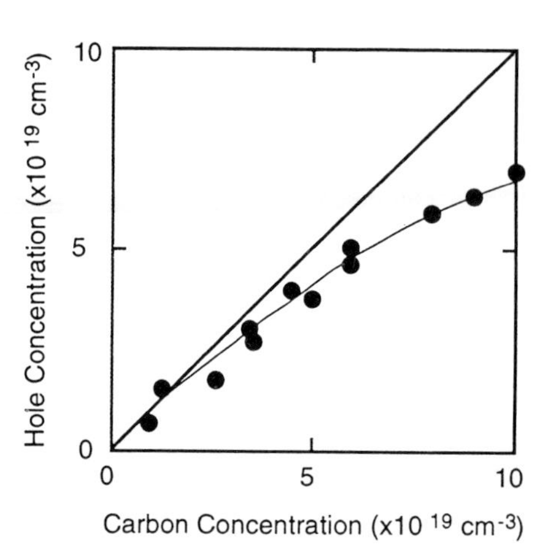

Figure 3 Hole saturation effects in C-doped GaAs

In the case of Be-related degradation, we have suggested the F-pair generation on the Ga sub-lattices to be a factor. Here, F-pair generation on the As sublattice is suggested. Although information on the As sublattices, some reports indeed support our suggestion: slow degradation of HBTs was accompanied by C precipitation by transmission electron microscopy (TEM) [18]. Precipitation needs a mobile C species, i.e., C interstitials Ci, since C on the As sublattices is a slow diffuser. Further studies are expected.

4. Discussion

We showed that the degradation factors are impurity-related and point-defect-related. The former example is hydrogen. C-H complexes are decomposed by CSE under MCI to release hydrogen, which leads to device degradation. However, it is an open question how H dissociated degrades devices at issue. It has been reported that {111} platelet defects are observed in a degraded HBT [5]. The platelets are similar to microdefects formed by H plasma-irradiation in Si [19]. If {111} platelets observed in degraded HBTs are precipitates containing H, the degradation should starts with the agglomeration reaction of H into H_2 of H_2* molecules or larger clusters. Weman, et al. proposed that oxide precipitates in Czochralski Si are one source of deep luminescence [20]. According to them, such precipitates compress the Si lattice to shrink the bandgap because of the negative deformation potential of Si and form quantum boxes. In GaAs the lattice compression induced by the {111} platelets should expand the bandgap. So, provided that the same argument is applicable to the platelets, the edges of the platelets may confine carriers via "quantum ring" formation, which may provide a recombination path between carriers.

We proposed that point-defect-related factors are self-interstitials which are generated as by-products in terms of Frenkel pair formation reactions on the group III and V sublattices. Since the number of self-interstitials to be generated is equal to that of vacancies, factors determining the vacancy concentration are very important. There are at least three factors to be considered: Fermi level E_F, stoichiometry, and temperature. Here we discuss the Fermi level effect.

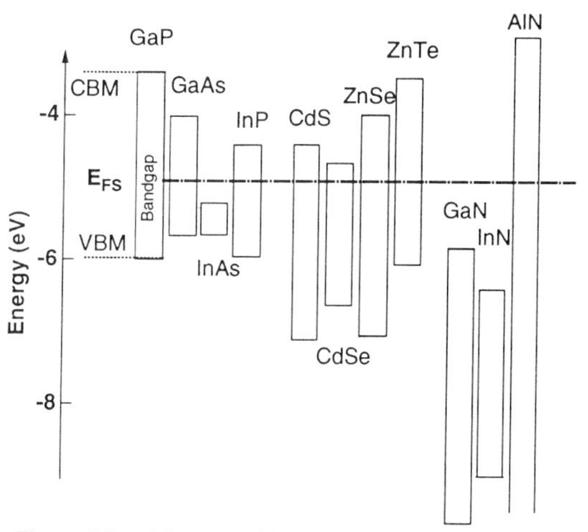

Figure 4 Band lineup and E_{FS} of compound semiconductors.

Following the amphoteric native defect model [17], the Fermi level stabilization energy E_{FS} is the energy where the formation energy of vacancy acceptors is equal to that of vacancy donors, and E_{FS} is always located at a certain depth from the vacuum level. Thus, both vacancies could be abundant if E_{FS} is located near the midgap. In other words, vacancy donors would be abundant in p-type while vacancy acceptors would be abundant in n-type in the materials with midgap E_{FS}. Figure 4 shows the relationship between band lineup [21] and E_{FS} of various compound semiconductors. Most of the semiconductors have E_{FS} located within their bandgap except InAs, GaN, and InN.

It is well-known that the LDs based on InGaAsP systems are robust against degradation. The band lineup for the alloy system is not shown in Fig. 4, but E_{FS} of materials for 1.55 μm emission would be resonant or close to resonant with the conduction band. Blue LEDs based on GaN and InN systems do not easily degrade. In clear contrast, II-VI blue LDs show rapid degradation. These can be explained by the position of E_{FS} in these materials. In p-type InAs, GaN, and InN, vacancy donors should be abundant, while in the n-type materials vacancy acceptors should be few because of the high formation energy. Consequently, it is expected that when E_{FS} is located near one of the band edges or even resonant with the bands, one type of vacancies becomes few, leading that both types of self-interstials are not available in such materials. This may be related to the robust properties of these material systems. Considering a model that the presence of one type of point defects can explain the degradation in the LDs [22], it can be fair to say that the location of E_{FS} in bandstructures should be one factor to be considered. In real device structures, we have to consider another factor induced by device processing as well.

5. Summary

The cunderstanding of HBT degradation processes is as follows:
1. Extrinsic and intrinsic point defects like hydrogen and self-interstitials are material factors of degradation. They are incorporated in heavily doped layers to form complexes in as-grown state.
2. MCI causes these extrinsic and intrinsic point defects to dissociate from these complexes via REDR and/or CSE. These point defects lead to HBT degradation.
3. Degradation of HBT without H is still unclear but inactive C, most likely a C interstitial, is probably a degradation factor.
Surface-Fermi-level-pinning-induced Frenkel pair formation is the most probable mechanism for incorporating intrinsic point defects in devices with heavily doped layers. A model is proposed for robust properties of compound semiconductors against device degradation.

References

[1] Y. Nanishi, H. Yamazaki, T. Mizutani and S. Miyazawa, 1981 Intern. Symp. GaAs and Related compounds, Oiso (Inst. Phys. Conf. Ser. 1982) p. 7.
[2] R. L. Hartman, L. B Schwartz, and M. Kuhn, Appl. Phys. Lett. 18, 304, 1971.
[3] O. Nakajima, H. Ito, T. Nittono, and K. Nagata, Technical Digest of Intern. Electron Device Meeting (IEEE, New York, 1990) p. 673.
[4] M. Uematsu, and K. Wada, Appl. Phys. Lett. 58, 2015, 1991.
[5] H. Sugahara, J. Nagano, T. Nittono, and K. Ogawa, GaAs IC Symposium (IEEE, 1993) p. 115.
[6] H. Fushimi and K. Wada, Mater. Sci. Forum (IVDS-18) in press.
[7] M. J. Tejwani, H. Kanber, B. M. Paine, and J. M. Whelan. Appl. Phys. Lett. 53,2411,1988.
[8] M. Uematsu, K. Yamada, and K. Wada, J. Appl. Phys. 72, 2520, 1992.
[9] U. Goesele, and M. Morehead, J. Appl. Phys. 52, 4617,1981.
[10] W. S. Hobson, S. J. Pearton, A. S. Jordan, Appl. Phys. Lett. 56, 1251, 1990
[11] Isolated H donors generated by the decomposition still compensate for C acceptors, but in reality the holes increase. This is explained by assuming that isolated H donors becomes inactive by forming H_2 molecules.
[12] S. J. Breuer, R. Jones, S. Oberg and P. R. Briddon, Mater. Sci. Forum (IVDS-18) in press.
[13] L. Pavesi, and P. Giannozzi, Phys. Rev. B-46, 4621, 1992.
[14] K. Watanabe and H. Yamazaki, Appl. Phys. Lett. 59, 434, 1991.
[15] G. Baraff and M. Schulter, Phys. Rev. B-33, 7346, 1986.
[16] K. Wada, Appl. Surf. Sci. 85, 246, 1995.
[17] W. Walukievicz, Appl. Phys. Lett. 54, 2094, 1989.
[18] T. Takahashi, S. Sasa, A. Kawano, T. Iwai, and T. Fujii, Intern. Electron Device Meeting, (IEEE, 1994) p. 191.
[19] J. N. Heyman, J. W. Ager III, E. E. Haller, N. M. Johnson, J. Walker, and C. M. Doland, Phys. Rev. B 45, 13363, 1992.
[20] H. Weman, B. Monemar, G. S. Oherlein, and S. J. Jeng, Phys. Rev. B-42, 3109, 1990.
[21] S. Sakai, Y. Ueta, and Y. Terauchi, Jpn. J. Appl. Phys. 32, 4413, 1993.
[22] P. M. Petroff, and L. C. Kimerling, Appl. Phys. Lett. 29, 1976, 461.

Inst. Phys. Conf. Ser. No 149
Paper presented at DRIP 95

Influence of Multi Wafer Annealing of LEC GaAs Substrates on the Quality of epitaxial Layers

J. Forker[1], M. Baeumler[1], J.L. Weyher[1], W. Jantz[1], D. Bernklau[2], H. Riechert[2] and T. Inoue[3]

[1] Fraunhofer Institut für Angewandte Festkörperphysik, Tullastr. 72, 79108 Freiburg, Germany
[2] Siemens AG, Otto-Hahn-Ring 6, 81730 München, Germany
[3] Japan Energy Corporation, Toda, Saitama 335, Japan

Abstract: The lateral distribution of dislocations and nonradiative recombination centers in bulk LEC GaAs is reproduced in the epilayer. Combined ingot/multi wafer annealing reduces the excess As concentration at dislocations and generates As matrix precipitates. We find that the presence of matrix precipitates does not adversely affect the epitaxial layer quality. On the other hand layer defects correlated with substrate dislocations are still visible. Hence either the ingot/wafer annealing does not sufficiently reduce dislocation-correlated substrate defects migrating into the epilayer. Alternatively, nonradiative recombination centers are generated during epitaxial growth near dislocations propagating from the substrate into the epilayer.

1. Introduction

The quality of epitaxial layers is strongly determined by the morphological and chemical characteristics of the substrate surface, but also by its bulk material properties, in particular by dislocations and nonradiative recombination centers (NRRC). The lateral distribution of these defects is reproduced in the epilayer [1,2]. The appearance of NRRCs in the epilayer may be due either to diffusion from the substrate or to gettering of deposition-induced defects at grown-in dislocations. Thus far the microscopic structure of the NRRC has not been established unambiguously, but is likely to be correlated with an intrinsic defect. In particular, excess As, known to be present in various forms in state-of-the art GaAs, has been suggested. In order to identify the role of the substrate in generating NRRCs in epitaxial layers, we studied GaAs layers grown on non-annealed or ingot annealed LEC GaAs crystals, in part further treated by two different wafer annealing procedures. Wafer annealing is known to reduce substantially the excess As content in a near-surface layer [3].

The substrates and epilayers were characterized by photoluminescence spectroscopy (PLS), topography (PLT) and microscopy (PLM), electron temperature topography (ETT) and light-assisted diluted Sirtl-like (DSL) etching. The influence of threading dislocations as well as decoration and matrix precipitates on the quality of the epitaxial layer will be discussed.

2. Experimental

Five LEC-GaAs wafers, cut from the same crystal but subjected to different annealing procedures, were investigated. Part of the crystal was ingot annealed (IA). Two wafers of each part were further treated by two different multi wafer annealing techniques (new MWA and old MWA). One wafer was only ingot annealed. The temperature versus time profiles of the annealing procedures are not known but should be similar to those described in Ref. [3] and [4].

The test structure consists of a 1.2μm MBE GaAs layer preceded by a 50nm GaAs buffer layer and a 10 period (5nm GaAs/5nm $Al_{0.2}Ga_{0.8}As$) superlattice. It is capped with a 20nm $Al_{0.2}Ga_{0.8}As$ barrier and a 5nm GaAs layer.

PLT as well as ETT were performed at 2K. The set-up and technique is described in Ref. [5]. For ETT the temperature of the thermalized electron gas T_e is measured optically by recording the exponential high energy tail of the band-to-band recombination. T_e has been evaluated topographical within 5x5 mm^2 sample areas. PLM images were obtained at about 80K with the PLM set-up used by Wang et al. [1]. The DSL etching procedure is described in Ref. [6]. Correlation between structural features in the substrates and epilayers was obtained using sequential etching and documentation. The images of the etched samples were acquired with a standard differential interference contrast (DIC) microscope.

3. Pre-characterization of the substrates

The photoluminescence (PL) spectra of the differently annealed substrates are similar with respect to spectral details and overall intensity. Only for the old MWA substrate the PL intensity is smaller by a factor of 4.

The PL intensity topograms reproduce the cellular structure typical for LEC GaAs with high intensity in the cell walls and low intensity in the center. The size of the cells is about the same for the five substrates, indicating comparable dislocation density, but the relative variation of the PL intensity across the cells differs. Two distinct homogenization effects were observed: a reduction of the intensity contrast amplitude and a broadening of

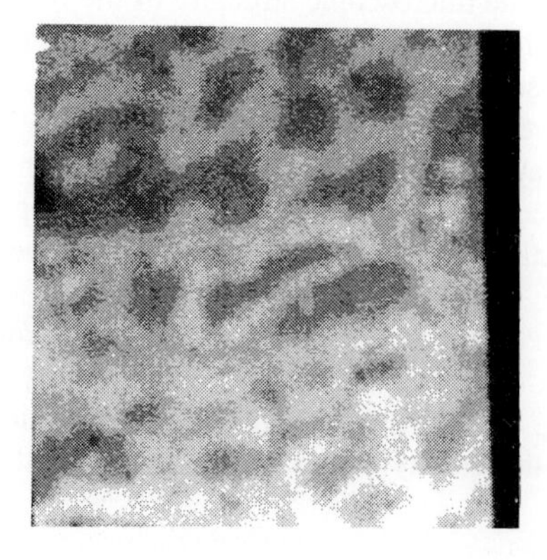

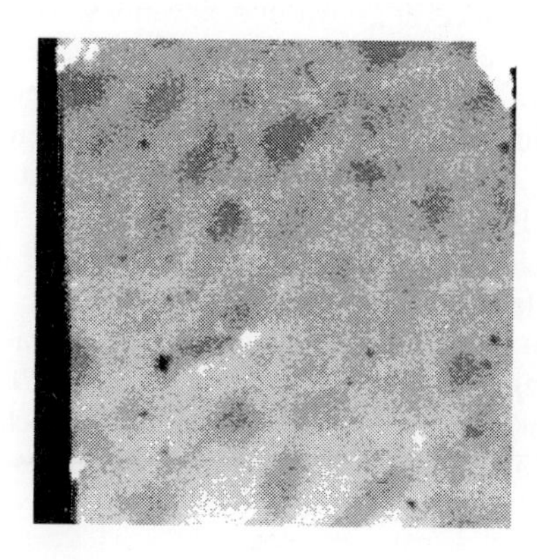

Fig. 1. Photoluminescence topograms of substrates subjected to new MWA (a) and combined ingot/new MWA (b).

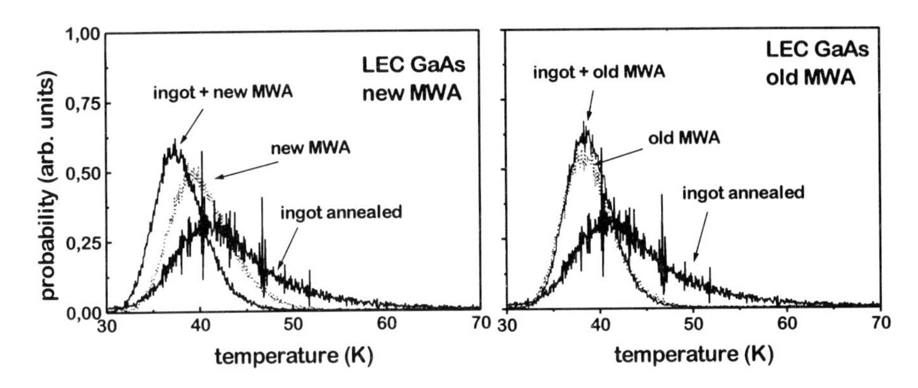

Fig: 2. Histograms of electron temperature for the substrates subjected to new MWA (a) and old MWA (b).

the cell walls at the expense of the interior cell area. As compared to samples that were only wafer-annealed, the intensity contrast is reduced by about a factor of two for wafers which had been cut from the ingot annealed crystal. Additional MWA does not result in a further reduction of the fluctuation amplitude. However, as can be seen from Fig. 1, homogenization of the material by broadening of the cell walls is more efficient for ingot annealing prior to new MWA (Fig. 1(b)). This is not observed for combined ingot/old MWA where the cell walls remain sharp.

The superiority of the combined ingot/new MWA is demonstrated by ETT. In Fig. 2(a) the electron temperature distribution is plotted histographically for the two new MWA substrates of Fig. 1. For comparison the histogram obtained for material with ingot annealing only is also shown. This material shows a wide distribution peaked at higher temperature, corresponding to a more inhomogeneous distribution as well as a higher average density of NRRCs [5]. Additional MWA results in a reduction and homogenization of NRRCs. The results in Fig. 2(a) and (b) point out that wafer annealing without ingot annealing is superior to ingot annealing without wafer annealing .

In order to estimate the degree of precipitation near the surface as well as in the bulk two samples of each wafer were photo-etched. The first sample was etched at the existing surface. The second sample was pretreated by mechanical polishing to remove 100µm prior to DSL etching.

Fig. 3(a) shows as one example the DIC image of the sample subjected to combined ingot/old MWA after removing 100µm and DSL etching. Well defined dislocation-related hillocks and ridges belong to the cell walls. Some of them have shallow etch pits (s-pits) on top which are due to decoration precipitates (DPs). The center of the cell is nearly free of dislocations. Instead many etch pits and hillocks appear which are due to matrix precipitates (MPs). They are concentrated in the center of the cell with decreasing concentration towards the cell walls. A detailed description of the different defects observed in annealed GaAs wafers is given in Ref. [6] and [7]. The images for the bulk did not differ much. The substrates subjected to ingot annealing prior to MWA show a slightly more homogeneous distribution of MPs.

Differences appear at the surface. The surface of the combined ingot/old MWA substrate (Fig. 3 (b)) still reveals the typical cellular structure of the substrate dislocations, but neither DPs nor MPs in the center of the cell are present. The same refers to the old MWA substrate without ingot annealing. For old MWA at least the last annealing step seems to

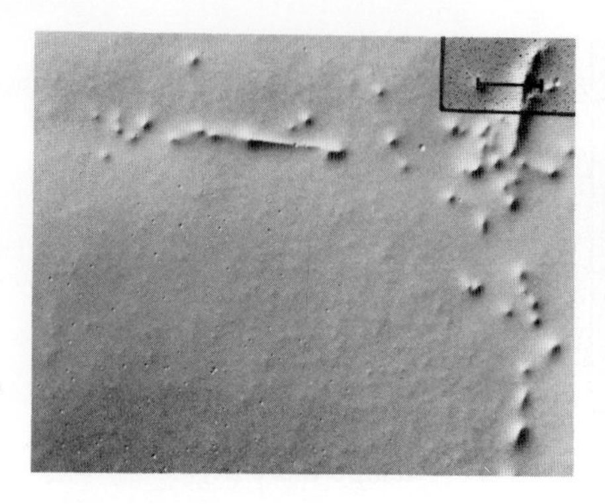

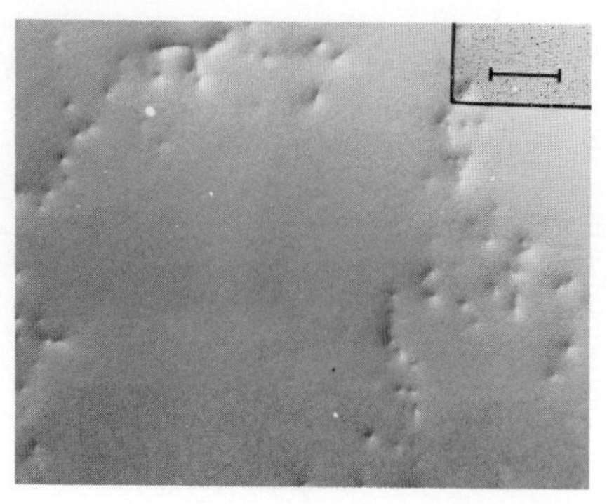

Fig. 3. Bulk (a) and surface (b) DIC image of the combined ingot and old MWA substrate after DSL photo-etching, the marker represents 50μm.

be performed without a controlled Arsenic atmosphere during annealing causing outdiffusion of Arsenic [3]. This is illustrated in Fig. 4 which shows the image of a DSL etched (110) cleavage plane; in the 60 μm thick sub-surface zone there are no MPs and the electrical properties differ as compared to the bulk of this material. The wafer subjected to ingot annealing and new MWA (Fig. 5), however, still shows some roughness due to very small MPs. For the substrate only subjected to new MWA this roughness is limited to the cell walls.

As indicated in Fig. 4 the surface area reaches about 60μm into the substrate. The photoluminescence results therefor are mainly due to this surface area. The differences observed by DSL etching for the surface of the two new MWA wafer can explain the PL topograms in Fig. 1(a) and (b). New MWA seems to cause precipitation mainly in the cell walls. The result is a high PL intensity contrast between cell wall and interior due to "cleaning" of this zone from NRRCs to form MPs. Ingot annealing prior to new MWA however seems to induce a more homogeneous precipitation and distribution of NRRCs. As a result in the PL topogram the cell walls seem to extend (Fig. 1(b)) and the electron temperature distribution is sharper and peaked at lower electron temperature (Fig. 2(a)).

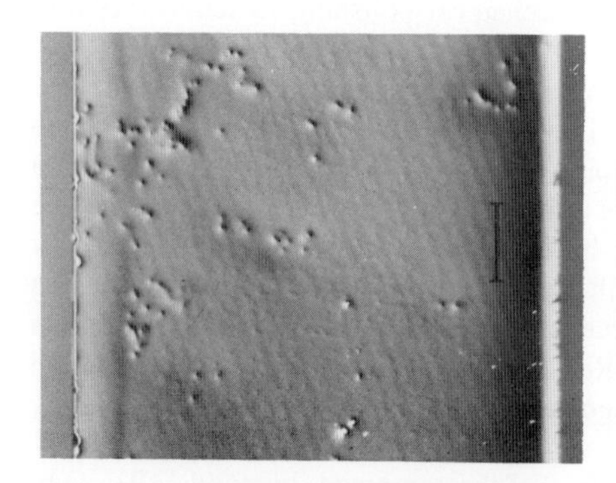

Fig. 4. DSL photo-etched (110) cleavage of a wafer after old MWA. The marker represents 100μm.

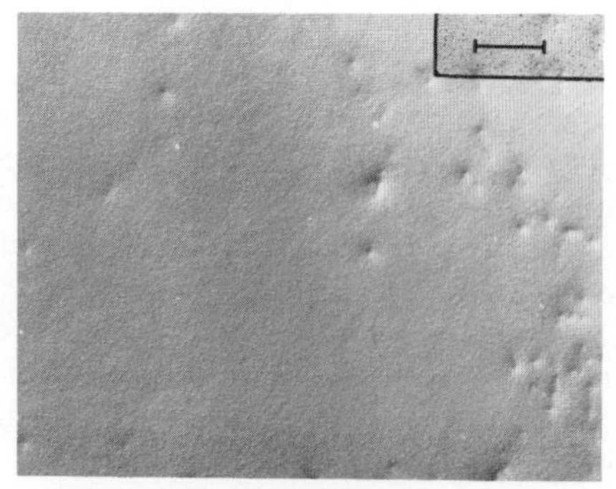

Fig. 5. Surface DIC image of the combined ingot and new MWA substrate after DSL photo-etching, the marker represents 50μm.

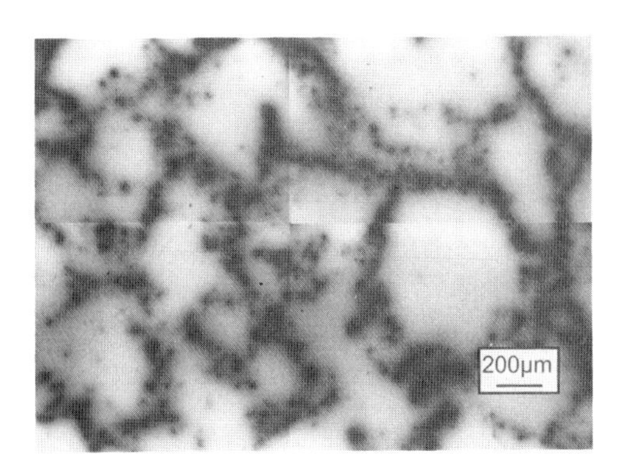

Fig. 6. PLM image of a epitaxial GaAs layer, grown by MBE.

The effect of ingot annealing prior to MWA is less pronounced for old MWA material (Fig. 2(b)).

4. Properties of epitaxial layers

The epitaxial test layer described in section 2 was deposited on the five differently pre-treated substrates and investigated with PL, PLT, PLM and DSL etching. Neither the PL spectra nor PL topograms did show significant differences. As a rule the PL topograms hardly revealed a cellular structure, partly due to competing intensity variations generated by contamination with residual donors, but also due to insufficient lateral resolution. PLM images, which are recorded with higher lateral resolution (about 6 µm as compared to about 100 µm for PLT) clearly show dark spots replicating the cellular arrangement of the substrate dislocations (Fig. 6). Comparison of the image of the DSL etched epilayer to that of the substrate (Fig. 7(a) and (b)) demonstrates that the substrate dislocations propagate into the epitaxial layer. However the PLM images as well as the images of the DSL etched epitaxial layers are practically identical for all five substrates. In particular the differences of substrate material properties induced by MWA, including matrix precipitation and excess As outdiffusion, do not result in differences of epilayer quality discernible by our investigative techniques.

Of course, this (negative) result can be interpreted in at least two different ways. One obvious explanation would be that MWA does not sufficiently reduce the defect species propagating from the substrate into the epilayer. This would mean that further improvement of the MWA procedure is needed to avoid the dark spot defects in the epitaxial layer.

Alternatively one might argue that the only significant interaction between substrate and epilayer is the propagation of substrate dislocations, whereas the pronounced variation of the excess As concentration (DPs,MPs and possibly fluctuation concentrations of interstitials) as well as NRRC agglomerations in the substrate do not affect the epilayer quality at all. To account for the occurence of dark spots in the epilayer one would further assume that NRRCs are generated during the epitaxial growth and are attracted by the dislocations. Indeed the comparison of PLM and images of the DSL photo-etched epilayers shows that the size of dark spots by far exceeds the size of isolated dislocations. Hence accumulation of NRRCs at the dislocations would be responsible for the dark spot defects.

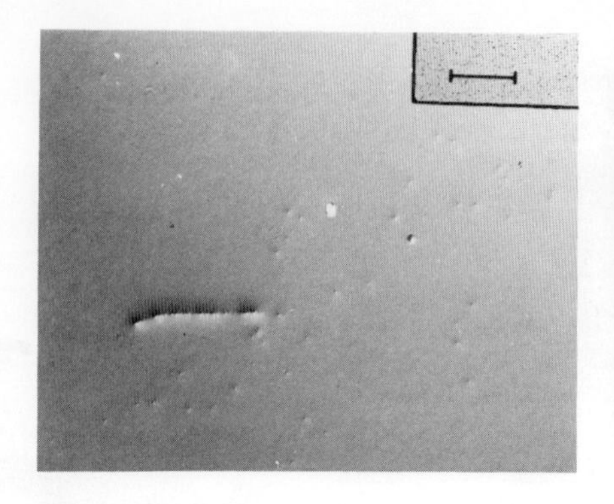

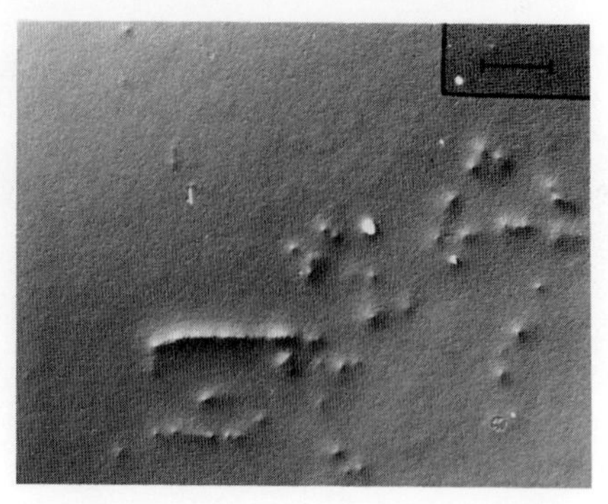

Fig. 7. Epilayer (a) and combined ingot/old MWA substrate (b) after removal of the epilayer after DSL photo-etching, the marker represents 50μm.

At present, no evidence is available to exclude either alternative. In view of the most prominent differences observed in the near surface substrate layer, in line with reported quantitative stoichiometry data, we are presently in favor of the second interpretation. However, one should keep in mind that clear correlations between substrate and epilayer properties *without* participation of dislocations have been observed unambiguously [1].

5. Conclusion

In conclusion we have shown that different combinations of ingot and wafer annealing procedures result in variations of the lateral distribution and depth variation of material inhomogeneities related to excess As. We have further demonstrated that these pronounced differences, in particular with respect to the concentration and distribution of matrix precipitates, do not influence the properties of epilayers as analyzed by DSL photo-etching and photoluminescence microscopy. The implications of alternative interpretations of these findings are discussed.

References

[1] Z.M. Wang, M. Baeumler, W. Jantz, K.H. Bachem, E.C. Larkins and J.D. Ralston, J. Crystal Growth 126 (1993) 205

[2] M. Baeumler, E.C. Larkins, K.H. Bachem, D. Bernklau, H. Riechert, J.D. Ralston and W. Jantz, Inst. Phys. Conf. Ser. 135 (1994) 169

[3] M. Mori, G. Kano, T. Inoue, H. Shimakura, H. Yamamoto and O. Oda, Semi-insulating III-V Materials, eds. A.G. Milnes and C.J. Miner, (Toronto, Canada, 1990), p. 155

[4] O. Oda, H. Yamamoto, M. Selwa, G. Kano, T. Inoue, M. Mori, H. Shimakura and M. Oyake, Semicond. Sci. Technol. 7 (1992) A215

[5] Z. M. Wang, J. Windscheif, D.J. As and W. Jantz, J. Appl. Phys. 73 (1993) 1430

[6] J.L. Weyher, P. Gall, Le Si Dang, J.P. Filliard, J. Bonnafé, H. Rüfer, M. Baumgartner and K. Löhnert, Semicond. Sci. Technol. 7 (1992) A45

[7] J.L. Weyher, C. Frigeri, P. Call and R. Kremer, Semi-insulating III-V Materials, ed. M. Godlewski, (Warsaw, Poland, 1994), p. 163

Inst. Phys. Conf. Ser. No 149
Paper presented at DRIP 95

Improved thermally stimulated current analysis in semi-insulating GaAs: new conclusions

Z-Q. Fang and D. C. Look
Physics Department, Wright State University, Dayton OH 45435, USA

Abstract. Measurements of $EL2^{o}$ and $EL2^{+}$ concentrations by IR absorption and temperature-dependent photocurrent (1.13 eV) and dark current at $80K < T < 300K$ in semi-insulating GaAs, allow a more accurate analysis of the thermally stimulated current spectrum. We conclude that trap T_2, at 220K, is related to As_{Ga} and controlled by both $EL2^{o}$ and $EL2^{+}$, and T_3, at 200K is most likely a V_{As}-related defect complex..

1. Introduction

The properties of undoped semi-insulating (SI) GaAs are mainly determined by the native deep donor EL2, believed to be an arsenic antisite (As_{Ga}) related complex, and shallow acceptors, like carbon (C_{As}) and the gallium antisite (Ga_{As}) [1]. However, more and more evidence indicates that, besides EL2, other point defects, such as arsenic and gallium vacancies (V_{As} and V_{Ga}), and their complexes with arsenic interstitials (As_i), and also with Ga_{As} and As_{Ga}, exist in SI GaAs (for example, see studies by positron lifetime measurements [2]) and play an important role in electrical compensation, carrier recombination and ion-implant activation efficiency. Actually, the activation energy of the dark conductivity for undoped GaAs is usually dominated by one of three different centers: an 0.75 eV center (due to EL2), an 0.42 eV center (unknown) or an 0.15 eV center (probably due to V_{As}-As_{Ga} [3]), depending on the crystal growth conditions and post-growth annealing. Thermally stimulated current (TSC) spectroscopy has been used to study deep centers in SI GaAs for nearly three decades [4]. The studies have included: 1) the determination of the trap parameters, i.e. the activation energy (E_T) and the capture cross section (σ) for, at least, six deep centers in the temperature range from 80K to 250K, which have been labeled by us as T_2 (0.50 eV) at 220K, T_3 (0.43 eV) at 200K, T_4 (0.29 eV) at 155K, T_5 (0.27 eV) at 140K, T_5* (0.23 eV) at 125K and T_6* (0.15 eV) at 96K for a thermal scan of 0.3 K/sec, although there are slight disagreements in the literature [4] on the actual trap depths; 2) the EL2-related infrared (IR) quenching and thermal recovery for a prominent TSC trap T_5 with $E_T = 0.27 - 0.31$ eV [5-7], as well as for several other TSC traps in the TSC spectrum; and 3) the association of relative peak height of TSC peaks, like T_2/T_3, with crystal stochiometry both among different SI GaAs wafers grown under various conditions or across the diameter of a given SI GaAs wafer [8, 9].

In this paper, we use a more accurate analysis of the TSC spectrum, based on input data from TSC, dark current (DC), photocurrent (PC), and absorption measurements on various SI GaAs samples, to reach the following new conclusions: 1) T_2 and T_3, along with T_5, are the dominant deep centers in SI GaAs; and 2) the peak height of T_2 is controlled by both $EL2^{o}$ and $EL2^{+}$, which also control the dark conductivity and the photoconductivity at $T > 200K$. We also note that T_3, with the largest capture cross section of the six traps, is most likely a V_{As}-related defect complex, and T_5 has been tentatively identified to be a As_{Ga}- or V_{Ga}-related defect [10,11]. Some examples of variations in the TSC spectra depending upon wafer uniformity, annealing history, and growth conditions will be presented to show how TSC spectroscopy can be used as a routine characterization of SI GaAs wafers.

2. Analysis of TSC spectrum for SI GaAs

A TSC spectrum for an undoped SI GaAs (see Fig. 1 below) basically consists of three portions: portion I at T > 240K, which mainly results from the DC rather than traps; portion II at 170K < T < 240K, which primarily consists of traps T_2 and T_3; and portion III at 80K < T < 170K, which includes traps T_4, T_5, T_5* and T_6*. The DC portion as well as the influence of the excess current due to sawing damage and poor polishing have been carefully studied in a previous paper [12]. If the excess current is negligible, then the bulk DC through a SI GaAs sample will be given by [13]

$$I_{DC}(T) = C_b \cdot \mu(T)\, T^{3/2} \left(\frac{N_D - N_A}{N_A}\right) \exp\left(-\frac{E_{D0}}{kT}\right) \tag{1}$$

where C_b is a constant, including the effective density of states in the conduction band, sample geometry, and applied electrical field; $\mu(T)$, the temperature dependent electron mobility; N_D, the donor (EL2) concentration; N_A, the net acceptor concentration; and E_{D0}, the donor activation energy (0.75 eV ± 0.02 eV). Experimentally, in the Arrhenius plot of the DC (for example see Fig. 2 below), we find an activation energy of 0.77 eV. Therefore, we believe that the DC, if dominated by EL2 centers, is mainly controlled by a ratio of $N_{EL2}{}^{\circ}/N_{EL2}{}^{+}$, since N_D-N_A = $N_{EL2}{}^{\circ}$, $N_A = N_{EL2}{}^{+}$ and the electron mobility does not vary as much as EL2, especially within a wafer. A given electron trap in TSC portions II and III, if fully filled at T_i, will emit electrons to the conduction band according to the following relationship [13]:

$$n = n_T \tau_n e_n(T) \exp\left(-\int_{Ti}^{T} \frac{e_n(T)}{a} dT\right) \tag{2}$$

where n is the free electron concentration; n_T, the trap concentration; τ_n, the carrier lifetime; T_i, the initial temperature; a, the heating rate; and e_n, the emission rate given by

$$e_n(T) = C_e \left(\frac{g_0}{g_1}\sigma_{\infty} \cdot e^{\frac{\beta}{k}}\right) T^2 e^{-(E_{T0}+E_{\sigma})/kT} \tag{3}$$

Here, σ_∞, E_σ , E_{T0} and β are defined by $\sigma = \sigma_\infty \exp(-E_\sigma/kT)$ and $E_T = E_{T0} - \beta T$, where σ is the capture cross section. With an applied field $\mathcal{E}$, the TSC produced by emitted electrons from the trap will be given by

$$I_{TSC}(T) = C\mathcal{E}e\mu_n(T) n_T \tau_n(T) e_n(T) \exp\left(-\int_{T_i}^{T} \frac{e_n(T)}{a} dT\right). \tag{4}$$

On the other hand, if we use a sub-bandgap light and neglect the contribution of holes (due to low hole mobility), the PC will be given by

$$I_{PC}(T) = C\, \mathcal{E} e\, \mu_n(T)\, I_0\, \alpha_n(T)\, \tau_n(T) \tag{5}$$

where C is a geometry related constant; I_0, the light intensity; and $\alpha_n(T)$, the absorption coefficient, which is related to the neutral EL2 concentration and the EL2 electron photoionization cross section ($\sigma_n{}^{\circ}$) by $\alpha_n(T) = N_{EL2}{}^{\circ}\, \sigma_n{}^{\circ}(T)$. Our new formalism is based on two improvements: 1) the use of PC vs. T to normalize out some of the unknown or poorly known factors in Eq. (4); and 2) exact, least-squares fitting to the basic TSC formula. PC measurements are useful because both PC and TSC depend upon mobility, carrier lifetime and the same geometric factors. In fact, the TSC can be written as

$$I_{TSC}(T) = \frac{I_{PC}(T)}{I_0 \alpha_n(T)} n_T \cdot e_n(T) \exp\left(-\int_{T_i}^{T} \frac{e_n(T)}{a} dT\right). \tag{6}$$

For undoped SI GaAs, both α (78K) and α (300K) have been accurately measured in the region $\lambda \cong 1.0$-1.5 μm, which makes possible the determination of EL2 and the net acceptor concentrations [14]. An approximate equation for α_n can be obtained using N_{EL2} and N_A. Thus, in principle, if we measure $I_{TSC}(T)$, $I_{PC}(T)$ and I_0, and determine N_{EL2} and N_A by near IR absorption (NIRA) measurements, then a least-squares fit of Eq. 6 will give the trap parameters n_T, E_T, and σ. Multiple traps are easily handled by adding the individual contributions, and this procedure is especially necessary for overlapping trap spectra. The fitting program was set up on a personal computer and made use of the commercial software package MATHCAD.

Experimentally, we have found that as long as the traps are completely filled at the initial temperature T_i, both portions II and III can be measured reproducibly. However, as compared to portion II, portion III is more affected by the excitation conditions at T_i, i.e. photon energy, light intensity, and illumination time. Besides, a thermal quenching in T_5 can be caused by a large bias due to the formation of high field domains [15]. Therefore, in this report, we focus on the TSC spectrum in portion II, i.e. traps T_2 and T_3.

3. Experiments, Results and Discussion

A total of 17 undoped SI GaAs samples were investigated. Five of them were cut from wafers grown by high-pressure liquid encapsulated Czochralski (HP-LEC) (from two vendors), low-pressure (LP) LEC, vertical gradient freeze (VGF), and vertical zone melting (VZM). The other twelve samples were cut from the center, the ring and the edge positions of four wafers (three vendors) for uniformity comparison. Two of the latter four wafers were grown by HP-LEC and HP-LEC respectively. The other two wafers were grown by HP-LEC, but they were treated differently, i.e., one had an ingot anneal (INA) and the other, a multiple wafer anneal (MWA). The electrical properties at 300K along with the total EL2 concentration (N_{EL2}), the net acceptor concentration (N_A) and the compensation factor (N_{EL2}/N_A-1) for the first five samples are summarized in Table I. The N_{EL2} ($=N_{EL2}^{\circ} + N_{EL2}^{+}$) and N_A were measured on a 6×6 mm^2 sample by NIRA at room temperature, using three wavelengths: 1.1 μm, 1.2 μm and 1.7 μm [14, 16]. In the measurement of PC vs. T at 80K < T < 300K, a rather weak 1.1 μm ($h\nu = 1.13$ eV) light of intensity 3.30×10^{14} phot/cm^2 sec was used, because: 1) it is subband-gap light, which can penetrate the whole sample even at 300K; 2) a strong 1.1 μm light will cause IR quenching of PC due to EL2 IR quenching; and 3) for light which is too weak, the PC will be dominated by TSC signals, as reported by Desnica et al. [17]. Under a reasonably high light intensity, say from about 10^{14} to 10^{16} phot/cm^2 sec, a proportional relation between PC and light intensity at 140K < T< 300K is observed. Experimental details for the TSC measurements can be found in Ref. [8, 10]. Here, we want to emphasize that the TSC spectrum in

Table 1. Summary of growth technique, electrical properties, EL2 and net acceptor concentrations for the tested samples.

Sample	Growth	Vendor	ρ 10^7ohm.cm	μ 10^3cm^2/Vs	n 10^7cm^{-3}	N_{EL2} 10^{16}cm^{-3}	net N_A 10^{15}cm^{-3}	$(N_{EL2}-N_A)/N_A$
H25	HP-LEC	A	3.4	6.9	2.7	1.56	1.35	10.5
V27	VGF	B	5.35	6.7	1.8	1.32	1.23	4.7
L28	LP-LEC	C	16.4	5.7	0.67	2.11	5.58	2.8
B22-17	HP-LEC	D	4.4	6.3	2.3	1.35	1.84	6.32
6-90B1	VZM	E	1	0.4	1×10^2	0.42	1.91	1.21

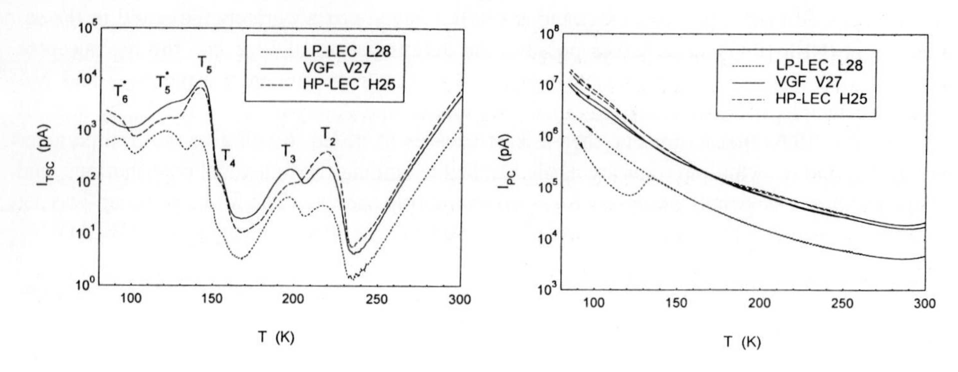

Fig. 1. TSC spectra for three samples.

Fig. 2. PC vs T for three samples.

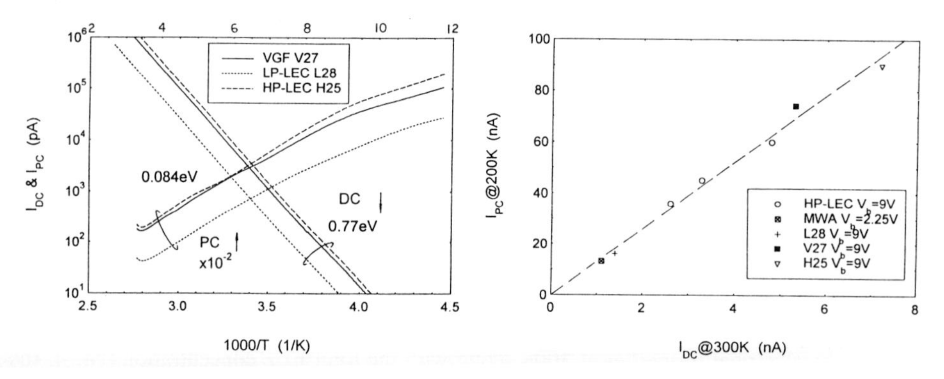

Fig. 3. Arrhenius plots of DC and PC.

Fig. 4. Relationship between PC and DC.

portion II, as a fingerprint, is not affected by the photon energy (say from 0.9 eV to 1.46 eV) for the excitation and only depends on the SI GaAs sample itself, as long as the TSC spectrum is not subject to IR quenching [7].

The TSC and PC spectra for samples grown by HP-LEC, VGF and LP-LEC, are shown in Fig. 1 and 2, respectively. The bias was 9V across a 6 mm sample. From the figures, we see that: 1) the relative peak height ratios T_2/T_3 and T_5/T_5^* for the three samples are quite different, indicating that the crystal stoichiometry is different, as discussed in our previous paper [9]; 2) the PC spectra measured during cooling and subsequent warming cycles, are basically identical, except for the LP-LEC sample, which experiences a decrease in PC at T < 140K, due to the IR quenching of EL2; and 3) by taking the PC into account, the normalized peak height of T_5, at 140K, is not much larger than that of T_2 or T_3. The Arrhenius plots of the DC and PC at 83K < T < 300K are shown in Fig. 3. From the figure we see that: 1) the activation energy for each DC plot is 0.77 eV, showing the control of EL2; 2) the average activation energy for the PC curves is 83.5 meV, which is very close to the thermal activation energy of the electron capture cross section for EL2 (66 meV), if the temperature dependence of electron mobility is considered; 3) the PC's are almost proportional to the DC's at T > 200K. These observations have been confirmed by measurements on other samples, cut from one HP-LEC and one LP-LEC wafer from different suppliers. Both wafers show poor uniformities in DC and PC, but the proportional relation between the DC and net PC at T > 200K still holds. Based on the data in this study, a linear relationship between DC at 300K and PC at 200K is found, as plotted in Fig. 4. On the other hand, from the TSC peak profile studies on HP-LEC, VGF and LP-LEC wafers [9], we have found another important relationship; that is the peak height of T_2 follows the DC at 300K very well. The linear relationship between the TSC of T_2 and the DC at 300K still holds for other wafers that are measured, as seen in Fig. 5.

These relationships lead to important conclusions. From Eq. (1), we see that I_{DC} (300K) $\propto$ $N_{EL2}°/N_{EL2}^{+}$ (7), a factor which is usually much more variable than μ (300K). According to Eqs. (4) and (5), we have I_{TSC} (220K) $\propto$ μ_n (220K) n_T τ_n (220K) (8) and I_{PC} (220K) $\propto$ μ_n (220K) I_0 $N_{EL2}°$ τ_n (220K) (9). We thus see that I_{PC} (220K)/I_{DC} (300K) = $K_1 I_0$ N_{EL2}^{+} τ_n (220K). Therefore, the experimentally observed linear relationship between the DC and net PC at T > 200K implies that the lifetime of the photoexcited carriers is proportional to $1/N_{EL2}^{+}$. From the experimental data a capture cross section at 200K of 1.7×10^{-17} cm^2 was found which is very close to the expected capture cross section of electrons for EL2. From Eqs. (7) and (8), and the constancy of N_{EL2}^{+} τ_n, we obtain I_{TSC} (220K)/I_{DC} (300K) = K_2 μ_n (220K) $n_T/N_{EL2}°$. Therefore, the nearly linear relation between the DC and TSC of T_2 for each wafer (if assuming the nearly same μ_n (220K) within a wafer) implies that the trap density of T_2 is proportional to $N_{EL2}°$. However, the TSC of T_2 is controlled by both EL2° and EL2$^+$ due to the constancy of N_{EL2}^{+} τ_n which was shown earlier.

Theoretical fitting of portion II of the TSC spectrum for sample V27 results in E_T = 0.60 eV,

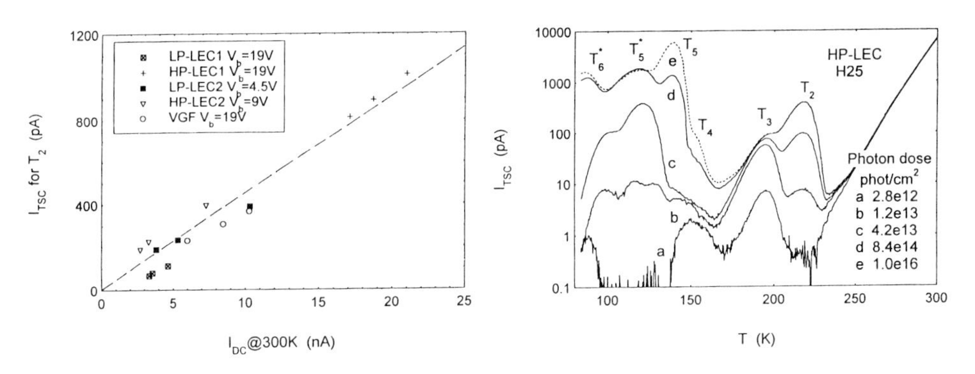

Fig. 5. Relationship between TSC of T_2 and DC. Fig. 6. TSC spectra as a function of photon dose.

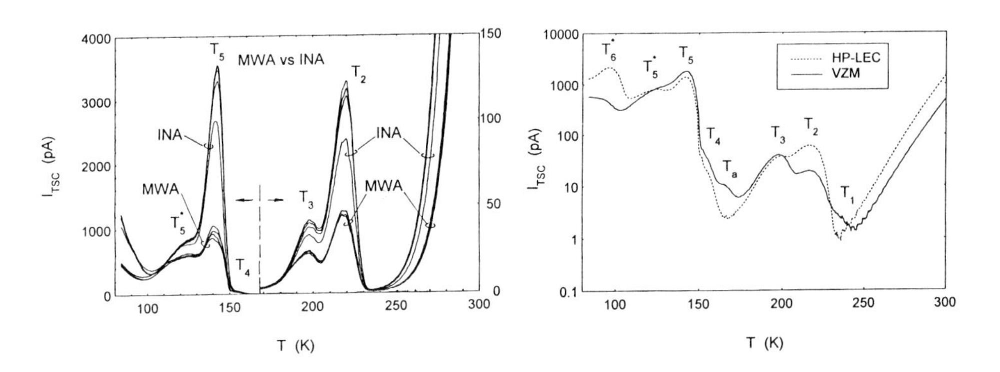

Fig. 7. TSC comparison between INA & MWA wafers. Fig. 8. TSC comparison between VZM & LEC samples.

$\sigma = 3.3 \times 10^{-13}$ cm^2 and $n_T = 5.5 \times 10^{14}$ cm^{-3} for T_2, and $E_T = 0.52$ eV, $\sigma = 1.3 \times 10^{-13}$ cm^2 and $n_T = 2.1 \times 10^{14}$ cm^{-3} for T_3. The numbers of E_T and σ are very close to those obtained by TSC measurements with heating rate ranging from 0.05 K/s to 0.4 K/s on the same sample and using an analytical equation of $E_T/kT_m = \ln T_m^4/a \ln 1.7 \times 10^{16}\sigma/E_T$ [13]. On the other hand, to compare are the capture processes for the various traps, TSC spectra were measured on sample H25 as a function of 1.13 eV photon dose, as shown in Fig. 6. From the figure, it can be clearly seen that T_3 is the first of the six traps to be completely filled (see curve b), indicating that it has the largest capture cross section. This discrepancy with the above fitting results has yet to be resolved; however, the concentrations should still be accurate.

We have earlier found that the peak height ratio T_2/T_3 is associated with crystal stoichiometry; i.e. an As-rich melt favors higher T_2 and a Ga-rich melt favors higher T_3 [8] and the profiles of T_2 and T_3 have an anticorrelation across the diameter of a LP-LEC wafer, i.e. a W vs M shape [see Fig 3 in Ref 9]. These facts, along with the present study, strongly suggest that T_2 is related to EL2 (with As_{Ga} as its core) and that the T_3 may be associated with V_{As}, most likely a V_{As} related defect complex. Of course, this latter statement needs further confirmation.

Finally, we present two more examples to show how TSC spectroscopy can be used as a routine characterization of SI GaAs. The first example concerns wafer anneal effects. To reduce the arsenic precipitates found in conventional INA HP-LEC grown SI GaAs, a MWA technology has been developed, in which wafers are annealed first at 1100°C and then at 950°C [18]. Comparative TSC spectra for samples cut from two wafers (INA vs MWA) are shown in Fig. 7. An improvement of the uniformity both in the DC and the TSC peaks can be clearly observed from the figure. The second example involves growth technique effects. Comparative TSC spectra for two samples, grown by HP-LEC and VZM (more Ga rich) are shown in Fig. 8. Once again, we see from the figure that the DC and T_2 (or T_2/T_3) are controlled by N_{EL2} and N_A. In addition, a new trap T_a was observed in the VZM sample; T_a was also found in VGF samples [8], which is not surprising since these two growth techniques have similarities. The work of Z-QF and DCL was supported under USAF Contract No. F33615-95-C-1619.

References

[1] Look D C 1993 Semiconductors and Semimetals **38** 91-116

[2] Saarinen K, Kuisma S, Hautojarvi P, Corbel C and LeBerre C 1993 Phys. Rev. Lett. **70** 2794-7

[3] Look D C, Fang Z-Q and Sizelove J R 1994 Phys Rev **B49** 16757-60

[4] Desnica D I 1992 J. Electron. Mater. **21** 463-71.

[5] Fillard J P, Bonnafe J and Castagne M 1984 Solid State Commun. **52** 855-9.

[6] Desnica U V, Desnica D I and Santic B 1991 Appl. Phys. Lett. **58** 278-80

[7] Fang Z-Q and Look D C 1991 Appl. Phys. Lett. **59** 48-50

[8] Fang Z-Q and Look D C 1991 J. Appl. Phys. **69** 8177-82

[9] Fang Z-Q and Look D C 1994 Semi-insulating III-V Materials, Warsaw (World Scientific) 143-6

[10] Fang Z-Q and Look D C 1993 J. Appl. Phys. **73** 4971-4

[11] Fang Z-Q and Look D C 1993 Appl. Phys. Lett. **63** 219-21

[12] Fang Z-Q and Look D C 1993 J. Electron. Mater. **22** 1361-3

[13] Look D C 1983 Semiconductors and Semimetals **19** 75-169

[14] Brierley S K and Lehr D S 1989 Appl. Phys. Lett. **55** 2426-8

[15] Fang Z-Q and Look D C 1995 Appl. Phys. Lett. **66** 3033-5

[16] Look D C, Walters D C, Kemerley R T, King J M, Mier M G, Sewell J S and Sizelove J S 1989 J. Electron. Mater. **18** 487-92

[17] Desnica U V and Santic B 1990 J. Appl. Phys.**67** 1408-11

[18] Mori M, Kano G, Inoue T, Shimakura H, Yamamoto H and Oda O 1990 Semi-insulating III-V Materials, Toronto (Adam Hilger) 155-160

Inst. Phys. Conf. Ser. No 149
Paper presented at DRIP 95

Detection of vacancy defects in gallium arsenide by positron lifetime spectroscopy

K. Saarinen[1], S. Kuisma[1], P. Hautojärvi[1], C. Corbel[2], and C. LeBerre[2]

1) Laboratory of Physics, Helsinki Univ. of Technology, 02150 Espoo, Finland
2) INSTN, Centre d'Etudes de Saclay, 91191 Gif-sur-Yvette CEDEX, France

Abstract. Vacancy-related native defects were studied in semi-insulating GaAs by positron lifetime measurements. Both gallium and arsenic vacancies are observed at concentrations of 10^{15} - 10^{16} cm^{-3}. The experiments in the dark after illumination manifest the vacancy nature of the metastable state of the EL2 center.

1. Introduction

The influence of intrinsic point defects on the electronic and optical properties of GaAs is particularly significant. However, there is a lack of experimental information on such basic defects as Ga and As vacancies. The well-known defect center EL2 in undoped GaAs has been the object of extensive experimental and theoretical investigations. This defect exhibits *metastable* properties, i. e. it is transformed persistently to another configuration by light illumination at low temperatures. Several models have been developed to explain these effects, but the atomic structure of EL2 is still under discussion [1]. In this work we have used photoexcitation together with positron spectroscopy to study the properties of native Ga and As vacancies [2] and the metastable EL2 defect [3] in semi-insulating (SI) GaAs.

2. Experimental method

Positron spectroscopy is a method to study vacancy defects in semiconductors. Positrons get trapped at neutral or negative vacancies. The trapping rate κ is proportional to the defect concentration c by $\kappa = \mu c$, where μ is the positron trapping coefficient. The electron density at vacancies is less than in the bulk and the lifetime of trapped positrons τ_v is thus longer than that of free positrons: $\tau_v > \tau_b$. The positron average lifetime can be defined as $\tau_{av} = \eta_b\tau_b + \eta_v\tau_v$, where η_b and η_v are relative amounts of positron annihilations in the bulk and vacancies, respectively. The increase of the average lifetime is a clear indication that a larger fraction of positrons annihilate as trapped at vacancies. The positron trapping rate and the defect concentrations can be estimated using the experimental average lifetime. The lifetime of trapped positrons τ_v can be obtained from the decomposition of the positron lifetime spectra and its value gives information on the open volume of the vacancy defect.

The positron lifetime experiments in this work were performed in undoped semi-insulating LEC GaAs crystals. The resistivity of the samples was about 10^8 Ω cm, and their

EL2 concentration was typically 10^{16} cm^{-3}. Two identical sample pieces were sandwiched with a 30 μCi positron source. The sample sandwich was mounted in a closed-cycle He cryocooler for positron experiments at 20 - 300 K. The illuminations were performed with 0.7 - 1.5 eV light obtained from a 250 W halogen lamp and a monochromator.

3. Metastable EL2 defect in semi-insulating GaAs

Positron average lifetime in SI GaAs in darkness at 300 K is very close to the bulk value of τ_b = 231 ps. At low temperatures the average lifetime increases to about 232 - 238 ps depending on the sample. After illumination at 25 K positron lifetime is longer than the reference levels obtained in darkness. Furthermore, the increase of positron lifetime is persistent, i. e. when the sample is kept at 25 K, the change of τ_{av} remains even if the illumination is turned off. In the isochronal annealing the higher values of the average lifetime obtained after illumination at 25 K remain up to 110 K, but at about 120 K an abrupt decrease is observed (Fig. 1). After this stage the positron average lifetime saturates at the level corresponding to the situation before illumination.

The increase of positron lifetime after illumination indicates that more positron trapping takes place at vacancies than before illumination. Hence, a vacancy defect is generated in the sample during the illumination. When the illumination is removed, the vacancy can be observed persistently, until thermal annealing is performed at about 120 K (Fig. 1). This behavior indicates that the vacancy possesses metastable character.

The positron trapping rate at the metastable vacancy is shown in Fig. 2 as a function of

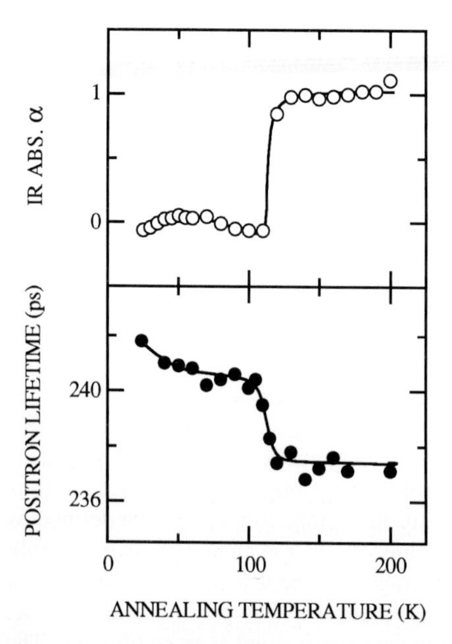

Fig. 1. The average lifetime as a function of the isochronal annealing temperature after 1.15 eV illumination at 25 K. The normalized infrared absorption coefficient is shown in the top panel. All measurements have been performed in darkness at 25 K.

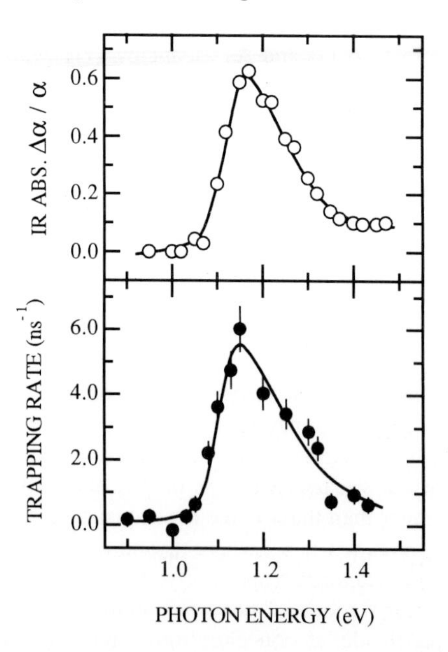

Fig. 2. Positron trapping rate at the metastable vacancy and the relative change of the IR absorption as functions of the photon energy. The data were measured at 25 K after illuminating the sample to the photon fluence of 4.5 x 10^{17} cm^{-2}.

the photon energy hν. There is no illumination effect at hν < 1.05 eV. However, at hν ≥ 1.05 eV the positron trapping rate increases and a local maximum is formed at hν = 1.15 eV. The metastable vacancy can thus be generated most efficiently with hν = 1.15 eV.

The infrared absorption experiments show that the absorption due to the EL2 defect becomes negligible after illumination at 30 K. This is due to the photoquenching of the EL2 defects: during illumination the EL2 defects transform from the stable state to the metastable state $EL2^*$, which does not absorb IR light [1]. The infrared absorption coefficient retains its original value after annealing at 120 K, when $EL2^*$ recovers (Fig. 1). The metastable EL2 can be induced most effectively with 1.15 eV light as indicated by the change in the IR absorption coefficient (Fig. 2).

In the positron experiments, the appearance of the metastable vacancy correlates well with the transformations of EL2 between the stable and metastable states. When the IR absorption shows the photoquenching of the EL2 defects, a metastable vacancy is detected in the positron annihilation experiments. Furthermore, when the metastable state of EL2 recovers at 120 K according to IR experiments, also the metastable vacancy defect disappears from the positron annihilation data. These correlations lead us to conclude that the metastable vacancy belongs to the structure of the metastable state of the EL2 defect. Using similar experiments we have also determined that (i) a defect with a vacancy in its metastable state can be introduced in the electron irradiation of GaAs [4] and (ii) the deep ground state of the DX center contains a vacancy which resembles in many ways that found in the metastable state of the EL2 defect [5].

It has been verified that the As antisite defect is a part of the stable state of the EL2 defect in GaAs. According to the vacancy-interstitial model, the As atom relaxes in the metastable state to the interstitial position leaving a Ga site vacant: $As_{Ga} \rightarrow V_{Ga} + As_i$ [6]. According to a similar model the DX center is simply an isolated donor atom [7]. When the DX level is occupied, the donor atom, e. g. Si, is not in the substitutional Ga lattice site but has relaxed to the interstitial position leaving a Ga vacancy behind: $V_{Ga} + Si_i$. The positron results give direct evidence that large lattice relaxations are associated in the transitions between the stable and metastable states of the EL2 and DX centers. They show further that a vacancy defect is involved in the structure of the relaxed configuration of these defects. The positron results are in a good agreement with the theoretical models involving the displacements of the donor atoms (As in case of EL2, group IV impurity in case of DX) from the substitutional lattice site to the interstitial position.

4. Native Ga and As vacancies in semi-insulating GaAs

At 300 K the positron lifetimes in darkness vary from 231 ps to 233 ps in undoped GaAs. When a sample is cooled down to low temperatures, the positron lifetime increases as shown for sample 1 in Fig. 3. The increase of τ_{av} at low temperatures shows that the samples contain vacancy defects and further that the charge of those defects is negative. The lifetime spectra at 20 - 80 K can be decomposed into two components. The second component is τ_v = 250 - 260 ps. This value is typical for monovacancies in GaAs [8-10].

According to theoretical calculations As vacancies are positive and Ga vacancies are negative in SI GaAs [11]. Since positrons repel positive centers we identify the defects as Ga vacancies or complexes involving V_{Ga}. Their concentration in SI GaAs is 10^{15} - 10^{16} cm^{-3}. In addition to V_{Ga}, ionic acceptors have also been detected in some SI GaAs samples [12].

Under illumination the positron average lifetime increases compared to the values in darkness (Fig. 3). The effect of illumination is the largest at 20 - 100 K and disappears com-

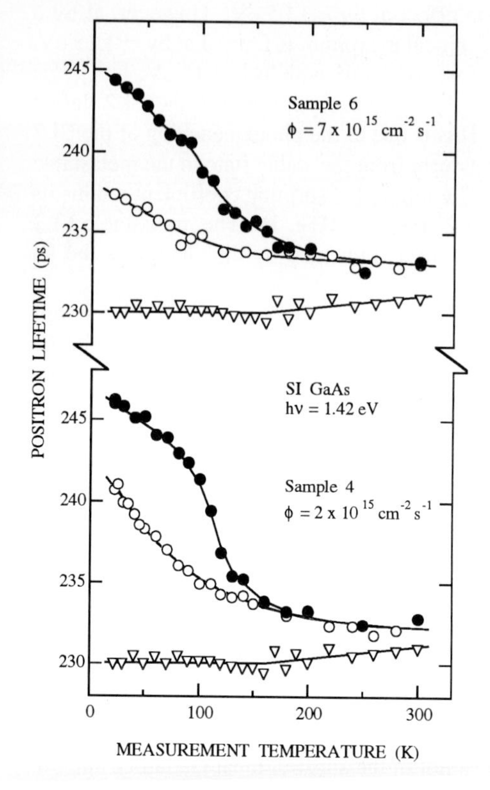

Fig. 3. Positron lifetime as a function of measurement temperature in darkness (○) and under illumination (●). The free positron lifetime corresponding to defect-free material is also given (▽).

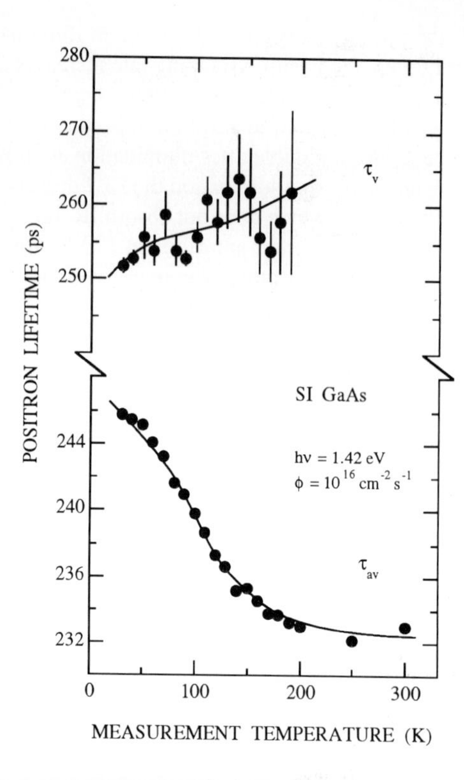

Fig. 4. Average positron lifetime τ_{av} and the lifetime component τ_v under 1.42 eV illumination as a function of measurement temperature.

pletely above 180 K. This increase is much larger than the persistent part of it, which is due to the EL2 defect (see Sec. 3). This indicates that most of the effect seen under illumination is not related to the EL2 defect.

The lifetime spectra measured under illumination can be decomposed into two components yielding τ_v = 250 - 265 ps (Fig. 4). Since this lifetime is the same as obtained in darkness, the defects trapping positrons under illumination can also be characterized with τ_v = 255 ± 5 ps. This value is typical for monovacancies in GaAs [8,9,10].

The increase of positron average lifetime under illumination indicates that some vacancies are converted to more efficient positron traps by capturing electrons under illumination. In a semi-insulating sample, the levels populated under illumination are above the mid-gap. As mentioned above, no ionization levels are expected in the upper half-part of the gap for Ga vacancies, whereas they are expected for As vacancies. It is thus natural to identify the defects seen under illumination to As vacancies or complexes related to the As vacancies. The decrease of τ_{av} at 20 - 70 K indicates further that the As vacancies are negatively charged. The concentration of the As vacancies is typically 10^{15} - 10^{16} cm^{-3}.

Fig. 5 shows the concentration of the detected As vacancies as a function of the photon energy under illumination with constant photon flux. When EL2 is in the stable state the negative charge state of the As vacancy can be generated with the photons of hν > 0.9 eV and the spectrum is almost flat at $h\nu > 1.2$ eV. In SI GaAs the absorption at $h\nu < 1.4$ eV is mainly due to the optical exchange of electrons between the EL2 defect and the conduction and valence bands [1]. These processes generate photoelectrons which can be captured by the defect levels in the band gap. Since the As vacancy becomes negatively charged under these illumination conditions, we associate the the generation process to the capture of photoelectrons emitted from the EL2 defect to the conduction band under illumination.

When EL2 is in the metastable state only the photons of $h\nu > 1.35$ eV are able to excite As vacancy to the negative charge state (Fig. 5). The photon flux needed for this excitation is much larger than that required in the case where EL2 is in the stable state. These observations indicate that a different optical process leads to the population of the negative As vacancy when EL2 is in the metastable state. We attribute this process to the direct excitation of electrons from the valence band to the ionization level of V_{As}.

Under illumination the concentration of negative As vacancies decreases strongly when temperature is increased above 70 K. (Fig. 6). The onset temperature for the decrease depends on the illumination intensity. This type of behavior shows that at $T > 70$ K the thermal emission of electrons from the ionization level of V_{As} starts to compete with the optical transitions. The ionization energy of 65 ± 15 meV can be determined for this process. This value correlates well with the maximum efficiency of the excitation at 1.4 - 1.5 eV (Fig. 5), since the GaAs band gap is 1.5 eV at 30 K.

In optical experiments on bulk GaAs, a strong absorption of monochromatic light

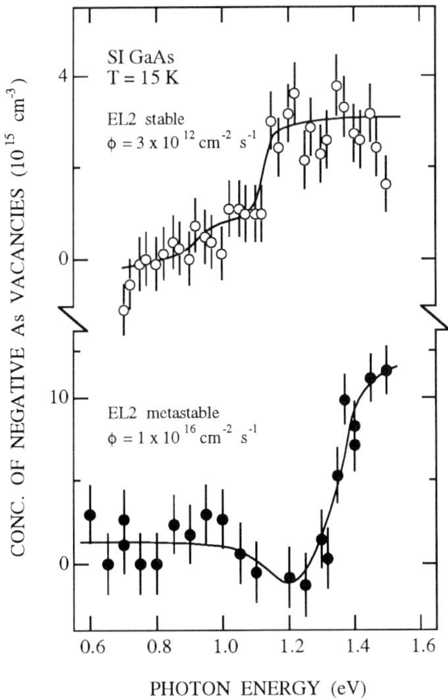

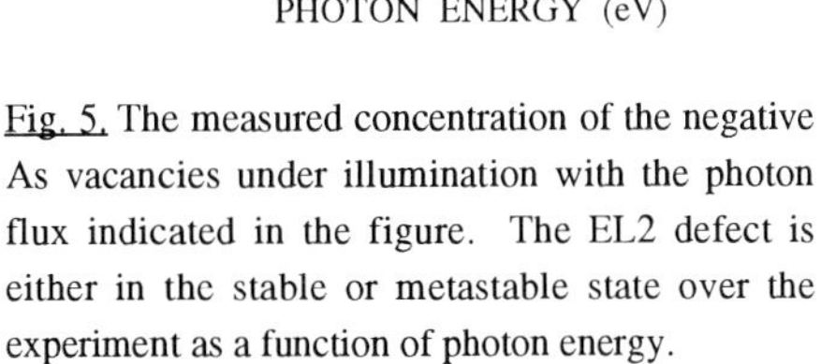

Fig. 5. The measured concentration of the negative As vacancies under illumination with the photon flux indicated in the figure. The EL2 defect is either in the stable or metastable state over the experiment as a function of photon energy.

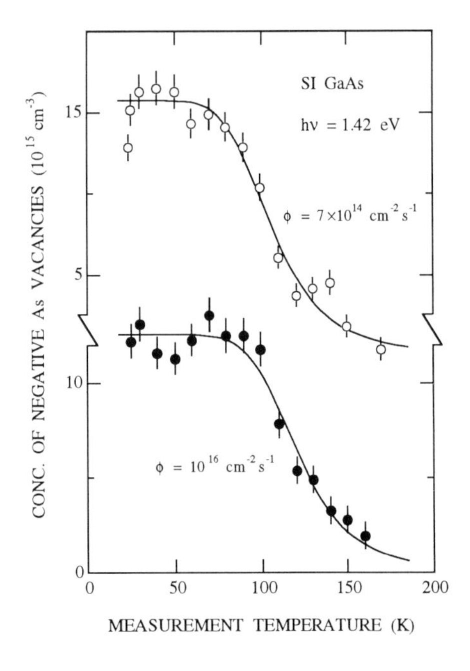

Fig. 6. The measured concentration of negative As vacancies as a function of temperature. The experiments have been performed under illumination with 1.42 eV photons and using the photon fluxes given in the figure.

within 50 meV of the conduction band edge is observed below 150 K [13]. This near-band-edge absorption ("RC absorption") has been attributed to an unidentified point defect, but it is not related to EL2. The present positron experiments indicate that the ionization level of the As vacancy is at about E_c - 50 meV and that this level can be most efficiently populated using 1.4 - 1.5 eV light. The concentration of the As vacancy correlates quantitatively with the magnitude of the RC absorption [14]. Therefore, our results provide a natural explanation to the near-band-edge absorption in GaAs: it results from the photo-induced electron transition from the valence band to the ionization level of the As vacancy.

5. Conclusions

The experiments of this work show that depending on the illumination conditions, three different point defects can be observed in semi-insulating GaAs by positron lifetime measurements. *In darkness*, positron trapping at native Ga vacancies is observed. *Under 1.4 eV illumination*, native As vacancies are converted to negative charge state and thus become detectable by positrons. Experiments in the dark *after 1.1 - 1.3 eV illumination* reveal the vacancy nature of the metastable state of the EL2 defect, i. e. the close V_{Ga} - As_i pair. The concentrations of these defects are typically 10^{15} - 10^{16} cm^{-3}. These results contribute to the understanding of some macroscopic properties of GaAs like the IR absorption or electrical compensation.

References

[1] S. T. Pantelides (Ed.) 1986 *Deep Centers in Semiconductors* (New York: Gordon and Breach)

[2] K. Saarinen, S. Kuisma, P. Hautojärvi, C. Corbel and C. LeBerre 1993 *Phys. Rev. Lett.* **70** 2794

[3] R. Krause, K. Saarinen, P. Hautojärvi, A. Polity, G. Gärtner and C. Corbel 1990 *Phys. Rev. Lett.* **65** 3329; K. Saarinen, S. Kuisma, P. Hautojärvi, C. Corbel and C. LeBerre 1994 *Phys. Rev. B* **49** 8005

[4] K. Saarinen, S. Kuisma, J. Mäkinen, P. Hautojärvi, M. Törnqvist and C. Corbel 1995 *Phys. Rev. B* **51**, 14152

[5] J. Mäkinen, T. Laine, K. Saarinen, P. Hautojärvi, C. Corbel, V. M. Airaksinen and P. Gibart 1993 *Phys. Rev. Lett.* **71** 3154; J. Mäkinen, T. Laine, K. Saarinen, P. Hautojärvi, C. Corbel, V. M. Airaksinen and J. Nagle 1995 *Phys. Rev. B* **52** 4870

[6] J. Dabrowski and M. Scheffler 1988 *Phys. Rev. Lett.* **60** 2183; D. J. Chadi and K. J. Chang 1988 *Phys. Rev. Lett.* **60** 2187

[7] D. J. Chadi and K. J. Chang 1988 *Phys. Rev. Lett.* **61** 873

[8] K. Saarinen, P. Hautojärvi, P. Lanki and C. Corbel 1991 *Phys. Rev. B* **44** 10585

[9] C. Corbel, M. Stucky, P. Hautojärvi, K. Saarinen and P. Moser 1988 *Phys. Rev. B* **38** 8192

[10] M. J. Puska and C. Corbel 1988 *Phys. Rev. B* **38** 9874; K. Laasonen, M. Alatalo, M. J. Puska and R. M. Nieminen 1991 *J. Phys.: Condens. Matter* **3** 7217

[11] G. A. Baraff and M. Schlüter 1985 *Phys. Rev. Lett.* **55** 1327; M. Puska 1989 *J. Phys.: Condens. Matter* **1** 7347; H, Xu and W. Lindefelt 1990 *Phys. Rev. B* **41** 5975

[12] C. LeBerre, C. Corbel, K. Saarinen, S. Kuisma, P. Hautojärvi and R. Fornari 1995 *Phys. Rev. B* **52** 8112

[13] S. Tüzemen and M. R. Brozel 1991 *Appl. Surf. Sci* **50** 395

[14] C. LeBerre, C. Corbel, R. Mih, M. R. Brozel, S. Tüzemen, S. Kuisma, K. Saarinen, P. Hautojärvi and R. Fornari 1995 *Appl. Phys. Lett.* **66** 2534

Inst. Phys. Conf. Ser. No 149
Paper presented at DRIP 95

Spectroscopy of Defects Induced by Ohmic Contact Preparation in LEC GaAs Particle Detectors

A. Castaldini*, A. Cavallini*, C. del Papa #, C. Canali§, F. Nava§ and C. Lanzieri$

*INFM and Dept. of Physics, Univ. Bologna, Italy, #Dept. of Physics, Univ. Udine, Italy, §Dept. of Physics, Univ. Modena, Italy, $ALENIA S.p.A. Roma, Italy; # §also at INFN.

ABSTRACT Semi-insulating LEC gallium arsenide particle detectors were realized with differently manufactured ohmic contacts to improve their performances and possibly avoid injection effects often experienced when the detectors work in full depletion conditions. I-V and C-V measurements on Schottky structures were carried out. Photo-induced current transient spectroscopy and also photo-deep level transient spectroscopy investigations, performed on both planar and Schottky structures, identified electron and hole traps. Detector performances were correlated to defects action.

1. Introduction

The recent use of semi-insulating (SI) liquid encapsulated Czochralski (LEC) gallium arsenide as bulk material for particle detectors gave rise to renewed interest in this material and aroused questions related to such kind of devices. Particle detector performances are measured in terms of charge collection efficiency, *cce*, i.e. the fraction of the charge released by incoming particles which gets collected. Till today the charge collection efficiency never reached 100% since most of the detectors reach breakdown as soon as they are fully active. Consequently, high electric field can not be obtained in the active region and severe trapping occurs in it. To this respect, it has been demonstrated [1] that a low injecting ohmic contact in the Schottky detector plays a key role in its performance, since it allows the application of high reverse bias without breakdown and, in turn, the increase of the collecting electric field strength to high values as well as of *cce* to almost 100%. It is, therefore, worth improving detector performance by avoiding breakdown to occur when the detector is fully active and, hence, increasing as much as possible its performances.

I-V and C-V measurements have been carried out. Besides, spectroscopic methods, namely photo-induced current transient spectroscopy (PICTS) and photo-deep level transient spectroscopy (P-DLTS), were applied to identify charge carrier traps induced by ohmic contact processing. Correspondingly, the detector *cce* was measured and related to the trapping processes identified by the spectroscopic analyses. It has been shown that: a number of traps exists, trap identity is the same independently of the ohmic contact preparation, whereas their content strongly depends on this. Some suggestions as to the reasons for their influence on the charge collection efficiency will be given.

2. Experimental

The detectors studied in the present paper were made by ALENIA S.p.A. on Hitachi as well as on Nippon SI LEC undoped <100> oriented substrates, with n-type resistivity $\rho \cong 2 \div 3 \times 10^7$ Ωcm. Detectors were 100 μm thick, with circular (ϕ=3 mm) Schottky pads metallized with Au/Pt/Ti . Two sets of detectors with different kinds of ohmic contacts were investigated. In set A a new non alloyed ohmic contact (NAOC) [2] has been realized on the whole back surface of the wafer by implanting Si^+ ions at two different doses and energies, a dose of 7×10^{12} cm^{-2} at 300 keV and a dose of 1×10^{13} cm^{-2} at 40 keV.

The wafers, with a reactively sputtered silicon nitride cap, were then fast annealed at 850°C for 30 s. Finally, the ohmic contact was achieved by alloying an e-beam deposited AuGeNi multilayer at 420°C for 30 s. Secondary Ion Mass Spectroscopy (SIMS) after Si^+ ion implantation showed that after the annealing 80% of Si atoms resulted electrically active to 0.80 μm from the surface [2]. In set B the ohmic contact was obtained [3,4] by an Au/Ge/Ni metallization (4250Å/500Å/500Å) followed by a thermal cycle [430°C; 20s] in N_2+H_2 (10%) atmosphere. A few structures with two ohmic contacts on opposite sides have been made, too, to perform PICTS measurements in planar configuration [5,6]. Both planar and Schottky structures produce similar spectra but the Schottky barrier should give sometimes better resolved peaks [7].

Current-voltage (I-V) characteristics have been analyzed, and ideality factor n and series resistance R_S of the detectors under test have been obtained. Capacitance-voltage measurements of the Schottky detectors were taken with modulating frequencies ranging from 40 Hz up to 100 kHz at different temperatures. Finally, photo-induced current transient spectroscopy (PICTS) [6] and photo-deep level transient spectroscopy (P-DLTS) [8] have been utilized to identify minority and majority charge carrier traps.

3. Results and discussion

Typical examples of forward I-V characteristics are reported in Fig.1, relevant to samples of both sets A and B, indicating that the forward currents are severely space charge limited. The current trend denotes that ideality factor and series resistance are high, suggesting that the pure thermoionic emission over the barrier is not the only dominant mechanism [9] and contributions of other current-transport mechanisms must be taken into consideration. The experimental data have been fitted by accounting for these additional contributions (generation-recombination, leakage and tunneling currents) [10]. From the fitting lines we get that the ideality factor is n_A=1.4 and n_B=1.8 and the series resistance R_S is 4.2×10^5 Ω and 1.3×10^5 Ω for these samples of set A and B, respectively. Although high if compared to values relevant to "good" metal-semiconductor barriers, the above values are very satisfying for a Schottky barrier on semi-insulating gallium arsenide [11] and certainly good enough

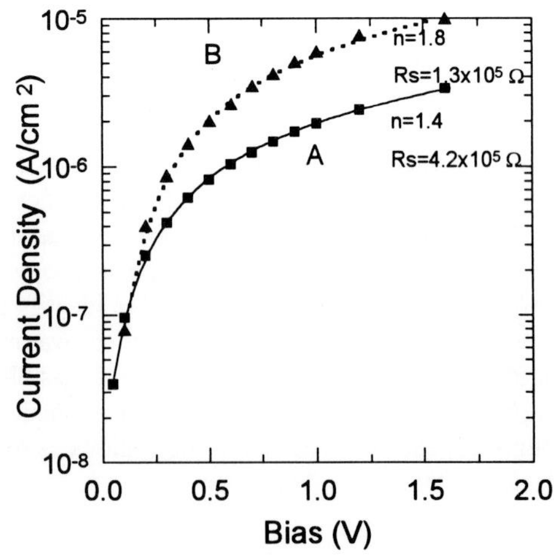

Fig.1 Examples of I-V characteristics for set A and B. Ideality factors n and diode series resistances R_S are also reported.

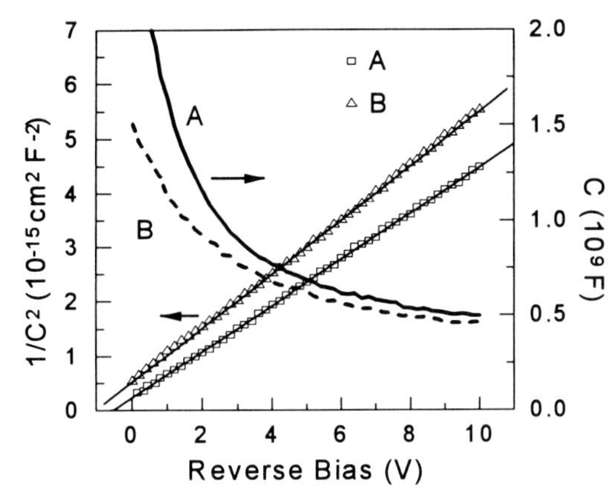

Fig.2 Reverse C-V characteristics (right hand scale) an $1/C^2$ (normalized to the diode area) vs bias (left hand scale) (f=40 Hz, T=420 °C).

for adopting hereafter the "depletion" or "one-sided abrupt junction" approximation about the metal-gallium arsenide barrier [5].

The measured C-V dependence of typical detectors A and B are shown in Fig.2 together with the calculated S^2C^{-2}=f(V) dependence,with S metallized pad area, at temperature T=420 K.

We assumed: (i) the parallel equivalent scheme of the structure [12], (ii) Berman's condition [13] been fulfilled, which enables the application of C-V analysis to SI materials, (iii) depletion width increase as √V, like in conventional Schottky barriers. Fulfilling Berman's condition has been verified by C versus T measurements [14] and checking the saturation of the measured capacitance, occurring in the present samples for T > 415 K. It can be shown [15] that

$$N_{ef}^{+} = \left(\frac{2}{q\,\epsilon\,S^2}\right)\left(\frac{\partial V}{\partial C^{-2}}\right) \tag{1}$$

where q is the electron charge, ϵ the semiconductor permettivity, and the terms N_i are the density of ionized centers $N_{ef}^{+} = (N^{+}_{EL2}-N^{-}_{net}+N^{+}_{TD}-N^{-}_{TA})$: deep centers EL2 ($N^{+}_{EL2}$), shallow impurities usually due to residual carbon (N^{-}_{net}), additional deep donors (N^{+}_{TD}) and acceptors (N^{-}_{TA}). It is noteworthy that in this expression the term N_{EL2} usually dominates. From the here shown slopes (N^{+}_{efA}= 2.5×10^{16}cm^{-3} and N^{+}_{efB} = 2.2×10^{16}cm^{-3}) N^{+}_{ef} results to be very nearly the same in both sets, which means that the overall defective state is approximately the same in the two sets even though differences in specific traps content are not to be excluded.

Trapping and emission processes from deep energy levels have been shown to play a key role in the electric field distribution across the detector [2] and, in turn, in the non-complete charge collection efficiency observed. Consequently, to explain set A and B performances in terms of *cce* it is necessary to have a thorough account of the traps in the starting material and to monitor changes possibly occurred during the detector preparation due to the differently processed ohmic contacts. To this aim spectroscopic methods, namely PICTS and P-DLTS, have been applied to the nuclear detectors under investigation.

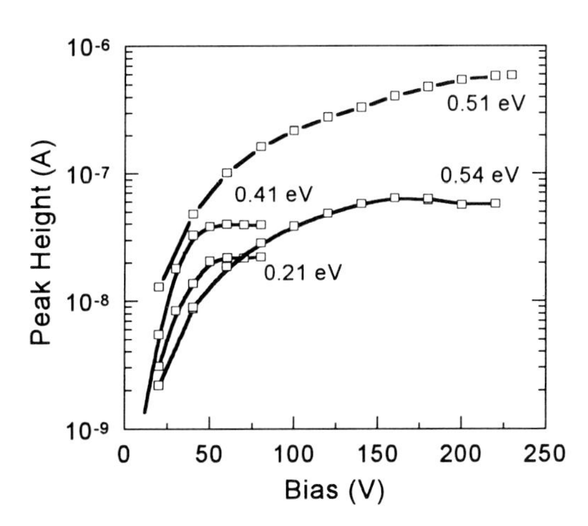

Fig.3 Variation of the amplitude of some peaks as a function of the applied bias.

To collect the all charges thermally released by the traps, PICTS spectra have been obtained at different bias voltages (Fig.3). They show that the amplitude of the peaks increases up to a plateau region, which has been used to determine the working conditions of the spectroscopic investigations. In what follows, PICTS spectra will be reported after normalizing the current signal i(t) to the net photocurrent $i_L(0)$, i.e. the value at the beginning of the transient following the light pulse. As a matter of fact, it is possible to gather information on the deep level trap content from the normalized signal $i_N(t)= i(t)/i_L(0)$ since it can be shown [6] that

$$N_t \propto \frac{\mu_0 \tau_0}{\mu_1 \tau_1} i_N(t_1, T_m). \tag{2}$$

where μ and τ are carrier mobility and lifetime, respectively, at the beginning of the pulse (subscript 0) and at t_1 (subscript 1). Thus, a comparison between sets A and B can be performed in terms of defective state.

By combining PICTS and P-DLTS investigations, electron and hole traps can be distinguished since PICTS, as well known, gives information on both trap kinds, while P-DLTS on Schottky barriers detects only majority carrier traps [8]. Figure 4 shows the Arrhenius plots of both PICTS and P-DLTS spectra. PICTS as well as P-DLTS spectra are reported in Fig. 5 (a) and (b), respectively. Comparing Fig. 5 a and b, it results that the traps at 0.15 eV (not detectable with the emission rate relevant to the present spectra), 0.21, 0.32, 0.54, and 0.81 eV are electron traps, while those at 0.41 and 0.51 eV are hole traps. In addition, comparing the A and B spectra brings into evidence that set A is more defective than set B about most of the traps observed, even though the difference is significant only for the traps located at 0.41 and 0.32 eV. It is worth noting that in A the PICTS peak at 0.41 eV is the convolution of two peaks, one actually located at 0.41 eV and one located at 0.32 eV which more clearly appears as a shoulder in the spectrum B. The trap at 0.81 eV appears with a very low current intensity in PICTS spectra for the following reasons: (i) the diode reverse current (also plotted in Fig. 5a with the relevant activation energy values) sharply increases for temperature T>300 K, (ii) the here shown spectra have been obtained with a bias voltage V=50 V, where the peak height of this trap is not yet in saturation.Onthe other hand, a further bias increase would induce an even more highreverse current and a noise-to-signal ratio too high for reliable current measurements in the whole

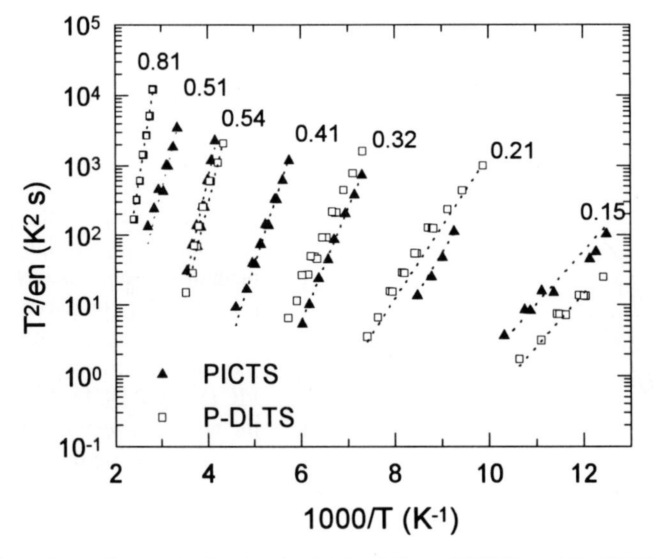

Fig.4 Arrhenius plots relevant to the set A obtained from PICTS and P-DLTS results.

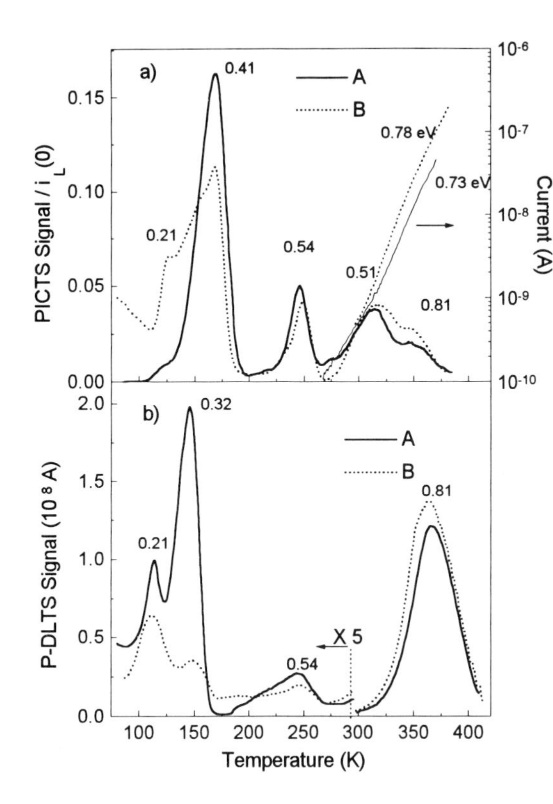

Fig.5 a) Representative PICTS spectra of samples of sets A and B. These spectra have been obtained with light wavelength $\lambda=940nm$, beam power $P=0.3W/cm^2$ and reverse bias V=50V. b) P-DLTS spectra of the same samples as in a). Trap depth (in eV) is reported on the relevant peaks.

temperature range explored. The actual comparison relevant to the trap located at 0.81eV is shown in the P-DLTS spectra (Fig.5b) .

The charge collection efficiency *cce* (fig.6) can be straightforwardly correlated with the deep levels found by PICTS and P-DLTS. Indeed, samples A have a *cce*, for both electrons and holes, lower than samples B but these, conversely, reach breakdown as soon as they are fully active, i.e. as soon as the electric field reaches the ohmic contact giving rise to an abrupt and high injection of current.

4. Conclusive Remarks

It is long established that most of the Schottky barrier nuclear detectors reach breakdown assoon as their active region equals the detector thickness due to current injection from the ohmic contact. To overcome this inconvenience, a new Si^+-implanted ohmic contact has been realized and compared to conventionally made contacts. It has been found that in both cases a non complete charge collection efficiency occurs ($cce < 100\%$), due to trapping effects, the realization of the implanted contact increases the detector defective content but, in spite of this drawback, it allows the detector to work at reverse bias voltages much greater than the one needed to make the detector fully depleted. In detail, electrical and spectroscopical analyses allowed to investigate the traps present in the as-grown material and their variations induced by processes. Namely, the increased content of traps in sample A can be attributed to the implantation process and/or to the subsequent high temperature treatment, as required to recovery the damage induced by the ion implantation.

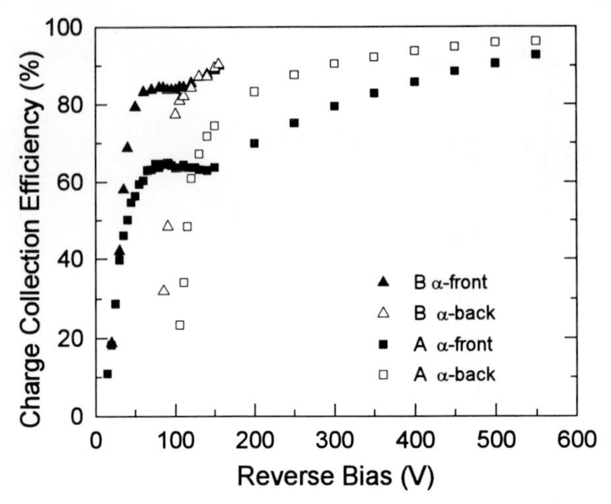

Fig. 6 Charge collection efficiency vs. reverse bias. Curves labelled *α-front* refer to irradiation with alpha-particles from ^{241}Am of the Schottky contact (hence relevant to electron collection), and *α-back* to irradiation of the ohmic contact (relevant to hole collection).

References

[1] Alietti M Canali C Castaldini A Cavallini A Cetronio A Chiossi C D'Auria S del Papa C Lanzieri C Nava F and Vanni P 1995 *Nucl. Instr. Meth. in Phys. Res.* **A362** 344-348

[2] Nava F Alietti M Canali C Cavallini A del Papa C Re V and Lanzieri C 1995 submitted to IEEE Trans. Nucl. Sci.

[3] Nava F in Proc. of the 20th Workshop of the I*NFN Eloisatron Project on GaAs Detectors and Electronics for High-Energy Physics*, edited by C. del Papa, P.G. Pelfer and K. Smith (World Scientific Singapore 1992) 121,129.

[4] Braslau N Gunn J B and Staples J L.1967 *Solid-State Electron.* **10** 38

[5] Look D C 1989.*Electrical Characterization of GaAs Material and Devices*, (John Wiley & Sons, Chichester)

[6] Tapiero M Benjelloun N Zeilinger J P El Hamd S and Noguet C, 1988 *J. Appl. Phys.* **64** 4006-4012

[7] Fang Z Shan L Schlesinger T E and Milnes A G 1989 *Solid-State Electron.* **32** 405-411

[8] Mooney P 1983 *J Appl. Phys.* **54** 208-213

[9] Sze S M 1981.*Physics of Semiconductors Devices* (John Wiley & Sons, New York)

[10] Donoval D Barus M and Zdimal M 1991 *Solid-State Electron.* **34** 1365-1373

[11] McGregor D S Knoll G F Eisen Y and Brake R Proc. of the 20th Workshop of the *INFN Eloisatron Project on GaAs Detectors and Electronics for High-Energy Physics*, edited by C.del Papa, P.G. Pelfer and K. Smith (World Scientific Singapore 1992) 30-43.

[12] Beaumont S P et al 1993 *Nucl. Instr. and Meth.* **A326** 313.

[13] Berman LS 1975.*Sov. Phys. Semicond.* **9** 406

[14] Dubecky F and Olejniová B J. 1991 *Appl. Phys.* **69** 1769-1771

[15] Dubecky F Darmo J and Betko J 1994 *Semicon. Sci. Technol* **9** 1654.

Inst. Phys. Conf. Ser. No 149
Paper presented at DRIP 95

Investigation of the Radiation Damage of GaAs Detectors by Protons, Pions, Neutrons and Photons

W.Braunschweig[1], Z.Chu, T.Kubicki, K.Lübelsmeyer, R.Krais, O.Syben, F.Tenbusch, M.Toporowsky, B.Wittmer, W.Xiao

I.Physikalisches Institut der RWTH Aachen, Sommerfeldstrasse 28, D-52056 Aachen, Germany

Abstract. Schottky diodes made of SI-GaAs have been irradiated with high doses/fluences of photons (mean energy 1 MeV), neutrons (1 MeV), pions (191 MeV) and protons (23 GeV). The detectors have been characterized in terms of macroscopic quantities like I-V characteristic curves and charge collection efficiencies for minimum ionizing as well as α-particles incident on the detector. Whereas no effect is seen after irradiation with photons, considerable degradation occurs after the irradiation with the neutral and charged heavy particles. Si-GaAs material with low carbon content is less affected than material with higher carbon concentration.

1 Introduction

At the forthcoming Large Hadron Collider LHC which will be built at Cern (Geneva, Switzerland) two proton beams with a kinetic energy of 7 TeV (7×10^{12} eV) will collide thereby producing events with energetic secondary charged and neutral particles at a rate of $\sim$ 800 MHz which emerge from a small interaction region. Our institute is participating in the (Compact Muon Solenoid)- CMS collaboration [1] which is one of the two big experiments which have been approved for the LHC. The physics goals of the experiment require a track detection system covering the full solid angle around the interaction region and providing spatial resolutions of O(μm). Therefore semiconductor microstrip and pixel detectors are foreseen, where as material for the baseline solution Silicon has been chosen. The innermost layers of these detectors have to be operated as close as $\sim 7.5cm$ to the primary proton beams, where the fluences of the above mentioned secondary particles are extremely high : per year one expects a γ-ray dose of $10 Mrad$ and heavy particle fluences of $\sim 10^{13}$ neutrons per cm^2 as well as $\sim 5 \times 10^{13}$ charged hadrons per cm^2 (mainly pions and protons). These fluences cause major radiation damage to the silicon detectors [1] degrading their performance within one year or two to unacceptable low levels. Gallium Arsenide is believed to be more radiation hard than silicon, mainly because of its larger bandgap. One of our main motivations to work with GaAs detectors is therefore to study their behaviour w.r.t. irradiation using artificial particle sources and if possible, to establish their longtime functionality under conditions as given at the inner tracking detector of CMS.

The paper is organized as follows : After a short description of the detectors and the parameters of the artificial irradiations we first present the characterization methods and then discuss the results.

[1]Presenting author

2 Detectors and Irradiation Specification

The detector wafers consist of SI-GaAs (provided by several commercial suppliers, see below) with Ni-Pt/Au Schottky contacts on both sides. The desired electrode structure (microstrips, pads) is applied on the reverse biased front side. The backside is forward biased resulting in an Ohmic behaviour. The processing was done in our institute at Aachen. In order to be operated as a tracking device in an high energy physics experiment one has to perform a series of additional processing steps providing a guard ring structure, the biasing of the readout electrodes, an insulating layer (acting as an integrated capacitance), a second metallic layer with the bond pads for the connection to the preamplifiers and eventually a passivation. Of course, at the end all parts of the detector have to be radiation hard. In this paper we concentrate on the behaviour of the bulk material and the metal-semiconductor contacts. We investigated the following material (supplier,thickness) : AXT ($370\mu m$), Freiberger-Low Carbon ($250\mu m$), Freiberger-Normal Carbon ($250\mu m$), MCP ($500\mu m$), Sumitomo ($600\mu m$), Showa Denko ($450\mu m$), Alenia-Sumitomo ($250\mu m$). The irradiation facilities together with the particle

Source	Particle	$< E_{kin} >$	Fluence	av.flux	av.flux at CMS
			cm^{-2}	$cm^{-2}s^{-1}$	$cm^{-2}s^{-1}$
ISIS at RAL	n	$\sim 1MeV$	$(.42 - 4.6) \times 10^{14}$	2.5×10^8	1.25×10^6
PSI	π	$191MeV$	$(0.5 - 0.6) \times 10^{14}$	4×10^8	$\sim 3 \times 10^6$
CERN	p	$23GeV$	$(0.6 - 1.7) \times 10^{14}$	5×10^9	$\sim 3 \times 10^6$
Fraunh.Inst.	γ	^{60}Co source	$100Mrad$		

Table 1: Irradiation facilities, particle species, energies and fluences

species and their energies are presented in table 1. The listed fluences were accumulated during periods ranging from 1/2 day to $\sim$ 3 weeks. Therefore also the flux (intensity) values are given and compared to the average flux expected at the tracker of CMS. It can be seen that the artificial fluxes are up to two orders of magnitude more intense than these average values.

3 Characterization

In this paper we characterize the detectors/detector material almost exclusively in terms of macroscopic quantities:

- I-V and C-V characteristic curves
- Analysis of pulse height spectra obtained from

  1. Minimum ionizing particles (mips) traversing the detector. We use electrons from a ^{90}Sr source which generate in SI-GaAs $\sim$ 13000 electron-hole pairs per $100\mu m$ thickness, uniformly distributed along their path across the detector. The charge collection efficiency, defined as the ratio of the measured charge and the charge deposited in the detector, $CCE = Q_{meas}/Q_{dep}$, hardly reaches unity, the reason being that the detectors are not fully active [2]. The active zone X_A, where the electric field is sufficiently large to cause a drifting of the charge carriers which leads to a detectable signal is smaller than the detector thickness but increases with reverse bias voltage. The limit of the CCE is

determined by the voltage limit (i.e. breakdown) and/or by the free mean paths of the charge carriers. Fig.1 shows an example: due to the increasing active zone the CCE increases towards higher bias voltage.

2. α-particles from a ^{241}Am source. Their range in SI-GaAs amounts to $\sim 12\mu m$. As a consequence, the pulse height analysis for injection from front- and backside, respectively provides information on the active zone of the detector and the different contributions from the drifting holes and electrons which constitute the signal. Fig.2 shows the CCE for α-particles injected from the front side (negatively biased). The signal is almost entirely due to the drifting holes. The CCE-value near to 90% indicates that the mean free path of the holes is larger than the detector thickness.

- Photon counting, i.e. measuring the rate due to photons from a ^{57}Co-source provides information on the thickness af the active zone X_A.
- Time structure : Signal shapes as well as pulse heights for different shaping times are analyzed; they probably provide information on the life times of deep level traps.
- Dark-current correlated noise, where the noise is either extracted from the width of the pulse height spectra without any particle incident (the so-called pedestal width σ_{ped}) or it is calculated from the total dark current or from its noise density spectrum, respectively.
- Behaviour at different temperatures.
- Performance of microstrip and micropad/pixel detectors in high energy particle beams with respect to efficiency and spatial resolution.

Concerning the microscopic characterization , as an institute for high energy physics we rely on the collaboration with outside solid state physics oriented institutes (universities of Berkeley, Bologna, Florence, Bratislava) and the material suppliers, who will analyze our probes using NIR, PL, PICTS, TSC and hall mobility measurements to get informations on deep level concentrations, carrier mobilities and carrier life times and the corresponding changes with irradiation.

From our data we were able to extract results on charge carrier life times (see below).

4 Results

As an example, Fig. 3 and Fig. 4 show the dark current versus the reverse bias voltage for two different materials before and after irradiation. The unirradiated detectors behave similarly : the initial sharp turn-on is followed by a slow increase with increasing voltage up to eventual breakdown. After rradiation the current rises continuously with the voltage showing for all materials a sudden increase of the slope at a certain value $U_{bias} = U_k$. It should be noted that the FEW-LC material is the only one for which the current decreases with increasing irradiation level. Once irradiated, all detectors can be operated up to very high reverse bias voltages without exhibiting the usual fast breakdown. The above mentioned change of slope occurs when the electric field reaches the backside contact (i.e. the detectors becoming fully active). This can be inferred from the signal as measured when α-particles are injected from the backside : The signal starts to be detectable as soon as the bias voltage reaches the value U_k.

The detailed structure of the dark current vs. U_{bias} as well as its origin, e.g. the current fraction due to generation-recombination of charge carriers are not understood. We have observed that the detector-initiated parallel noise (σ_{ped}, see above) does not scale with the dark current I_{dark} in the usual way : one expects its contribution to the equivalent noise charge to obey the relation $ENC^2 = C \times I_{dark}$ with C being a calculable factor depending on the shaper characteristics, especially its peaking (integration) time. The noise density spectrum of I_{dark} for the FEW-LC material shows the expected (see fig. 3) noise reduction after irradiation. By taking only the flat white noise above $\sim 10^4 Hz$ of I_{dark} leads to a good agreement with the measured value for the noise as extracted from σ_{ped}.

As a consequence of the irradiated detectors to be fully active for bias voltages $> U_k$ the signal response to mips saturates towards high U_{bias}. Before irradiation the signals increase nearly linearly with U_{bias}, whereas they start to level off at $U_{bias} \sim U_k$. The saturation value decreases with increasing irradiation level, indicating decreasing free mean path lengths of the charge carriers. Fig. 5 summarizes the results concerning the dependence of the measured charges for incident mips on the various materials for proton irradiation. The results look similar for pion and neutron irradiation. Due to the saturation it is sufficient to give the data at $U_{bias} = 200\ volts$ (beyond U_k). All materials show a degradation with increasing irradiation level; it seems that the damage induced by pions is largest followed by protons and neutrons. It should be mentioned that the damage induced by photons turned out to be negligible for doses up to $100 Mrad$ (details can be found in [3]).

We have observed that the time structure of the signals changes with the irradiation. Unirradiated detectors show a slow time component of $O(\mu s)$, probably due to detrapping of deep levels, which dissappears even after rather low levels of irradiation. The size of the effect is dependent on the material.

The temperature dependence of the response to mips and the ENC has been studied in the range $-5^0C \leq\ t \leq 30^0C$. The signal height and noise charge are stable within this temperature interval for an integration time of the shaper of $40ns$.

The degraded detectors could be partly recovered by annealing them at 226^0C for 1/2 hour. For the FEW-LC detector an increase of the signal by $\sim$ 20% has been observed; the dark current decreases. A second annealing under the same conditions did not further improve on the behaviour.

5 Results-Microscopic Characterization

Based upon an empirical model for the electric field distribution and the charge carrier transport we could obtain from a combined fit to the data lifetimes for electrons and holes separately for different irradiation levels. They are presented in Fig 6 together with earlier data from [4]. The two measurement agree quite well and show that electron lifetimes are stronger affected by irradiation than hole lifetimes.

6 Conclusion

All detectors work well after irradiationwith high particle fluences, with no dependence of signal height and detector-induced noise on the ambient temperature in the range between $-5°C$ and $+30°C$. The life times of the electrons show a stronger decrease with increasing irradiation level than those of the holes.

References

[1] Technical Proposal CMS Collaboration, CERN/LHCC 94-38, LHCC/P1 December 1994

[2] O.Syben et al, 1994 Aachen Preprint PITHA 1994/67
Physikalische Institute RWTH Aachen, 52056 Aachen, Germany

[3] W.Braunschweig et al, 1994, Aachen Preprint PITHA 1994/43
Physikalische Institute RWTH Aachen, 52056 Aachen, Germany

[4] E.Weber UC Berkeley, Erice 1992 and private communication

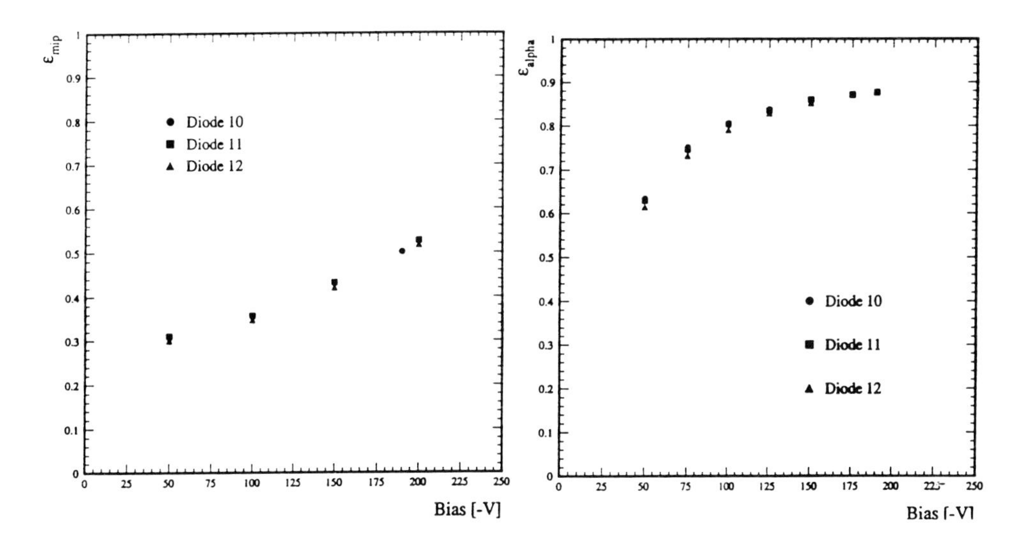

Figure 1: The charge collection efficiency for a minimum ionizing particle vs. the reverse bias voltage for three different diodes on the same wafer.

Figure 2: The charge collection efficiency for an α-particle injected from the frontside for three different diodes on the same wafer.

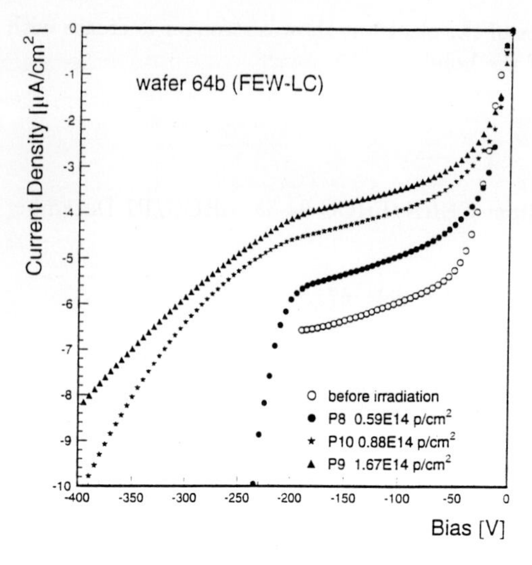

Figure 3: Characteristic I-V curve for FEW-LC material before and after irradiation.

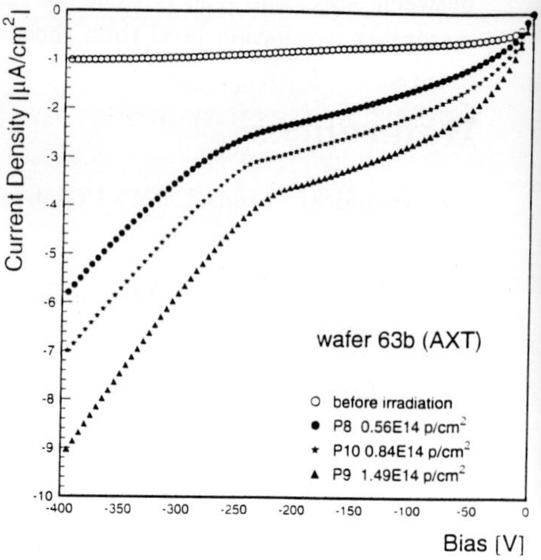

Figure 4: Characteristic I-V curve for AXT material before and after irradiation.

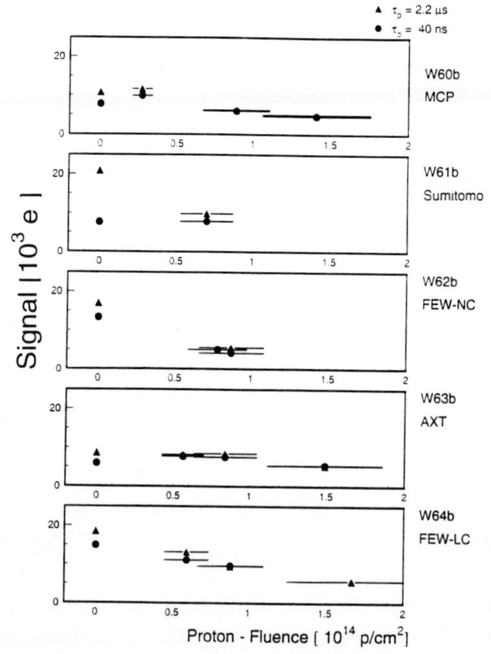

Figure 5: The mean signal height for mips for the indicated materials versus the irradiation fluence of protons at $U_{bias} = 200 volts$.

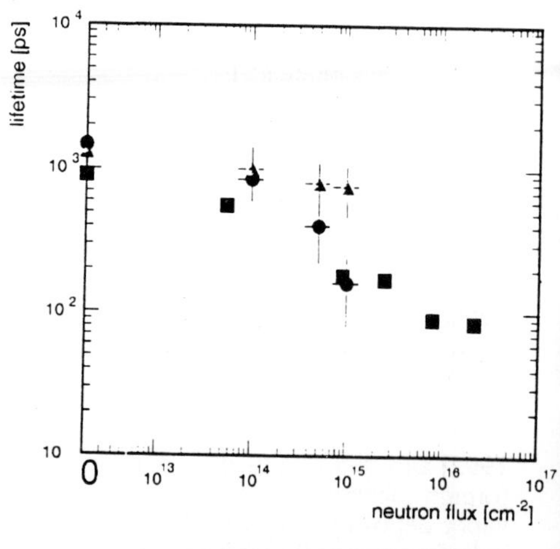

Figure 6: Life times of electrons (•) and holes (△) in SI-GaAs (FEW-LC material) as a function of the neutron fluence. The data marked as (□) give the life times for holes in a LT-GaAs msterial and are from ref[4].

Inst. Phys. Conf. Ser. No 149
Paper presented at DRIP 95

Surface Morphology of Silicon Carbide Layers Deposited by Cyclotron Resonance Plasma

F.J. Gómez [1], M.L. Rodríguez-Méndez [2], J. Piqueras [3], J. Jiménez [4], J.A. De Saja[5]
1) Dpt.Ingeniería Informática, Universidad Autónoma de Madrid, Cantoblanco, 28049 Madrid. Spain.
2) Química Inorgánica. ETS Ingenieros Industriales. 47011 Valladolid. Spain.
3) Dpt. Física Aplicada, Universidad Autónoma de Madrid, Cantoblanco, 28049 Madrid. Spain.
4) Física de la Materia Condensada. ETS Ingenieros Industriales. 47011 Valladolid. Spain.
5) Física de la Materia Condensada. Facultad de Ciencias. 47011 Valladolid. Spain

Abstract . The surface morphology of Silicon Carbide layers grown by ERC (Electron Cyclotron Resonance) on (100) and (111) Si substrates has been studied by ex-situ Atomic Force Microscopy (AFM). The morphology was observed to depend on the deposition temperature and the CH_4/SiH_4 flow ratio. RTA (Rapid Thermal Annealing) was also found to induce changes in the surface morphology. These changes are discussed in terms of the different growth parameters and the stoichiometry of the layers controlled by the flow ratio, as deduced from Fourier Transform Infrared (FTIR) spectroscopy and Spectroscopic Ellipsometry (SE) measurements

1.Introduction

Silicon Carbide is a wide gap semiconductor, whose energy gap ranges from 2.3 eV, for the β-SiC cubic phase, to 3.4 eV for the 2H α-SiC phase; thus, the energy gap and its electronic and optical properties can be continuously changed controlling the composition and the hydrogen content. This makes SiC an attractive material for high-temperature electronic devices and for potential applications in optoelectronics.

Thin film preparation of SiC has received a continued effort. Sputtering [Mafagas, 1992] and pulsed laser ablation [Rimai, 1993] have been used to obtain thin SiC layers on foreign substrates. Among the different techniques, those based on plasma enhanced chemical vapor deposition (PECVD) are becoming widely used for SiC:H obtention. Both, standard radiofrequency (RF) [Gat,1992] and electron cyclotron resonance (ECR) [Chayahara, 1993] are being extensively investigated. The deposition conditions: substrate temperature and the flow ratio, r= CH_4/SiH_4, play a major role in the characteristics of the films, both the stoichiometry and the structure. Post growth treatments are also considered in view of the improvement of the film processing.

2.- Experimental

A "low profile" ECR plasma system from Plasma Quest was used to deposit the a-SiC:H layers. In this low profile reactor one of the electromagnets is placed in the neighborhood of the sample holder. A load lock chamber and a high speed turbomolecular pump enable to get background pressures below 10^{-7} Torr.

The different gases are introduced by mass flow controllers. A flow of 100 sccm. of pure argon is introduced in the resonance region to form the plasma column while the reactive gases, pure CH_4 and 5% argon diluted SiH_4, are introduced by a second gas inlet in the proximity of the substrate holder. Pure silane flow was varied between 1.25 and 5 sccm, while CH_4 flow was 5 sccm; thus, flow ratio, r= CH_4/SiH_4, ranged from 1 to 4.

Prior to loading in the vacuum chamber, the (100) and (111) p-type Si substrates were degreased and the native oxide removed with ethanol diluted HF. Once the samples were loaded, a hydrogen and argon plasma is formed for 180 seconds, to assure that the starting Si surface was completely clean, finally, the flows of both precursors were set to the desired values. Different SiC layers were deposited varying the substrate temperature between 150 and 600 ^{o}C. The pressure in the work chamber was maintained between 1 and 3 mTorr in all the experiments. Under these conditions, reactions between the precursors gases and nucleation inside the plasma are greatly prevented, resulting very compact layers.

The deposition time was 120 s long to get a thickness between 500 Å an 1000 Å. For this thickness range the ellipsometric curves are free from interferences and it is possible to get a good estimation of the layer composition that may be compared with that obtained from infrared measurements [Gómez, 1995]. After deposition several samples were annealed for 3 minutes in a RTA (Rapid Thermal Annealing) system under vacuum at temperatures between 700 and 1100 ^{o}C.

3.- Results

3.1- Infrared and ellipsometric characterization.

For the CH_4/SiH_4 flow ratios used only small traces of the SiH_n and CH_n stretching bands are observed by infrared spectroscopy, being the deposited amorphous silicon carbide close to stoichiometry [Gómez, 1996]. Fig.1 shows typical FTIR spectra of four samples deposited at 550 ^{o}C with different flow ratios. The main difference between these spectra lays in the in the

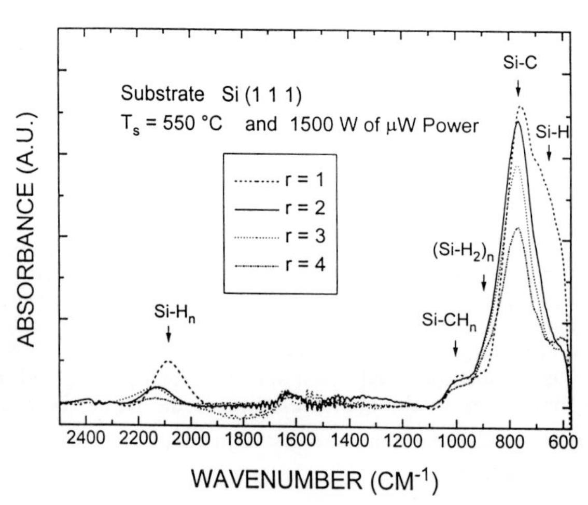

Fig.1. Infrared spectra of SiC layers as deposited on (111) Si at 550°C and 1500 W microwave power for several CH_4/SiH_4 flow ratios.

band around 2100 cm^{-1} assigned to SiH_n stretching vibrations [Brodsky, 1977]. The smaller the flow ratio the stronger the SiH_n band, as we have previously reported [Gómez, 1995]. It must be noted the nearly complete absence (not shown in the figure) of the CH_n stretching band at around 2900 cm^{-1} [Guivarc'h, 1980].

The main band at 780 cm^{-1}, common to all the spectra, is related to the Si-C stretching vibration [Guivarc'h, 1980]. Shoulders clearly appeared at high and low wavenumber sides of this band. In the low side, the shoulder at 670 cm^{-1} is generally assigned to the Si-H rocking or wagging modes [Basa, 1990]. In the high side, two additional shoulders appear depending on the deposition conditions. The shoulder at 900 cm^{-1} may be due to $(SiH_2)_n$ bending modes, whereas the mode at 1000 cm^{-1} should be related to wagging vibrations of CH_n bonds in Si-CH_n groups [Gat,1992; Basa, 1990; Katayama, 1991]. As previously reported [Gómez, 1996], we have tried to qualitatively follow the composition evolution of the films across the changes of the 780 cm^{-1} band and its shoulders. For this purpose, we have deconvoluted this band into four gaussian components corresponding to the Si-C, $(SiH_2)_n$, Si-(CH_n) and Si-H related bands. Fig.2 shows the normalized area of these bands as a function of the flow ratio for a series of samples deposited at 550 °C and 1500 W microwave power. For comparison we have plotted as well the compositions obtained for these samples from ellipsometric spectra measurements. The agreement between both data is excellent. On the other hand, the changes of the $(SiH_2)_n$ and Si-CH_n normalized band areas seem to be related to the void content. Furthermore, the a-Si concentration deduced from ellipsometry follows a parallel evolution of the IR band areas associated with the Si-H rocking or wagging modes. Figure 3 shows the normalized area and the ellipsometric composition changes with the annealing temperature for a sample deposited with r=4 and 1500 W microwave power. The SiC band area increased from 70% in the as-deposited sample up to nearly 90% after annealing at 1100°C.

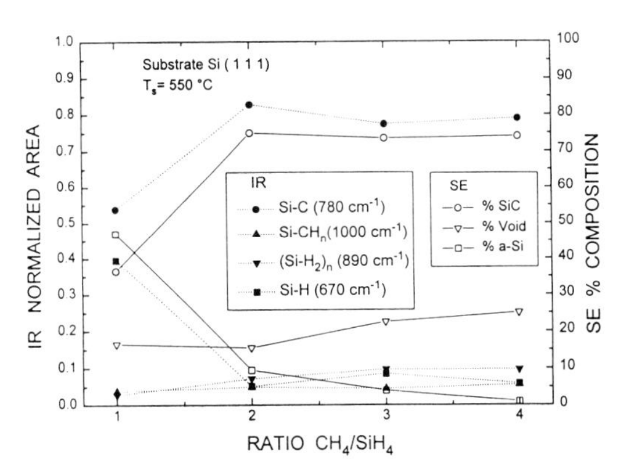

Fig.2. Normalized infrared band areas and composition deduced from ellipsometry measurements as a function of the flow ratio.

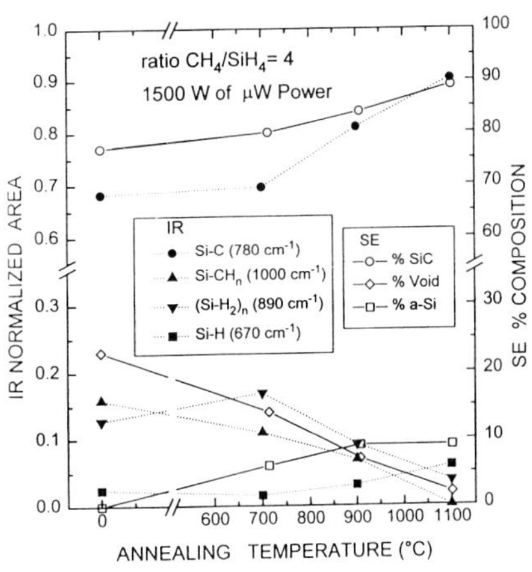

Fig.3. Compositional changes with the annealing temperature.

3.2.- Surface morphology.

The morphology of the surface was studied by optical microscopy, SEM (Scanning Electron Microscopy) and AFM (Atomic Force Microscopy). First of all, optical microscopy revealed point like inhomogeneities in samples grown at high substrate temperature and intermediate flow ratios, r= 2 and 3; the density of points appear enhanced for (100) substrates and r=3. The presence of these points was found to be dependent on the substrate temperature, the threshold for the observation of such inhomogeneity features was 500°C for both substrate orientations. The other samples grown on different conditions exhibit an homogeneous aspect under optical microscope observation. Samples grown below this threshold and samples grown at flow ratios 1 and 4 exhibited an homogeneous aspect under optical microscope observation.

These samples were subsequently studied by AFM, figs. 4a shows AFM image of the surface of a sample grown on an (111) substrate at 600°C and a flow ratio of 3. The points initially observed appear as hillocks, fig.4b, with an average height of 1000 A. We did not appreciate any regular geometry shape for these hillocks, though all of them were truncated at

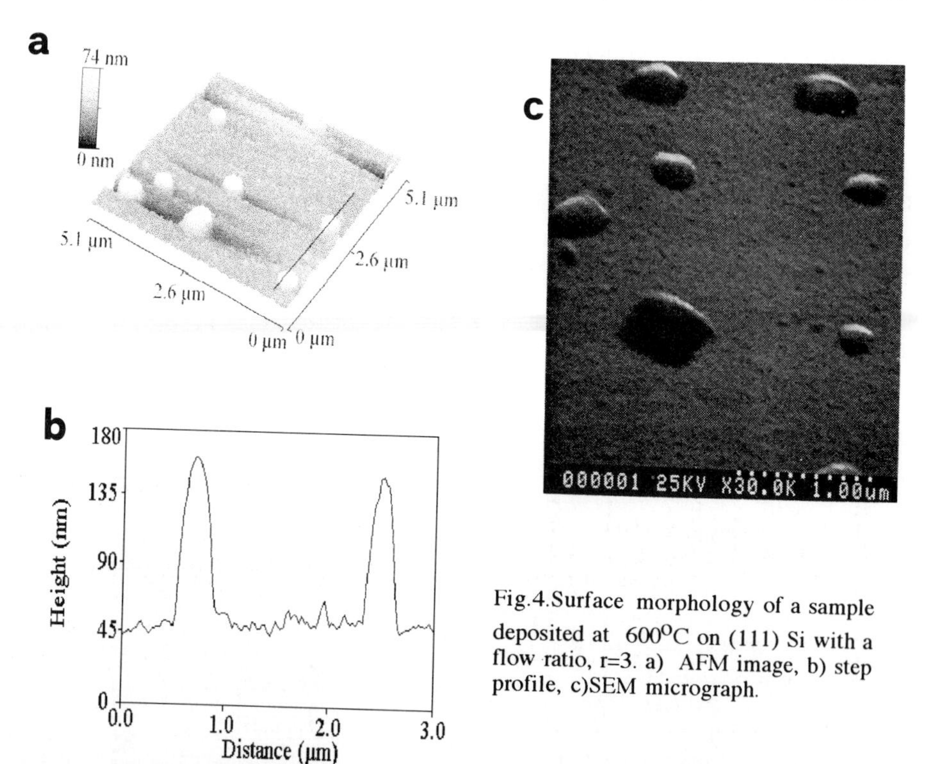

Fig.4.Surface morphology of a sample deposited at 600°C on (111) Si with a flow ratio, r=3. a) AFM image, b) step profile, c)SEM micrograph.

the top. SEM micrograph also evidenced the existence of randomly distributed columnar grains, fig.4c. In order to ascertain that these columnar grains were not generated by silicon segregation from neighbor areas, the surface was etched with a HNO_3, HF, CH_3-COOH solution; the grains were not etched and only areas presenting deposition failures were found to etch. This observation suggests that the columnar grains are not formed by segregated Si. Samples deposited under the same conditions but on (100) substrate exhibit as above mentioned a more marked columnar granularity, Fig. 5, the columnar grains pointed out a romboedral geometry. In order to complete the study of the surface morphology, we carried out

AFM measurements in the columnar free areas. We have compared the results above described with those obtained on a sample that did not exhibit columnar granularity (substrate temperature 600°C and flow ratio 4). Both kind of samples present a smooth granular texture in such areas, which looks more ordered in the samples grown on high flow ratios, without columnar grains, fig.6.

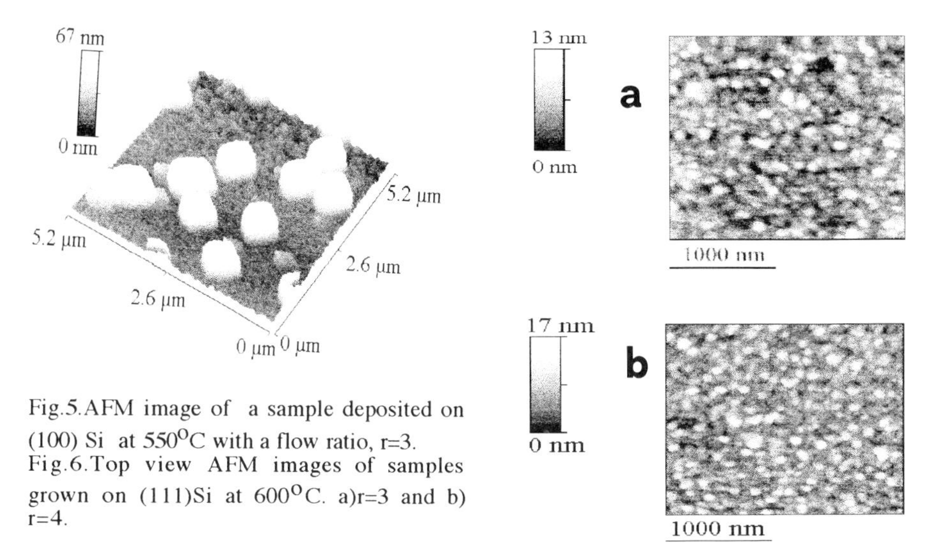

Fig.5.AFM image of a sample deposited on (100) Si at 550°C with a flow ratio, r=3.
Fig.6.Top view AFM images of samples grown on (111)Si at 600°C. a)r=3 and b) r=4.

3.3 .-Rapid Thermal Annealing

A study of the evolution of the columnar grain morphology under annealing in vacuum at 1100°C during 3 minutes was carried out. After such a treatment all the samples exhibited a granular like surface. SEM imaging of these defects revealed triangle shaped structures for (100) substrates and rectangle shaped structures for (111) substrates, Figs. 7a and 7b.

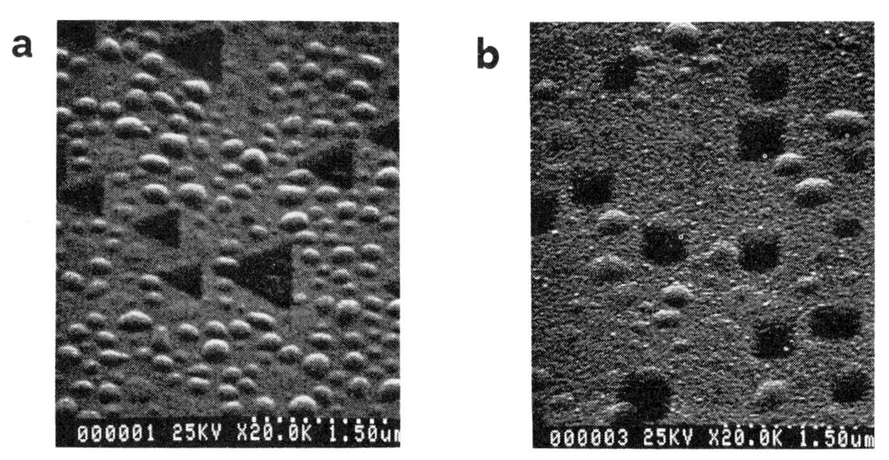

Fig.7.- SEM micrographs of samples deposited at 550°C with r=4 after RTA at 1100°C. a) (111) Si substrate, b) (100) Si substrate.

The size of these defects increases with the flow ratio. AFM demonstrated that such defects are depressions and that the granular texture extends all over the sample surface. RTA was also performed on a sample grown on a *cold* substrate (150°C) with a flow ratio of 4, without significant consequences on the surface morphology; only a few defects were observed. The texture of the defect free regions is smooth and uniform, the size of the grains is slightly larger than that observed for the as-grown specimen, but smaller than the grains in samples grown on substrates above 500°C.

4. Discussion

The generation of triangle and rectangle shaped defects on (111) and (100) substrates during RTA at 1100°C may be caused by a fast hydrogen release together with a microcrystallization process. The geometry of the defect should be determined by the orientation of the substrate. A more difficult question to answer is the formation of columnar grain defects in samples grown on substrates heated above 500 °C and only for intermediate flow ratios, 2 and 3. A tentative hypothesis would be that these are seed points for crystallization, which should start at temperatures depending on the flow ratio, 500°C for r=3 and > 600°C for r=4. For small flow ratio, r=1, the probability of symmetric Si-C bond formation is very low, since it needs excess CH_4. The fact that this eventual crystallization forms localized columnar structures instead of an homogeneous epitaxial layer, might be associated with the short surface migration length of C. The granular texture after RTA could be due to the formation of seed points, that result in a microcrystallization pattern dependent on the substrate orientation.

5. Conclusion

The morphology of the SiC films grown by ERC on Si substrates presents different features, that appear controled by the growth conditions, i.e. the flow ratio and the substrate temperature. The orientation of the substrate seems to influnce the geometry of such structures rather than their formation. The morphology can be changed by RTA treatment, resulting in a granular background texture and some features, which the geometry, either triangles or rectangles, depends on the substrate orientation.

References

Basa DK and Smith FW 1990 Thin Solid Films 192 121
Brodsky MH, Cardona M and Cuomo C 1977 Phys.Rev.B 16 3556
Chayahara A, Masuda A, Imura TI and Osaka Y 1986 Jpn.J.Appl.Phys. 25 L564
Gat E, El Khakani MA, Chaker M, Jean A, Pepin H, Kiefer JC, Durand J, Cros B, Rousseaux F and Gujrathi S 1992 J.Mater.Res. 7 2478
Gómez FJ, Martínez J, Garrido J, and Piqueras J 1995 J. Electrochem. Soc. (in press)
Gómez FJ, Martínez J, Garrido J, Gómez-Aleixandre C and Piqueras J, 1996 J. Non-Cryst. Solids (in press)
Guivarc'h A, Richard J, Le Contellec M, Ligeon E and Fonteneille J 1980 J. Appl. Phys. 51 2167
Katayama Y, Usami K and Shimada T 1991 Phil.Mag. B 34 283
Mafagas L, Georgeoolas N, Girginoudi D and Thanailakais A 1992 J. Non-Cryst. Solids 139 146
Rimai L, Ager R, Hangas J, Logothetis EM, Abu-Ageel N and Asland M 1993 J. Appl. Phys. 73 8242

Inst. Phys. Conf. Ser. No 149
Paper presented at DRIP 95

Morphological and MicroRaman Study of MBE $In_xGa_{1-x}Sb$ / $In_yAl_{1-y}Sb$ Heterostructures Grown on (100) GaAs Substrates

P.Martín*, J.J. Pérez-Camacho(+), J.Ramos*, J.Jiménez*, F.Briones****
*Física de la Materia Condensada, ETS ingenieros Industriales, 47011 Valladolid, Spain
**CNM, CSIC, Serrano 144, 28006 Madrid, Spain
(+) Present address: Optronics Ireland, Physics Department, Trinity College, Dublin 2, Ireland

Abstract. The morphology and structure of $In_xGa_{1-x}Sb$ / $In_yAl_{1-y}Sb$ heterostructures grown by MBE on (100) GaAs substrates under either low or high Sb_4 flux are studied by means of optical microscopy and Raman spectroscopy. The texture of the surface appear, as obtained by PSM imaging, to be different for both specimens, being the surface rougher for low Sb_4 flux. Raman spectroscopy reveals different cation composition for both Sb_4 flux conditions. These observations are discussed in terms of In and Ga incorporation. Some growth defects are studied.

Introduction

Antimonide based ternary alloys are used as photodetectors and light emitters in optical fiber systems [1]. The demonstration of antimonide based mid-infrared semiconductor lasers operating at room temperature [2] has increased the interest of these materials grown by molecular beam epitaxy, which might be the key for the next generation of atmospheric monitoring devices. The narrow gap ternary $In_xGa_{1-x}Sb$ is very attractive for mid-infrared optoelectronics, while $In_yAl_{1-y}Sb$ is a wide gap material that can be lattice matched to $In_xGa_{1-x}Sb$, providing the barriers for quantum well confinement. Physical information concerning $In_xGa_{1-x}Sb$ and $In_yAl_{1-y}Sb$ alloys is up today very limited, in particular only a few reports concerning the structural aspects are available [3,4]. Structural changes are often accompanied by compositional changes that modify the optical properties of the alloy [5]. In particular the growth conditions have a high influence on the surface morphology and hence the composition and structure are expected to present local fluctuations [6]. In these conditions the study of the surface morphology and the surface defect structure presents a high interest in order to control the growth parameters. Information about stoichiometry changes and surface texture are essential to the understanding of the incorporation of antimonium and cation diffusion during growth. We present herein a study of the morphology of the surface of $In_xGa_{1-x}Sb$ / $In_yAl_{1-y}Sb$ epilayers grown by MBE(Molecular Beam epitaxy) on GaAs (100) substrates under either high or low Sb_4 flux. The morphology study was carried out combining Nomarski microscopy, SEM (Scanning Electron Microscopy) and PSM (Phase Stepping Microscopy). Structural and composition analysis is done by means of microRaman spectroscopy. This technique allows to study small areas (slightly submicronic), allowing the assessment of local crystal order changes and composition fluctuations. These results are correlated to the morphology and the Sb flux during growth.

2. Experimental and Samples

PSM (Phase Stepping Microscopy) is a technique that combines optical interferometry and image processing for fringe analysis. It provides a profile of the studied surface with nanometer vertical resolution and slightly submicronic lateral resolution. The image of the surface can be reconstructed, providing thus a complete view of the surface topography. For a more detailed description of the technique see ref. 7

MicroRaman measurements were carried out with a DILOR X-Y Raman spectrometer attached to a metallographic microscope (Olympus BHT). The excitation is made with an Ar^+ laser. Both the excitation and the scattered light collection are made through the microscope objective, conforming thus a nearly backscattering geometry. The diameter of the laser beam at the focal plane is diffraction limited, $D= 1.22\ l /\lambda\ NA$, according to the Rayleigh criterion, where λ is the laser wavelength and NA is the Numerical Aperture of the objective; in our standard experimental conditions, λ = 514.5 nm and the objective was a 100X with NA= 0.95. Thus the lateral resolution is better than 1 μm and the spectral resolution was better than 2 cm^{-1}.

The samples were grown on (001) GaAs substrates using a conventional solid source MBE. After standard oxide desorption (at 600 °C under an As_4 flux), a 0.2 μm thick GaAs buffer layer was grown at 580 °C, in order to improve the surface flatness. Then the substrate temperature was lowered, the arsenic cell cooled down and the antimony cell was heated up, in order to avoid the incorporation of Sb in the buffer layer and of As in the antimonide layers. The antimony cell provided a beam of Sb_4, as the cracker was not in operation. Group III elements fluxes were previously calibrated by standard procedures. The substrate temperature was 420°C during the antimonide deposition. We will focus our study on samples grown in non optimum conditions, either low ($2.5*10^{-6}$ torr , sample 1) or high ($3*10^{-6}$ torr , sample 2) Sb_4 pressure.

3. Morphology

SEM micrographs are shown in fig.1. The morphology appears different for each sample. It is known that the flux ratio Φ_{V}/Φ_{III} plays a crucial role in the surface morphology of the layers. Unlike As over GaAs, the vapor pressure of Sb (mostly Sb_4) over GaSb is very low, which results in stoichiometry changes under excess Sb flux and the formation of cation droplets under a low Sb flux [6]. The presence of these metallic droplets leads to a strong rugosity since they are nucleation sites for tridimensional growth. In addition to this surface texture, growth under non optimal flux ratio conditions can favour the presence of extended surface defects, such as oval defects, which have been extensively studied in GaAs and GaAlAs epilayers [5,8,9]; showing that they grow at surface contaminant particles or at Ga droplets. In order to assess the complete morphology of our specimens PSM images of the surface were obtained, fig. 2. 3-D views of the surface morphology and transverse cross sections are shown. The period of the surface profile differs from

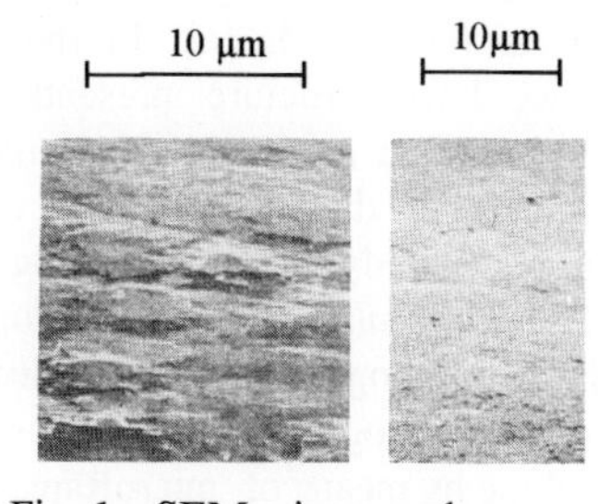

Fig. 1.- SEM micrographs.
a) Sample 1, b) Sample 2.

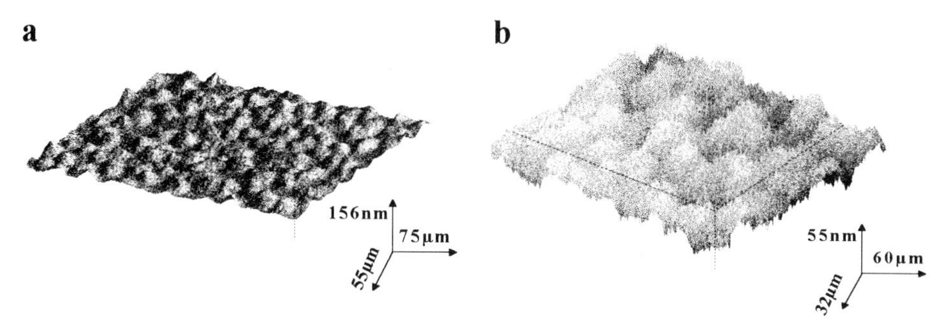

Fig 2.- PSM images (3D), a)Sample 1,b)Sample 2.

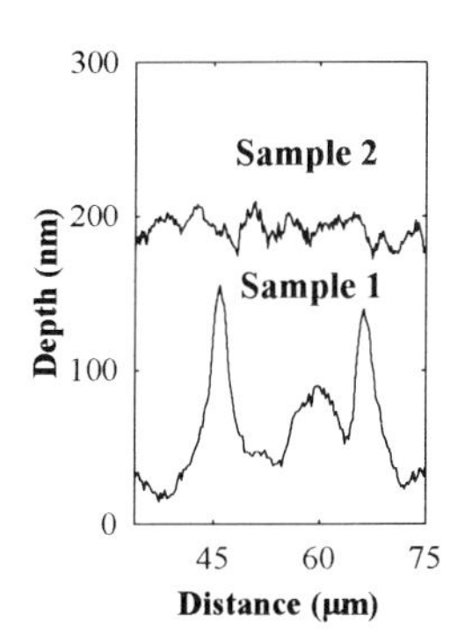

Fig.3 Cross sections from PSM images.

each other, being the texture much strongly marked for Sb poor (sample1) than for Sb rich (sample2) specimens, fig.3. In samples grown under low Sb_4 flux the formation of cation droplets should be responsible for the overgrowths, while in the case of samples grown under excess Sb flux, the stoichiometry fluctuations could account for the observed texture.

Some oval defects were also imaged, fig.4. The size of these defects in low Sb flux samples is significantly larger than in Sb excess samples. On the other hand, they present a truncated shape in low Sb flux. Also the formation of terraces is observed, suggesting the presence of successive metallic droplets, that will favour the overgrowth in successive stages. Such terraces were also observed in oval defects in GaAs epilayers [8].

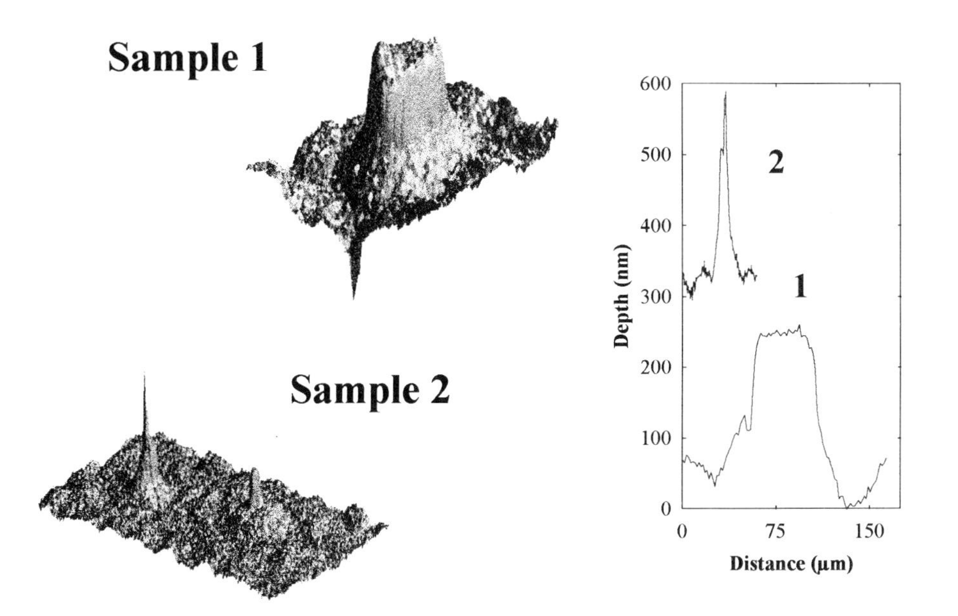

Fig. 4.- Oval defects (3D image) and cross section.

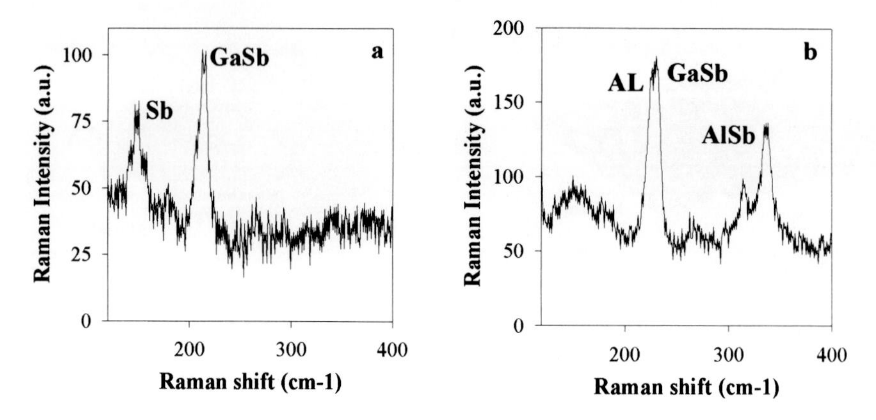

Fig. 5.- Typical Raman spectra a) Sample 1, b) Sample 2.

4. Raman Spectroscopy

The heterostructures herein studied are formed of two layers, $In_xGa_{1-x}Sb$ / $In_yAl_{1-y}Sb$. According to the masses of the atoms forming these alloys the MREI (Modified Random Element Isodisplacement) [4,10] theory predicts a two mode behavior for $In_yAl_{1-y}Sb$ and a partly two mode behavior for $In_xGa_{1-x}Sb$. The frequency of the different modes should give information about the alloy composition. On the other hand the Raman backscattering selection rules should inform about the crystal order. The Raman microprobe was used to study the characteristics of different defect structures.

First of all, typical Raman spectra obtained in both samples are shown in fig5. These spectra differ from each other. The most significant difference between them is the absence of $In_yAl_{1-y}Sb$ like modes in sample 1, that only exhibited modes corresponding to the top layer. This should imply that this top layer is thicker for this sample than it is for sample 2. This might be due to a faster growth associated with the formation of cation droplets.

Other feature of the Raman spectra to be pointed out is the different frequency maximum of the $In_xGa_{1-x}Sb$ like modes in the two specimens. Only the GaSb like modes were observed for the composition of our samples. This Raman band is centered at about 213 cm^{-1} for sample 1 and 230 cm^{-1} for sample 2; this frequency difference should correspond to a significant composition difference between both samples ($x = 0.7$ for sample 1 and $x = 0.3$ for sample 2). We can assume that sample 2 presents a higher Ga amount than sample 1, that gives the expected composition ($x=0.7$) [10]. Paradoxally, some points of sample 1, low Sb_4 flux, exhibit a Raman band at 150 cm^{-1}, that is rather probably related to solid Sb [11]. It is interesting to remark that in such points the In content is enhanced, the frequency of the Raman band shifts to the low frequency, which can be related to the enormous vapor pressure of Sb in InSb. This is consistent with the astonishing fact that in sample 2, grown under a high Sb flow, we did not observe Sb precipitation, since it is Ga rich.

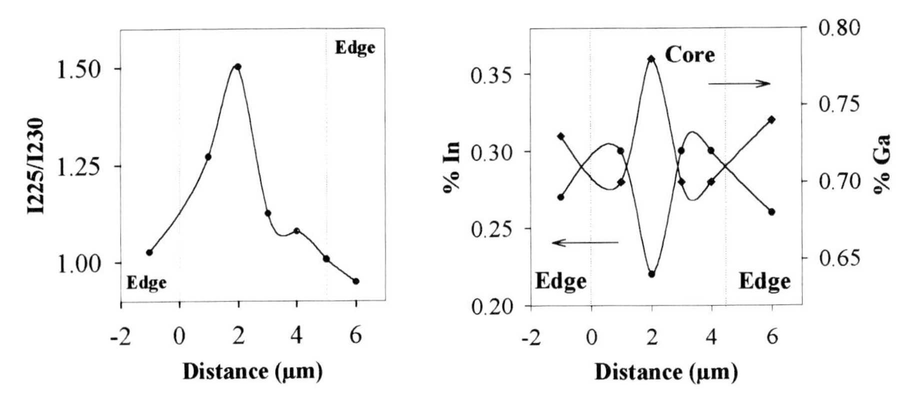

Fig. 6.- 6a) I_{255} / I_{230} ratio along a scan lline crossing an oval defect on sample 2, 6b) Cation composition along the scanning line.

Sample 2 presents other Raman bands that need to be studied, among them the Raman bands associated with the underlying layer $In_yAl_{1-y}Sb$ [4], which will not be considered here and the two peaks exhibited by the GaSb like band of the top layer. This complex structure merits special attention. The peak at 230 cm^{-1}, as it was previously discussed, is related to the LO GaSb like band. The other peak can be the TO GaSb like, which is selection rule forbidden. The high intensity of this Raman band suggests that the origin of this band can be different. The other possibility is that this peak can be partially associated with an acoustic localized (AL) mode [12], corresponding to the motion of Sb atoms about Ga in In position (Ga_{In}). The non observation of such a mode in sample 1, In rich, supports this hypothesis, ruling out plane tilting effects, as the main cause for such a Raman peak. This should imply that the excess Sb should favor the incorporation of Ga. It is rather probably that the metallic droplets, either In or Ga, are solved by Sb_4, being the Ga droplets more easily solved than the In droplets due to the different vapor pressures.

Following these general considerations about the Raman spectra of both samples we can analyze the local changes observed at some defects observed in both samples. In particular an oval defect (sample 2) was scanned by the Raman microprobe. The main results are summarized in fig.6a. The relative intensity, I_{225} / I_{230} is enhanced in the oval defect. This can obey two causes, an enhancement of the AL mode or a plane tilting effect resulting in the observation of the forbidden TO mode. The first hypothesis is supported by the alloy composition deduced from the frequency of the GaSb like LO peak, that reveals a Ga rich composition, fig.6b. Nevertheless, misorientation effects are typical of oval defects [5,8,9], All this suggests that both effects can contribute to the local enhancement of the 225 cm^{-1} peak. Taking into account the rich Ga composition at the oval defect, it should rather probably nucleate at an In droplet.

Other defects were studied in sample 1. Different points of the textured surface were probed by the Raman microprobe. The alloy composition deduced from the frequency of the GaSb like LO peak, fluctuates at both sides of $x = 0.7$. This would suggest that metallic droplets originating the tridimensional growth could be formed indistinctly of In or Ga, depending on the point probed and in agreement with the very strong rugosity determined by PSM in this sample, figs. 2 and 3.

5. Conclusion

$In_xGa_{1-x}Sb$ / $In_yAl_{1-y}Sb$ heterostructures have been studied by optical microscopy and microRaman spectroscopy. Samples grown under low Sb_4 flux present a strong rugosity, due to the formation of metallic droplets. Samples grown at high Sb_4 flux, result in Ga rich alloys and a smooth texture; the formation of In droplets appears as the main cause of such texture. The structure of some growth defects is studied and it was found to agree with the growth features of the samples.

Acknowledgements samples were grown at CNM -CSIC Madrid (ESPRIT III Project No. 6374). P.M., J.J and J.R. were supported by DIGICYT (MAT 94-0042)

6. References

[1] Eglash S J and Choi H K 1994 Appl. Phys. Lett. **64** 833
[2] Lee H, York P K, Menna R J, Martinelli R U, Garbuzov D Z, Narayan S Y and Connolly J C 1995 Appl. Phys. Lett. **66** 1942
[3] Cerdeira F, Pinczuk A, Chiu T H and Tsang W T 1985 Phys. Rev. B **32** 1390
[4] Genezdilov V P, Lockwood D J, Webb J B and Maigné P 1993 J. Appl. Phys. **74** 6883
[5] Sapriel J, Chavignon J and Alexandre F 1988 Appl. Phys. Lett. **52** 1970
[6] Chyi J I, Kalem S, Kumar N S, Litton C W and Morkoc H 1988 Appl. Phys. Lett. **53** 1092
[7] Montgomery P C, Fillard J P, Castagné M and Montaner D 1992 Semicond. Sci. Technol. **7A** 237
[8] Martín P, Ramos J, Jiménez J, Sanz L F, González M A (to be published) Mater. Sci. Technol.
[9] Dobal P S, Bist H D, Mehta S K and Jain R K 1995 J. Appl. Phys. **77** 3934
[10] Brodsky M H, Lucovsky G, Chen M F and Plaskett T S 1970 Phys. Rev. B **2** 3303
[11] Farrow R L, Chang R K, Mroczkawski S and Pollak F H 1977 Appl. Phys. Lett. **31** 768
[12] Parayanthal P and Pollak F H 1984 Phys. Rev. Lett. **52** 1822

Inst. Phys. Conf. Ser. No 149
Paper presented at DRIP 95

Imaging of the Si-Buried Oxide Interface Charges by Surface Photovoltage for the SOI CMOS Applications.

K. Nauka and M.Cao

Hewlett-Packard Company, Palo Alto, CA 94304.

Abstract. Surface Photovoltage has been employed to image the distribution of charges at the Si - buried oxide interface in Si-on-Insulator wafers. Substrates fabricated by oxygen ion implantation and wafer bond and etchback have been compared. Oxygen implanted wafers exhibited larger and less uniformly distributed interfacial charges than bonded wafers with comparable buried oxide thicknesses. Interfacial charges were stable at temperatures below the 1000° C.

1. Introduction

Silicon-on-Insulator (SOI) substrates (Figure 1.A) offer opportunities for extending CMOS IC technology beyond the boundaries imposed by the bulk Si substrates. SOI MOSFETs can consume less power and switch faster than the corresponding bulk Si transistors [1]. Additionally, presence of buried oxide simplifies insulation between the PMOS and the NMOS devices in the CMOS circuits [2]. SOI wafer can be fabricated either by implanting Si wafer with a large dose of oxygen, followed by a high temperature anneal that restores crystallinity of the top Si layer (SIMOX wafer) and forms buried oxide layer, or by etching back one side of two thermally bonded Si wafers (BESOI wafers).

Application of SOI wafers for sub - 0.25 μm CMOS devices requires an in-depth undertanding of the material properties that might impact the device quality. Electrical properties of the Si - buried oxide interface are of particular interest, because this interface becomes an integral part of the device. Interfacial defects frequently manifest their presence as interfacial charges. Their impact on device performance has been evaluated by simulating properties of partially depleted SOI MOSFET as a function of Si - BOX interface charge density (Q_i) [3]. NMOS transistor with 0.25 μm channel length, top Si thickness between 100 nm - 150 nm, BOX thickness = 400 nm, and retrograde implanted dopant concentration in the channel region was assumed. It was found that the off-state leakage current was severely degraded when the Q_i exceeded $2 * 10^{12}$ cm^{-2} (Figure 2). Further reduction of device dimensions decreased the device neutral body and lowered the maximum acceptable Q_i to below the 10^{12} cm^{-2}. In the case fully depleted device, where the channel neutral region disappeared, Q_i related leakage could be even more severe.

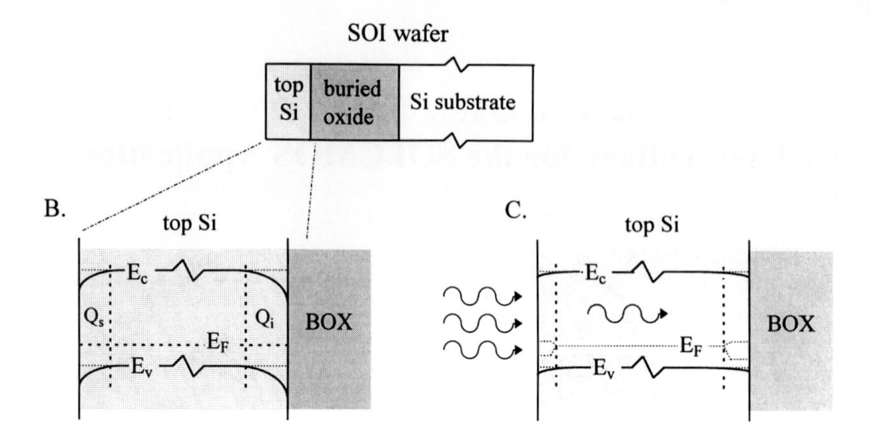

Fig.1. A.) schematic SOI Si wafer; B.) band diagram of the p-type thin top Si under equlibrium; C.) band diagram of the p-type thin top Si under strong illumination ($h\nu > E_g$).

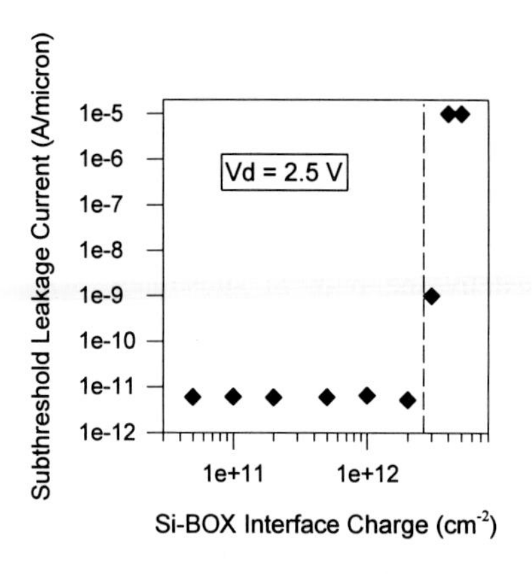

Fig.2. Off-state leakage current for a partially depleted 0.25 μm SOI NMOS transistor.

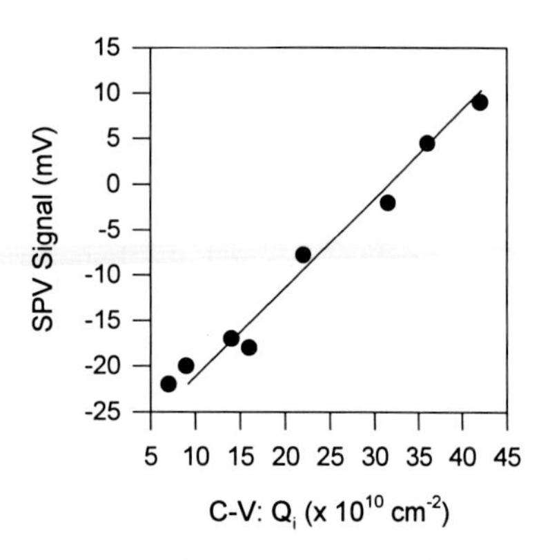

Fig.3. Correlation between the interfacial charge data obtained by C-V and SPV for bulk Si MOS capacitors.

2. Experimental

SPV technique employed for the evaluation of Q_i was based on previously described [4] Si surface charge measurements. The SPV signal was obtained by illuminating one side of a wafer with monochromatic photon flux having energy higher than the Si bandgap, while the other side remained in darkness. Nonequlibrium carriers generated by absorbed photons caused a decrease of both the surface and the interface potential barriers (Figure 1 B. and C.). At high photon flux intensities both potential barriers disappeared, and the SPV signal reached the maximum. Under these conditions the measured SPV signal can be expressed as follows:

$$V_{SPV} = V_s + V_i = A * Q_{SPV} = A * (Q_s + B * Q_i) \quad (1)$$

where V_s and V_i are respective surface photovoltage components due to the Si surface (Q_s) and Si-BOX interface (Q_i) charges. A is a calibration constant that has been obtained by comparing SPV data obtained for bulk Si wafers with the corresponding interface charge density results obtained from the C-V measurements of MOS capacitors. (Figure 3). B is a constant describing attenuation of the photon flux before it reaches the Si-BOX interface; it has been calculated using the known Si absorption and reflection coeficients [5]. Since only a small portion of the photon flux reaches the BOX-Si substrate interface, its contribution to the measured SPV signal has been neglected.

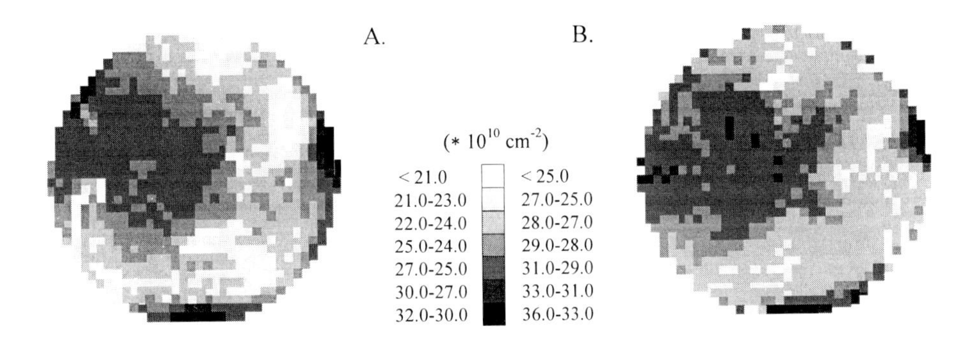

Fig.4. A.) Q_{SPV} map of a SIMOX wafer after the RCA clean (SC2 last); B.) Q_{SPV} map of the same SIMOX wafer after the HF dip. Top Si thickness = 100 nm, BOX thickness = 370 nm.

Additionally, V_s has been evaluated by applying Si surface cleans that affected only the Si surface charges. Since cleaning changed only V_s while V_i remained constant, comparison of the SPV results obtained for the SOI and the bulk wafers allowed to evaluate the contribution of V_s to the measured V_{SPV} signal. It was found that V_s was only a small part of the overall V_{SPV} signal, and it was constant throughout the wafer's surface. Thus, the observed variations of the V_{SPV} signal across the wafer were, in most cases, due to the local variations of the top Si - BOX interfacial charges. Figure 4 demonstrates maps of the Q_{SPV} obtained for a SIMOX wafer cleaned using chemical treatments leaving distinctly different types of surface charges: first, mixture of H_2O, HCl, and H_2O_2 was employed (SC2 clean [6]) and the wafer was measured (Figure 4.A); then the wafer was washed with HF and measured again (Figure 4.B). SC2 clean formed thin, chemical oxide, while HF dip removed the oxide and left Si surface terminated with H atoms. Corresponding experiment conducted on bulk Si wafers demonstrated that both treatments left Si surface uniformly covered with surface charges; lower charges were seen in the case of a wafer covered with the chemical oxide. Thus, distributions of the SPV signal shown in Figure 4, similar for both cleans, are due to the top Si - BOX interfacial charges that remain unaffected by the surface cleans. Q_s appears to be a small, constant, additive component of the measured Q_{SPV} signal.

SPV measurements were conducted using a commercial [7] SPV monitoring system facilitating non-contact measurements that did not require any wafer processing. Clean room compatibility and fast operation allowed to employ it for the in-line qualification of incoming SOI wafers.

3. Characterization of Interfacial Charges

Figure 5 demonstrates interfacial charge maps obtained for a SIMOX wafer (A) and a BESOI wafer (B) with similar top Si and BOX thicknesses. Interfacial charge variability, similar to the one shown in Figure 5.A has been observed for all measured SIMOX wafers.

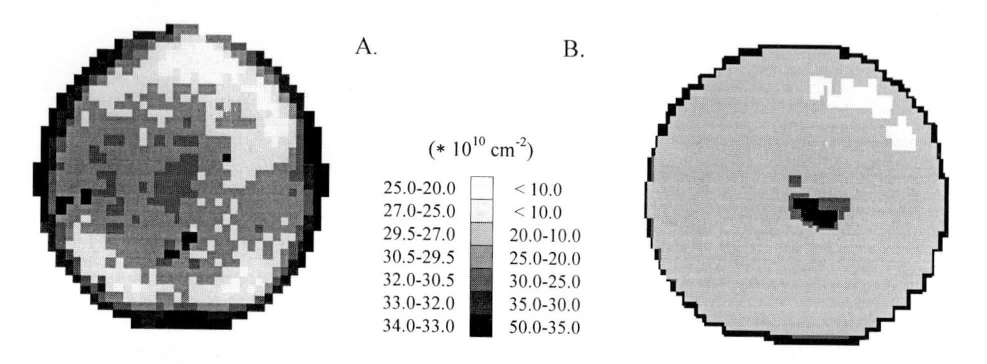

Fig.5. A.) Q_i map of a SIMOX wafer (top Si thickness = 120 nm, BOX thickness = 380 nm); B.) Q_i map a BESOI wafer (top Si thickness = 110 nm, BOX thickness = 400 nm).

Although pattern of charge distribution across the wafer varied between the wafers, the relative magnitude of variations remained constant for a given O dose. An increase of the O dose increased both, the magnitude of interfacial charge variations across the SIMOX wafer and the average charge values. For example, change of the O dose from $3 * 10^{17}$ cm^{-2} to $1.8 * 10^{18}$ cm^{-2} caused 6 times increase of the average value of Q_i and 35 times increase of the standard deviation of Q_i calculated for an average SIMOX wafer.

Interfacial charge densities in BESOI wafers were lower than in the SIMOX wafers with similar top Si and buried oxide thicknesses. The only exception were the areas where interfacial voids were present. High interfacial charge region observed in the center of Figure 5.B is due to the presence voids at the top Si layer - buried oxide interface, as confirmed independently by the cross-sectional TEM observations.

Also, it had been seen that high temperature heat treatment had no impact on the interfacial charge distribution in SIMOX wafers. Experiments, in which high temperature annealing steps imitating the furnace operations of a 0.25 μm CMOS device manufacturing process were used, left SPV measured Q_i distribution virtually unchanged. Thus, interfacial charge nonuniformities present in SIMOX substrates cannot be annealed out.

4. Conclusion.

It has been demonstrated that Surface Photovoltage can be employed to monitor charges at the top Si layer - buried oxide interface in SOI wafers. SIMOX wafers exhibited non-uniform interfacial charge distribution while BESOI wafers had smaller and more uniformly distributed interfacial charges except for the regions where interfacial voids were present.

Interface charge distribution and magnitude in SIMOX wafers were dependent on the O implant dose; wafers fabricated using large O doses exhibit large and less uniformly distributed interfacial charges. High temperature annealing did not impact Q_i distribution in SIMOX wafers. The proposed measurements could be used for the initial monitoring of SOI wafers.

Acknowledgment.

Collaboration of F. Assaderaghi in SOI MOSFET modelling is acknowledged.

References:

[1] Shadini G.G., Ning T.H., Dennard B., Davari B., in Extended Abstracts of the *1994 Solid State Device and Materials Conf. (SSDM)*, p.265.
[2] Suma K. et al., *Proc. of the ISSCC*, p.138 (1994).
[3] Nauka K., Cao M., Assaderaghi F., *1995 Internat. SOI Conference Proc.*, IEEE Publ., p.52
[4] Edelman P., Lagowski J., Jastrzebski L., "*Surface Charge Imaging in Semiconductor Wafers by Surface Photovoltage*", presented at the Materials Research Meeting, San Francisco, CA, April 1992.
[5] *Data in Science and Technology - Semiconductors, Group IV*, Springer-Verlag 1991.
[6] Kern W., Puotinen, *RCA Review*, June '70, p.187.
[7] Model CMS A/R, Semicond. Diagnostics, Inc.

Inst. Phys. Conf. Ser. No 149
Paper presented at DRIP 95

Effect of Temperature on implant isolation characteristics of AlGaAs/GaAs multilayer heterojunction structure

Hong-Da Xu, Sen Wang, Dong Zheng, Huan-Zhang Liu and Li-Shou Zhou
Institute of semiconductors, Chinese Academy of Sciences, P. O. Box 912, Beijing 100083, PRC.

Abstract. Thermally stable high-resistivity regions have been formed using hydrogen ion implantation at three energies (50, 100, and 180 keV) with three corresponding doses (6×10^{14}, 1.2×10^{15}, and $3\times10^{15}cm^{-2}$), oxygen implantation at 280keV with $2\times10^{14}cm^{-2}$ as well as subsequent annealing at about 600℃ for 10-20s, in AlGaAs/GaAs multiple epitaxial heterojunction structure. After anncaling at 600℃, the sheet resistivity increases by six orders more of magnitude from the as-grown values. This creation of high resistivity is different from that of the conventional damage induced isolation by H or O single implantation which becomes ineffective when anneal is carried out at 400—600℃ and the mechanism there of is discussed.

1. Introduction

Ion implantation is widely used in GaAs based system semiconductor device processing for the purpose of selectively altering the resistivity at the surface, e. g., in metal-semiconductor field effect transistors (MESFETs), the major application of ion implantation is the formation of n and n^+ regions for channel and source/drain contact regions, respectively. Alternatively, ion implantation can also be used to create high resistivity regions for device isolation. For device structure comprising conductive epitaxial layers on semi-isolation substrate, high resistivity region between devices can be confirmed by implantation of protons(H) or oxygen(O). However, H or O implant isolation is thermally stable only at the temperature of about 400℃ or of 700℃, respectively. [1,2]. Although the isolation induced by implanting O into AlGaAs is due to chemical activity to be thermally stable at $>$ 700℃, unfortunately, a similar process has been shown to be unsuccessful in n type GaAs[1]. Hence, in device processing, any high temperature process steps must be carried out prior to the isolation implant that introduces more complications there into. So, A higher temperature-stable isolation technique is very attractive for device processing flexibility.

In this study, we report on the activation, diffusion and isolation characteristics of implanted H and O in AlGaAs/GaAs multilayer heterojunction structure material grown by molecular beam epitaxy (MBE). To the best of our knowledge, this is the first report on the implant properties of H at three energies and doses and O at single energy and dose (multi-bombarded) in AlGaAs/GaAs multilayer heterojunction structure.

2. Experiment

All the $Al_x Ga_{1-x} As$/GaAs structures studied had $x=0.3$, which are the typical values used in the growth of heterojunction bipolar transistors (HBTs). The structure consists of from the substrate up, 4000 Å GaAs ($Si=2\times10^{18}cm^{-3}$) buffe layer, 5000 Å GaAs ($Si=5\times10^{16}cm^{-3}$) collector, 800 Å GaAs ($Be=1\times10^{19}cm^{-3}$) Base, 300 Å undoped GaAs spacer, 1500 Å AlGaAs ($Si=5\times10^{17}cm^{-3}$) emitter, with 200 Å graded $Al_xGa_{1-x}As$ ($x=0$ to 0.3) both sides and a 1500 Å GaAs ($Si=2\times10^{18}cm^{-3}$) emitter cap, totaling a thickness of 1.35μm and grown on a semi-insulating GaAs substrate by MBE. For all samples, hydrogen ions (H) were implanted at three energies (50, 100, and 180 keV) and three corresponding doses (6×10^{14}, 1.2×10^{15}, and 3×10^{15} cm^{-2}) to produce an approximately uniform hydrogen concentration throughout the $Al_{0.3}Ga_{0.7}As$/GaAs multiple epitaxial layers. The doses at 100 keV and 180 keV were 2 times and 5 times the dose at 50 keV, respectively. The peak hydrogen concentration for the dose at 50 keV is approximately 10 times the grown-in sillicon concentration in the $Al_{0.3}Ga_{0.7}As$/GaAs multilayer.

After the hydrogen implantation was carried out, the oxygen ions (O) was subsequently implanted at the energy of 280 keV and the dose of $2\times10^{14}cm^{-2}$. the peak oxygen concentration for the dose of $2\times10^{14}cm^{-2}$ is approximately 6 times the grown-in beryllium concentration in the GaAs layer.

All the ion implantations were performed in a nonchanneling direction (7° tilt) at room temperature.

The samples were followed by rapid thermal annealing (RTA) at 400, 450, 500, 550, 600, 650, 700 and 800℃ for 10-20s in a pure nitrogen ambient. After anneals, the sample was monitored by recording O, H, Al, and Be depth distributions using Camera IMS 4f SIMS machine and its sheet resistivities, concentrations and mobilities were determined by Van der Pauw measurements using indium ohmic contacts.

3. Resutts and discussion

The dependence of the sheet resistivity on RTA temperture is shown in Fig. 1 for samples implanted with H at multiple energies (50, 100, and 180 keV) and corresponding doses (6×10^{14}, 1.2×10^{15}, and $3\times10^{15}cm^{-2}$), and O at an energy of 280 keV and a dose of $2\times10^{14}cm^{-2}$, and subsequently annealed at temperatures of 400-800℃. For temperature ranging 400-800℃ except 600℃, a lower sheet resistivity is obtained. During the annealing process the sheet resistivity appears to be reversed from normal operation in the vicinity of 600℃. We found that this reverse annealing effect observed on the implantation with H and O at two different sets of doses is almost the same.

The salient features of the annealing process which can be extracted from Fig. 1 are following: a) as the temperature is increased from 550 to 600℃, the activation increases, indicating that a thermal budget of 550℃ for 10-20s is not sufficient to fully activate the implantation; b) as the temperature increases from 600 to 650℃, the activation decreases, indicating that the onset of reverse annealing occurs at temperature Tr≈600℃.

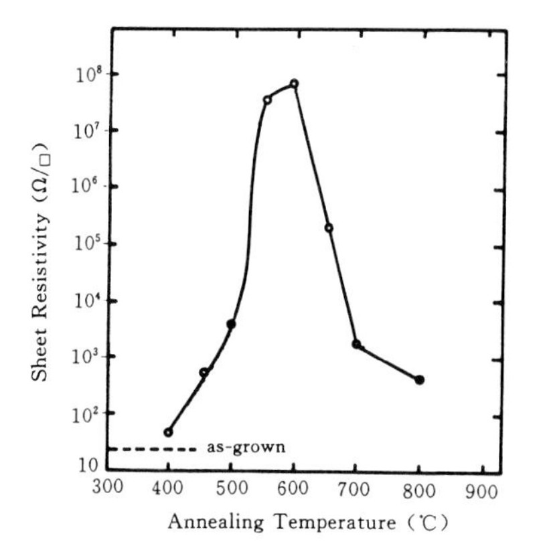

Fig. 1. Sheet resistivity vs anneal temperature for H and O implantation.

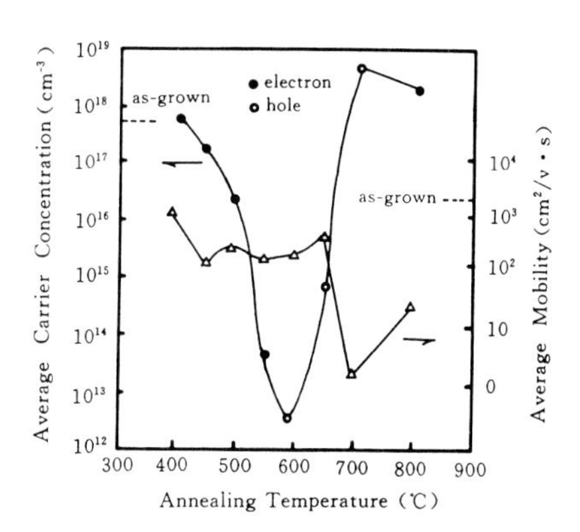

Fig. 2. Average carrier concentration or average mobility vs anneal temperature for H and O implantation.

The change in average carrier concentration and mobility in the implanted multiple epitaxial layers is shown in Fig. 2 as a function of the post-annealing temperature. As shown in Fig. 2, the average carrier concentrations or mobilities are decreased with increasing post-annealing temperature until reaching the temperature of 600℃, at which the minimum concentration is confirmed. At about 550℃, the electrical characteristic of the implanted epitaxial layer changes from electron conduction into hole conduction, but the decrease of the mobility value is not obvious. After the anneal at the critical temperature of 600℃, the hole concentration is increased from 5×10^{12} to $8\times10^{18}cm^{-3}$and the hole mobility is further decreased to a minimum value of 1.9 $cm^2/v.s$ with increasing annealing temperature until 700℃, the original electron conduction characteristic is restored, but the average electron concentration is higher than its initial value, and the mobility value is on the contrary until an annealing at 800℃. This fact implies that due to implantation with H and O and subsequent high temperature anneal, the shallow donor level and the electron scattering center are finally increased in the multiple epitaxial layer, resulting in an increas of the total electron concentration and a decrease of the total electron mobility.

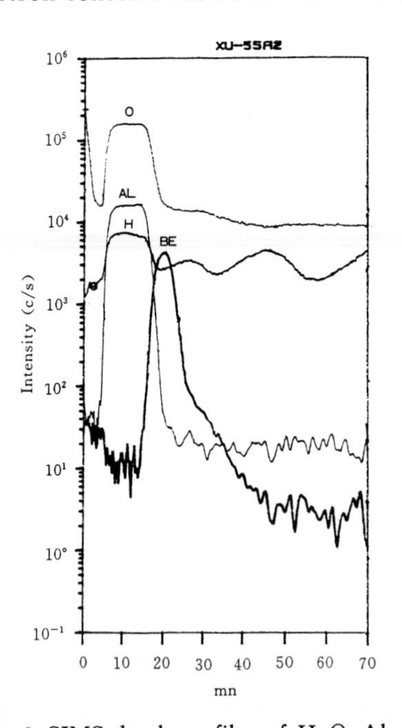

Fig. 3. SIMS depth profiles of H. O. Al and Be recorded on H and O implantation before annealing treatment.

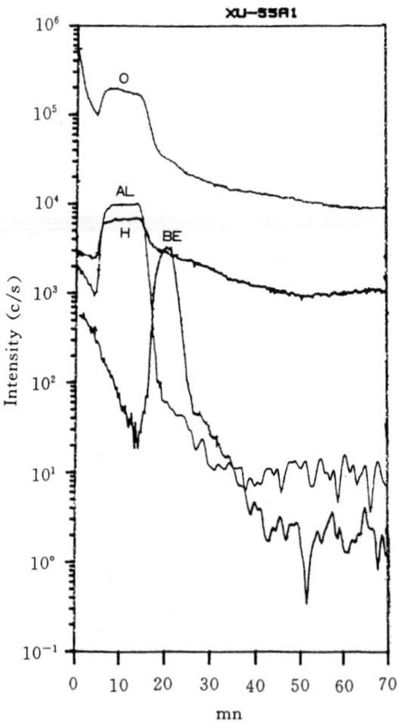

Fig. 4. SIMS depth profiles of H. O. Al and Be recorded on H and O implantation after annealing at 600℃.

Figs. 3 and 4 represent the SIMS-recorded depth distributions of H, O, Al and Be on the implanted sample before and after annealing at 600℃.

As shown in Figs. 3 and 4, an O concentration is higher than that in the Be or Si-doped GaAs counterparts, O concentration is probably over 10^{18} cm^{-3}. In O-implanted n-type GaAs and AlGaAs, it is found that oxygen-related deep double-electron traps are primarily responsible for the the high resistivity[3]。This means that the chemical doping makes contribution to the high-resistivity behavior rather than the radiation damage. Temperature stable donor compensation can be attributed to the formation of an Al-O complex that forms a deep acceptor level in n-type AlGaAs[4] and also in Be-doped GaAs, the oxygen may form a complex with the Be impurity atoms reaulting in thermally stable acceptor compensation[5]. Before the anneal, the SIMS depth profile of hydrogen on implanted multiple layers is shown in Fig. 3. There are three hydrogen peaks in it. After anneal, only one hydrogen peak at the position of AlGaAs layer remains. the other two hydrogen peaks become a uniform hydrogen distribution in the subsequent Be and Si-doped GaAs layers as shown in Fig. 4. However, about SIMS depth profile of oxygen as shown in Fig. 3 and 4, which display that two oxygen peaks is located at Al and Be peaks, respectively. After anneal the shape of the peaks is not sharp as before, that imply the implanted O has been diffused into Be and Si-doped region. Upon implantation in the begining, the average electron concentration of AlGaAs/GaAs multiple layers in almost the same as the initial concentration. As the subsequent annealing temperature is increased, the electron concentration is suddenly decreased with the reduction in hopping conduction until at a temperature (600℃), the H-implanted damage-related deep level density rises to a little over the initial eleetron concentration. After that, the average hole concentration still increases to the value of $8.36 \times 10^{14} cm^{-3}$. This behavior is different from that of H or O implanted GaAs or AlGaAs[6], which indicates that the deep levels associated with H or O bombardment have been annealed out between 400-600℃。During above 650℃ annealing, H-implanted damage related defect sites, which act as acceptors to trap the electrons in the multiple layers, are annealed out along with the temparature increaments, thus, the average electron concentration rapidly rises up to over the initial value. Although the high-resistivity region of Be-doped GaAs layer by O-implantation can reach about 10^8 $\Omega/\square$ and be maintained at 900℃, but in this work, the thickness of the Be-doped GaAs layer inserted between Si-doped AlGaAs and GaAs is too thin (~800 Å) to affect sufficiently the total resistivity of all the multiple layers. It is important to note that the high-resistivity of the multi-P-N type heterojunction epitaxial layers implanted with H and O and subsequently annealed at 600℃ can be maintained in any of the incoming heat treatments below 600℃.

4. Summary

we have shown that high-resistivity implant isolation of the AlGaAs/GaAs multiple epitaxial heterojunction structure can be achieved by H and O implantation at higher doses and energies followed by high-temperature annealing and sheet resistivity of $8 \times 10^7 \Omega/\square$ cna be maintained in the subsequent heat treatment below 600℃. The thermal stability of the high resistivity is attributed to both the activated damage-related deep levels by the H-implantation and the O-implantation leading to two complexes: one is aluminum complex and the other is beryllium complex.

The authors gratefully acknowledge the technical assistance of Yi Huang for MBE growth and Zhu-Qi Xuang for ion implantation. This work was supported by the Foundation of Academia Sinica.

References

[1] Pearton S J 1990 *Mater. Sci. Rep.* **4** 315

[2] Short K T and Pearton S J 1988 *J. Electrochem. Soc.* **135** 2835

[3] Xiong F and Tombrello T A 1989 *Nucl. Instr. and Meth. in phys. Res.* **B40/41** 526

[4] Pearton S J, Iannuzzi M P, Reynolds Jr C L and Peticolas L 1988 *Appl. Phys. Lett.* **52** 395

[5] Von Neida A E, Peatton S J, Hobson W S and Abernathy C R 1989 *Appl. Phys. Lett.* **54** 1540

[6] Cummings K D, Pearton S J and Vessa-Caleiro G P 1986 *J. Appl. phys.* **60** 163

Inst. Phys. Conf. Ser. No 149
Paper presented at DRIP 95

High Speed, High Density SPV Mapping of Iron Contamination in Silicon Wafers

J. Lagowski

University of South Florida, 4020 East Fowler Avenue, Tampa, FL 33620

P. Edelman, R. Erickson

Semiconductor Diagnostics, Inc., 6604 Harney Road, Suite F, Tampa, FL 33610

Abstract. The refined surface photovoltage, minority carrier diffusion length technique provides, for the first time, the fast mapping capability of iron contamination and other recombination center defects in p-type silicon wafers. The method is non-contact and uses an IR-light as the only medium to reach the wafer. Mapping with a density of up to 6000 points/wafer is achieved in about 20 minutes with an Fe detection limit below 10^{10} atoms/cm^3.

1. Introduction

Iron is an omnipresent, and most dangerous, heavy metal contaminant frequently introduced to silicon integrated circuits (ICs) from processing tools, chemicals, and wafer handling equipment. For most advanced circuits, the Fe concentration threshold for yield reduction is currently in the 10^{11} atoms/cm^3 range and is constantly shifting to lower values [1]. In-line iron monitoring is, therefore, a necessity for silicon IC manufacturing. The leading, quantitative Fe monitoring methodology relies on the minority carrier diffusion length, L_D, measured in p-type silicon by the non-contact, surface photovoltage (SPV) technique [2, 3]. The separation of the iron contribution is done by measuring L_D before and after dissociation of iron-boron pairs, increasing iron recombination efficiency by about ten times. In commercial SPV apparatus, pair dissociation is done optically at room temperature [3].

Existing methods have been very successful in measuring iron in discrete spots on the wafer and also in obtaining low density Fe concentration, N_{Fe}, maps up to about 180 spots per wafer. High density, Fe mapping was not possible due to pairing of the dissociated iron during long measurement. For boron concentration, $N_B = 10^{16}$ cm^{-3}, 300K pairing time is 30 minutes and decreases to 3 minutes for $N_B = 10^{17}$ cm^{-3}.

Refinements of the SPV technique in recent years have accelerated the measurements from about 1 point per 5-12 minutes in Goodman's original method [4] to 1 point per 2 seconds in modern, commercial SPV apparatus. This constituted the first step toward Fe mapping, making feasible 15 x 15 maps. However, high density mapping with 5 to 6 thousand points per wafer still took 3 to 4 hours.

In this paper, we present fast SPV diffusion length measurements with a speed of up to 1200 points/minute, compatible, for the first time, with the needs of whole-wafer iron mapping. This new capability is combined with a rapid photo-dissociation of iron-boron pairs, enabling the completion of the entire mapping cycle in 10 to 20 minutes per wafer.

2. Fast SPV Diffusion Length Measurement

In the constant photon flux SPV method, a steady-state surface photovoltage signal, V, (i.e., the change of the surface potential under illumination) is measured for different light penetration depths, z. L_D is then determined from V(z) using procedures given in Ref. [5]. The wafer is illuminated with a chopped, monochromatic light through a small, transparent conducting electrode, positioned about 0.25 mm above the wafer, which forms an air capacitor with the wafer. A very low illumination level assures that the recombination in the bulk and on the wafer surface is not affected by this illumination. A required linear SPV range, where V is proportional to the incident photon flux, Φ, is usually achieved when V is 1 mV or less [6]. Measurement of this low signal with an accuracy of about 1 μV is required for precise determination of L_D in a clean silicon wafer.

This is a difficult task, considering the open-circuit type capacitive coupling of the pick-up electrode to the wafer. For 1 mV, 510 Hz signals, a desired 10^3 signal-to-noise ratio requires a long time constant of 100 ms. The use of a substantially higher light chopping frequency is not possible due to minority carrier lifetime limitations.

In the fast SPV technique, a solution was found by permitting the SPV signal to increase from 1 mV to about 10 mV. This is still a very low injection level region, not affecting recombination parameters. However, the signal is large enough to cause a slight SPV sub-linearity $V \sim \Phi^k$ where $k < 1$. This sub-linearity must be determined and corrected for in order to reliably measure L_D. In SPV apparatus, k is measured and the photon flux is adjusted to a achieve a pre-set value, typically, k = 0.92. Then, using a semi-empirical procedure incorporated in the software, the signals $V_\ell(z_\ell)$, measured for different z_ℓ are corrected for non-linearity. The linearized SPV signals, $V_\ell^*(z_\ell) = R_\ell(V_\ell) \cdot V_\ell(z_\ell)$ (where $R_\ell(V_\ell)$ is the semi-empirical correction factor), are then used instead of $V_\ell(z_\ell)$ to determine L_D using Eqs 2-4, in Ref. 5.

With an increased SPV signal, the signal-to-noise ratio increases without resorting to long-time constants. This enables actual "in flight" readout of V_ℓ while the wafer is rotating. Mapping of L_D in (r, Θ) coordinates is then done by repeating each circle for two penetration depths. Data $V_\ell^*(z_\ell, r, \Theta)$, and $V_2^*(z_2, r, \Theta)$ are processed by a computer giving L (r,Θ). High density maps, over 6000 points per wafer, can be obtained in only 6 minutes. Such a map is

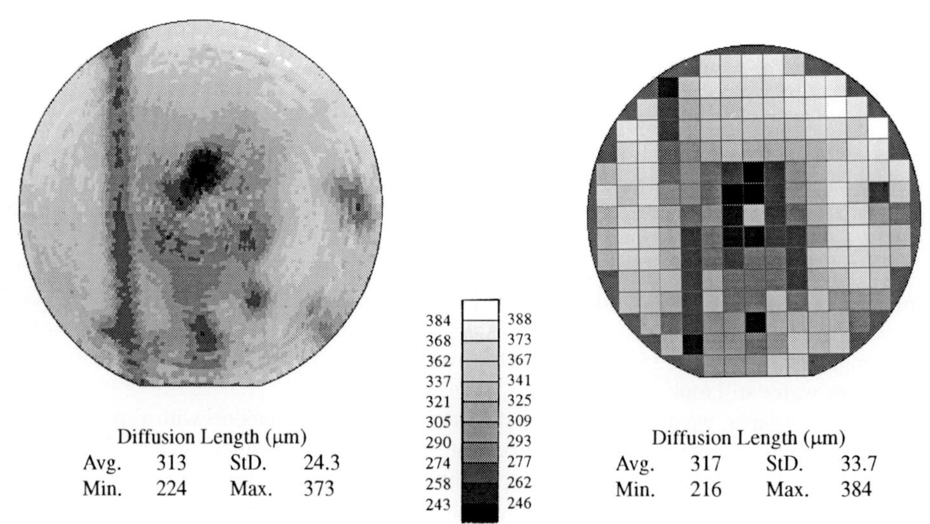

Figure 1. A comparison between: a.) a fast high-density L map obtained with a new approach; and b.) a 177-point map measured with the old procedure.

shown in Figure 1, together with a corresponding, standard, 177-point map, that was also done in 6 minutes with the old technique. In spite of an accelerated measurement by about 30 times, the average, minimum, and maximum L_D values are very close in both cases. This confirms the reliability of the new, fast mapping procedure.

3. Mapping of Iron and Other Recombination Centers

Determination of iron concentration involves two measurements: the initial one when all iron is paired and L_D is at its maximum, L_0; and the second one when all pairs are dissociated and L_D is at its minimum, L_1. An example of the fast, L_0 and L_1 maps of a 200 mm wafer (p-type, 10 Ωcm) is given in Figure 2a and 2b, respectively. L_0 was measured after 24 hour storage of the wafer at room temperature but a complete pairing state may also be obtained after a 20 minute annealing at 80°C [2].

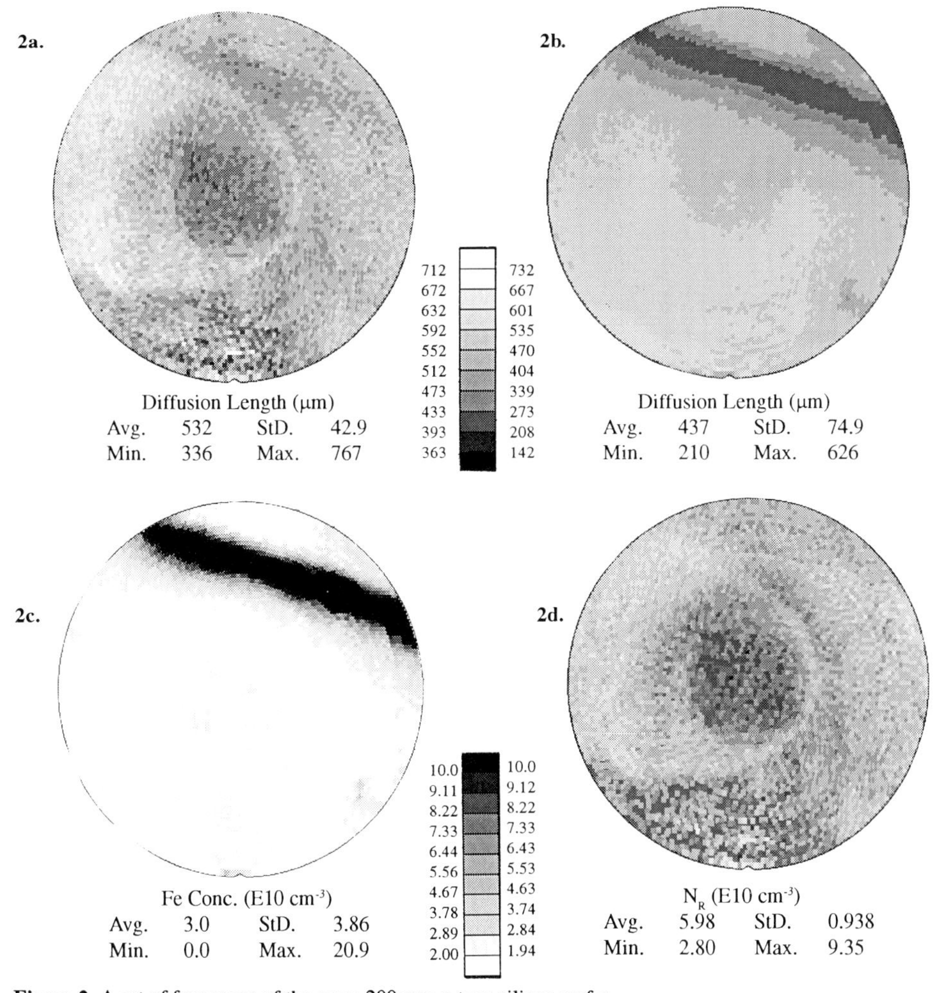

Figure 2. A set of four maps of the same 200 mm p-type silicon wafer.
a.) diffusion length before and b.) after iron-boron pair dissociation.
c.) iron contamination; and d.) other than iron recombination centers.

The pairs were then dissociated by illuminating the wafer with a strong halogen light with a removed, visible-UV range of $\lambda < 0.7$ μm. The elimination of these short wavelengths is important, otherwise the SPV signal can be significantly reduced by photo-decomposition of the organic residue on the surface. The whole-wafer photo-activation station used in this study consisted of nine 250 watt halogen bulbs fitted with cold-mirror filters and positioned above the slowly rotating wafer in a way which provided uniform exposure of the entire 200 mm wafer. A computer-controlled station assured over 95% dissociation of the Fe-B pairs in about 60 to 90 seconds, depending on the initial diffusion length and surface treatment. Pair photo-dissociation is a recombination-enhanced process with a rate dependent upon all recombination centers and not only iron [3].

Using a well-known relationship for diffusion length, one gets:

$$L_0^{-2} = D_n^{-1} (C_{FeB} N_{Fe} + C_R N_R) \quad \text{1a}$$

$$L_1^{-2} = D_n^{-1} (C_{Fe} N_{Fe} + C_R N_R) \quad \text{1b}$$

where D_n is the electron diffusion constant, C_{FeB}, C_{Fe}, and C_R are the electron capture rate of the FeB pair, the isolated Fe and the other recombination centers, respectively. N_R is the concentration of bulk recombination centers other than iron and $C_{Fe} = P \cdot C_{FeB}$ where the capture rate ratio, P, is typically between 10 and 20, depending on the Fermi energy.

As discussed in Ref. [2, 3], N_{Fe}, derived from Eqn 1, is:

$$N_{Fe} \sim (D_n/C_{Fe})(L_1^{-2} - L_0^{-2}) \quad 1.1 \times 10^{16} (L_1^{-2} - L_0^{-2}) \quad \text{2a}$$

N_{Fe} is in atoms/cm^3 and L is in μm. The corresponding expression for N_R is:

$$N_R = (D_n/C_R)(P-1)^{-1}(PL_0^{-2} - L_1^{-2}) \quad \text{2b}$$

The maps of iron and other recombination centers obtained from L_1 and L_0 maps in Figure 2a and 2b are show in Figure 2c and 2d. The iron map reveals a characteristic Fe contamination trail with a concentration up to 2 x 10^{11} atoms/cm^3. The remaining part of the wafer is clean with Fe in the low 10^{10} atoms/cm^3 range. Other recombination centers show a completely different distribution than that of Fe. In the calculation of N_R, the electron cross-section, C_n, was taken, the same as that of the iron interstitial. The specific choice of C_n affects the absolute value of N_R, but not the distribution pattern on the wafer.

4. Examples of Iron Contamination

Iron contamination mapping is a necessity for silicon IC manufacturing. First of all, it often occurs in localized parts of the wafer and can be missed in commonly utilized 5 or 9 point measurements. Secondly, high resolution mapping gives an footprint of contamination and, thus, it identifies the contaminating tool. Finally, the mapping procedure discussed here enables the separation of the omnipresent Fe contribution from that of other recombination centers such as stacking faults, oxygen precipitates, and slip-induced defects. This offers the possibility of reliable mapping of these recombination centers. Examples illustrating these cases in wafers after various stages of IC processing are given in Fig. 3, 4, and 5. A brief description is provided in the figure captions.

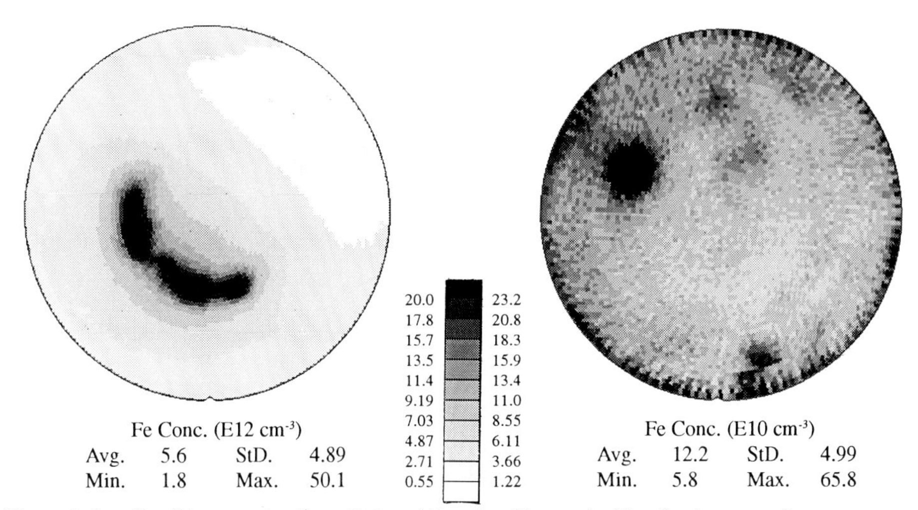

Figure 3. Localized Fe contamination, which could be missed by standard 5 or 9 point-per-wafer measurements. The pattern on the left (3a) shows N_{Fe} as high as 5×10^{13} atoms/cm^3; while that on the right (3b), shows N_{Fe} at about 7×10^{11} cm^{-3}.

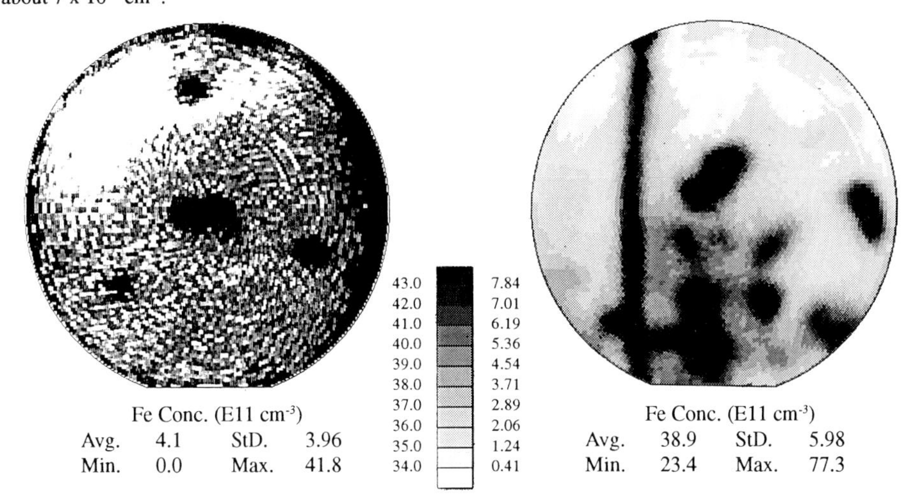

Figure 4a. Fe concentration maps with a characteristic "dot" pattern caused by wafer contact with supporting pins.

Figure 4b. A pattern composed of: an "iron trail" (seen in Fig. 2), typically caused by wafer transport; and "dots" caused by wafer supporting pins (the same wafer as that in Figure 1).

5. Summary

Our experimental results demonstrate that fast SPV measurements are uniquely suited for whole-wafer mapping of iron contamination introduced during silicon IC manufacturing. The entire measurement procedure is non-contact. Iron identification is based on optical iron activation as a recombination center. The wafer is touched only from the back side for transport and motion during mapping. We expect that further refinement of the technique should increase sensitivity of Fe detection to the low 10^9 atoms/cm^3 range and further decrease the mapping time to perhaps meet the IC requirement of measuring 25 wafers per hour.

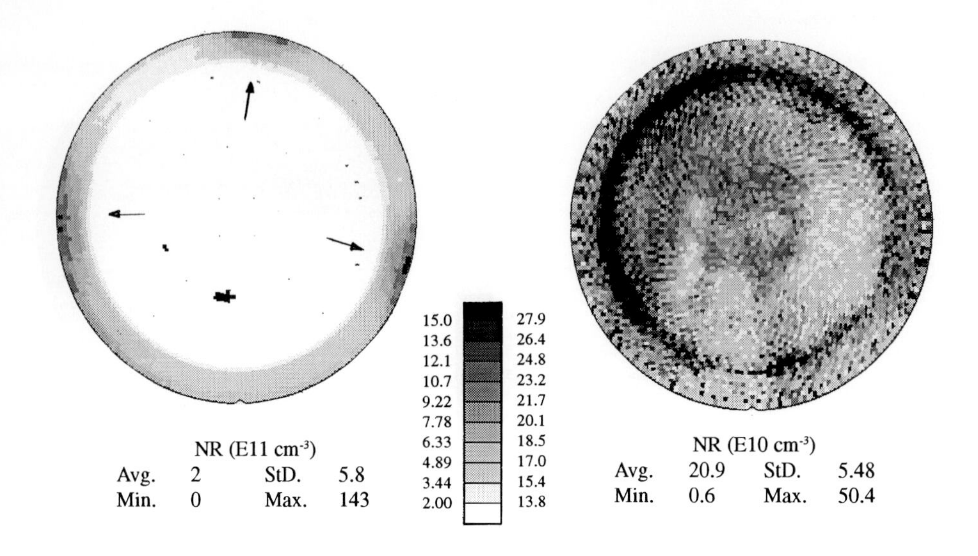

Figure 5a. A map of other (than Fe) recombination centers, revealing three regions of slip-induced defects near the edges of a 200 mm wafer.

Figure 5b. A characteristic ring of oxidation-induced defects in a 200 mm wafer, revealed by the map of other recombination centers.

6. Acknowledgment

The authors are indebted to Semiconductor Diagnostics, Inc. for making available the prototype apparatus used in this study.

References

[1] Armour, N.; *Monitoring Metallic Contamination in the ULSI Era*; SPIE Vol. 2337; October 20, 1994.

[2] Zoth, G and Berholz, W.; *A Fast, Preparation-free Method to Detect iron in Silicon*; 1990 *J. Appl. Phys.* **67** (11).

[3] Lagowski, J. et al; *Iron Detection in the Part per Quadrillion Range in Silicon using Surface Photovoltage and Photo-dissociation of Iron-Boron Pairs*; 1993 *Appl. Phys. Lett.* **63** (22).

[4] Goodman, A.M., *J. Appl. Phys.*; **32**, 2550 (1961).

[5] Lagowski, J. et al; *Method for the Measurement of Long Minority Carrier Diffusion Lengths Exceeding the Wafer Thickness*; 1993 *J. Appl. Phys.* **63** (21).

[6] Lagowski, J. et al; *Semiconductor Sci. Technology*; 7, A185 (1992).

Inst. Phys. Conf. Ser. No 149
Paper presented at DRIP 95

Raman and Photoluminescence Imaging of the Gan/Substrate Interface

H. Siegle, P. Thurian, L. Eckey, A. Hoffmann, and C. Thomsen

Institut für Festkörperphysik, Technische Universität Berlin, 10623 Berlin, Germany

B. K. Meyer
Technische Universität München, Physik-Department E16, 87547 Garching, Germany

T. Detchprohm, K. Hiramatsu
Nagoya University, Nagoya, Japan

H. Amano, I. Akasaki
Meijo University, Nagoya, Japan

Abstract. GaN exhibits apart from the near-bandgap excitonic and donor-acceptor-pair luminescence a broad "yellow" photoluminescence between 2.0 and 2.5 eV. We performed spatially-resolved photoluminescence and Raman experiments on the substrate interface region of wurtzite GaN layers. We found that the broad photoluminescence band is strong only near the interface. Our investigations reveal that both the substrate interface and a region of structural reorientation of the layer near the interface act as source of the photoluminescence. The Raman-scattering experiments show that at least a portion of the GaN layer near the substrate interface is oriented in such a way that the c-axis of the layer is parallel to the substrate interface. At a distance about 30 μm away from the interface the layer reorients by turning the c-axis by 90° into a direction perpendicular to the substrate interface.

1. Introduction

The wide-bandgap semiconductor GaN has attracted considerable attention over the last years because of its application as a basic material for optoelectronic devices working in the blue and UV spectral region, such as blue laser diodes [1]. Apart from the near-bandgap excitonic and donor-acceptor-pair luminescence GaN shows an unwanted broad "yellow" photoluminescence band between 2.0 and 2.5 eV at low temperatures [2]. Recently, intensive work has been done to clarify its origin, and some authors connect this luminescence band with the inherent property of GaN to be automatically n-type conductive [3-5]. However, this issue is still controversial.

Similar luminescence bands are also known from several II-VI semiconductors like ZnS [6]. They are often interpreted as a recombination between shallow donors and deep acceptors in which the donors or the acceptors were build by anion or cation vacancies. The creation of such intrinsic vacancies is very probable near surfaces, interfaces or grain boundaries.

In order to clarify wether the broad “yellow“ photoluminescence band is an intrinsic property of GaN or caused by defects located near the interface to the substrate, we performed spatially-resolved photoluminescence and Raman measurements on hexagonal GaN samples which were grown on sapphire. We found that the luminescence band is strong only in a region near the interface and hence not an intrinsic property of GaN. Our investigations reveal that the photoluminescence originates from both the substrate interface and a region of structural reorientation near the interface where the c-axis of the GaN layer rotates by 90° from a direction parallel into a direction perpendicular to the substrate interface.

2. Experiment

The samples investigated were undoped wurtzite GaN layers grown on [0001] sapphire using hydride vapor phase epitaxy (HVPE) with thicknesses of 220 μm, 230 μm and 400 μm and a free carrier concentration of about $1 \cdot 10^{17}$ cm^{-3}. The spatially-resolved photoluminescence and Raman experiments were carried out using a Dilor XY800 triple-grating spectrometer with a charge-coupled device (CCD) detector and confocal optics. The sample was excited either parallel (in-plane) or perpendicular (on-plane) to the substrate surface using the 488 nm (2.54 eV) line of an Ar^{+}-Kr^{+} mixed-gas laser and the 632.8 nm (1.96 eV) line of an He-Ne Laser. By passing the laser through a microscope objective (x80) the laser beam was focused on a point spot with a diameter of about 1 μm and a power of 2 mW. The scattered light was detected in backscattering geometry which corresponds to an $x(..)\bar{x}$ configuration for in-plane excitation and a $z(..)\bar{z}$ configuration for on-plane excitation (under the assumption, that the z is parallel to the c-axis). The samples were cooled down to 4.2 K using an Oxford microscope cryostat. With this setup we obtained a spatial resolution of about 1 μm and a spectral resolution better than 1 cm^{-1}.

3. Results

The photoluminescence spectra of a 400 μm thick GaN layer taken at 4.2 K and room temperature after excitation at 3.75 eV are compared in Fig. 1. While the spectrum at low temperature is dominated by the the excitonic transitions and the donor-acceptor-pair luminescence, the broad photoluminescence band with an intensity maximum at 2.4 eV becomes the dominating part of the room-temperature spectrum.

Figure 2 shows a 40 μm long linescan across the GaN-substrate interface of a 400 μm thick layer where we have taken a spectrum every 1 μm. The region of the substrate is marked by the presence of the A_g sapphire mode at 419 cm^{-1} [7]. The transition to the GaN layer is indicated by the appearance of the A_1(TO) and the E_2 modes at 534 cm^{-1} and 569 cm^{-1}, respectively. A photoluminescence band with an intensity maximum at 2.4 eV appears directly at the interface with the substrate; it is seen as constant in Raman shift background, since it is broad compared to the spectral range observed here. The spatial

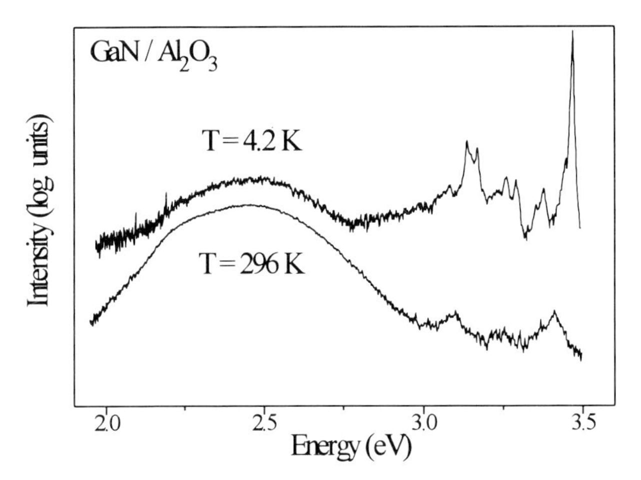

Fig. 1: Photoluminescence spectra of a 400 μm thick GaN layer grown on sapphire taken at 4.2 K and at room-temperature after excitation at 3.75 eV.

width of the region from which this photoluminescence occurs is about 3 μm. The GaN region is dominated by the abruptly increasing A_1(TO) mode. At a distance d of about 30 μm away from the substrate interface, in a region several μm wide, a second broad photoluminescence band appears peaking at the same spectral position as the first photoluminescence band. Simultaneously with the increasing photoluminescence the intensity of the A_1(TO) Raman signal decreases but the modes continues to be visible.

The spatial continuation of Fig. 2 is shown in Fig. 3 on an expanded spectral and spatial scale. The decrease of the A_1(TO) Raman signal (near d = 20 μm) and the broad

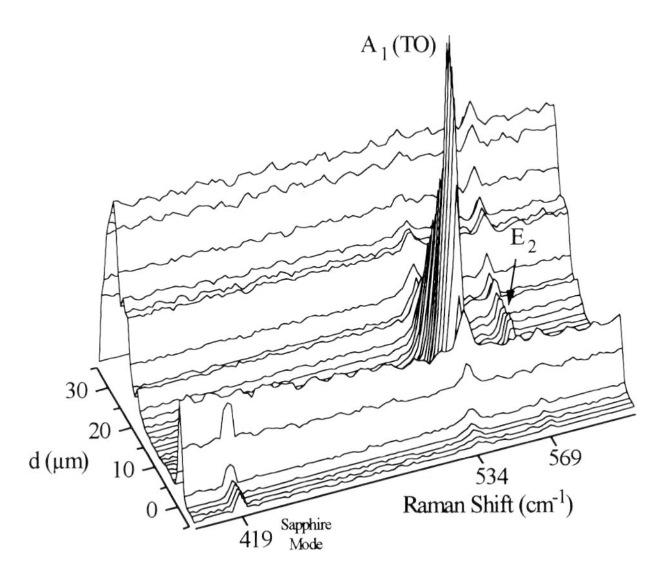

Fig. 2: Linescan across the GaN-substrate interface of a 400 μm thick layer taken at 4.2 K. In order to emphasize the sapphire Raman mode it was multiplied by a factor of 2.

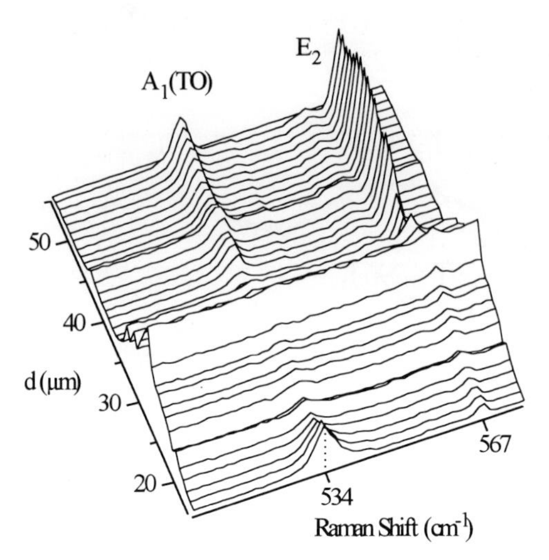

Fig. 3: Spatial continuation of Fig. 2. Linescan across the inner interface caused by a reorientation of the GaN layer taken at 4.2 K. For clarification the spectral scale was enlarged.

photoluminescence band (at d = 30 μm) can be seen. Of particular importance is the change in scattering intensity of the different phonon modes. While both, the E_2 mode and the A_1(TO) mode appear in this region the ratio of their intensity inverts with increasing distance from the substrate interface. We have clarified this in Fig. 4 where we show individual spectra taken at 15, 25 and 50 μm. The strong luminescence was subtracted in the middle spectrum of Fig. 4. The change in intensity starts where the second photoluminescence band appears. While near the interface with the substrate ($0 \leq d \leq 20$ μm) the spectra are dominated by the A_1(TO) mode the intensity of the E_2 mode increases with the appearence of the second photoluminescence band and becomes the most intensive phonon mode (d > 35 μm). The inversion of the relative intensities can be explained by a structural reorientation of the GaN layer. It is interesting to note that the frequencies of both, the A_1(TO) mode and the E_2 mode decrease by about 2 cm^{-1} for distances larger than 25 μm (Fig. 3), which is likely to be caused by strain relaxation. This is consistent with the results of Kozawa et al. [8] who observed a decrease of ~ 1 cm^{-1} for increasingly relaxed GaN layers.

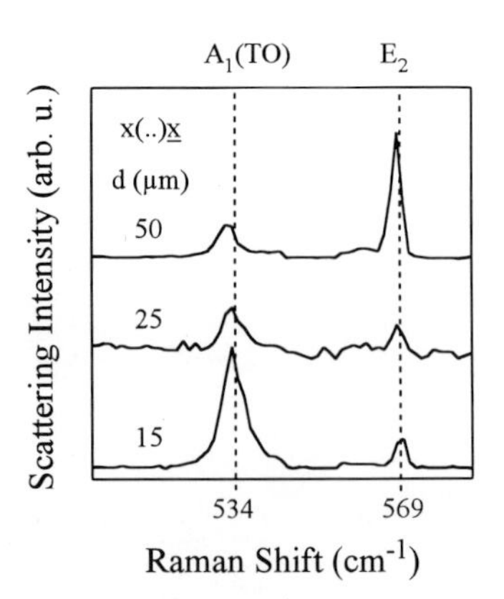

Fig. 4: Raman spectra of the 400μm thick GaN layer taken at various distances d from the interface with in-plane excitation. The red shift can be explained by strain relaxation.

Table 1: Raman selection rules for backscattering configuration used in this work. The c-axis is parallel to the z direction.

Scattering Configuration	Allowed Modes
$z(yy)\bar{z}$	A_1(LO), E_2
$x(zz)\bar{x}$	A_1(TO)
$x(yz)\bar{x}$	E_1(TO)

In Table 1 the selection rules for first-order Raman scattering of hexagonal material and in-plane excitation are listed [9, 10]. Comparing Table 1 with the corresponding Raman spectra in Fig. 4 reveals that for the region near the substrate interface ($0 \leq d \leq 20$ µm) where the spectra are dominated by the A_1(TO) mode the GaN layer is oriented in such a way that the scattering geometry corresponds to $x(zz)\bar{x}$. For d > 35 µm the E_2 mode is the strongest mode and the A_1(TO) is still observable. This combination corresponds to $x(yy)\bar{x}$. Since the incident polarization of the exciting laser remained parallel to the substrate interface and constant throughout the experiment our results show that the c-axis of the GaN layer near the substrate is parallel to the interface and turns by about 90° at a larger distance (d > 20 µm) from the interface. This becomes more clear when considering

Fig. 5 in which a depth profile of room-temperature Raman spectra taken in $z(yy)\bar{z}$ configuration on a 220 µm thick layer are plotted. By using the 632.8 nm line of a He-Ne laser no photoluminescence was excited. The confocal optics allows to record Raman spectra at several depths, i. e. distances to the substrate interface. All spectra are dominated

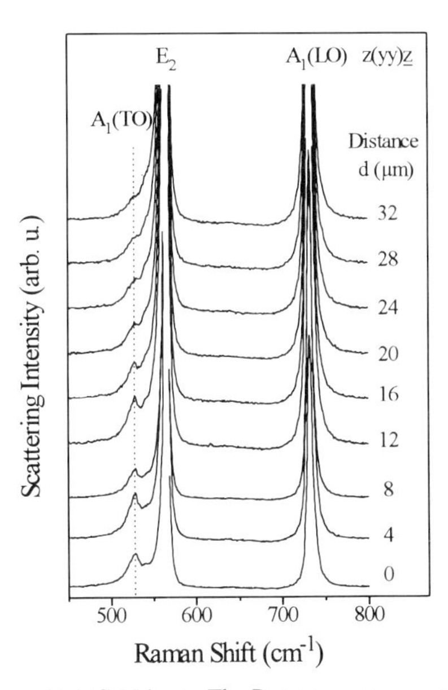

Fig. 5: Depth profile of a 220 µm thick GaN layer. The Raman spectra are taken at room temperature using the 632.8 nm line of a He-Ne laser.

by the E_2 and the $A_1(LO)$ modes but near the substrate interface the $A_1(TO)$ mode can clearly be seen. The scattering intensity decreases with increasing distance to the interface. According to the selection rules listed in Tab. 1 the $A_1(TO)$ is only excitable if the c-axis lies perpendicular to the incident wave vector. Hence it follows from both in- and on-plane Raman measurements that at least a portion of the GaN layer near the interface to the substrate is oriented parallel to it. This reorientation yields a further interface which like the substrate interface acts as source of the broad luminescence band.

4. Conclusions

We have shown that the origin of the broad photoluminescence band with an intensity maximum at 2.4 eV is not homogeneously distributed in our GaN layers. Instead, we found that the luminescence is strong only near the interface region. A first luminescence band appears directly at the interface with the substrate, a second one appears approximately 30 μm away from the interface in a region which is several μm wide. Simultaneously performed Raman scattering experiments allowed us to analyze layer orientation and strain and revealed that the second band appears where a significant reorientation of the wurtzite GaN c-axis from parallel to perpendicular to the substrate surface occurs. This observation suggests that the 2.0 - 2.5 eV luminescence is not intrinsic to GaN but occurs primarily near structural defects.

5. Acknowledgments

This work was in parts supported by the Stifterverband für die Deutsche Wissenschaft.

References

[1] R. F. Davis, Physica **185B**, 1 (1993); S. Strite and H. Morkoç, J. Vac. Sci. Technol. **B10**, 1237 (1992)

[2] J. I. Pankove and J. A. Hutchby, J. Appl. Phys. **47**, 5387 (1976); T. Ogino and M. Aoki, Jap. J. Appl. Phys. **19**, 2395 (1980)

[3] D. M. Hofmann, D. Kovalev, G. Steude, B. K. Meyer, A. Hoffmann, L. Eckey, T. Detchprom, A. Amano, and I. Akasaki, Phys. Rev. B, in print

[4] E. R. Glaser, T. A. Kennedy, H. C. Crookham, J. A. Freitas jr., M. Asif Khan, D. T. Olson, and J. N. Kuznia, Appl. Phys. Lett. **63**, 2673 (1993)

[5] P. Perlin, T. Suski, H. Teisseyre, M. Leszczynski, I. Grzegory, J. Jun, S. Porowski, P. Boguslawski, J. Bernholc, J. C. Chervin, A. Polian, and T. Moustakas, Phys. Rev. Lett. **75**, 296 (1995)

[6] J. R. James, J. E. Nicholls, B. C. Cavenett, J. J. Davies, D. J. Dunstan, Solid State Commun. **17**, 969 (1975)

[7] For a survey of Raman spectra on sapphire see for example: S. P. S. Porto and R. S. Krishnan, J. Chem. Phys. **47**, 1009 (1967)

[8] T. Kozawa, T. Kachi, H. Kano, H. Nagase, N. Koide, and K. Manabe, J. Appl. Phys. 77, 4389 (1995)

[9] C. A. Arguello, D. L. Rousseau, and S. P. S. Porto, Phys. Rev. **181**, 1351 (1969)

[10] H. Siegle, L. Eckey, A. Hoffmann, C. Thomsen, B. K. Meyer, D. Schikora, M. Hankeln, K. Lischka, Solid State Commun. **96**, 943 (1995)

Inst. Phys. Conf. Ser. No 149
Paper presented at DRIP 95

Thermal Treatment of ZnSe Crystals Studied by Light Scattering and Photoluminescence Tomography

Toshihide Tsuru, Minya Ma and Tomoya Ogawa

Department of Physics, Gakushuin University
Mejiro, Tokyo, 171, Japan
FAX +81-3-3590-2602

Abstract Some crystals emit strong photoluminescent light when they are illuminated by a laser beam for light scattering tomography.

The luminescence is only noise for precise measurement of light elastically scattered by the defects to be studied. However, luminescence caused by impurities and/or defects is useful information to characterize the crystals.

An instrument to individually observe luminescence and elastically scattered light is developed for tomographic studies on laser materials.

1. INTRODUCTION

Heat treatment of materials is an important and popular method to use in study on the structure and nature of materials, so ZnSe crystals were heated by an incandescent lamp in flowing dry nitrogen gas. After the treatment the crystals emitted a fairly bright and red luminescence under illumination of 488 nm radiation from an Ar ion laser for light scattering tomography (LST) (Ma and Ogawa,1995).

The luminescence is only noise from the viewpoint of ordinary LST observation of defects but can provide important information from the viewpoint of light emission as laser materials. The luminescence should, therefore, be separated from the signal caused by elastic scattering from defects in the crystal: use of both images allow more effective characterization of the crystals.

2. EXPERIMENTAL PROCEDURES AND PRELIMINARY RESULTS

A (111) ZnSe plate with 1.2 mm thickness was optically

polished and then cleaved along {110} into several small parallelepipeds to use as specimens of heat treatments by an incandescent lamp. The temperature of each specimen was systematically different from the other within range of 500 to 800°C by 50°C step and kept constantly for 1 to 7 hrs in a flowing dry nitrogen gas environment. Before and after the heat treatment, geometrical structures of the dislocation walls and nets were studied by LST using 488 nm radiation from an Ar laser. After the treatment, the red luminescence from the crystals was so strong that it disturbed the light elastically scattered by defects in the ZnSe crystals, which made our evaluations difficult.

To achieve this separation, an optical filter was tentatively used to cut the luminescent light off for clear observation of the elastic scattering images. By elimination of 488 nm radiation we could get the luminescent images where another filter was also used. By these filtering the pictures shown in figs. 1 and 2 were obtained.

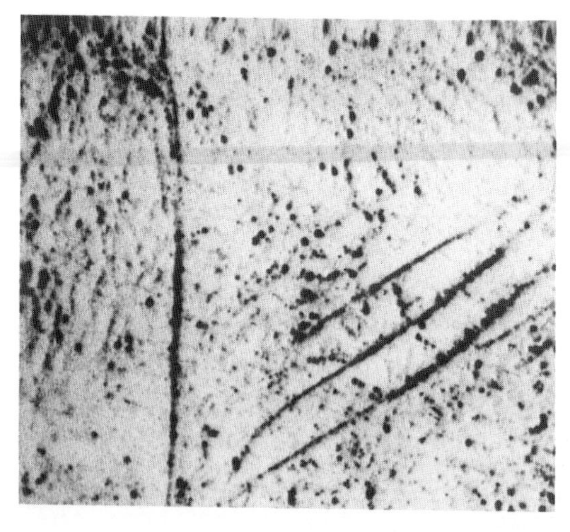

Fig.1
Light scattering tomograph of a heat-treated ZnSe crystal obtained by a beam of 488 nm radiation from an Ar laser

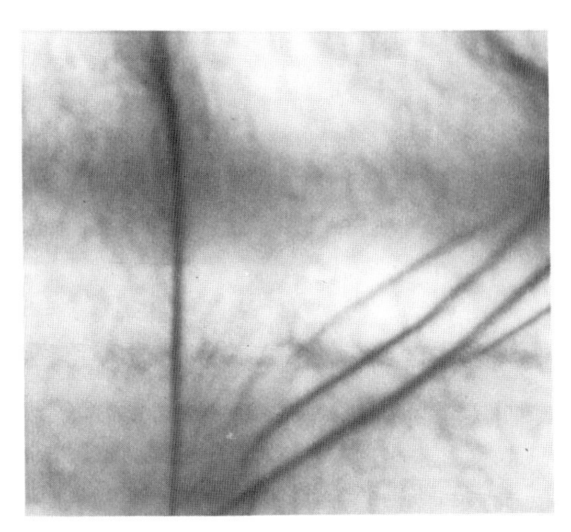

Fig.2
Photoluminescence tomograph obtained by red SA (self-activated) emission from the heat-treated ZnSe crystal

3. LIGHT SCATTERING AND PHOTOLUMINESCENCE TOMOGRAPHY

For individual and simultaneous observation of elastically scattered light and photoluminescence, a new instrument has been developed as schematically shown in fig.3; the scattered light and luminescence are separated by a dichroic mirror and received by two independent vidicons. Hereafter, the light scattered by defects is indicated as the signal "S" and the luminescence as the signal "L".

In the figure these two signals are individually transferred to the terminals R and G of a frame memory for color TV images which has a three-fold frame system to separately receive three fundamental color components R, G and B. The signals passing through the terminal R, G and B are stored as digitized data in its memory sections, where the data acquisition system is very similar to this reported previously (Ogawa and Nango, 1986).

Intensity of the scattered light will be very widely changed according to size of the scatterers and by difference between the characteristic frequency Ω of the scatterers and frequency ω of the incident laser beam, because the scattered light intensity is proportional to square of the scatterer volume and is nearly and inversely proportional to $(\Omega^2 - \omega^2)^2$. The switch shown in fig.3 is, therefore, closed during data acquisition to receive the scattered intensity since the frame memory acts as (8 + 8) = 16 bits per pixel due to addition of the data from the terminal G and B, which is good enough for the data acquisition of the S signal. By image processing due to the computer both the image signals are adjusted into 8 bits.

Since the magnification factor of optical lenses is dependent upon wavelength of an optical beam passing through

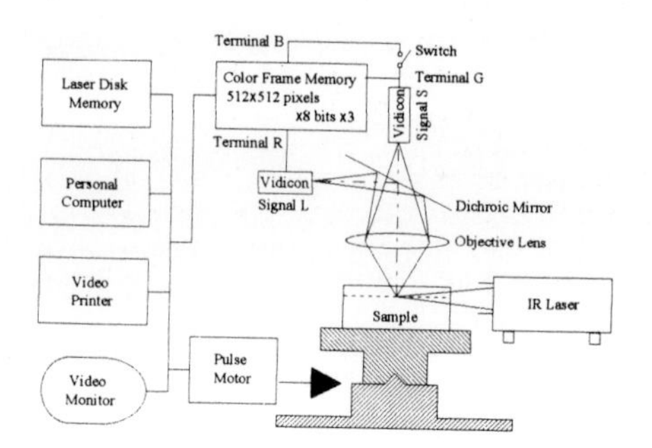

Fig.3
Schematic drawing of the instrument for light scattering and photoluminescence tomography

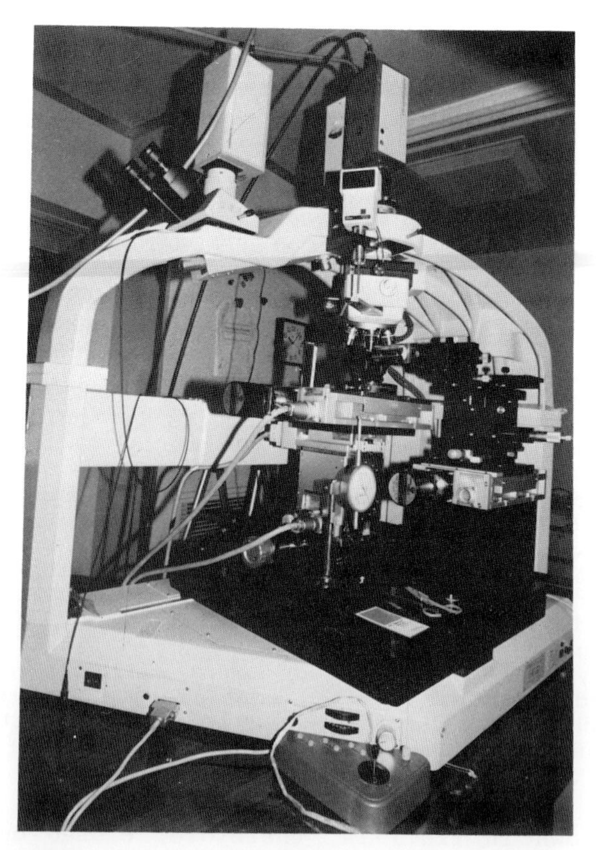

Fig.4
Mechanical appearance of the instrument

the lenses, the factor for the L signal is optically adjusted to be equal to that for the S signal by a sort of zoom lens system. The pictorial coincidence between the L and S signals will be automatically made by an ordinary process of color picture construction while only two components are involved. Of course, the picture of only the S or the L signal will be individually obtained. When both the signal intensities are received by 8 bit systems, one more signal from the specimen can be acceptable by the terminal B of the color frame memory.

The mechanical appearance of this instrument is shown in fig.4 but its software is still being refined.

4. CONCLUSION

An instrument for light scattering and photoluminescence tomography is being developed for characterization and evaluation of semiconducting materials for laser diodes and oxide ones for laser rods. This will also be applicable to characterization and evaluation of nonlinear optical crystals for harmonics generation.

5. REFERENCES

Ma M and Ogawa T, 1995, Phil. Mag. A 72 113.
Ogawa T and Nango N, 1986, Rev. Sci. Instrum. 57, 1135.

Inst. Phys. Conf. Ser. No 149
Paper presented at DRIP 95

Composition fluctuations in (In,Ga)(As,P) single layers and laser structures based on GaAs

I. Rechenberg[1), A. Knauer[1), U. Zeimer[1), F. Bugge[1), U. Richter[2), A. Klein[1) and M. Weyers[1)

1) Ferdinand-Braun-Institut für Höchstfrequenztechnik Berlin, Rudower Chaussee 5, D-12489 Berlin, Germany,
2) Labor für Elektronenmikroskopie e.V., Weinbergweg 23, D-06120 Halle/Saale, Germany

Abstract. GaInAsP layers lattice matched to GaAs were grown by metal organic vapour phase epitaxy (MOVPE) and characterized by cathodoluminescence (CL), high resolution x-ray diffraction (HRXRD), transmission electron microscopy (TEM) and transmission electron diffraction (TED). Depending on growth conditions and composition TEM images show a modulated contrast indicating spinodal decomposition which also leads to broad CL emission peaks and broad X-ray rocking curves. The decomposition leads to columnar growth with significant anisotropy. In the growth of buried structures additionally the composition depends on the orientation of the growing surface. The resulting composition fluctuations give rise to defect formation in overgrown trenches.

1. Introduction

The semiconductor alloys $In_{1-x}Ga_xP$ and $In_{1-x}Ga_xAs_yP_{1-y}$ (hereafter called InGaP and InGaAsP) lattice matched to GaAs attract attention mainly for their application in optoelectronic devices. These materials are an alternative to AlGaAs in GaAs based heterostructures and laser diodes, because they promise an improved long-term stability due to a smaller amount of mirror facet overheating /1/. However, the Al-free alloys show a tendency for atomic-scale ordering and also for spinodal-like decomposition. Both effects have to be controlled to allow for the production of high quality material for device applications. Ordering is well studied for GaInP but is also reported for GaInAsP /2-4/. The quaternary additionally shows a tendency for phase separation. Thermodynamic calculations /5/ predict a miscibility gap (MG) the effect of which up to now has primarily been studied for InP-based structures due to their importance in optoelectronics /6/. In dependence on composition and deposition temperature samples suffer microscale variations of the alloy composition as a result of spinodal-like decomposition /7/. It is assumed, that this phase separation occurs at the surface while the layer is growing /6,8/. Direct experimental evidence for decomposition in ternary and quaternary layers is obtained from TEM images where coarse-scale and fine-scale contrast modulations are observed /6,7/. An irregularly spaced modulation is observed with a wide variation in spacings (few nanometers up to ≈ 100 nm and more) in the growth plane. The decomposition in the growth plane is direction

dependent and leads to a columnar-like growth /9/. Along the growth direction modulated contrast was not observed /4/.
Phase separation in heterostructures for laser applications potentially limits the long-term stability of laser diodes. To eliminate or to minimize the decomposition we have investigated the relationship between MOVPE growth parameters (especially growth temperature) as well as misorientation of the (001) GaAs-substrate on the occurence of phase separation. We have studied InGaAsP single layers lattice matched to GaAs with y=0; 0.12; 0.48-0.52; 0.76. These compositions are interesting as cladding, confinement and active layer materials in laser diodes. These layers were characterized by X-ray diffraction taking rocking curves and area scans in different azimuths. Additionally cathodoluminescence spectra and monochromatic images were taken. On the basis of these results samples were selected for TEM analysis. TEM was also used to characterize heterostructures for broad-area laser diodes. Finally, a preliminary study of the impact of the observed phenomena on the growth of buried real index selfaligned structures (RISAS) for monomode lasers was performed.

2. Experimental

$Ga_xIn_{1-x}As_yP_{1-y}$ undoped layers were grown lattice matched to GaAs in a horizontal MOVPE system using a rotating 2" susceptor at a reactor pressure of 70 hPa with trimethylindium, trimethylgallium, pure arsine and phosphine as precursors in a hydrogen carrier gas. Temperatures of 650, 680, 700 and 750°C were used for the growth of InGaAsP and InGaP on GaAs substrates with exact orientation and misorientation of 2° to {111}B. The growth rate was 0.5-1 μm/h for InGaAsP and 2.4 μm/h for InGaP. Single layers and laser structures were examined by scanning electron microscopy (JEOL 840A) in the backscattered electron (BSE) mode and in the monochromatic CL mode (Oxford Instruments MonoCL-system). CL spectra were taken at 110 K and the full width at half maximum (FWHM) was correlated to growth temperature and substrate orientation. Additionally, the composition of the grown layers was determined by electron probe microanalysis (EPMA). HRXRD was performed in a five-crystal diffractometer (Philips MRD) to check the lattice matching and to determine the FWHM of the rocking curves of the (004) reflection. Furthermore, area maps of the (-2-24) reflection were recorded to distinguish between mosaicity and composition fluctuations. For TEM investigations samples were glued together and thinned mechanically and by ion beam etching.
The investigated laser structures were broad area lasers with a lattice-matched GaInAsP (y=0.76, 20 nm thick), active layer embedded in waveguide layers with y=0.5. The cladding layers are made of GaInP. Additionally, RISAS laser structures were investigated where GaInAsP (y=0.12) is filled into a trench etched through a 300 nm thick GaInP layer.

3. Results and discussions

3.1. *Single layers*

Table 1 summarizes the results of the analysis of GaInAsP layers of different composition grown at different temperatures and on different substrate orientations. All layers are lattice-matched excluding strain relaxation as possible reason for the observed trends. Since the layers are of different thickness the theoretically expected half width $FWHM_{theor}$ is different.

For a valid comparison thus the normalized difference $\Delta FWHM_{norm}$ of the experimentally obtained rocking curve width $FWHM_{exp}$ is listed for two orthogonal azimuths ($\Delta FWHM_{norm} = (FWHM_{exp} - FWHM_{theor})/FWHM_{theor}$).

Table 1 Layer properties depending on composition and growth conditions

sample No.	substrate orientation	T_d (°C)	x	y	$\Delta a/a$ x 10^{-6}	HRXRD $\Delta FWHM_{norm}$ [1-10]	[110]	CL peak (eV)	CL FWHM (meV)
1a	0°	680	0.54	0.12	460	2.57	3.93	1.796	32.3
1b	2°B	680	0.54	0.12	180	0.43	1.36	1.799	28.7
2a	0°	680	0.75	0.5	-386	0.86	1.48	1.674	20.7
2b	2°B	680	0.75	0.5	-878	1.05	1.48	1.686	42.2
3	2°B	680	0.9	0.76	-853	0.81	1.38	1.607	13.5
4a	0°	700	0.748	0.5	277	0.19	1.32	1.675	19.9
4b	2°B	700	0.753	0.48	-696	2.35	2.55	1.703	30

At a growth temperature of 680°C layers with y=0.12 prefer the misorientation to {111}B as can be seen from the FWHM in HRXRD which is closer to the expected one than on exact substrates. Also the CL peak is narrower on the off-orientation. This is also consistent with a smoother surface observed on the off-orientation and shows that at low As-content the material behaves similar to GaInP /10/. However, it is striking that the rocking curves show a considerably different FWHM for the two orthogonal azimuths.
For y=0.5 from the FWHM of the CL spectra and the rocking curves a preference of this composition for the exact substrate orientation can be concluded. This preference also shows up in the surface morphology which is much smoother on the exact substrate.
Since in the laser structure thick layers of GaInP and GaInAsP with y=0.12 both prefering the off-orientation to (111)B are present while only thin layers with a higher As-content (y=0.76) are used in the active region, the off-orientation was chosen for the complete laser structure. At 680°C the quaternary layer with y=0.76 shows a narrow CL peak and a reasonable rocking curve width (sample 3) on the off-orientation although the crystalline perfection is better on the exact substrate /11/.
The anisotropy in the rocking curve width was investigated more closely. Rocking curves in all four <110> azimuths were taken and compared. While no difference is observed when the sample is rotated by 180° a 90° rotation yields the different $\Delta FWHM_{norm}$ listed in table 1. Broadening of rocking curves due to defects can be excluded, because in no sample misfit dislocations or extended defects were observed. The main feature in cross section TEM images was a modulated contrast parallel to the growth direction for samples 1a,b and 2a,b. Fig. 1a shows the image for sample 2b. The modulated contrast is interpreted as being due to spinodal-like decomposition leading to columnar growth. The circular form of TED spots indicates an only small difference in lattice parameters between the different phases.

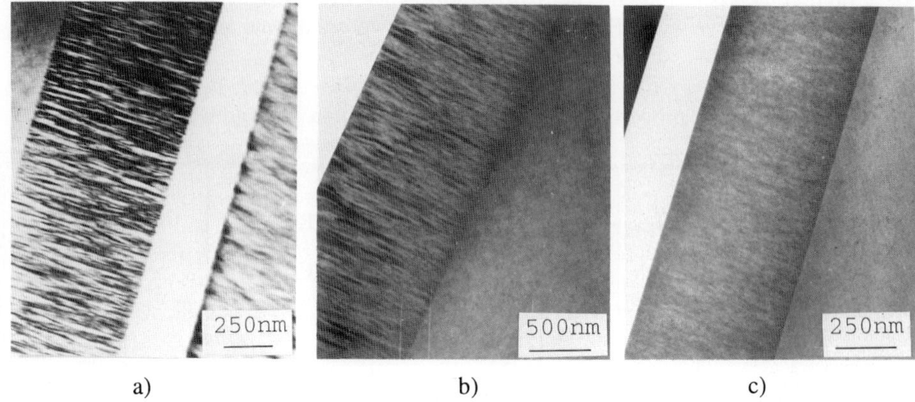

Fig. 1 TEM bright field images from InGaAsP with y=0.5 grown at a) 650°C , b) and c) 700°C, a) and b) on a misoriented substrate, c) on a exact oriented substrate

Anisotropy in the size of the growing columns is one reason for the anisotropic rocking curves. From (-2-2 4) area maps it can be concluded, that in addition to decomposition mosaicity leads to the broadened rocking curves for the sample 1a,b and 2a,b while only mosaicity but no decompositions is observed for sample 3 /11/ which also shows no modulated TEM contrast. For y=0.12 an additional phenomenon is observed. TED patterns from samples 1a and 1b show 1/2 {11n} superlattice spots indicating the presence of ordering in this material with a composition close to GaInP. This ordering is more pronounced on the off-orientation /3/. The occurence of partial ordering leads to a broadened CL spectrum for sample 1b while the smaller width of the rocking curves indicates that the crystalline quality is superior to sample 1a.

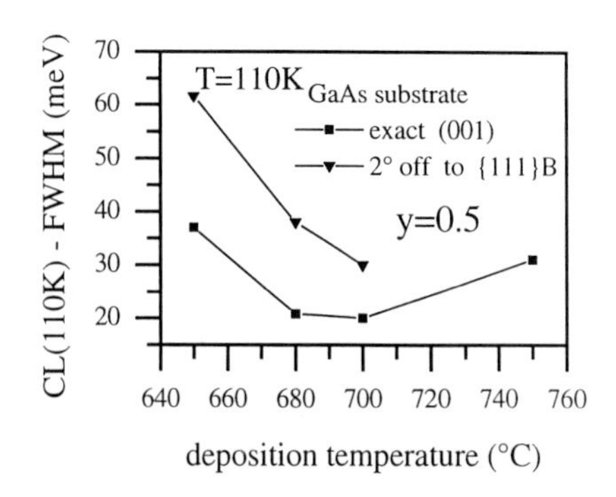

Fig. 2 Dependence of CL-FWHM at 110 K on the growth temperature

The broadest CL spectra are obtained for y=0.5, the composition which lies closest to the miscibility gap. For this composition the influence of growth temperature has been studied. The FWHM of the CL emission strongly decreases upon increasing the growth temperature to 700°C, a further increase to 750°C again leads to broader CL emission and also a stronger contrast modulation in TEM. The peak is also narrower on the exact than on the off-orientation again showing that the composition lying on the GaAs-side of the MG prefers exact substrate orientation.
This can also be seen from the HRXRD data in table 1 (samples 4a and 4b) and fig. 1b and 1c. While a strong contrast modulation in TEM is observed for sample 4b (fig.1b), only a slight residual contrast is visible for sample 4a (fig. 1c). which not necessarily has to be due to decomposition /12/. The strong anisotropy of the rocking curve width for sample 4a is due to mosaicity.

3.2. Laser structure

Broad-area laser diodes processed from layer structures grown at 680°C show good laser properties /10/ despite some indications for decomposition in layers with y=0.12. For buried

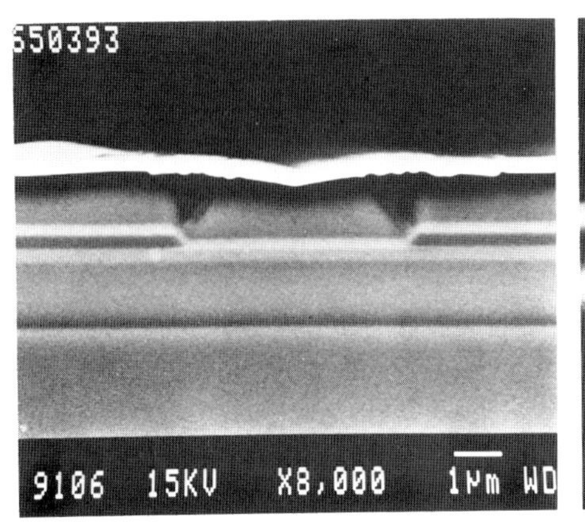

a)

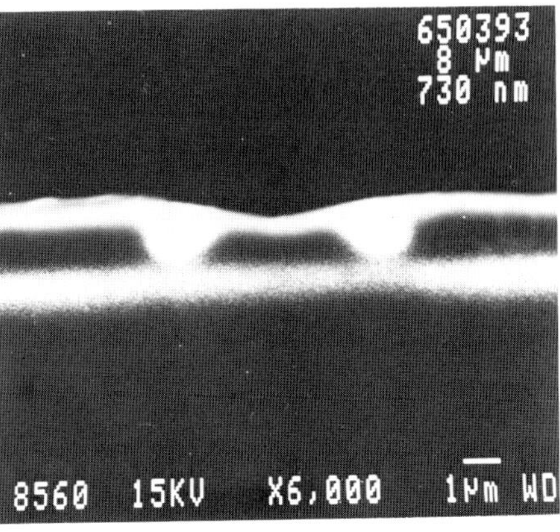

b)

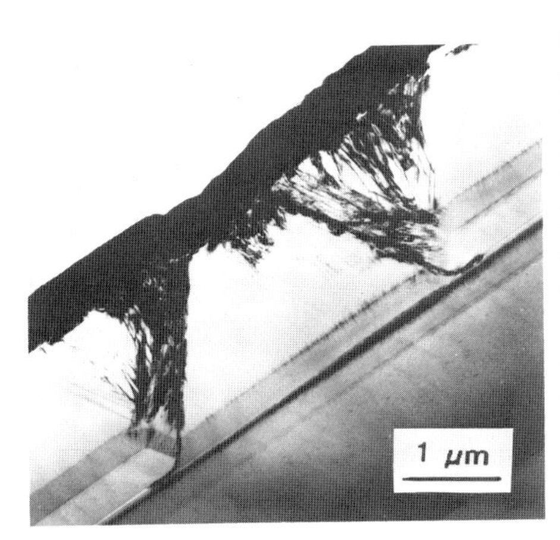

c)

Fig. 3 RISAS structures
a) BSE/compo-mode image
b) Monochromatic CL image
c) TEM image

laser structures it can be expected that the presence of different growth facets gives rise to additional effects or will promote those discussed in the previous section.

The laser structure under study was a RISAS-laser where an etched trench has to be filled in a second epitaxial growth step. First studies indicate that indeed a number of problems has to be overcome to successfully produce such a structure. Already from SEM/BSE micrographs of a facet cleaved through such a trench (fig. 3 a) it can be seen that considerable composition fluctuations between the flat bottom (bright) and the channel sidewalls (dark) occur. EPMA shows that on the sidewall the material with a nominal As-content of y=0.12 is Ga-rich and As-poor. CL spectra and CL images taken with the wavelength of the emission maximum confirm the inhomogeneous composition (fig. 3b). TEM (fig. 3c) reveals that in the sidewall region corresponding to the dark contrast in SEM/BSE a complicated defect structure exists. Dislocations, antiphase boundaries and composition fluctuations contribute to the observed contrast. Apparently, strain resulting from the compositional inhomogeneity leads to the formation of dislocations originating from the overgrown GaInP sidewall. Due to this disturbed region, the CL intensity from the active layer is lower in the trench than outside for a trench width below 5 μm. Only for wider trenches the intensity in the channel exceeds the one outside.

4. Conclusions

GaInAsP lattice matched to GaAs shows a strong dependence of the obtained crystallographic perfection on the MOVPE growth conditions and the substrate misorientation. The observed phenomena are different for different compositions. For a small As-content (y=0.12) some degree of spinodal decomposition is observed together with ordering. Off-orientation to {111}B is preferred over the exact one although it promotes ordering like in GaInP. For compositions close to GaAs (y=0.76) the exact orientation is the preferred one and no indication of decomposition but some mosaicity is found on the off-orientation.

The composition with y=0.5 is the most critical one in our study since it lies closest to the MG. Here an increase of the growth temperature from 680°C to 700°C significantly improves the crystalline quality with the exact orientation being the favoured one. While 680°C spinodal-like decomposition is observed on both orientations it is no longer visible in the TEM images taken from samples grown at 700°C on exact substrate.

The observed spinodal-like decomposition leads to columnar growth with an anisotropy of the column size distribution in the [110] and [1-10] azimuth.

While the above phenomena can be controlled by a proper choice of the growth parameters for broad area laser diodes, buried laser structures suffer from additional problems. The dependence of the composition on the growing facet leads to strain that promotes decomposition and extended defect formation. For such buried structures the geometry of the sidewalls also has to be optimized in order to suppress these effects.

References

[1] Zhang G, Ovtchinnikov A, Näppi J, Asonen H and Pessa M 1993 *IEEE J. Quantum Electron.* **29** 1943

[2] Plano W E, Nam D W, Major Jr J S, Hsieh K C and Holonyak Jr N 1988 *Appl. Phys. Lett.* **53** 2537

[3] Oster A, Knauer A, Gutsche D and Weyers M 1995 *J. Crystal Research and Techology* in press

[4] Rechenberg I, Oster A, Knauer A, Richter U, Menniger J and Weyers M 1995 *Material Research Society, Fall Meeting*, Boston

[5] de Cremoux B, Hirtz P and Ricciardi J 1981 *GaAs and Related Compound, Inst. Phys. Conf., IOP Conf. Proc.* **56** 115

[6] McDevitt T L, Mahajam S, Laughlin D E, Bonner W A and Keramidas V G 1992 *Phys. Rev. B* **45** 6614

[7] Norman A G and Booker G R 1985 *J. Appl. Phys.* **57** 4715

[8] Launois H, Quillec M, Glas F and Treacey M M J 1992 *GaAs and Related Compounds, Inst. Phys. Conf., IOP Conf. Proc.* 64 537

[9] McDevitt T L 1990 PhD Dissertation, Carnegie Mellon University, Pittsburgh

[10] Knauer A, Erbert G, Gramlich S, Oster A, Richter E, Zeimer U and Weyers M 1995 *J. Electron. Mater.* **24** 1653

[11] Zeimer U, Rechenberg I, Oster A, Knauer A, Bugge F and Weyers M to be published

[12] Glas F 1993 *Microscopy of Semiconducting Materials, Inst. Phys. Conf. Ser.* **134** 269

Inst. Phys. Conf. Ser. No 149
Paper presented at DRIP 95

Defect energy levels in Cd-based compounds

A.Castaldini, A.Cavallini, B.Fraboni, J.Piqueras*and L.Polenta

Dipartimento di Fisica, Università di Bologna, Bologna - Italy
*Departamento de Fìsica de Materiales, Facultad de Fisicas,Universidad Complutense, Madrid - Spain

Abstract. The influence of deep levels on the electrical and optical properties of semiconductors is widely acknowledged. We have utilized several complementary spectroscopic techniques to investigate the deep traps in undoped CdTe, CdTe:Cl and $Cd_{0.8}Zn_{0.2}Te$. The electrical activity of the defects has been studied by DLTS, PICTS and P-DLTS while their optical properties have been characterized by cathodoluminescence, CL. Various deep levels have been found and by critically comparing the results obtained with the different techniques in different samples, we were able to achieve a better understanding of the nature of the defects.

1. Introduction

Semiconducting II-VI binary and ternary compounds have recently arisen a great interest due to their medical and optoelectronic applications. Many of these applications are limited by the lack of understanding and control of carrier traps whose energy levels lie deep in the forbidden gap. A thorough investigation of these deep levels can only be carried out by combining several characterisation techniques, which can provide complementary information both on their electrical and optical properties.

We have investigated by capacitance junction spectroscopy, i.e. DLTS (Deep Level Transient Spectroscopy) [1], the electrically active defects in semiconducting CdTe. As their role becomes even more significant in compensated semi-insulating materials, we have analyzed by current transient spectroscopy methods, i.e. PICTS (Photo-Induced Current Transient Spectroscopy) [1] and P-DLTS (Photo-DLTS) [2], semi-insulating II-VI materials which cannot be studied by capacitance methods. The radiative properties have been investigated by cathodoluminescence, CL.

We have also monitored the evolution of the existing traps after annealing treatments, by using the same techniques. A critical comparison of the deep levels found in the various materials analyzed with different techniques allows us to achieve a better understanding of the nature of the defects. By comparing our results with those reported in the literature, we have tentatively assigned an origin and a character (donor or acceptor) to some of the deep traps.

2. Experimental

We have investigated 4 different sets of samples: CdTe undoped ($\rho \approx 30\ \Omega$cm), CdTe:Cl ($\rho > 10^7\ \Omega$cm), CdTe:Cl annealed for 5 hours at T=600°C, and $Cd_{0.8}Zn_{0.2}Te$ ($\rho > 10^{11}\ \Omega$cm). The CdTe samples have been grown by the travelling heater method (THM) while the $Cd_{0.8}Zn_{0.2}Te$ ones have been grown by the high pressure Bridgman method (HPB). The samples were chemo-mechanically polished with a bromine-methanol solution in concentration from 1% to 0.1% followed by rinsing in deionized water. A further etch in HCl:H_2O (1:1) for 1 min. was then necessary to remove the newly formed V_{Te} rich superficial layer.

DLTS analyses have been carried out only on the semiconducting samples, i.e. CdTe undoped, after a Au Schottky barrier had been deposited on the sample. The applied reverse bias has been varied from V=-5V to V=-9V and the pulse width from 100μs to 10ms.
PICTS analyses have been utilised to study the semi-insulating samples, CdTe:Cl and $Cd_{0.8}Zn_{0.2}Te$, in two configurations: planar and Schottky. The planar one requires two ohmic contacts, prepared with an InHg alloy, while the Schottky configuration consists of an Au Schottky barrier and a backside ohmic contact. They both produce similar spectra, but the Schottky configuration should allow to achieve better resolved peaks [3]. We employed two different excitation lights, λ=880nm and λ=670nm corresponding respectively to 1.41eV (below band gap) and to 1.85eV (above band gap), to vary the carrier excitation conditions. The applied bias was V=30V. The P-DLTS method has been applied on semi-insulating samples in Schottky configuration to reveal the majority carrier traps (hole traps). The applied reverse bias varied from V=-5V to V=-9V and we used both wavelengths λ=880nm and λ=670nm as external excitation source. In all spectroscopic experiments the heating rate was 0.2 K/sec and the emission rates varied from 5 to $2\times10^4\ s^{-1}$.

Cathodoluminescence measurements have been performed in an Hitachi S-2500 scanning electron microscope at temperatures between 80K and 300K with an accelerating voltage of 25kV. Emission was measured with a North Coast E0-817 germanium detector.

3.Results

The electrical characterisation of deep levels in the four investigated sets of samples revealed the existence of ten different traps labelled A_0, A, B, C, D, E, F, G, H and I. They are reported in Table I, together with the results of DLTS, PICTS, and P-DLTS analyses, where applicable, and with the results of CL investigations for each set of samples. In CL measurements deep levels often manifest themselves as luminescence bands. In some cases the broad bands have been determined by deconvolution to be composed of two or more sub-bands.
A PICTS spectrum in the planar configuration for a CdTe:Cl sample is reported in Fig.1,with e_n=256.41 s^{-1}, pulse width 30 ms and V_b=10 V .The existing peaks are labelled according to Table I.

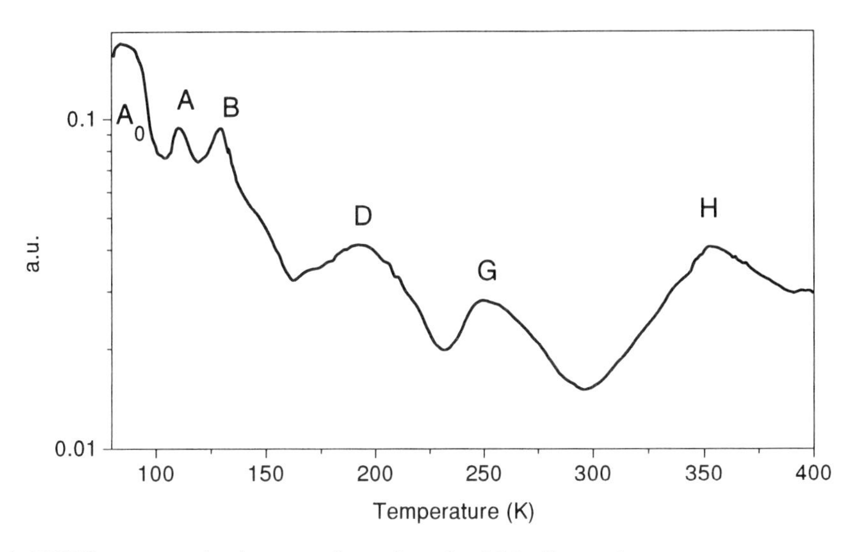

Fig. 1: PICTS spectrum in planar configuration of a CdTe:Cl sample

4. Discussion

In order to discuss our experimental findings, it is worth stressing that while DLTS and P-DLTS reveal only majority carrier traps, i.e. hole traps being the material p-type, PICTS spectra show both majority and minority carrier traps at the same time, but does not allow easily to distinguish between them [1]. Therefore, care has to be taken in deciding if a trap is not present in the sample studied or if it is simply not detectable. Moreover, the energy band gaps of $Cd_{0.8}Zn_{0.2}Te$ and CdTe are different, being respectively 1.65eV and 1.5eV at room temperature [4,5] and this has to be taken into account when comparing the various levels in different materials. A critical comparison of the results obtained with the different spectroscopic method used allows us to advance hypotheses on the character of each trap (donor or acceptor) and on whether its energy level has to be calculated from the bottom of the conduction band or from the top of the valence band. The increase in the band gap in $Cd_{0.8}Zn_{0.2}Te$ with respect to CdTe mostly affects the conduction band, whose bottom rises in energy. The top of the valence band, on the contrary, remains almost constant and so do the energy levels of the acceptor traps associated to E_V [6].

Levels A_0 and A: in all samples a level at 0.15eV has been detected. A similar trap has been reported in the literature by DLTS measurements [7,8] and associated to the so called centre A. Luminescence measurements also report on a 1.4 - 1.5eV band attributed to centre A in CdTe and $Cd_{0.8}Zn_{0.2}Te$ respectively [4,5,9]. It has been suggested that centre A consists of a V_{Cd} and a donor an Te site (Don_{Te}) and that it should behave as an acceptor [4,8-10]. This last hypothesis is confirmed by the presence of the level A in DLTS and P-DLTS spectra and, thus, we can locate A at E_V +0.15eV. In Cl doped samples there is also the A_0 level, detected both

Table I: Deep levels found in the investigated samples

	Energy level	$Cd_{0.8}Zn_{0.2}Te$		CdTe undoped		CdTe:Cl			CdTe:Cl annealed	
	(eV)	PICTS	CL	DLTS	CL	P-DLTS	PICTS	CL	PICTS	CL
A_0	0.12					▨	▨	▨	▨	▨
A	0.15	▨	▨	▨	▨	▨	▨	▨	▨	▨
B	0.20			▨	▨	▨	▨	▨	▨	▨
C	0.25	▨	▨							
D	0.32					▨	▨			
E	0.40			▨					▨	
F	0.57	▨								
G	0.64						▨		▨	
H	0.78	▨	▨	▨	▨	▨	▨		▨	
I	1.1	▨	▨		▨		▨	▨	▨	▨

in PICTS and P-DLTS spectra. It has, therefore, an acceptor character and is located at E_V + 0.12eV. Its nature is not clear but, as it is present only in Cl doped samples, it could be related to the presence of Cl. Moreover, as centre A usually manifests itself as a band of defects, level A_0 could be one of the components of this band in CdTe:Cl.

Level B: all CdTe samples exhibit a level at 0.20eV. Since their impurity contents and conductivity differ significantly, this trap cannot be ascribed to any defect related to the above mentioned properties. It has been reported that different growth techniques influence the formation of the defects lying close to the E_V+0.15eV band [11] and, significantly, the Arrhenius plot of one of the levels associated to the THM method coincides just with our level B. As our CdTe samples are all THM grown and the $Cd_{0.8}Zn_{0.2}Te$ ones, which are grown with HPB method, do not show level B, it seems plausible to assign the origin of the 0.20eV level to the growth method.

Level C: this level at 0.25eV is only found in $Cd_{0.8}Zn_{0.2}Te$ samples. A similar level is reported in the literature, found in CdZnTe:In and assigned to a Zn related defect [12]. The Zn attribution is supported by our results. Conductance measurements [12] suggest that its activation energy has to be measured from the valence band, i.e. E_V+0.25eV.

Level D: The level at 0.32eV has only been found in CdTe:Cl before the annealing treatment. As it is also present in P-DLTS spectra, it must be a majority carrier trap and its activation energy results E_V+0.32eV. Some researchers attribute the defects located in the band 0.27-0.35eV to Cu contamination [7,8], but in our case such contamination should be present in all of the investigated CdTe samples, grown in identical conditions. Other possible attributions of this defect are to a Te_{Cd}^- complex introduced during growth [7]. This hypothesis could be supported by our findings as level D is not present in $Cd_{0.8}Zn_{0.2}Te$ (different growth method). Its absence in undoped CdTe could be due to the dominant nearby lying level E, that may mask any effect due to level D. The disappearance of this trap in CdTe:Cl annealed samples could be interpreted as a transformation of the Te_{Cd}^{-} defect into V_{Cd}^{-} and Te_i, an effect supported by the significant concentration of level E (attributed to V_{Cd}^{-}, see below) in the annealed samples.

Level E: the 0.40eV level is present in CdTe undoped and in CdTe:Cl annealed samples. It has been previously found in CdTe and attributed to the isolated V_{Cd} [8,13]. Cadmium vacancies are not easily revealed in CdTe:Cl as they tend to form the centre A complex with a Cl donor, leaving the concentration of isolated V_{Cd} very low [14]. On the contrary, in CdTe undoped crystals which are not compensated, the V_{Cd} is present in significant concentration [14]. In CdTe:Cl annealed samples, both the annealing treatment and the transformation of the Te_{Cd}^{-} defect (level D) may account for the appearance of a relevant cadmium vacancy concentration.

Level F: the 0.57eV PICTS level is only present in $Cd_{0.8}Zn_{0.2}Te$ and cannot be related to any of the major traps found in CdTe. A similar trap has been found in PICTS measurements of $Cd_{0.9}Zn_{0.1}Te$ [13], but the Arrhenius plots of these two traps do not have much in common. The CL analysis does not straightforwardly suggest the existence of a band associated with this energy. However, it has been suggested [15] that the defect located at 0.55 - 0.65eV has an acceptor character and is related to a Zn vacancy.

Level G: this trap has only been found in PICTS measurements in CdTe:Cl and CdTe:Cl annealed samples. A level with similar characteristics has been reported in the literature and attributed to Cd_i^{++} [8,13]. DLTS measurements on n-type CdTe reveal the presence of a E_C-0.66eV donor level which has also been attributed to Cd_i^{++}, possibly associated with an extrinsic impurity [16]. Our findings confirm these hypotheses: DLTS and P-DLTS measurements, carried out on CdTe undoped and CdTe:Cl respectively, only show hole traps and, thus, level G cannot be revealed.

Level H: the 0.78eV trap is detected in all of the investigated samples with every technique used. Such level has been reported in the literature studied both with electrical [7,8,13,15] and optical characterisation techniques [4,9,17,18]. It has been associated to a deep acceptor formed by V_{Cd}^{--}, and it is supposed to play a crucial role in the compensation mechanism and in the pinning of the Fermi level [14,18,19]. Some researchers attribute the level to a V_{Cd}^{--} associated with an impurity [16,17] and photocapacitance measurements have provided indication for its behaviour as a recombination centre with $e_p \approx 5e_n$ [16]. This explains why level H has been detected with DLTS or P-DLTS both in our p-type CdTe:Cl and CdTe undoped samples and in n-type CdTe by other researchers [16]. As our samples are all p-type, the majority carrier cross section is bound to control the process and we can locate the level at E_V+0.78eV.

Level I: the 1.1eV level is present in the CL spectra of all samples while it is not revealed in DLTS and P-DLTS measurements of CdTe undoped and CdTe:Cl samples, respectively. The 1.1eV trap has been found in the past in CdTe and $Cd_{1-x}Zn_xTe$ and has been attributed to V_{Te}^{+} [5,6,9]. This defect is supposed to behave as a donor [10] and, in fact, the trap is not present in the majority carrier spectra (DLTS, P-DLTS). We can therefore locate the level at E_C-1.1eV.

5. Conclusions

We have investigated various CdTe and $Cd_{0.8}Zn_{0.2}Te$ samples with complementary spectroscopic techniques (DLTS, PICTS, P-DLTS and CL) to characterise the existing defects. Ten different deep levels have been found and some are common to all samples. By comparing the results obtained in different samples we were able to advance hypotheses on the nature and character (donor or acceptor) of the traps.

This research has been partially supported by the Cooperation Programme "Azione Integrata" between Italy and Spain and by DGICYT (Project PB 93 - 1256). The authors are indebted to Prof. Casali for providing the $Cd_{0.8}Zn_{0.2}Te$ samples and to Japan Energy Corporation for the undoped and Cl doped samples.

References

[1] Look D.C. 1989 *Electrical Characterization of GaAs Materials and Devices* (New York: Wiley)
[2] Mooney P.M. 1983 *J.Appl.Phys* **54**, 208
[3] Fang Z., Shan L.,Schlesinger T.E. and Milnes A.G. 1989 *Solid State Electron.* **32**, 405
[4] Hofmann D.M., Omling D., Grimmeiss H.G., Meyer B.K., Benz K.W. and Sinerius D. 1992 *Phys.Rev.B* **45**, 6247
[5] Barnett Davis C., Allred D.D., Reyes-Mena A., Gonzalez-Hernandez J., Gonzales O., Hess B.C. and Allred W.P. 1993 *Phys.Rev.B* **47**, 13363
[6] Hofmann D.M., Stadler W., Oettinger K., Meyer B.K., Omling P., Salk M., Benz K.W., Weigel E. and Müller-Vogt G. 1993 *Mat.Sci.Eng.* **B16**, 128
[7] Samimi M., Biglari B., Hage-Ali M., Koebel J.M. and Siffert P. 1987 *Phys.Stat.Sol.(a)* **100**, 251
[8] Hage-Ali M. and Siffert P. 1992 *Nucl.Instr. and Meth.* **A322**, 313
[9] Stadler W., Hofmann D.M., Alt H.C., Muschik T., Meyer B.K., Weigel E., Müller-Vogt G., Salk M., Rupp E., Benz K.W. 1995 *Phys. Rev.B* **51**, 10619
[10] Allen J.W. 1995 *Semicond.Sci.Technol.* **10**, 1049
[11] Eiche J.W., Maier D., Sinerius D., Weese J., Benz K.W. and Honerkamp J. 1993 *J.Appl.Phys.* **74**, 6667
[12] Suzuki K., Inagaki K., Kimura N., Tsubono I., Sawada T., Imai K. and Seto S. 1995 *Phys.Stat.Sol.(a)* **147**, 203
[13] Fiederle M., Ebling D., Eiche C., Hofmann D.M., Salk M., Stadler W., Benz K.W. and Meyer B.K. 1994 *J.Crystal Growth* **138**, 529
[14] Hoschl P., Grill R., Franc J., Moravec P. and Belas E. 1995 *Mat Sci.Eng.* **B16**, 215
[15] Larsen T.L., Varotto C.F. and Stevenson D.A. 1972 *J.Appl.Phys.* **43**, 172
[16] Takebe T., Saraie J. and Matsunami H. 1982 *J.Appl.Phys.* **53**, 457
[17] Pal U., Fernandez P., Piqueras J., Suchinski N.V. and Dieguez E. 1995 *J.Appl.Phys.* **78**, 1992
[18] Agrinskaya N.V. and Arkadeva E.N. 1989 *Nucl.Instr. and Meth.* **A283**, 260
[19] Moravec P., Hage-Ali M., Chibani L. and Siffert P. 1993 *Mat Sci.Eng.* **B16**, 223

Inst. Phys. Conf. Ser. No 149
Paper presented at DRIP 95

Determination of three-dimensional deep level defect distribution by Capacitance-Voltage Transient Technique (CVTT)

S Dueñas, R Pinacho, L Quintanilla, E Castán and J Barbolla

Dpto Electricidad y Electrónica. Facultad de Ciencias, Universidad de Valladolid, 47011 Valladolid (SPAIN)

Abstract. In this work we present the application of the CVTT technique to obtain three-dimensional distributions of deep level defects in boron ion implanted silicon wafers which were rapid thermal annealed just after the implantation process. The surface temperature is not homogeneous along the wafer during the RTA and, in consequence, the resulting damage varies from one point to another. Our results show the existence of a strong dispersion in the damage profiles. Moreover, the total amount of damage as well as its preferential location in the bulk vary with temperature showing a clear tendence. As temperature increases, the height and the depth of the damage peaks decrease, and the total amount of the damage increases. The fact that the deep level damage is non homogeneous over the wafer is very important because of its impact on the performance and yield of the electronic devices which will be later fabricated.

1. Introduction

Present days, ion implantation is the preferred doping technique in microelectronics processing because of its advantages over dopant diffusion: better control of dose and depth, ability to self-align junctions and to produce very shallow pn junctions, etc. The main problem of the ion implantation is the damage produced in the crystalline lattice of the semiconductor. Therefore, a thermal annealing step (either in conventional furnace or rapid thermal annealing, RTA) becomes as necessary to repair the induced damage as well as to electrically activate the dopant impurities. In recent years much interest in ion implantation research has focused on minimizing the amount of dopant redistribution while activating the implanted dopants and to remove the damage. There is a good evidence that this is best done by rapid thermal annealing. Many investigators are interested in the study of RTA annealings in the range of seconds to minutes time because the fact that the activation energy for the elimination of secondary defects is greater than the activation energy for impurity diffusion [1]. In order to produce defect-free pn junctions with the minimum amount of diffusion, the above results seems to suggest the use of shorter time and higher temperature anneals. However, the fact that much of the initial diffusion is anomalously fast because of the high point defect concentrations in the implanted layers appears as a serious inconvenient. Thus, RTA is also seen as a method of limiting anomalous diffusion by concentrating on the range of seconds.

Nuclear stopping processes of ions implanted into Si are responsible for producing displacement damage. The types and amount of damage produced depend on implant species, energy, dose, wafer temperature and orientation, dose rate, and materials covering the substrate. The as-implanted defect morphologies then change during wafer annealing depending on temperature, time, furnace ambient, and ramp-up/ramp-down rates. Much of the point-defect generation (vacancies and self-interstitials) occurs during the changes from as-implanted damage to stable or dissolved damage structures as a result of annealing [2, 3]. Thus, a large part of the experimental work involving ion implantation of semiconductors has been aimed at the development of characterization techniques that are highly sensitive to the residual defects in the lattice which remain after annealing.

DLTS [4-6] has been proved to be very useful to study the electrically active damage induced by ion implantation in semiconductors because of its high sensitivity and easy interpretation. However, some times it seems not to be sufficient to do this study, for example when emission transients become non-exponential because of different effects that can take place in the space-charge region of the junctions, as the Poole-Frenkel effect, refilling effects, and so on. In order to achieve a more complete characterization we have recently set up a new technique named Capacitance-Voltage Transient Technique (CVTT) [7], which allows to determine the spatial distribution of the concentration and of the emission rates of the deep levels. This technique basically consists on recording the instantaneous C-V curve just after an emission pulse is applied to a pn or Schottky junction. The experimental set up is basically the same of standard C-V. The only difference is that the reverse bias consists on emission pulses whose duration is scanned in order to modify the time in which the deep levels are emitting. As this duration is varied, the charge state of the deep levels at the end of these pulses is modified also. From the experimental C-V curves we can extract information about the deep levels. From the slopes of these curves it can be obtained the transients corresponding to the concentration of emptied deep levels as a function of the time, for every point into the space charge region and the experimental temperature. In this way, recording the amplitude of these transients as a function of the spatial coordinate, we can obtain the profiles of the deep levels that emit at this temperature. Deep levels with emission energies distant enough can be easily separated by choosing the adequate temperatures. Carrying out these measurements at several temperatures, we can obtain the variation of the emission rates of the deep levels as a function of temperature, and then the thermal emission energy, for each point into the space charge region. This result is very important since it will allow to determine if the activation energy of a deep level presents some variations in this zone. These variations can be due, for instance, to the existence of several deep levels with non homogeneous distributions, as is the case of the damage associated to an ion implanted and RTA-annealed sample.

In this work we present the aplication of the CVTT technique to obtain three-dimensional distributions of deep level defects due to ion implantation and RTA annealing in the whole semiconductor wafer. Indeed, if the semiconductor wafer is segmented as a two-dimensional array of pn junctions, the application of CVTT to all samples in the wafer will allow us to obtain the three-dimensional distribution of the deep level damage. We have used this method to study the damage distribution induced in boron ion implanted silicon wafers which were rapid thermal annealed just after the implantation. As the surface temperature is non homegeneous along the wafer during RTA, the damage effects vary from one point to another in the wafer. Finally, we have correlated the deep level distribution so obtained with the variation in the temperature along the surface of the wafer.

2. Samples description

We have used 100 mm diameter p-type (100) silicon wafers for sample fabrication. The background doping level in the 12-18 Ω cm wafer was ≈9 x 10^{14} cm^{-3}. The samples were 1 mm^2 area squares n^+p junctions distributed along all the wafer. The fabrication process was as follows: First, for the back ohmic contact, p^+ layers were made in the rear side of the wafers by implanting 150 keV boron ions with a dose of 1 x 10^{15} cm^{-2}. The layers over we have carried out our study were p^+ layers made by implanting the front face of the wafers with a dose of 3 x 10^{12} cm^{-2} of boron ions. The implantation energy was 120 keV and the tilt angle was 7°. The p^+n junctions were completed by implanting very low energy (20 keV) phosphorus ions in a very high dose (1x$10^{14}$$cm^{-2}$) to obtain n^{++} layers very much higher and shallower than the underlying p^+ layers. After this the wafers underwent a 1100 °C RTA process for 10 s. Finally, metallizations were made for the contacts and the pads.

The surface temperature is not homogeneous along the wafer during RTA. In order to obtain the temperature distribution, we have carried out in the RTA chamber an oxidation process of silicon wafers under the same conditions of time and temperature (1100°C for 10s) that we have used for the RTA annealing of the implanted wafers. In a first approximation, we can assume that the grown oxide width exponentially depends on the temperature at every position in the wafer. Figure 1 shows the mapping of temperatures obtained from ellipsometry measurements of the oxide width. We can see that the highest value is located at the center, and the temperature decreases and reachs its lowest value near of wafer edge. As the termocouple that controls the RTA system temperature is located at a distance of approximately half of a radius from the center, the RTA temperature nominal value corresponds to an intermediate position. For the system we used we found that there is a temperature dispersion of about 30°C for 100 mm diameter wafers. One can expect greater dispersions for bigger wafers (p.e. 6 or 8 inchs diameter wafers).

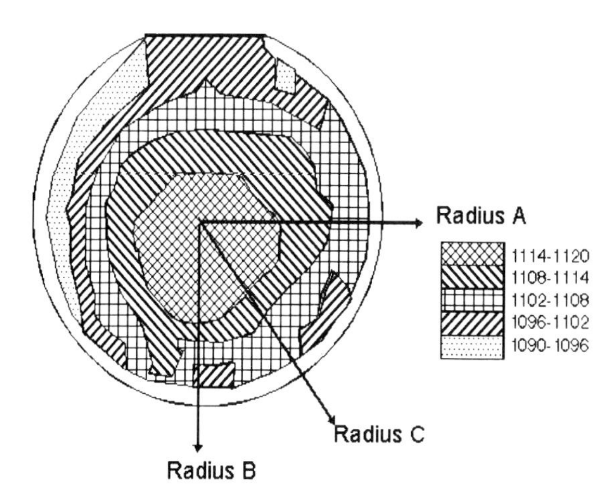

Figure 1. Wafer temperature distribution in the RTA annealing system

3. Experimental results and discussion

In this paragraph we present the deep level profiles obtained by using the CVTT technique as a function of the annealing temperature for several positions in the wafer. Figure 2 shows three-dimensional view of the damage profiles along the three radii (A, B and C) marked in Figure 1.As we can see, the profiles are deeper as the distance from the center

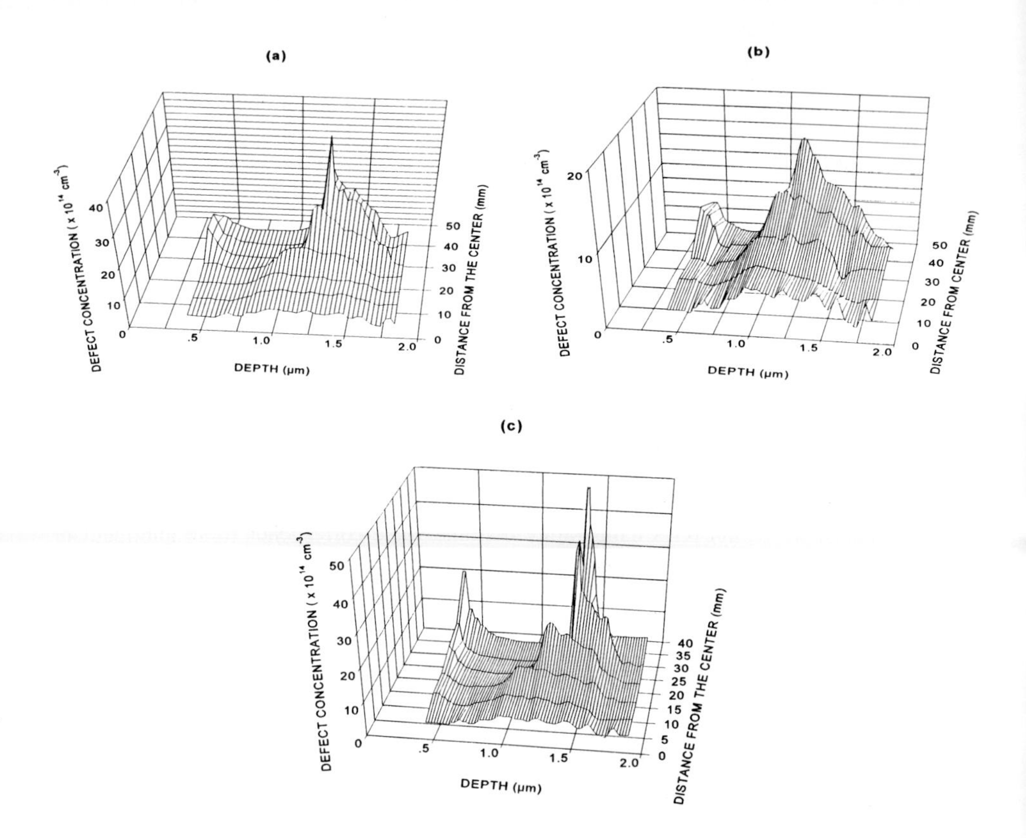

Figure 2.- Three-dimensional damage profiles along the three radii indicated in Figure 1 : (a) Radius C, (b) Radius B and (c) Radius A.

increases, that is, as temperature decreases. Near the center, where temperature is higher, the damage profiles are continuum-shaped and the damage distribution have not so much preferential location as for the lower temperatures annealed samples. These results indicate that when temperature increases a process like a gettering takes place: the damage moves towards the plane of the junction and the total amount of deep levels decreases. One could think that some temperature-activated process works moving the damage and repairing the lattice.

We calculate the total amount of damage in a sample by integrating the profile curves. Figure 3 shows the variation of the total damage as a function of the RTA temperature. We can see that when temperature decreases the damage becomes bigger. Figure 4 shows the damage mapping for a quadrant of the wafer. When comparing with Figure 1 we clearly observe the correlation between temperature and damage.

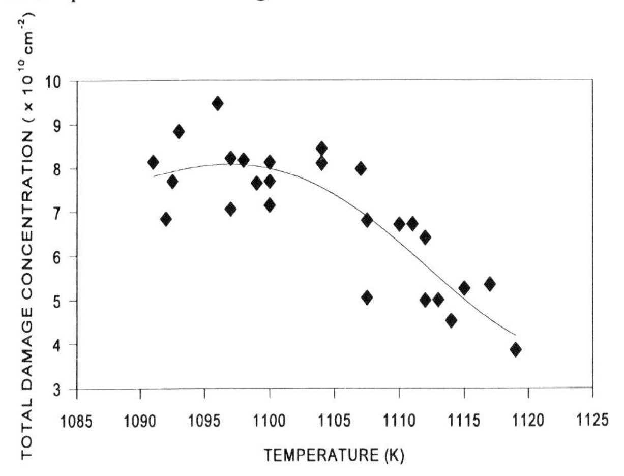

Figure 3. Variation of the damage with the temperature

A detailed observation of the damage profiles shows the existence of two deep levels.These levels are more clearly observed in the samples where the damage is higher (i.e. the more external samples). We have seen already these deep levels in a previous work [7]. In that work, we used CVTT to determine the activation energy of both deep levels: E_v+90 meV, for the center that has its maximum closer to the junction plane and E_v+350 meV for the deeper one.

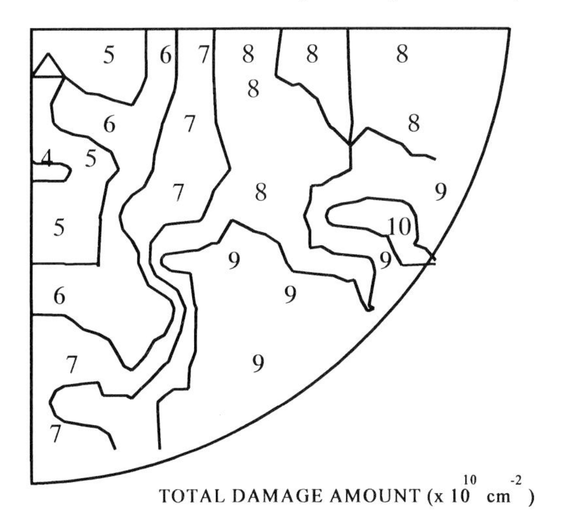

Figure 4. Total damage amount in the measured quadrant of the wafer

Although it is not easy to establish the origin of these deep levels because of the extremely complex reactions taking place during RTA annealing of the implanted layers, we related the 350 meV deep level with an interstitial silicon-shallow dopant complex, and the 90 meV deep level with a dislocation center, according the references dealing with this subject that we have found in the literature [8, 9].

4. Conclusions

In order to complete a previous work in which we studied the damage produced by boron implanted and RTA annealed silicon substrates, we present the evolution of this damage with RTA temperature. Because of surface temperature is not homogeneous along the wafer during RTA, we have obtained the three-dimensional distribution of the deep level damage by using CVTT technique. Our results show the existence of a strong dispersion in the damage profiles for different coordinates over the wafer. Moreover, the total amount of damage as well as its preferential location in the space charge region vary as temperature does. The fact that the deep level damage is non homogeneous over the wafer should be taken into account because of the impact of the damage on the performance and yield of the electronic devices that will be later fabricated.

References

[1]. Seidel T E , Lischner D J,Pai C S, Knoell R V, Maher D M and Jacobson D C 1985 Nucl. Instr. Meth.Phys. Res. **B7/8** 251

[2].Fair R B, Wortman J J and Liu J 1984 J.Electrochem.Soc. **131** 2387

[3].Kim Y, Massoud H Z and Fair R B 1989 J.Electron Mater **18** 143

[4].Lang D V 1974 J.Appl.Phys. **45** 3023

[5].Johnson N M, Regolini J L and Bertelink D J 1980 Appl. Phys. Lett. **36** 425

[6].Beck S, Jaccodine R J and Clark C 1992 J. Electron. Mater. **21** 125

[7].Dueñas S, Castán E, Enríquez L, Barbolla J, Montserrat J and Lora-Tamayo E 1994 Semicond. Sci. Technol. **9** 1637-1648

[8].Kimerling L C and Patel J R 1979 Appl.Phys.Lett. **34** 73

[9].Peaker A R and Marco M D 1988 Properties of Silicon (EMIS Datareviews Series **4**) ed INSPEC (London: IEE) 232

Inst. Phys. Conf. Ser. No 149
Paper presented at DRIP 95

Recognition and recombination strength evaluation by LBIC of dislocations in FZ and Cz silicon wafers

I. Périchaud, J.J. Simon, S. Martinuzzi

Lab. de Photoélectricité des Semiconducteurs, E. A. 882 - DSO
Fac. des Sciences et Techniques de Marseille - St Jérôme
13397 Marseille cedex 20 - FRANCE

Abstract. In the present work, dislocation networks are investigated in Czochralski (Cz) and in float zone (FZ) grown silicon wafers by the Light Beam Induced Current (LBIC) mapping technique at different wavelengths, which appears to be able to recognize and to detect these networks and to evaluate their recombination strength. Dislocations are found to be more recombining in Cz than in FZ. It is shown that in Cz wafers, a four hours phosphorus diffusion at 900°C, realized before dislocations creation, impedes the formation of oxygen precipitates during subsequent annealings. In FZ dislocated wafers, a phosphorus diffusion at 850°C for 30 min cancels the LBIC contrast of dislocations. Electrical activity of these defects which are still physically present as shown by X-Ray topography, seems to disappear.

1. Introduction

Dislocations are frequently generated in monocrystalline silicon wafers during processing steps needed to make electronic devices. This is the reason why the evaluation of their recombination strength and the knowledge of its evolution during high temperature treatments is of a great interest. The nature of the dislocation electrical activity in silicon is a problem under discussion for many years. First, Schockley [1] has suggested that only the electronic states of the dislocation core determine the electrical and optical properties of this defect. B. Pichaud et al. [2] has emphasized that the recombination activity is mainly due to geometrical defects distributed along the dislocation cores : double kinks, or their associated dangling bonds or solitons.

Other authors have shown the importance of the dislocation-point defect and dislocation-impurity interactions wich can play a determinative role in the evolution of the dislocation electrical activity [3,4]. Working with Czochralski (Cz) grown silicon in which interstitial oxygen concentration overpasses 10^{17} cm^{-3}, K. Sumino [5] has shown the major role of oxygen on dislocation motion and on dislocation electrical activity in this material.

One convenient way to try to clear this problem is to create individual dislocations by plastic deformation in float zone (FZ) grown and in Cz silicon and to modify, by different treatments, the impurity atmosphere around these defects.

Such a treatment could be phosphorus diffusion from $POCl_3$ which is well known to produce an external gettering when it is applied for duration overpassing 30 min at temperature around 900°C [6], and to inject a strong density of interstitial silicon atoms in the bulk of the wafers [7,8]. These self-interstitials in excess are able to shrink metallic or oxyde precipitates or impurity aggregates and then favour the transfer of substitutionnal metallic atoms to interstitial positions. Therefore, phosphorus diffusion treatment is a very efficient way to modify the distribution of impurities around crystallographic defects in the material.

In the present work, dislocation networks are investigated by the Light Beam Induced Current technique at different wavelengths, from which a minority carrier diffusion length (L) can be deduced. This technique appears to be able to recognize, to detect and to evaluate the recombination strength of these networks. It is shown that in Cz wafers, oxygen controls the electrical activity of dislocations and that a preliminary phosphorus diffusion impedes the formation of oxygen precipitates. In FZ wafers, phosphorus diffusion reduces drastically the LBIC contrast of dislocations.

2. Experimental

Rectangular samples 3 x 1 cm^2, 380 μm thick, were cut from p type Czochralski (Cz) and float zone (FZ) silicon wafers. They were boron doped ($\leq 10^{15}$ cm^{-3}) and two orientations were used : (100) and (111), the length of the rectangular sample being always along <110> direction. Dislocation sources were nucleated by scratching the samples in a direction parallel to their length with a diamond. Then, the samples were elastically bent at room temperature in the cantilever mode along the transversal axis, and finally they were heated under stress for 6 hours at 750°C in a pure argon atmosphere. Dislocation networks were characterized by X-ray transmission topography and by chemical etchings. Infrared spectroscopy was used to determine the interstitial oxygen concentration $[O_i]$.

Semi-transparent diodes were obtained by thermal evaporation of a thin aluminium layer (100 Å) after cleaning the surface with CP4 etch. These barriers, realized on dislocation containing and on dislocation free regions, were used for Light Beam Induced Current (LBIC) and for minority carrier diffusion length (L) mappings.

Phosphorus diffusions were carried out in an open tube furnace, using a $POCl_3$ source and nitrogen flow, at 850°C for 30 min or at 900°C for 4 hours. As phosphorus diffusion occurs by the two faces, one N^+ region was removed by chemical etching and covered by an aluminium layer which gave an ohmic contact after annealing at 450°C for 30 min (back surface). On the other N^+ region (front surface), arrays of mesa diodes were realized by classical photolithography techniques.

LBIC mappings were done with a monochromatic light spot less than 10 μm in diameter, given by a monochromator and the focusing system of a metallographic microscope, while an x-y stage moved the sample. This photocurrent mapping technique allows to follow easily the evolution of defect electrical activity when the samples are submitted to different treatments. It is possible to investigate large areas (4 cm^2) and then to compare the behaviour of defect containing and of defect free regions. The lateral resolution is sufficient to characterize precisely the recombination strength of precipitates, dislocations, or grain boundaries in silicon. By changing the wavelength, one can study regions near the surface (λ < 800 nm) as well as the bulk of the wafers (λ > 900 nm). Thanks to LBIC scans at different wavelengths ($0,8 < \lambda < 1$ μm), diffusion length maps were obtained from the spectral variation of the local quantum efficiency in the near infrared range correlated with that of the optical absorption coefficient. Details and performances of the technique have been already published [9,10].

After scratching and deformation annealing, the samples were first characterized by LBIC at 940 nm in order to obtain a view map of the electrical activity of defect-free and defect containing regions. Then, a part of the sample was selected and a diffusion length map was computed.

3. Results and discussion

Figure 1a shows the diffusion length map obtained for a (100) oxygen rich Cz sample in which $[O_i] \cong 10^{18}$ cm^{-3}. The selected region is choosen at the beginning of scratches where the dislocation density reaches 10^6 cm^{-2}. In this highly dislocated region, diffusion length is found to collapse to values around 2 μm. The L map is well correlated with X-ray topography and microphotographs obtained after chemical etches [11]. In oxygen poor FZ wafers, for the same dislocation density, diffusion length remains around 40 μm. Figure 1b gives a diffusion length map obtained in a similar region to that of figure 1a, for a (100) dislocated FZ sample. Notice the different diffusion length scales.

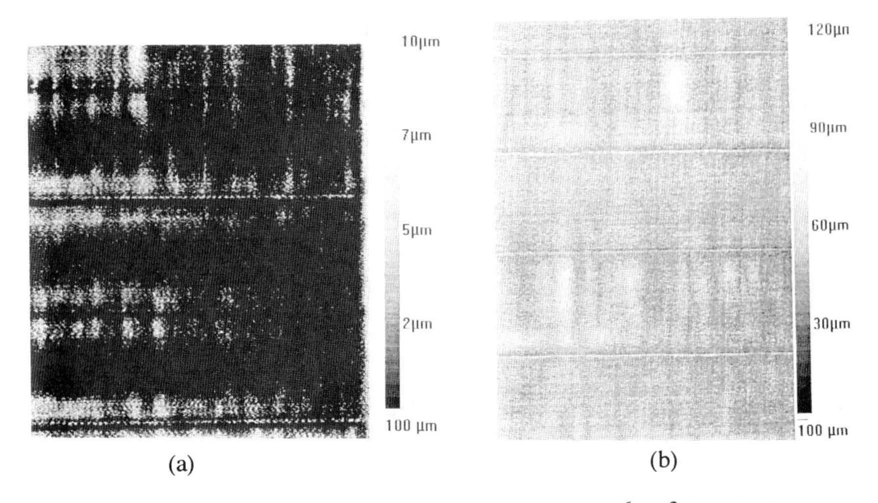

Figs. 1 : Diffusion length maps obtained in a highly dislocated region ($N_{dis} = 10^6$ cm^{-2}) for (a) a (100) Cz dislocated sample and for (b) a (100) FZ dislocated sample.

So, thanks to L maps, it is observed that the recombination strength of dislocations in Cz is more marked than in FZ and it is reasonnable to assume that this is due to the difference in oxygen concentration between the two types of wafers. When Cz dislocated samples are submitted to a short phosphorus diffusion (850°C - 30 min), the LBIC map changes and the figure 2 shows such a map for a (100) Cz sample. The obtained N^+ region was used as collecting barrier. In these samples, the recombination activity of dislocations appear clearly, depending on their density.

Fig. 2 : LBIC map at 940 nm of a (100) dislocated Cz sample after a phosphorus diffusion at 850°C for 30 min.

In the dislocation free regions, at the end of the scratches, ring like patterns begin to appear, probably due to oxygen microprecipitates and associated defects which are formed during thermal treatments (especially for temperatures higher than 700°C which are used for the growth of oxygen precipitates in silicon). These defects create recombination centers which can hide more or less those due to dislocations. Consequently, in order to evaluate what occurs at dislocations, it is needful to eliminate or to impede the formation of these precipitates. This could be done by a preliminary long phosphorus diffusion, carried out before the creation of the dislocations.

Some samples (cut in the same wafer used for the sample which the LBIC map is given by figure 2) were submitted to such experimental sequences and the figure 3 shows the obtained LBIC map. Notice that after the long phosphorus diffusion, the N^+ regions were removed, then the dislocations were created and, endly, a collecting structure is obtained by a new phosphorus diffusion at 850°C for 30 min.

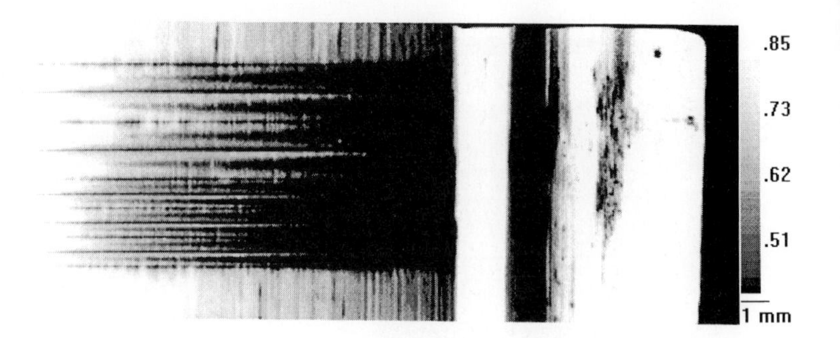

Fig. 3 : LBIC map at 940 nm of a (100) dislocated Cz sample which was preliminary longly phosphorus diffused.

The LBIC scan can be compared with that of figure 2 because the two scan maps were obtained in the same conditions : same wavelength, same illumination level and the grey level scales are identical in the two pictures.

A dramatic signal improvement is obtained in the dislocation free regions of the previously diffused sample. Diffusion length is around 100 μm and even if we change the grey level scale to investigate these regions more precisely, ring like patterns never appear.

Consequently, a preliminary phosphorus treatment of Cz wafers seems to prevent the formation of ring like contrast i.e. the formation and the growth of oxygen precipitates and associated defects. This behaviour could be due to the strong injection of self-interstitials [12] during the phosphorus diffusion, which shrinks the precipitates [8,13].

The influence of a phosphorus diffusion on dislocated FZ wafers is completely different. Before phosphorus diffusion, dislocations appear clearly electrically active like in figure 1b and the alone difference with Cz samples is that these defects are less recombining as previously shown. If such a dislocated sample is submitted to a short phosphorus diffusion at 850°C for 30 min, we obtain the LBIC map of figure 4 : the contrast due to dislocations disappears !. One can compare this map with that of figure 2 made on (100) Cz dislocated and diffused sample. The electrical behaviour, after phosphorus diffusion, of Cz and FZ dislocated silicon is drastically different, certainly due to the higher oxygen content in Cz samples.

Fig. 4 : LBIC map at 940 nm of a (100) dislocated FZ sample after a phosphorus diffusion at 850° C for 30 min.

As shown by figure 5, in highly dislocated regions (N_{dis} = 10^6 cm^{-2}) of (100) dislocated and diffused FZ sample, diffusion length reaches values around 100 μm, suggesting that the recombination strength of the defects has strongly decreased (comparison can be made with figure 1b). Electrical activity of dislocations seems to disappear because X-ray topography of such regions indicates that these defects are still physically present, as shown by figure 6. A similar effect of phosphorus diffusion is observed on (111) dislocated FZ samples.

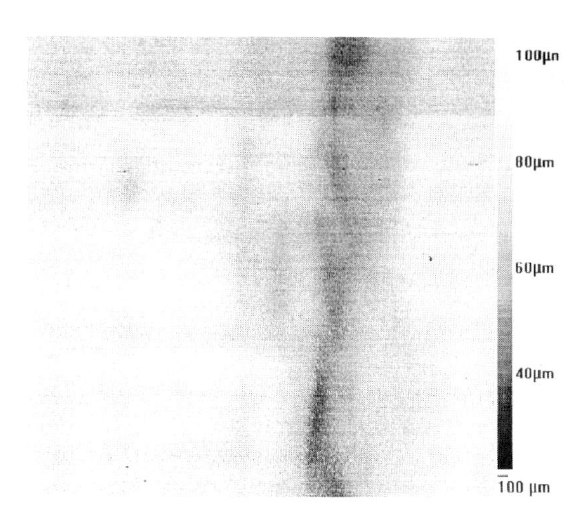

Fig. 5 : Diffusion length map in a highly dislocated region (N_{dis}= 10^6 cm^{-2}) of the sample of figure 4 ((100) dislocated and diffused FZ sample)

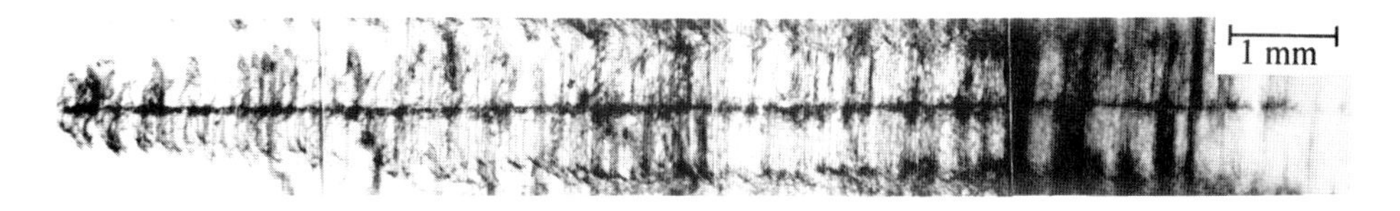

Fig. 6 : X-Ray topography of a (100) FZ dislocated and phosphorus diffused sample (sample of fig. 4). The selected region was choosen around a scratch.

Several explanations could be proposed to understand this last result. First, one can suppose that a slight dislocation reconstruction occurs due to the strong self-interstitial injection during phosphorus diffusion. This reconstruction would be able to modify the electrical activity of the defects but would not be sufficient to be detected by X-ray topography. If such an explanation is right, a four hours diffusion at 900°C would be more efficient than a 30 min one. We didn't observed such result : a long phosphorus diffusion doesn't seem to give as good results as the short one because some dislocated regions remain slightly recombining.

Another possible explanation is that the diffusion coefficient of phosphorus atoms is higher along the dislocation core than in the defect free regions. Such phenomena would give rise to a larger collecting surface. The electrical activity of the dislocation would be masked by this effect. But, like previously, a longer phosphorus diffusion would be more efficient, as it increases the heterogeneity of the junction structure. Such an increase is not observed. Therefore, the best explanation we can propose is that a gettering effect is developped during the phosphorus diffusion, which clean the defects by the removal of impurities (fast diffusers) aggregated at or around the dislocation cores. These impurities can use the dislocation pipes to diffuse rapidly to the heavily doped N^+ layer.

Some preceding results could be partly explained by the model of El Ghitani and Martinuzzi [14] who have computed the variation of L in dislocation containing silicon wafers and have shown that the value of L depends on the dislocation density N_{dis}, on the recombination strength of these defects S_d and also on the value of the diffusion length L^* in the dislocation free regions. L^* depends on the presence of recombining impurities and/or precipitates. In annealed Cz wafers, L^* values are degraded by the oxygen related precipitates and associated defects. As shown in [14], L tends to the value of L^* when N_{dis} and S_d are very low. The LBIC map contrast, which depends on the ratio L^*/L, is marked if L^* is high for given values of N_{dis} and S_d. Conversely, if L^* is small, a poor contrast is obtained. This is what is shown by the comparison of the figures 2 and 3. Note that a long phosphorus diffusion is needed in order that self-interstitials diffuse through the entire wafer.

4. Conclusion

In the present paper, it is shown that the LBIC mapping technique and the associated minority carrier diffusion length scan maps are able to detect and to recognize dislocations in silicon wafers as well as X-ray topography. In addition, the technique can evaluate the recombination strength of the defects. This recombination strength is higher in Cz than in FZ samples due to the aggregation of oxygen atoms at dislocations.

A long phosphorus diffusion before the creation of the dislocation networks by plastic deformation, impedes the formation of oxygen related precipitates and the LBIC contrast of the defect containing regions is drastically improved. This is a consequence of the injection in the bulk of an excess of self-interstitials.

When a phosphorus diffusion is applied to FZ wafers after the formation of the dislocation networks, the recombination strength of the defects is practically cancelled, probably due to an external gettering of impurities.

The results tend to demonstrate that impurities aggregated or segregated at dislocations are the main source of the recombination centers associated to these defects.

The authors would like to thank Dr N. Burle from MATOP Laboratory of Marseilles for X-Ray topographies.
This work was supported by CNRS - ECOTECH France and by ADEME France.

References

[1] W. Shockley 1953 Phys. Rev. 91, 228
[2] B. Pichaud, F. Minari and S. Martinuzzi 1991 Journal de Physique IV, 1, C6-187
[3] I.E. Bondarenko and E.B. Yakimov 1990 Phys. Stat. Sol. (a) 122, 121
[4] I.E. Bondarenko and E.B. Yakimov 1988 Solid State Phenomena, 182, 59
[5] K. Sumino 1983 Journal de Physique C4, 44
[6] I. Périchaud and S. Martinuzzi 1992 J. de Phys. III, 2, 313
[7] J.J. Kang and D.K. Schroder 1989 J. Appl. Phys., 65, 2974
[8] C. Clays and J. Vanhellemont 1994 in : Advanced and Semiconducting Silicon Alloy Based Materials and Devices, p 35, ed. by J. Nijs (IOP Pub. Bristol)
[9] S. Martinuzzi and M. Stemmer 1994 Mat. Science and Eng. B 24, 152
[10] M. Stemmer 1993 Applied Surface Science 63,213
[11] J.J. Simon and I. Périchaud 1995 communication to EMRS 95, Strasbourg, to be published in "Materials Science and Engineering B"
[12] A. Ourmazd and W. Schröter 1984 Appl. Phys. Lett., 45, 781
[13] S. Martinuzzi and I. Périchaud 1995 Solid State Phenomena, 47-48, 153
[14] H. El Ghitani and S. Martinuzzi 1989 J. of Appl. Phys. 66, 1717

Inst. Phys. Conf. Ser. No 149
Paper presented at DRIP 95

Extended defect related excess low-frequency noise in Si junction diodes

E Simoen, J Vanhellemont, G Bosman*, A Czerwinski** and C Claeys

IMEC, Kapeldreef 75, B-3001 Leuven, Belgium
*Dept. Electr. Engineering, Univ. of Florida, Gainesville (FL), USA
**Institute of Electron Technology, Warsaw, Poland

Abstract. This paper investigates the correlation between the density of oxygen-precipitation induced extended defects and the low-frequency (LF) noise spectral density of Cz Si junction diodes. It is shown that the excess LF noise intensity increases with increasing starting oxygen concentration and with increasing extended defect density, as revealed for instance by infrared Laser Scanning Tomography (LST). From the correlation with the bulk minority carrier lifetime, it is derived that the increase in excess 1/f noise originates from carrier recombination fluctuations associated with the SiO_x precipitates, or the associated secondary defects.

1. Introduction

It has been reported recently that the starting material has a pronounced impact on the low-frequency (LF) noise behaviour of Si junction diodes in forward operation [1],[2]. An example is given in Fig. 1, showing that in the low-frequency part, typically below 1 kHz, a clear 1/f-like noise component is observed, followed by the frequency independent shot noise. The 1/f excess noise strongly depends on the substrate type (e.g. Float-Zone FZ, Czochralski Cz, epi,...) and on the applied initial heat treatment (internal gettering IG, nuclation step, ...).

Furthermore, the experimental evidence points towards a defect-related generation-recombination (GR) origin. In this paper, the correlation with the starting material is further explored in detail, on a large number of diodes, fabricated in different Cz, FZ and epitaxial substrates. As is shown, the excess noise for the Cz substrates is correlated with the density and distribution of oxygen-precipitation related extended defects, revealed by TEM and LST. Finally, it is reported that in some cases individual processing-induced extended defects show up in the LF noise spectra, giving rise to Random Telegraph Signals (RTS).

2. Experimental

Diodes are processed on both p- and n-type 125 mm diameter wafers. The p-type wafers have a B-doping density in the range 6 to 20×10^{14} cm^{-3}; the n-type wafers are $\approx 10^{15}$ cm^{-3} P-doped. For the p-type substrates, Cz wafers with various initial interstitial oxygen contents ($[O_i]$) are compared with FZ and epi substrates, as described in Table I. Three different pre-treatments have been applied, which have been described in detail elsewhere [3]: either no pre-anneal (no) or a low-temperature nucleation step to create a large density of SiO_x precipitates and associated secondary defects (nucl.) and, finally, a three-step internal gettering sequence, resulting in a defect-lean zone of around 10 μm, revealed by TEM and LST [3],[4]. N-type FZ diodes are compared with medium-oxygen (MO) Cz devices. Finally, also (111) p- and n-type wafers have been processed. The diodes have been fabricated in a CMOS compatible technology, using ion implantation, furnace anneal and standard aluminum metallization. At the periphery of the devices a LOCOS isolation oxide is grown, to separate active device regions.

Diodes with different geometry (area/perimeter ratio) have been characterized in detail [2], using a large variety of electrical and analytical techniques. The minority-carrier recombination

Table I. Relevant material parameters of the p-type Si substrates studied. S_{I1} is the wafer average excess noise intensity at a forward current of 1 μA. LO: low oxygen; MO: medium oxygen; HO: high oxygen.

Si-Material	Initial $[O_i]$ (10^{17} cm^{-3})	τ_{rec} (μs)	LST density (10^9 cm^{-3})	S_{I1} (10^{-24} A^2/Hz)
LO Cz no	8.0	32	0.63	1.1
LO Cz IG	7.3	40	0.088	2.3
MO Cz no	9.3	1.6	1.0	1.9
MO Cz IG	9.2	7.2	0.76	1.8
HO Cz no	11.1	0.062	13.0	4.1
HO Cz IG	10.8	0.75	7.4	2.6
FZ	---	45	--	0.95
p/p^+ epi	---	--	--	0.32

lifetime τ_{rec} reported here has been derived from the forward I-V characteristics [5]. The details of the noise measurements have been described previously [1],[2]. Unless otherwise mentioned, the noise is measured between the top junction contact and the bottom ohmic contact, on square 900 μmx900 μm diodes. Per wafer type, a minimum of 10 diodes has been characterised, to study the statistical variation of the noise across the wafer.

3. Results

3.1. Impact of the Si substrate on the excess 1/f noise

The LF noise spectrum of the diodes generally consists of a 1/f-like part in the lower frequency range, followed by shot noise at higher frequencies (Fig. 2). The latter equals the fundamental relationship:

$$S_I = 2\,q\,I_F \qquad (1)$$

with S_I the LF noise spectral density, I_F the forward diode current and q the elementary charge. Qualitatively similar results are obtained for the different p-type substrates, whereby a higher average noise spectral density is obtained for the Cz wafers (Table I), compared with FZ (Fig. 2) or epi diodes (Fig. 3). This also follows from the data in the last column of Table I, representing the average excess noise intensity at f=10 Hz and I_F=1 μA. This value is obtained by a least-square fit to the wafer average noise data, after subtracting the back-ground system and the shot noise, given by equation (1). The n-type diodes generally have a larger average noise level, compared with their p-type counterparts, while, again, the Cz devices are more noisy than the FZ ones (Fig. 4). The larger noise of the n-type devices is most likely due to the occurrence of single-defect related Random Telegraph Signal (RTS) noise, which will be treated in more detail in the next section.

The LF excess noise can empirically be described by [1]:

$$S_I = C\,\frac{I_F^{\beta}}{f^{\gamma}} \qquad (2)$$

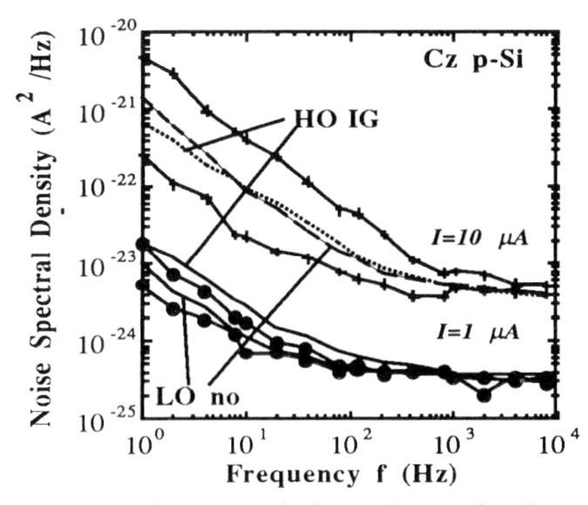

Fig. 1. LF noise spectra in forward operation for low oxygen (no) and high oxygen IG Cz diodes. The wafer averages at 1 μA and 10 μA correspond to the indicated line spectra. The other spectra correspond to the minimum and maximum for the LO Cz diodes.

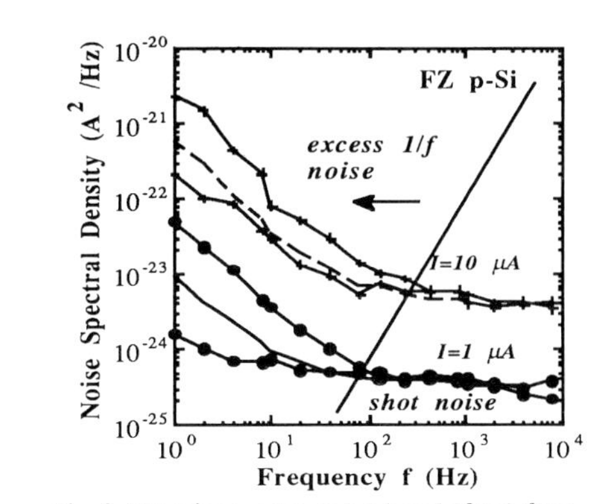

Fig. 2. LF noise spectra at 1 μA and 10 μA for p-FZ Si n$^+$p junction diodes. Circles and + correspond to the mimimum and maximum spectra. Wafer averages are represented by the lines.

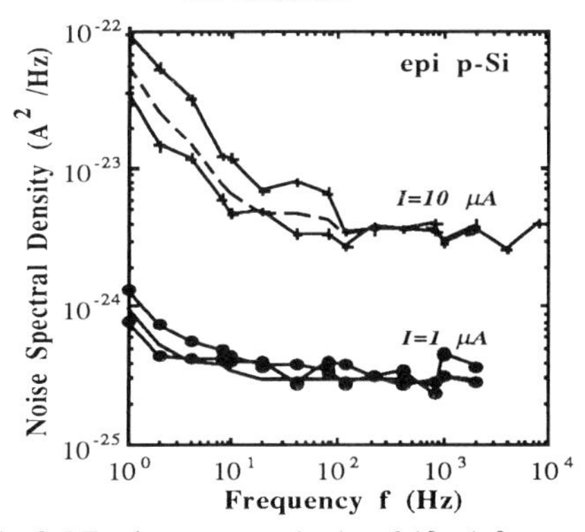

Fig. 3. LF noise spectra at 1 μA and 10 μA for p-epi Si n$^+$p junction diodes. Circles and + correspond to the mimimum and maximum spectra. Wafer averages are represented by the lines.

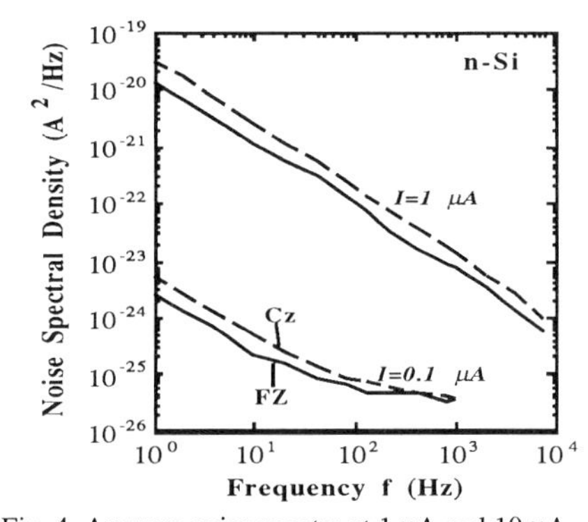

Fig. 4. Average noise spectra at 1 μA and 10 μA for a Cz and FZ n-type wafer.

with the frequency index γ between 0.7 and 1.3 and the current index β between 1.5 and 2.2 [2]. The fact that γ is not constant, but generally increases with increasing I_F, indicates a carrier-fluctuation or defect-related generation-recombination (GR) origin and a McWhorter type of model should therefore be applied [6],[7]. This is further supported by the β-values, which are not in agreement with the 'mobility-fluctuations' theory, where β=1 is expected [8].

3.2. Origin of the excess 1/f noise

From structural characterisation by TEM [3] and LST [4], it is known that a large density of oxygen-precipitation induced extended defects exists in the Cz substrates (see also Table I), so that the increase in excess noise could be partly associated with deep-level GR centres in the bulk of the material. This is illustrated in Fig. 5, representing the correlation between S_I and the LST defect density N_{LST}. While for low bulk defect densities the noise is independent from N_{LST}, a clear trend develops for the highly defective p-type Cz substrates. In other words, for the HO-Cz substrates an additional source of LF excess noise is present, which is

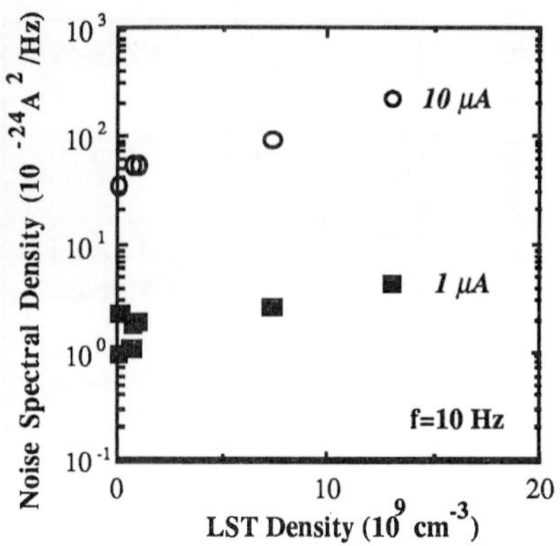

Fig. 5. Correlation between N_{LST} and S_I for the p-type substrates studied. I_F=1 μA (squares) and 10 μA (circles).

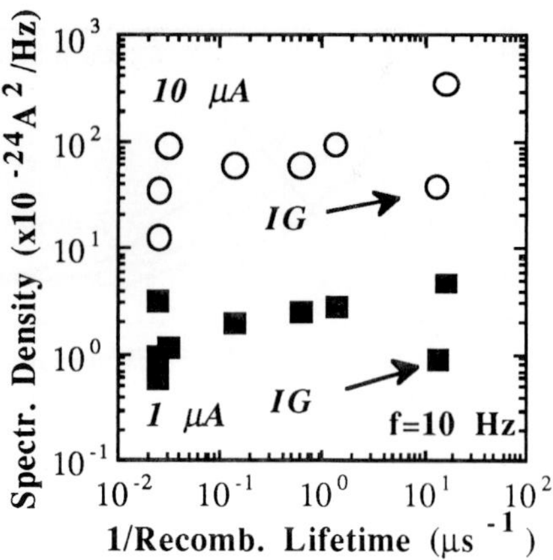

Fig. 6. Correlation between the inverse recombination lifetime and S_I for the p-type substrates studied.

associated with the extended defects in the substrate. This is the more pronounced at higher forward currents ($I_F \approx 100$ μA), where the series resistance of the diode becomes of the same order as the dynamic diode resistance $r_d=\partial V_F/\partial I_F$. In this case, a drastic increase in the 1/f noise occurs [1],[2], which is believed to be due to the 1/f noise generated in the series/contact resistance of the diode. Again, a GR origin is suspected, pointing towards the recombination activity of the oxygen-precipitation induced extended defects.

On the other hand, the origin of the 1/f like noise in the LO and FZ diodes is most likely related to the "surface" or peripheral diode region [7]. In that case, the noise is generated mainly by carrier exchange with slow oxide traps close to the Si-SiO_2 interface in the LOCOS regions surrounding the diode. This also follows from a systematic study of the LF noise in different geometry diodes.

Further proof for the electrical activity of the extended defects is provided by DLTS studies, which reveal the presence of two electron traps at E_c-0.2 eV and E_c-0.4 eV in the HO no p-Cz diodes [3]. The DLTS density profile follows closely the TEM distribution of the extended defects and the leakage current profile [3]. While the near midgap trap is thought to be responsible for the leakage current generation, temperature-dependent lifetime measurements reveal that the minority carrier recombination is governed by the E_c-0.2 eV electron trap. Additionally, a clear correlation between N_{LST} and the recombination lifetime for the different n- and p-type substrates has been demonstrated [4].

When plotting S_I versus $1/\tau_{rec}$ a similar picture as in Fig. 5 emerges: for low-defect density substrates, the noise is not correlated, while a trend develops for the low-lifetime Cz substrates (Fig. 6). In fact, a close relationship has between τ_{rec} and S_I has been proposed on several occasions in the past, both theoretically [8] and empirically [9]. In both cases, it is shown that S_I is proportional to the inverse recombination lifetime. This is to some extent also found in Fig. 6, for the diodes with a low bulk recombination lifetime compared to the surface recombination lifetime, which is estimated to correspond to 10 μs. Note, however, that IG reduces clearly the excess 1/f noise for the HO Cz diodes, which is an additional benefit of internal gettering, not reported in the past. The displacement of the HO-IG data point in Fig. 6 can be explained qualitatively as follows: due to the presence of the defect-lean zone, the recombination lifetime is higher in the near-surface layer and reduces considerably in the bulk of the wafer. For not too high forward currents, the excess LF noise will mainly be determined by the GR centers close to or in the forward depletion region - this has also been demonstrated for Ge pre-amorphisation induced end-of-range damage in p^+n Si diodes [10]. In other words, the LF noise intensity corresponds to a near-surface effective recombination lifetime which is much larger than the bulk value, represented in Fig. 6. Finally, another advantage of IG is that the sample-to-sample noise dispersion in IG diodes is less than in non-gettered devices [2], so

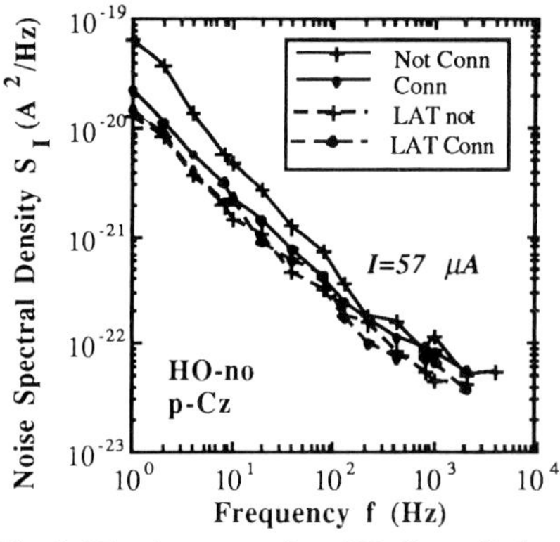

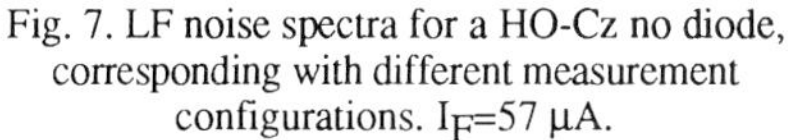

Fig. 7. LF noise spectra for a HO-Cz no diode, corresponding with different measurement configurations. I_F=57 μA.

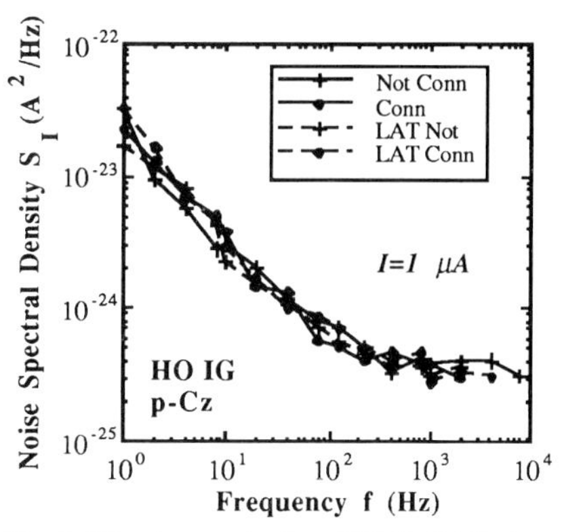

Fig. 8. LF noise spectra for a HO-Cz IG diode, corresponding with different measurement configurations. I_F=1 μA.

that there is a better control of the electrical device properties.

3.3. Excess noise generated in the bulk

As mentioned above, a drastic increase in the excess 1/f noise can be observed for increasing I_F, particularly for the HO-Cz diodes (Fig. 7). This increase could be due to the 1/f noise generated by the extended defect related GR centres in the bulk of the wafer. To investigate this further, the LF noise has been measured for different contact geometries : 1/ the current flowing from the biased top junction to the grounded bottom ohmic contact (Not Conn.); 2/ same as 1, but with a closeby top ohmic contact grounded to the bottom contact (Conn); 3/ the current flowing between a top junction and a grounded top ohmic contact, leaving the bottom contact floating (LAT Not) and 4/ the same as 3, with the bottom contact grounded (LAT Conn). By changing the electrode configuration, the current path will be changed. In case of a top ohmic contact, the current flow will be more parallel to the top surface (lateral flow), while for a bottom contact, the charge will drift in a vertical direction, through the defective bulk of the wafer. By adding an additional grounded substrate electrode, the current flow is further affected in one sense or another. With respect of the noise, it is expected that it will be largest for the vertical flow (case 1), which is indeed observed in Fig. 7 for the HO-Cz diode at I_F=57 μA.

On the other hand, for low forward current, little effect is expected if the above picture is valid, which is indeed observed in Fig. 8. Additionally, for an FZ diode, which contains no measurable extended defects, no effect is anticipated, as confirmed by Fig. 9.

4. Single defect related RTS

It is well-known from the past that in bipolar devices so-called burst or RTS noise can occur [11]-[14]. In its most simple form, the current through the device switches stochastically between two discrete levels, giving rise to the random telegraph signal waveform. When measuring the corresponding LF noise, a Lorentzian GR spectrum is obtained, which consists of a plateau at low frequencies, followed by a $1/f^2$ roll-off. From this follows that when $f \times S_I$ is plotted, a peak-shaped function is obtained, like in Fig. 10, whereby the peak maximum corresponds to the characteristic GR centre time constant $\tau=1/\pi f$. The GR centre in Fig. 10, observed in a (111) oriented FZ n^+p junction diode has a time constant of ≈0.1 s at room temperature, which increases slightly with increasing I_F. In this case, the RTS is associated with a LOCOS-edge 60° dislocation, penetrating the depletion region, in agreement with previous observations [13]-[14]. Similar type of RTS can be generated by ion-implantation induced end-of-range defects located in the diode depletion region, as is believed to be the case for some of the p^+n diodes studied, or by other extended defects placed at a strategic site near

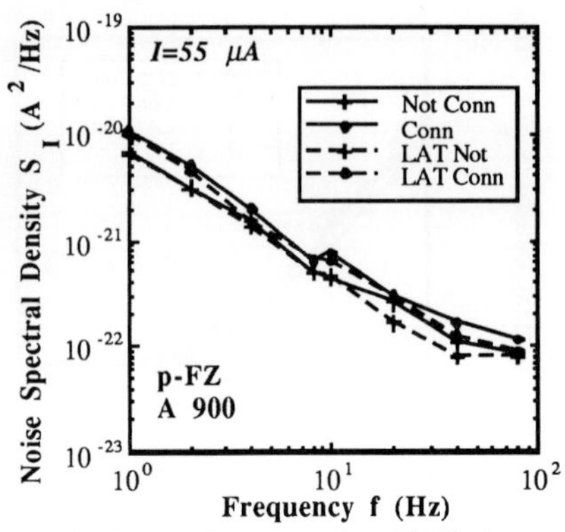

Fig. 9. LF noise spectra for a FZ diode, corresponding with different measurement configurations. I_F=55 μA.

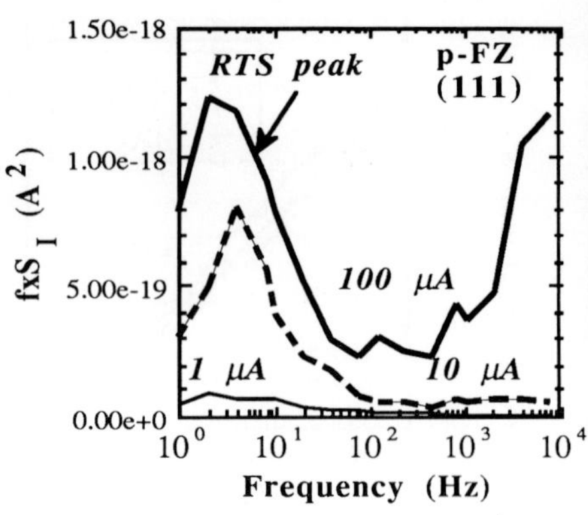

Fig. 10. LF noise spectrum for a (111) n^+p junction diode showing RTS, which is related to LOCOS edge dislocations.

the junction. It should finally be remarked that RTSs have also been observed after high-energy proton irradiations.

5. Conclusions

In summary, it has been clearly demonstrated that there exists a close relationship between the excess low-frequency noise spectral density of forward biased Si junction diodes on the one hand, and oxygen-precipitation induced extended defect density on the other. For high-oxygen Cz material, internal gettering can be useful to control the diode noise, particularly if the current is forced in a vertical direction. In addition, it has been demonstrated that other processing induced extended defects can also deteriorate the LF noise performance, through the ocurrence of RTSs.

Acknowledgements

The authors would like to express their gratitude to P Clauws, G Kissinger, A Kaniava, E Gaubas and A L P Rotondaro for the use of co-authored results and for many stimulating discussions. Part of this work has been financed by the European Space Agency (ESA) under Frame Contract 8615/90/NL/NB(SC) and in the framework of the Flemish-Polish bilateral Scientific Collaboration Agreement.

References

[1] Simoen E, Bosman G, Vanhellemont J and Claeys C 1995 *Appl. Phys. Lett.* **66** 2507-2509
[2] Simoen E, Vanhellemont J, Rotondaro A L P and Claeys C 1995 *Semicond. Sci. Technol.* **10** 1002-1008
[3] Vanhellemont J, Simoen E, Kaniava A, Libezny M and Claeys C 1995 *J. Appl. Phys.* **77** 5669-5676
[4] Vanhellemont J, Kissinger G, Clauws P, Kaniava A, Libezny M, Gaubas E, Simoen E, Richter H and Claeys C 1996 *Proc. GADEST '95*, Eds. Richter H, Kittler M and Claeys C (Switzerland : Scitec Publications) *Solid State Phenomena* **47-48** 229-234
[5] Vanhellemont J, Simoen E and Claeys C 1995 *Appl. Phys. Lett.* **66** 2894-2896
[6] McWhorter A L 1957 *Semiconductor Surface Physics* , Ed. Kingston R H (Philadelphia : University of Pennsylvania Press) pp. 107-208
[7] Hsu S T 1970 *Solid-State Electron* **13** 843-855
[8] Kleinpenning T G M 1980 *Physica B* **98** 289-299
[9] Lukyanchikova N B 1993 *Physics Letters A* **180** 285-288
[10] Kamarinos G, Mounib A, Minondo M and Jaussaud C 1995 *Proc. 13th Int. Conf. on Noise in Physical Systems and 1/f Fluctuations*, Eds. Bareikis V and Katilius R (Singapore : World Scientific) pp. 589-592
[11] Hsu S T, Whittier R J and Mead C A 1970 *Solid-State Electron* **13** 1055-1071
[12] Cook K B Jr and Brodersen A J 1971 *Solid-State Electron* **14** 1237-1250
[13] Nishida M 1973 *IEEE Trans Electron Devices* **ED-20** 221-226
[14] Blasquez G 1978 *Solid-State Eelectron* **21** 1425-1430

Inst. Phys. Conf. Ser. No 149
Paper presented at DRIP 95

Laser Scanning Tomography Studies of Lithium Niobate Crystals

D. Benhaddou and A. R. Mickelson

ECE department, University of Colorado, Boulder, CO 80309-0425, USA

Abstract. Some preliminary studies of light scattering in Czochralski grown lithium niobate wafers will be described. Motivation for this effort will be discussed as well as some plans for future studies. It is felt by the authors that integrated optics fabrication technology has progressed to the point that material purity is beginning to be the limit on device performance. The dearth of literature on defects in lithium niobate is therefore a limitation on ever considering how to improve the correct material characteristics.

1. Introduction

Lithium niobate is the material of choice when it comes to applications of electro-optic integrated optical (IO) devices. There are still problems hindering widespread commercial application of lithium niobate IO. A prime one is high cost. This high cost, however, is not material cost but processing cost associated with low manufacturing yield. It is our position that the high loss and low yield are related issues and that both relate to defects in the Czochralski grown substrates, and how these defects affect the results of subsequent processing steps.

Future progress in integrated optics technology is likely to be due to better control of the quality of optical material. A number of techniques that reveal defect structures (dislocations, impurity clusters, micro-precipitates etc.) in wafers has been developed. Light Scanning Tomography (LST) has been used to detect defects in quartz crystals [1,2] as well as used in semiconductors such as Si, GaAs, etc. [3,4].

Defect detection in lithium niobate is of outmost importance for researchers because the light loss in such material is significantly larger than in optical fibers. This loss limits the widespread application of $LiNbO_3$ in integrated optics.

We present preliminary results of LST studies of $LiNbO_3$ and trace the future application of LST in this material.

2. Experiment

Crystal imperfections, such as precipitates and dislocations, are generally visible in the LST images as sparkling spots, but sometimes forming larger structures with straight or meandery lines. The image represents the spatial distribution of scattering centers, which may include precipitates, grain boundaries and stacking faults.

The experimental setup is shown in figure 1. A red HeNe laser beam is focused with microscope objective on the lateral face of the sample. Owing to the transparency of LiNbO3 in visible domain and its refractive index, the laser beam enters the material as a quasiparallel beam illuminating the particles located along the path. This linear image is detected by a CCD camera through the upper face of the sample and it is digitized in a computer. The sample is moved in xy directions by stepper motor and images are then combined in computer [5]. The thickness of the tomographic plane is directly related to the size of the laser spot.

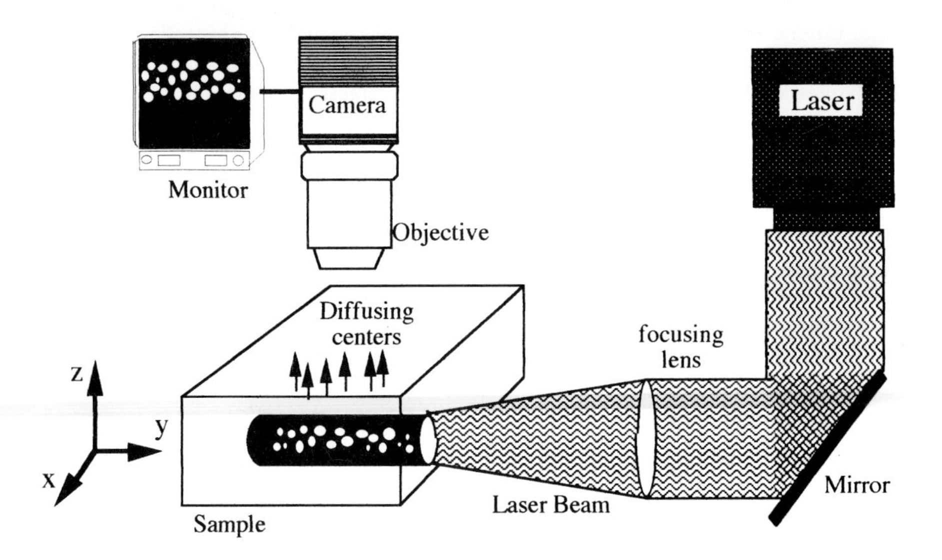

Characteristics :

COHU CCD Camera : 754 (H) X 488 (V)
Maximum of resolution : 0.3 (H) X 0.4 (V)
Sensitivity : High sensitivity due to the technique of dark field observation and high sensitivity of the camera (0.007 fc/ 0.07 lux)
Maximum density of defects detected is about 9 $10^{12}/cm^3$

Fig. 1 : LST experimental setup

LST images of a LiNbO3 wafer are shown in figures 2 and 3. Both pictures are obtained from the same sample in different locations. These defects appear as sharp sparkling spots and reproduce the cellular structure of the substrate dislocation network.

Figure 2 : First result of LST on LiNbO3

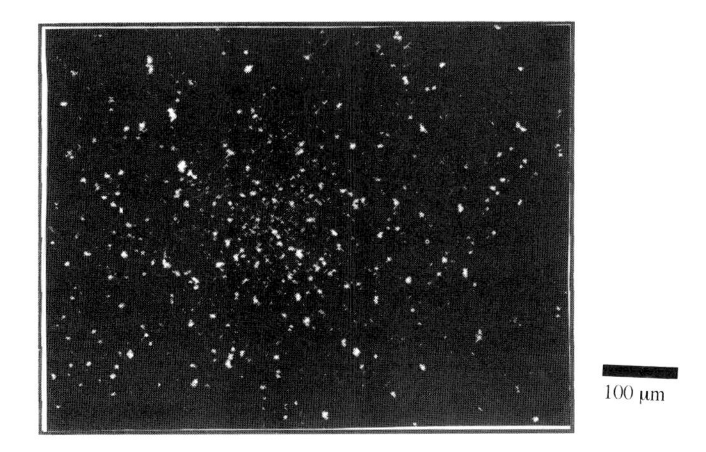

Figure 2 : First result of LST on LiNbO3

3. Interpretation

Typical objects that are observed by LST are micoprecipitates ; their size is usually below the micron range. The scattering is considered as Rayleigh scattering from scatterers which are generally associated with the dislocations. It is difficult to determine the shape of particles because of the diffraction limit. Nevertheless counting grain localization and their classification are of great interest.

Cells structures show individual precipitates on dislocations. Also, high density of defects can explain the high propagation losses. The yield of scattering is high, multiple scattering occurs which broadens the beam trace.

From a given image, the number of scattering centers can be calculated and therefore the intensity scattered by a cross section. Then the effect of this intensity on crosstalk can be estimated. However, since the light scattering mechanism is not well understood, studies must be undertaken to clarify the mechanism to obtain information from the scatterers as well as the defects.

4. Conclusion

The LST experiment will find applications in optical materials since the distribution of the scattering centers affect directly the interaction between light and these materials. It gives valuable information on the precipitates in transparent materials, no doubt they must play a decisive role in the specification of optical electro-optic IO. LST is effective and highly dependable for the analysis of LiNbO3.

5. references

[1] K. Morya and T. Ogawa, J. Crys. Growth 44 (1978) 53

[2] Lu Taijing, K. Toyoda, N. Nango and T. Ogawa, J. Crys. Growth 108 (1991) 482

[3] Proc. DRIP I, Montpellier, 1885 Ed J.P.. Fillard (Elsevier, Amsterdam, 1985)

[4] J.P. Fillard, P.C. Montgomery, A, Baroudi, J. Bonnafe and P. Gall, Japan, J. App. Phys 27 (1988) L258

[5] J.P. Fillard, P. Gall, J. Bonnafe, M. Castagne and T. Ogawa, Semcond. Sci. Technol. 7 (1992) A283-A287.

Inst. Phys. Conf. Ser. No 149
Paper presented at DRIP 95

Photon tunneling transfer through semiconductors with atomic force microscopy

P. Gall-Borrut, A. Bakro, M. Castagne, J. Bonnafé, J.P. Fillard

CEM- Université Montpellier II - 34095 Montpellier Cedex5 -France

Abstract We present the coupling of an Atomic Force Microscope with the photon tunneling effect in order to characterize semiconductor based devices. The experimental apparatus as well as the sample are described. In order to be sure that the photon tunneling effect effectively takes place the variation of the transmitted light through the AFM cantilever is registrated according to the distance of the tip to the sample. By this means topographic and photonic images are achieved. A good agreement is found between the topographic and the photonic results that open the field for spectroscopic studies.

1. Introduction

Nanotechnology applied to semiconductors has made it possible to achieve devices in the range of dimensions of 10 to 100 nm. Now it is necessary not only to characterize the topography of these devices but also to provide information about the subsurface. No universal method exists that provides all this information at the same time, so we have to use different techniques and/or to combine them.
Nowadays, Atomic Force Microscopy is, with S.T.M. [1] [2], one of the most popular techniques for obtaining topographic images at the nanoscale. On the other hand, the characterization of the subsurface not to be destructive must of the optical type and workers have successfully used photons in the place of electrons as tunneling particles [3][4][5].
As useful information will be provided by the combination of topographic and optical techniques, we have coupled an Atomic Force Microscope with an optical near field experiment.

2.Experimental

2.1. Experimental set-up

An atomic force microscope (Autoprobe CP from PSI) is used in the contact mode with a silicon tip on a cantilever of high stiffness (50 nN.m); the backside of the cantilever is not gold coated in order for the light transmitted through the tip not to be reflected inside the tip.

As shown in figure 1, an optical set up is adapted on the Autoprobe : a mirror and a microscope objective allows the focusing of a YAG laser (1,06 μm) beam onto the bevel of the sample. The resulting beam in the sample is quite parallel with a diameter of around 80 μm.

After tunneling, the infrared light transmitted at the back of the cantilever is collected by a microscope objective and an image (of the backface of the cantilever and of the sample's surface beneath) is formed on the target of a CCD camera.

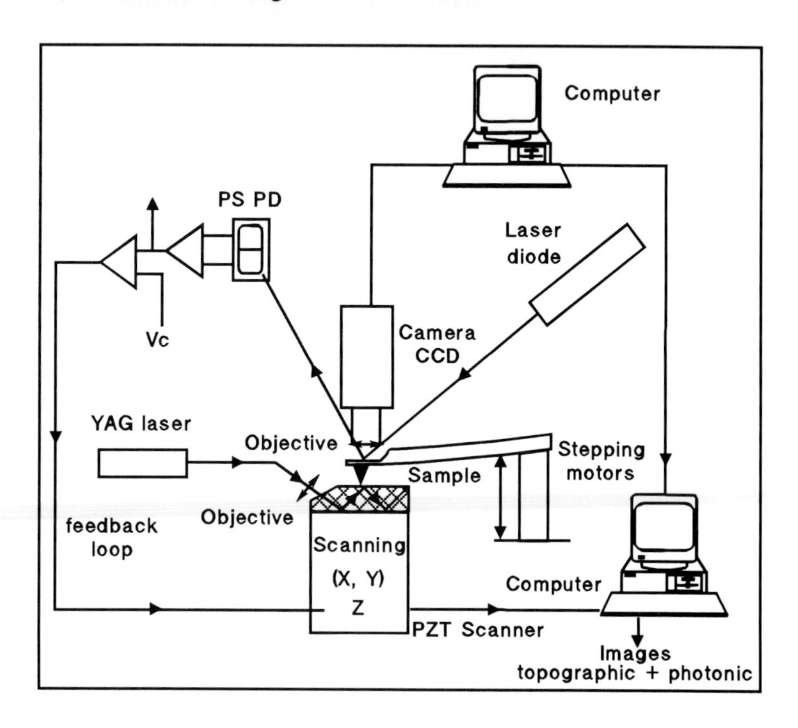

figure 1 : Experimental set-up

2.2. Sample

The sample is iron doped InP. It is cut in a 4 mm thick wafer . An array of gold patterns (25 nm thick - lateral dimensions of a few microns) has been evaporated onto the surface. A bevel has been made on one edge of the wafer with an angle of 20°. This makes it possible to inject the light in the sample in order to provide a succession of total internal reflections (the sample behaves as a light guide). The incident angle of the beam in the sample is 25°.

2.3. Experimental principle and procedure

The surface of the sample is illuminated from the inside with an incidence angle larger than the limit for total reflection (17° in InP); the classical theory of refraction shows that an evanescent wave stands at the surface in free space.The intensity of the wave decreases exponentially with the distance to the surface [6].

With the help of the camera, the tip is brought into the area of total reflection at a very small distance to the surface ; the bound electrons of the atoms at the tip are excited by the evanescent wave and their collective vibrations give rise to a propagating field in the direction of the incident photon . This field and the corresponding photons propagate in the tip and are transmitted at the back of the cantilever [7].

The video signal provided by the camera is digitized and the area of interest, corresponding to the illuminated pixels corresponding to the back of the cantilever, is defined: the average value of the grey levels of the pixels gives rise to the optical signal .

The z position of the tip is controlled using the force acting beween the tip and the sample surface (atomic force), the sample is moved back and forth according x and y with the piezoactuator (scanning). During the scanning, the classical A.F.M. topographic image is obtained at constant force as well as the photonic image.

3. Experimental results

In order to be sure that the light detected is not spurious scattered light and really comes from the near field photon transfer, we have looked for the decay of the light according to the distance tip-sample. The curves shown in figure 2 are obtained during extension and contraction according to the z of the piezoactuator. They show the exponential-like decrease of the transmitted light T, typical of photon tunneling tranfer [6]. As the cantilever is very stiff, there is no effect of snapping due to the aqueous layer on the surface. The classical d_C parameter (around 150 nm) we find is consistent with values found in previous works [6].

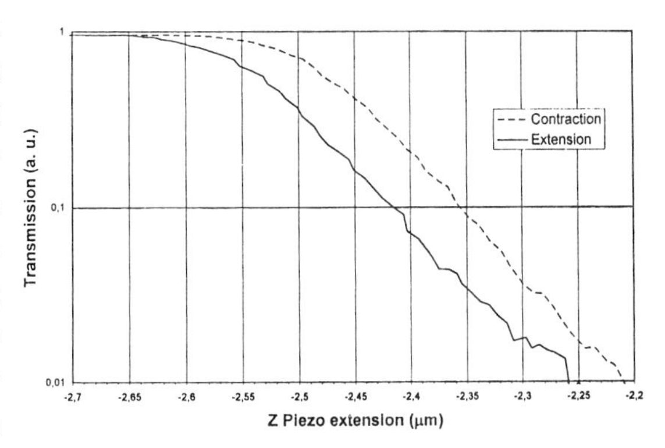

Figure 2 : Transmitted light versus piezo (sample) z position

Topographic and photonic images of figure 3 are obtained at the same time with a scanning frequency (fast scan direction) of 0,2 Hz. The tip is in contact with the sample, so the photonic image is obtained at z=0 and maps T_0[7]. We can see a clear correlation between the two images. The gold layer is higher and absorbs the light, so the corresponding zone appears brighter on the topographic image and darker on the photonic image. We can note that the gold layer is not homogeneous and the edges are not steep.

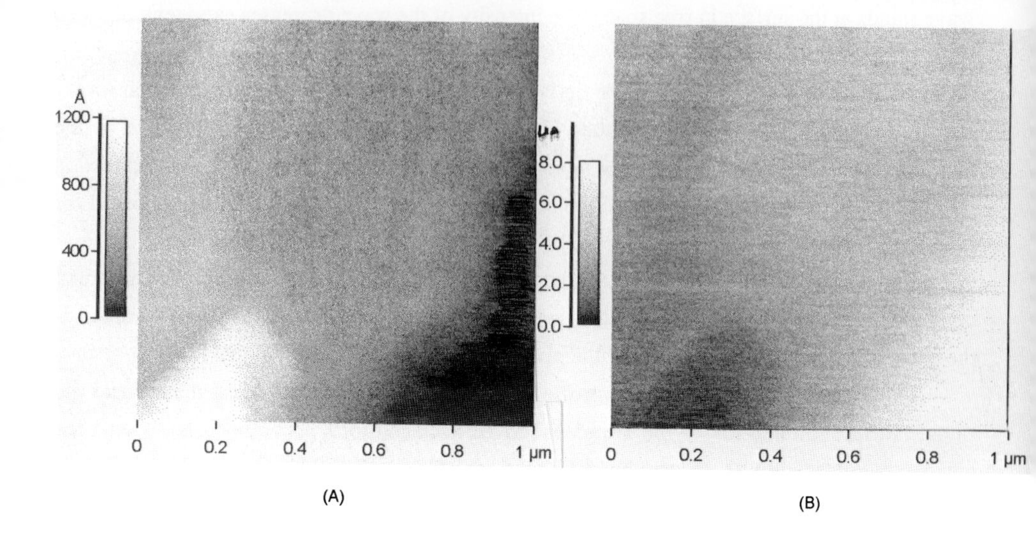

figure 3 : Topographic (a) and photonic (b) image of a gold layer on InP substrate

4. Conclusions

There is a good agreement between the topographic and photonic results. As the dimensions of the gold pattern are rather high, the limit in lateral definition of the photonic image has not been reached. We will have to work with nanometric patterns.

As we do not work in the conventional PSTM configuration, with a distance tip to sample controlled with the optical signal, but with a distance z=0, we minimize the contribution of the effect of topography of the surface in the images. This should lead more easily to spectroscopic results.

5. Acknowledgments

The authors thank R. Coquille (CNET Lannion) for providing the sample, J.C. Cano for the fabrication of the mechanical parts and also P. Mante.

References

[1] Binnig G and Röhrer H 1984 *Physica* **127 B** 37-45

[2] Binnig G, Quate C F and Gerber C H 1986 *Rev Lett* **56** 930

[3]Van Hulst NF, Segerink FB and Bolger B 1992 *Opt. Com.* **87** 212

[4] Reddick R C, Warmack R J and Ferrell T L 1989 *Phys Rev* **B39** 767

[5] Courjon D, Sarayeddine K and Spajer M 1989 *Opt. Com.* **71** 23

[6] Castagne M, Prioleau C and Baudry E1995 *SPIEConf "Sacnned Probe Microscopy III" San Jose* **2384** 186

[7] Castagne M, Prioleau C and Fillard JP 1995 *Appl. Optics* **34** n°4 703

Inst. Phys. Conf. Ser. No 149
Paper presented at DRIP 95

Comparative study of the topography of submicron opto-electronic device structures using coherence probe microscopy, SEM and AFM

P C Montgomery and P Gall-Borrut

Laboratoire LINCS, Centre d'Electronique de Montpellier, Université Montpellier II, Place Eugène Bataillon, 34095 Montpellier, France.

Abstract. Non-destructive analysis and measurement control in opto-electronic device fabrication requires new techniques that are fast, accurate and easy to use. In this paper we show how coherence probe microscopy, making use of the short coherence length of white light in an interference reflection microscope, can be used to this end for many applications requiring nanometric axial resolution, sub 0.5 µm lateral resolution and multi micron depth of field in a comparitive study with SEM and AFM.

1. Introduction

Non-destructive three dimensional geometrical analysis of modern microelectronic and opto-electronic devices requires the measurement of bars, blocks, grooves, epilayers and other more complicated micron sized structures with submicron and and even nanometric resolution (Chisholm). This is necessary both for optimising fabrication processes as well as in end product quality control. The combination of photonic and electronic elements on the same substrate increases the complexity of future telecommunication, signal processing and calculating devices, requiring a very demanding performance from the characterisation methods used. Analysis by SEM and AFM can provide the necessary resolution but often present limitations and constraints in certain practical fabrication situations.

Coherence probe microscopy, or CPM, (Davidson et al, Montgomery et al 1992) provides a new solution for high speed optical measurement, which can attain nanometric axial resolution and better than 0.3 µm lateral resolution, making it particularly useful in surface shape measurement. The method uses automatic interpretation of white light fringes in a reflection interference microscope for profiling surface topography.

This paper presents some preliminary results of the measurement of the topography of laser strucures, grooves and gratings with CPM and the comparison with the results of SEM and AFM. Some of the advantages and disadvantages of each technique are discussed.

2. Coherence probe microscopy

During the 1980's, phase shifting interferometry provided a convenient solution for the automatic interpretation of interference fringes in reflection microscopy using monochromatic and quasi-monochromatic illumination (Creath). While giving nanometric axial resolution, it suffered from a fundamental limitation in vertical range due to the periodicity of the fringes (Montgomery et al 1993). CPM, using white light interferometry, has virtually superseded this technique and greatly extended the applications because of the much larger axial range.

2.1 Profiling with white light interferometry

In white light interferometry, the temporal coherence of the light used is very short due to the wide spectrum of the source. For polychromatic light, the pattern of fringes can be described by the sum of the intensities produced by each wavelength component λ:

$$I_{\Delta}(x,y) = \sum_{\lambda 1}^{\lambda 2} 2I_o(\lambda)[1 + \cos(2\pi/\lambda.\Delta)]$$

where $I_o(\lambda)$ is the spectral distribution of the source and Δ is the optical path difference between the two beams. Some typical white light fringes are shown in fig 1, the shape of the envelope being described by the coherence function.

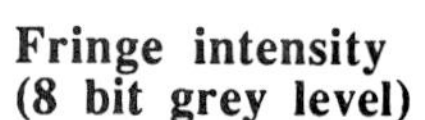

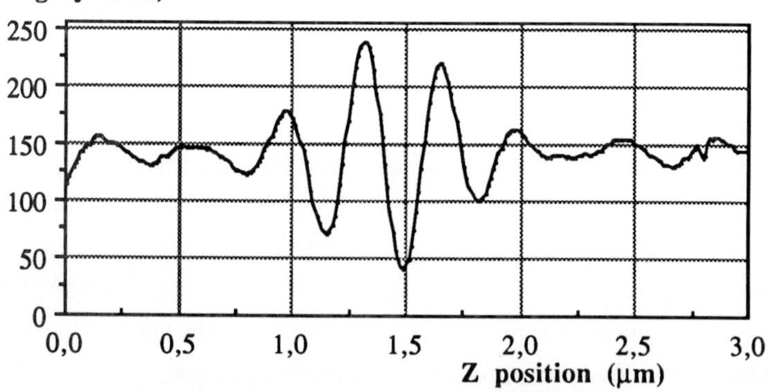

Figure 1 Profile of typical white light interference fringes

Using a specially designed algorithm, the peak of the coherence function can be determined with a high precision (Davidson et al, Montgomery et al 1992, De Groot 1995). If this is carried out at every pixel in an image, it is possible to reconstruct the 3D relief.

2.2 High resolution mode

The Linnik interferometer (figure 2) allows the use of large numerical aperture objectives (x100, 0.95) giving a lateral resolution of 0.3 µm with bluish light from a monochromator. For certain types of surface features, the CPM displays super-resolution (Davidson et al, Montgomery et al 1995) giving better than 0.3 µm lateral resolution.

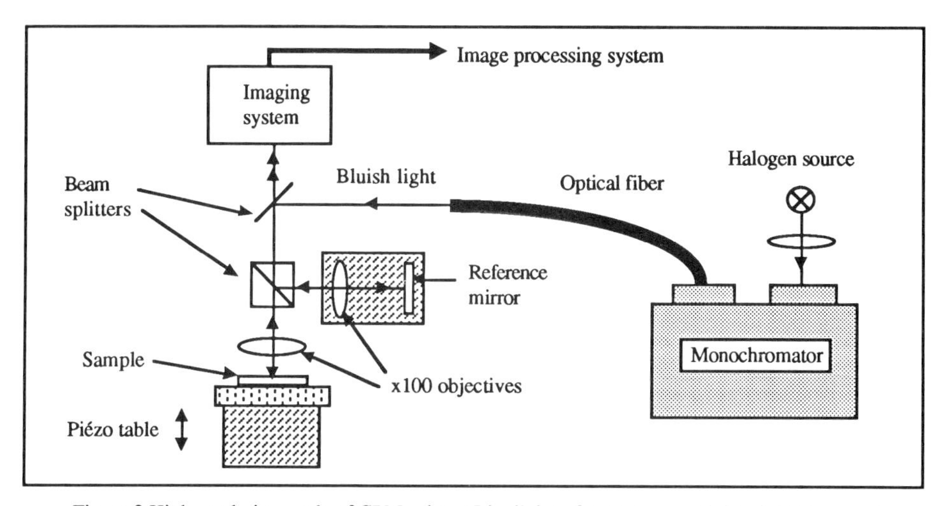

Figure 2 High resolution mode of CPM using a Linnik interferometer and bluish light illumination

Experiments have shown that it is possible to measure complicated surface shapes at high speed. With our own peak fringe scanning microscopy (PFSM) technique (Montgomery et al, 1992) measurement of typical opto-electronic samples of a few microns deep takes from a few seconds to a few minutes depending on the size of the image required.

3. Comparison of results on electro-optic structures

A particularly difficult problem in surface measurement is the profiling of deep narrow grooves. A typical example is the requirement for calibrating etching processes in the fabrication of ultra high speed lasers for frequency multiplexed optical telecommunication systems (Kazmierski et al). The "V" on "U" groove technique using (AP) MOVPE has been used successfully to give a very low RC-product value of 2 ps. It is important to be able to selectively etch the 2 µm wide "U" shaped groove down to the active layer stack through the quaternary mask.

Analysis of the grooves by SEM (figs 3(a) and 4(a)) gives good results but can only be achieved by cleaving the sample in order to image the cross sectional view. Using mechanical probes or near field scanning probes is difficult, in the latter case requiring great skill and slow scanning speeds. CPM on the other hand can be used quickly to give the etch depth required (fig 3(b)).

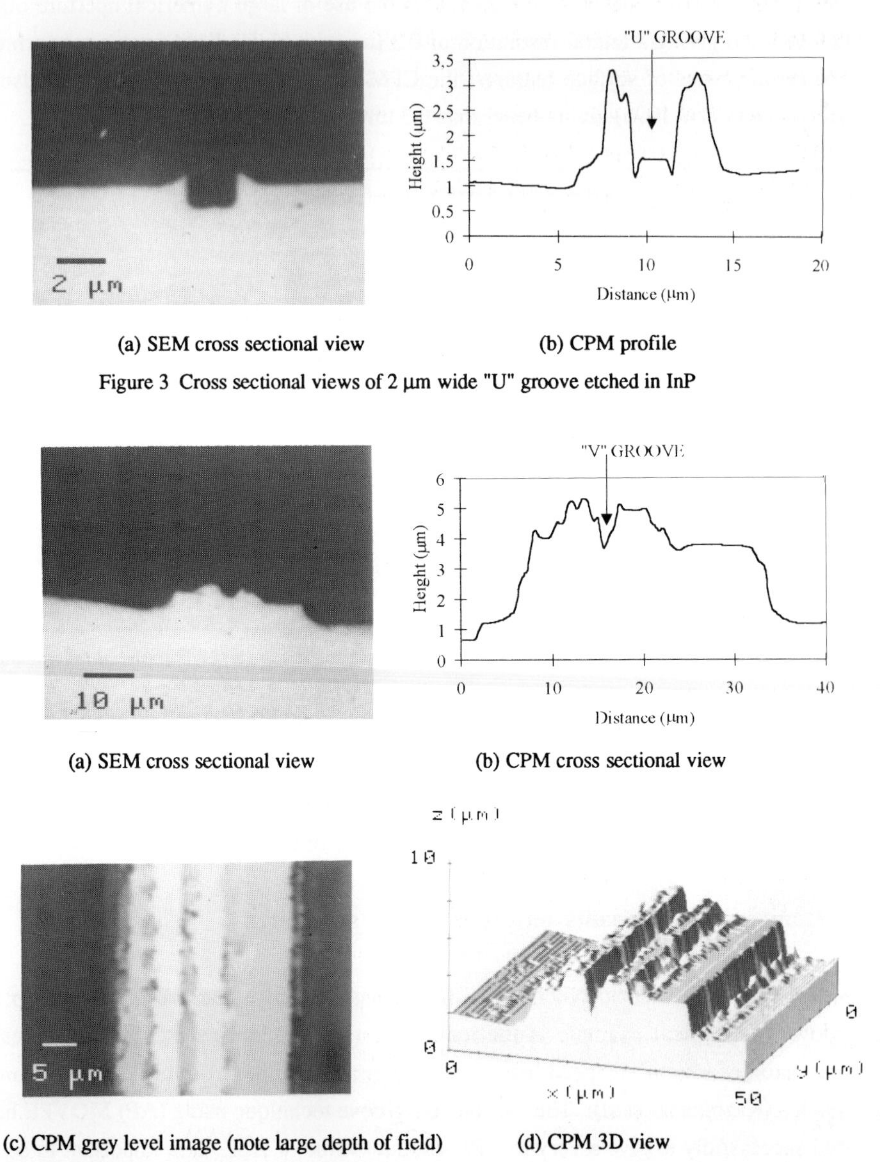

(a) SEM cross sectional view (b) CPM profile

Figure 3 Cross sectional views of 2 µm wide "U" groove etched in InP

(a) SEM cross sectional view (b) CPM cross sectional view

(c) CPM grey level image (note large depth of field) (d) CPM 3D view

Figure 4 Comparison of results between SEM and CPM of "V" groove in InP in ultra high speed laser

A qualitative appreciation of the etched relief is also important. This is difficult with classical high resolution optical reflection imaging because the large numerical aperture of the objectives limits the depth of field to less than 0.2 µm which is far less than the depth of the sample. For the laser cited, which is over 4 μm in depth, in any one image most of the details are out of focus. With CPM this problem is overcome because of the improved depth of field, making analysis of the top layer V shaped groove possible (see figs 4(c) and (d)); this shape is important to ensure a low series resistance and better thermal dissipation in the structure.

Finally we show the results on a 4800 Å period DFB (direct feedback) grating in InP in fig 5; this was done with the phase shifting technique (Montgomery et al, 1993). Although the grating is visible, the trapezoidal shape which is observable with AFM (fig 5(b)) is rounded because the lateral resolution is comparable to that of the period.

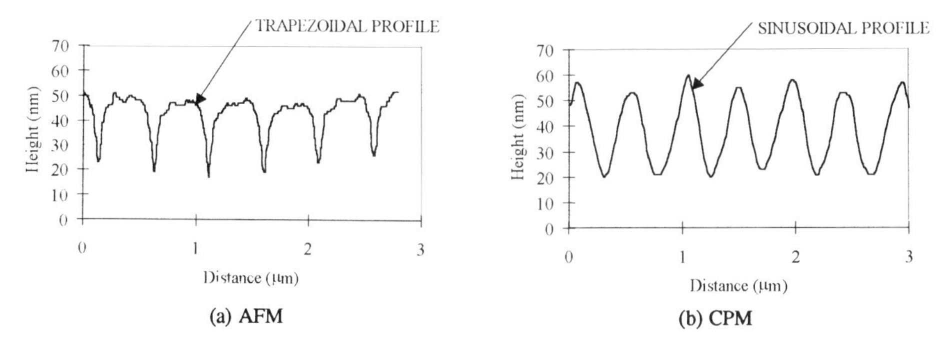

(a) AFM (b) CPM

Figure 5 Comparison of profiles of a 4800 Å period DFB grating in InP

4. Discussion of results

Table 1 gives a summary of the comparison between these different techniques. The main advantages of SEM and AFM are the greater lateral resolution, but as has been shown, CPM is suitable for many applications where 0.3 μm is sufficient.

TECHNIQUE	SEM (aerial view)	SEM (cross sectional view)	AFM	CPM
LATERAL RESOLUTION	<1 nm	<1 nm	<1 nm	<0.3 μm
AXIAL RESOLUTION	0.2 μm	<1 nm	<1 nm	<3 nm
EFFECTIVE TIME PER SAMPLE	2 to 3 hrs	2 to 3 hrs	10 to 20 mins	1 to 5 mins
DESTRUCTIVE-NESS	Sample metallized	Sample cleaved and metallized	None	None
SKILL LEVEL	High	High	Medium	Low
ARTEFACTS	Secondary emission from edges	None	Piezo non-linearities, tip shape convolution	Diffraction, low reflectivity
ADVANTAGES	Good qualitative view; high resolution	Good qualitative view of cross section	High resolution	Ease of use
DISADVANTAGES	Non-quantitative	Calibration required	Possible artefacts	Possible artefacts

Table 1: Comparison of techniques for analysing surface relief of opto-electronic devices

SEM gives very good qualitative views, but is non-quantitative where high precison is required. The other disadvantages of SEM are the destructiveness and the time involved. AFM and CPM have the advantage of being non-destructive. They are quantitative techniques but require careful interpretation because of the possible presence of artefacts, AFM being subject to non-linearities of the scanning system and the shape of the stylus probe, and CPM being prone to the effects from diffraction and low reflectivity due to high sample slope or the presence of inhomogeneous materials. CPM is easier to use than AFM, because of its similarities to a normal optical microscope and is quicker because of the parallel nature of the scanning probe compared with the stylus probe used in AFM.

6. Conclusions

In this paper we have compared the use of CPM, SEM and AFM in the relief measurement of electro-optic samples. SEM gives excellent qualitative results but cannot be used to quantify in high precision applications. AFM and CPM on the other hand give quantitative results, but which can contain the presence of artefacts, requiring careful interpretation. CPM is well adapted for many 3D measurement applications on opto-electronic devices at the sub-micron scale, because of its very high speed and ease of use. Further work is being carried out to better understand image formation and interpretation.

Acknowledgements

Thanks are extended to C Kazmierski of CNET (Bagneux) for providing the samples and the SEM results.

References

Chisholm B Yansen D and Morrill C 1989 Proc. Integrated circuit metrology, inspection, and process control III, SPIE **1087** p 360

Creath K 1988 Progress in optics **XXVI** (Elsevier science publishers) p 350

Davidson M Kaufman K Mazor I and Cohen F 1987 Proc. Integrated circuit metrology, inspection and process control, SPIE **775** p 233

De Groot P and Deck L 1995 J. Mod. Opt. **42** p 389

Kazmierski C 1992 J. Lightwave Tech. **10** p 1935

Montgomery P C and Fillard J P 1992 Proc. Interferometry: techniques and analysis, SPIE **1755** p 12

Montgomery P C Vabre P Montaner D and Fillard J P 1993 J. de Physique III **3** p 1791

Montgomery P C Lussert J M Vabre P and Benhaddou D 1995 SPIE **2412** p 88

Inst. Phys. Conf. Ser. No 149
Paper presented at DRIP 95

Linking Interface Quality to the Current-Voltage Characteristic of a Resonant Tunneling Diode

J S Hurley

Department of Engineering, Clark Atlanta University, Atlanta, Georgia USA

Abstract. The effect of interface quality on the current-voltage characteristic of an MBE-grown double barrier resonant tunneling (DBRT) diode is examined. The current density of a real diode is defined in terms of the current density of an ideal DBRT structure with planar interfaces and a normalized probability distribution which specifies the probability of finding at an arbitrary location on the front surface of the diode, a DBRT structure directly beneath the surface characterized by geometry, i. Differences between layer quality at inverted and normal heterointerfaces, neglected in a previous study, are now determined to significantly affect the form of the normalized distribution and ultimately the current-voltage characteristic of the real diode. Charge carriers injected into the DBRT structure are assumed to propagate coherently, and insensitive to irregularities in the structure's interfaces.

1. Introduction

Commercial and defense industries continue to demand faster and more efficient electronic devices to extend present data and information transmission and processing limits. Earnest efforts to address this technological challenge will require a better understanding of the nature of carrier movement through quantum well structures and devices. Investigations have shown that $Al_xGa_{1-x}As/GaAs/Al_xGa_{1-x}As$ quantum well systems experience significant deviations in transport characteristics due to numerous factors. For example, consequential reduction in mobility due to factors such as the: (i) order in which layers are grown [1]; impurity buildup during the growth of thick layers of $Al_xGa_{1-x}As$ [2,3]; (iii) impurity buildup during the growth of Al containing layers [3]; (iv) Si migration from the doped $Al_xGa_{1-x}As$ region into the GaAs layer either by diffusion or by segregation [4,5]; (v) strain resulting from the difference in lattice parameters of GaAs and $Al_xGa_{1-x}As$ [6]; and (vi) scattering which arises from band-edge discontinuity fluctuations at the $Al_xGa_{1-x}As$ heterojunction, have been identified [7].

In a previous study, a coherent transport model was used to determine the effect of interface roughness on the current-voltage (I-V) characteristic of a double barrier resonant tunneling (DBRT) structure and to evaluate whether interface roughness contributions could account for the approximately ten percent underestimation of the current at voltages exceeding the valley voltage, V_v, [8-12]. Based on the assumption that the charge carriers injected into the DBRT structure have a coherence length, l_c, that is much larger than the thickness of the structure's active region, W, but much shorter than the average linear dimension of regions composed of only one type of compound (e.g., GaAs or AlAs), Λ, it was concluded that interface roughness was not likely to change predicted characteristics in a significant way [12]. Hence, no substantial contributions to the I-V characteristic at voltages exceeding V_v, should be anticipated.

Notwithstanding, interface roughness is a suspected source of potential fluctuations at the inverted interface. The fluctuations are interpreted as carrier trapping centers that can reduce transport characteristics. Localization occurs when the magnitude of potential fluctuations exceeds the Fermi energy, μ, of the carriers. Recent studies have noted a significant difference between the quality of layers for "inverted" and "normal" heterointerfaces. An inverted heterointerface, the result of a GaAs layer grown on top of an $Al_xGa_{1-x}As$ layer, is found to be of lesser quality than a normal heterointerface, the result of an $Al_xGa_{1-x}As$ layer grown on top of a GaAs layer. Both interfaces exhibit asymmetrical properties attributed to the quality of the $Al_xGa_{1-x}As$ layer [3]. Differences in layer quality at the inverted and normal heterointerfaces, not incorporated into a previous study, were found to significantly alter the expression for the probability of outgrowth occurrence [12]. Thus, the negligible contributions attributed to interface roughness effects on the I-V characteristic may no longer be accurate.

2. Device Transport

Transport within the device is based on the assumption that charge carriers injected into the DBRT structure are governed by two conditions: (i) W (structure's active region) << l_c (coherence length); and (ii) l_c << Λ (average linear dimension of regions composed of only one compound, e.g., GaAs or AlAs). Conditions (i) and (ii) establish coherent propagation through the structure and insensitivity to interface outgrowths, respectively. It is additionally assumed that if τ (the characteristic lateral dimension of a planar DBRT structure on the order of magnitude of Λ), satisfies $l_c << \lambda$, then a real diode structure can be modeled in terms of a series of ideal diodes of varying geometries and planar interfaces connected in parallel [12]. Equations (1-3)

$$j_i(V)=\frac{em^*}{2\pi^2\tilde{h}^3}\{\theta(\tilde{\mu})\int^{\tilde{\mu}} dE_z T_i(E_z)(\tilde{\mu}-E_z) - \int^{\mu} dE_z T_i E_z(\mu-E_z)\} \tag{1}$$

$$j(V)=\sum_i P_i V_i \tag{2}$$

$$P_i=\lim_{A\to\infty}\frac{a_i}{A} \tag{3}$$

remain valid under the assumptions made in [12,13]. The parameters in (1-3) represent $\tilde{\mu} \equiv \mu - eV$, wherein μ is the Fermi energy of electron gas in the emitter contact, m* is the effective mass of the electron, $\theta(y)$ =1,0 for y > 0 and y < 0, respectively, V is the externally applied bias voltage, E_z is the longitudinal component of the electron's energy, $T_i(E_z)$ is the electron transmission coefficient, j(V) is the current density of the real diode, P_i is the normalized probability distribution function, j_i (V) is the current density in an ideal (perfectly planar interfaces) DBRT structure at T= 0°K, a_i is the section of the diode's cross sectional area that has geometry i , and A is the cross sectional area of the diode.

Equation (4) gives the binomial distribution for the probability of finding k deviations in the structure without consideration of the inverted and normal heterointerface distinctions [12]. The parameter, k, is used to describe the number of interfaces deviating from their nominal positions at a given transverse position of the diode, i.e.:

$$A_k = \frac{4!}{k!(4-k)!}(2p)^k(1-2p)^{4-k}, \qquad (4)$$

given that deviations on different interfaces are uncorrelated. Interface fluctuations or deviations from the ideal structures are limited to ± 1 monolayer ($\sim$ 2.83 $\mathring{A}$), in order to retain the validity of (2). The assumption that interface deviations could be described completely in terms of parameter p for $\pm$ deviations provided the foundation for (4)--fluctuations above and below the front interface could be assigned the same probability.

A "modified" A_k which takes into consideration the differences between "normal" and "inverted" heterointerfaces is now given by

$$MA_k = \frac{4!}{k!(4-k)!}(p+q)^k[1-(p+q)]^{4-k}, \qquad (5)$$

wherein p is used to represent the $Al_xGa_{1-x}As \rightarrow GaAs$ transition, while q is used to represent the $GaAs \rightarrow Al_xGa_{1-x}As$ transition.

3. Summary and Conclusion

The probabilities A_k and MA_k , can be significantly different even for the same values of p as shown in tables 1 and

2,; q variations must be included in the overall system assessment.

Table 1. Probability of Finding k Interface Deviations
(Layer Quality not Considered)

k	p	A_k
0	0.1	0.4096
	0.3	0.0256
	0.5	0
1	0.1	0.4096
	0.3	0.1586
	0.5	0
2	0.1	0.1536
	0.3	0.3456
	0.5	0
3	0.1	0.0256
	0.3	0.3456
	0.5	0
4	0.1	0.0016
	0.3	0.1296
	0.5	1

Obvious advantages exist if it is possible to define q in terms of p. Several authors have proposed methods to deal with interface fluctuations ranging from fractals to specifically designed scaling factors [14]. To date, the latter approach is most promising because of the need to replicate the material system's profile from a small region of the system [15]. Future consideration will be given to finding q in terms of p through a free energy-based term such as the surface tension (ν) which is related to h(**x**), deviation in height as a function of d-1 coordinates **x** in d-dimensional space, and a roughness term which connects the heterointerfaces [15,16]. Methodology will focus on determining small interface outgrowths about the perfectly planar interface (which is considered to be in equilibrium or the system's lowest energy configuration).

4. Acknowledgements

This work is supported in part by NASA contract NAG3-1394. I wish to take this time to thank J D Bruno for past communication, A. Sa'Di and A. Bassey for support in putting the document together, and C D Parker for suggestions.

References

[1] Airaksinen V M et al July/Aug 1988 J. Vac. Sci. Technol.B**6** No. 4 151
[2] Morko*c* H et al 1982 J. Electrochem. Soc. **129** 824
[3] Alexandre F et al 1986 Surf. Sci. **168** 454
[4] Inoue K et al 1985 Appl. Phys. Letts. **46** 973
[5] Gonzales L et al 1986 Appl. Phys. Letts. **41** 237
[6] Drummond T J et al 1983 Appl. Phys. Letts. **42** 615
[7] Cho N M et al 1987 Appl. Phys. Letts. **51** 1016
[8] Drummond T J et al 1982 J. Appl. Phys. **53** 3321; Miller R C et al 1982 Appl. Phys. Letts. **41** 374
[9] Tsu R and Esaki L 1973 Appl. Phys. Letts. 22 562
[10] Collins S et al 1988 J. Appl. Phys. **63** 142
[11] Sollner T C L G et al 1983 Appl. Phys. Letts. **43** 588
[12] Hurley J S and Bruno J D 1992 Superlattices and Microstructures **11** No. 1 23
[13] Bruno J D Unpublished notes
[14] Mandlebrot B 1982 The Fractal Geometry of Nature (San Francisco: Freeman)
[15] Mehran K and Hermann H and Roux S 1990 Disorder and Fracture (New York: Pleaum Press)
[16] Kardar M 1987 J. Appl. Phys. **61** 3601

Table 2. Probability of Finding k Interface Deviations
(Layer Quality Considered)

k	p	q	MA_k	k	p	q	MA_k
0	0.1	0.1	0.4096			0.3	0.1536
		0.3	0.1296			0.5	0
		0.5	0.0256	3	0.1	0.1	0.0256
	0.3	0.1	0.1296			0.3	0.1536
		0.3	0.0256			0.5	0.3456
		0.5	0.0625		0.3	0.1	0.1536
	0.5	0.1	0.0256			0.3	0.3456
		0.3	0.0625			0.5	0.4096
		0.5	0		0.5	0.1	0.3456
1	0.1	0.1	0.4096			0.3	0.4096
		0.3	0.3456			0.5	0
1	0.1	0.5	0.1536	4	0.1	0.1	0.0016
	0.3	0.1	0.3456			0.3	0.0256
		0.3	0.1536			0.5	0.1296
		0.5	0.0256		0.3	0.1	0.0256
	0.5	0.1	0.1536			0.3	0.1296
		0.3	0.0256			0.5	0.4096
		0.5	0		0.5	0.1	0.1296
2	0.1	0.1	0.1536			0.3	0.4096
		0.3	0.3456			0.5	1
		0.5	0.3456				
	0.3	0.1	0.3456				
		0.3	0.3456				
		0.5	0.1536				
	0.5	0.1	0.3456				

Inst. Phys. Conf. Ser. No 149
Paper presented at DRIP 95

Electrooptic sampling for measuring proton (H^+) exchanged induced defects in $LiNbO_3$

Paul D. Biernacki and Alan R. Mickelson

Department of Electrical & Computer Engineering, University of Colorado, Boulder, Colorado 80309-0425

Abstract. Electrooptic coefficients are sensitive to local crystal structure and therefore macroscopic defects. The electrooptic sampling technique can therefore determine some information concerning precipitates that may be formed during material processing. In past work, we can infer that electrooptic coefficients can be reduced by defect influenced diffusion of Ohmic contact material in GaAs. In this work, the mechanism of proton exchange (indiffused H^+ ions) via diffusion results in a disruption of the electrooptic coefficient in the LiNbO3 crystal which can then be detected by an electrooptic sampling probe.

1. Introduction

Although defect recognition has played an important role in the development of the semiconductor processing industry, not as much attention has been given to the processing of integrated optics devices in crystals such as $LiNbO_3$. Much of the high cost associated with integrated optics devices employing channel waveguides is due to the poor yield of such devices occurring at the processing level. An example of the key role defects have in channel waveguides could be the explanation of the reasonably high measured crosstalk found in fabricated 3 dB couplers.

Even though it is well known that H^+ indiffusion (proton exchange) destroys the electrooptic coefficient in $LiNbO_3$ [1], various processing methods have been employed to restore the electrooptic coefficient [2,3]. These researchers have shown that by reducing the proton concentration, restoration of the electrooptic coefficient can be achieved. It has not been emphasized, however, that defects present in the bulk $LiNbO_3$ could affect the restoration of the electrooptic coefficient or that these defects could eventually manifest themselves in overall poor device performance. In this paper electrooptic sampling is proposed as a technique for accurately measuring the reduction in the electrooptic coefficient in $LiNbO_3$ due to indiffused H^+ ions. Defects present in the crystal will also affect the overall electrooptic coefficient. The mechanism of proton exchange (indiffused H^+ ions) via diffusion results in a disruption of the electrooptic coefficient in the $LiNbO_3$ crystal which can then be detected by an electrooptic sampling probe.

2. Electrooptic Sampling For Defects Recognition

Since electrooptic sampling measurements are sensitive to the electrooptic coefficient, a lattice disruption (disordering) severe enough to affect the electrooptic coefficient should provide a significant perturbation in the modulated test optical pulse to be detected. This was seen in previous work [4]. This work concentrated on detecting the influence of diffusion induced defects due to the process of Ohmic contact formation.

Presented for the first time was the observation that the reverse case of using electrooptic sampling to provide accurate electrooptic coefficient measurements could be accomplished if the electric fields are known a priori. Usually the coefficients are assumed to be constant and the electric fields are mapped.

The test device was a coplanar waveguide fabricated on GaAs (an electrooptic substrate) as shown in fig. 1.

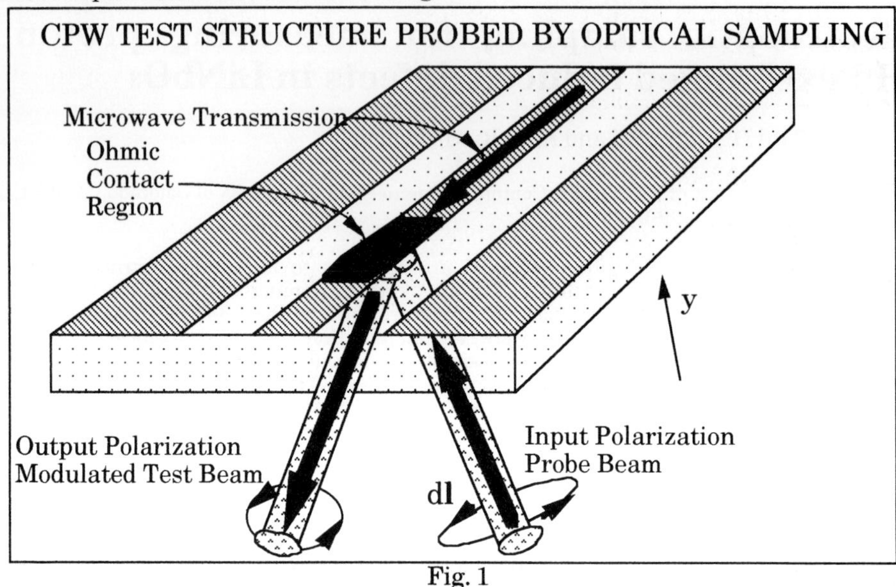

Fig. 1

The local electrical field along the coplanar wave guide (CPW) test circuit induces a birefringence in the electrooptic substrate. Electrooptic sampling then detects the modulated signal due to the rotation of the plane of polarization caused by this induced birefringence. More precisely, the electrooptic signal is proportional to the integrated electric field between the front and back surfaces of an electrooptic crystal (when a microwave field is applied along the CPW) weighted by the electrooptic coefficient. Therefore, any deviation in the electrooptic coefficient from its nominal value will affect the results of the electrooptic sampling measurements. This deviation was detected in the GaAs sample in certain regions where Ohmic contact material was diffused into the crystal substrate.

The detected voltage signal determined by the integrated field along a path normal to the surface when the electrooptic coefficient is not uniform throughout the crystal for GaAs can be expressed as

$$I_{det} \approx V \approx \int r_{41}(x,y,z) * \mathbf{E}_y(x,y,z) \cdot d\mathbf{l}.$$

Here the presence of a now spatially varying electrooptic coefficient is indicative of the defects influence on the coefficient.

The case of $LiNbO_3$ will however prove to be more complicated since many electrooptic coefficients are present causing both polarization and phase changes in the test optical probe beam. Generally the index ellipsoid for determining the change of the indices induced by an applied electric field is given for x, y, and z cuts of $LiNbO_3$ as

$$\left(\frac{1}{n_o^2} + r_{22}E_y + r_{13}E_z\right)y^2 + \left(\frac{1}{n_e^2} + r_{33}E_z\right)z^2 + 2r_{51}E_z yz = 1 \qquad \text{x cut}$$

$$\left(\frac{1}{n_o^2} - r_{22}E_y + r_{13}E_z\right)x^2 + \left(\frac{1}{n_e^2} + r_{33}E_z\right)z^2 + 2r_{51}E_z xz = 1 \qquad \text{y cut}$$

$$\left(\frac{1}{n_o^2} - r_{22}E_y + r_{13}E_x\right)x^2 + \left(\frac{1}{n_o^2} + r_{22}E_y + r_{13}E_z\right)y^2 - 2r_{22}E_x xy = 1 \qquad \text{z cut.}$$

However, for the transverse sampling geometry of the CPW test structure under the center conductor and for a y cut $LiNbO_3$ crystal, the sampling signal is mostly due to the r_{33} electrooptic coefficient ($r_{33}=31 * 10^{-12}$ m/V). Gap field information is obtained through the coefficients r_{22} and r_{13} [5]. Optical sampling for detecting the disruption of the electrooptic coefficients due to processing or bulk defects will now be explored for $LiNbO_3$.

3. Defects in $LiNbO_3$

Optical sampling for defect recognition is now discussed for the important integrated optic material $LiNbO_3$. Although a polarimeter was used in the final detection of GaAs, it is generally possible to do a mapping of more than one electrooptic coefficient in $LiNbO_3$ by employing an interferometer in the detection scheme. It should be noted that the optical sampling experiment requires a calibration of the detected signal but generally provides a precise relative measurement of the electrooptic coefficient. Figure 2 demonstrates the general optical sampling experimental arrangement for detecting the electrooptic coefficient influenced voltage amplitude signal.

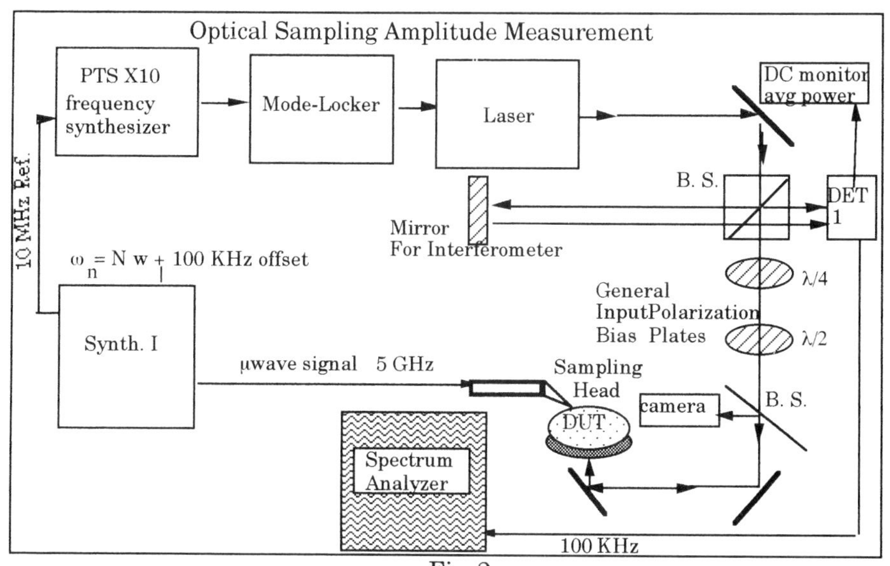

Fig. 2

The test device is a coplanar wave guide (CPW) structure consisting of alternating stripes 1 mm wide, separated by a 1mm spacing of H^+ indiffused regions transverse to the waveguide's propagation direction. This is shown in fig. 3. Since six hours of proton exchange was used we expect a destroyed electrooptic coefficient extending approximately 5 microns below the surface. However, defects will further influence the diffusion process of H^+ ions into the $LiNbO_3$ crystal. A simple phenomenological model based on a diffusion mechanism will then be used to explain the proton exchange process as well as the bulk defects' influence on the diffusing H^+ ions. Future work will include a precise laser scanning tomography technique to provide corroborating evidence of the size and location of the diffusion process.

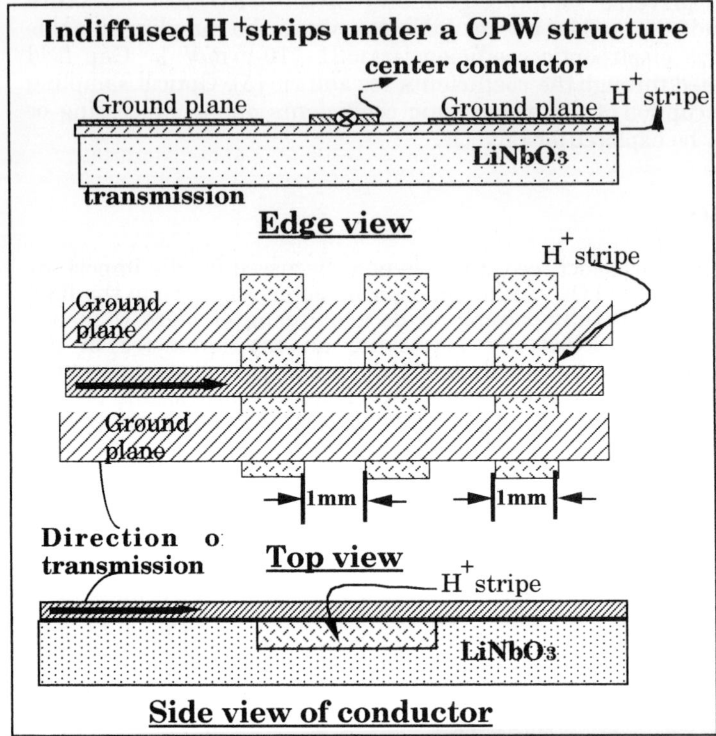

Fig. 3

Acknowledgments

The authors are grateful for the assistance of Wei Feng and Raghu Narayan for providing the mask and proton exchange for the $LiNbO_3$ CPW.

References

[1] R. A. Becker, "Comparison of Guided-Wave Interferometric Modulators on LiNbO3 via Ti Indiffusion and Proton Exchange," *Appl. Phys. Lett.*, **43**(2), pp. 131-, (1983).

[2] P. G. Suchowski, T. K. Findakly, and F. J. Leonberger, "Stable Low-Loss Proton Exchanged LiNbO3 Waveguide Devices With No Electro-Optic Degradation." *Opt. Lett.*, **13**(11), pp. 1050-1052, (1988)

[3] A. Loni, R. M. De La Rue, "Very Low Loss Proton Exchanged LiNbO3 Waveguides with a Substantially Restored Electro-Optic Effect," *Integrated and Guided Wave Optics Technical Digest Series,* **vol. 5**, Paper MD3, pp. 84-87, (1989).

[4] Paul D. Biernacki, Henry Lee, and Alan R. Mickelson, "Evaluation of Defect Related Diffusion in Semiconductors by Electrooptic Sampling," accepted IEEE J. Quantum. Electronics.

[5] D. R. Hjelme, A. R. Mickelson, "Voltage Calibration of the Direct Electrooptic Sampling Technique," IEEE J.Trans. Microwave Theory Tech., 40, no. 10, pp. 10 1941-1950, 1992.

Inst. Phys. Conf. Ser. No 149
Paper presented at DRIP 95

Investigations on Epitaxial and Processed InP by Optical Photoreflectance Spectroscopy

S Hildebrandt, J Schreiber, R Kuzmenko, A Gansha, W Kircher and L Höring

Martin–Luther–Universität, Fachbereich Physik, Friedemann–Bach–Platz 6, D–06108 Halle, Germany

Abstract. Photoreflectance (PR) modulation spectroscopy is a widely used optical technique on GaAs but it has been applied much more rarely on InP being an equally important optoelectronic compound semiconductor. Typical PR spectral lineshapes in the fundamental gap region of various InP materials are investigated. Spectral components such as Franz–Keldysh oscillations, low–field features, and epilayer interference phenomena are analyzed. Current applications concern the determination of the surface electric field, investigations of ion etching and hydrogenation processing, and the characterization of homo- and strained heteroepitaxial layers.

1. Introduction

Photoreflectance (PR) modulation spectroscopy is being widely applied as a non–contact, non–destructive quantitative optical technique on semiconductors and is well established on GaAs [1]. However, in the past it has been used much less for indium phosphide [2,3,4,5,6] being an equally important compound semiconductor for a variety of promising micro– and optoelectronic applications such as laser diodes, high–speed transistors, integrated devices, quantum wells, and solar cells based on the InP, InGaAsP, and InAlAs material systems [7].

Therefore, the implementation of the PR tool with relatively high experimental certainty and spectral resolution under room temperature conditions is becoming increasingly important for the characterization of InP epitaxial layers and microstructures.

In this paper, we will give a survey of results of our current experimental and theoretical investigations on various InP bulk and epilayer samples in their as–grown as well as processed state. We will show which typical spectral lineshapes are observed and how critical material parameters can be obtained by the use of a suitable interpretation scheme based on a multi–component non–linear fitting procedure, pump power, frequency, and wavelength dependent investigations as well as the synchronous phase analysis [8,9].

2. Results and analysis

2.1. Typical PR lineshapes from InP

As principal spectral lineshapes observed in E_0 PR spectra from InP are similar to those

Table 1. Representative lifetime broadening parameter values determined from various InP samples

Sample	Γ (meV) (E_0)	Γ (meV) (E_0+Δ_0)
n–InP:Sn (n = 3.8 · 10^{17} cm^{-3})	66.6	71.6
p–InP:Zn (p = 1.0 · 10^{18} cm^{-3})	49.7	46.1
SI–InP:Fe	15.1	–
n–InP/SI–InP (n = 4 · 10^{15} cm^{-3})	3.0	–
n–InP/SI–InP (n = 8.2 · 10^{16} cm^{-3})	8.7	14.8
InP/Si (n = 7.2 · 10^{17} cm^{-3})	8.0	19.8
InP/Si (n = 9.7 · 10^{16} cm^{-3})	10.4	–
InP/Si (LF spectra)	23 ± 5	–

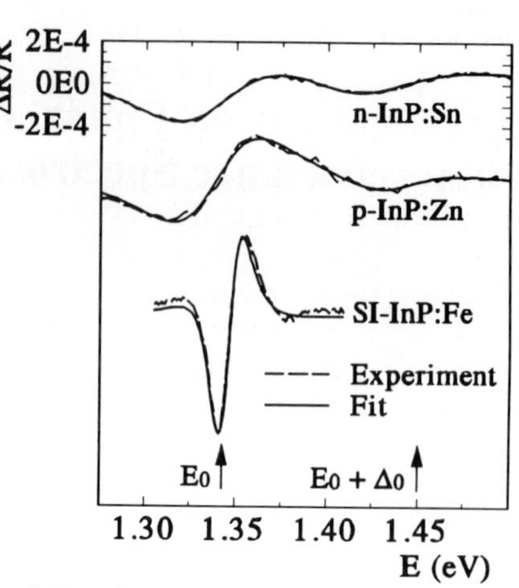

Fig. 1. Experimental and fitted LF PR spectra from bulk InP. Top: n–InP:Sn (n = 3.8 · 10^{17} cm^{-3}), middle: p–InP:Zn (p = 1 · 10^{18} cm^{-3}), bottom: SI–InP:Fe (n = 3 · 10^{8} cm^{-3})

from GaAs, a general complication is the small spin–orbit (s–o) splitting Δ_0 = 0.11 eV compared to Δ_0 = 0.34 eV for GaAs. Thus, spectral parts from the fundamental gap E_0 = 1.34 eV may often be superimposed with a feature originating from the $E_0 + \Delta_0$ = 1.45 eV transition.

A further problem of the application for the determination of the surface electric field F_s consists in the fact that often only a rather weak low–field (LF) signal is observed. This might point at smaller F_s values and/or larger spectral lifetime broadening Γ and less effective modulation than in comparable GaAs samples.

Fig. 1 shows typical PR spectra from n– and p–type as well as semiinsulating (SI) InP bulk material. LF lineshapes for both E_0 and $E_0 + \Delta_0$ are visible which can be analyzed using Aspnes' TDFF (third–derivative functional form) theory [10]. Representative Γ values are compiled in Table 1; transition energies were reproduced as given above. The rather narrow lineshape and small Γ from the SI sample corresponds to the vanishing free carrier concentration in this sample. The character of this spectrum is primarily excitonic and could hence be fitted with a two–dimensional TDFF at room temperature [11]. Generally, it has been observed that as–grown n–type substrates in the doping range n = $10^{16...18}$ cm^{-3} yield rather little LF PR spectra or even no detectable signal. This is in contrast to n–GaAs where medium field (MF) spectra with moderately damped Franz–Keldysh oscillations (FKO) are visible at least up to the 10^{17} cm^{-3} range [8,12].

The p–InP:Zn (p = 10^{18} cm^{-3}) material also shows LF behaviour which, however, changes into medium–field (MF) spectra for lower carrier concentrations (see 2.3. and [13]). The larger signal amplitude on p–InP hints at a higher surface field as on n–InP corresponding to a surface Fermi level position near to the conduction band edge [14].

On the other hand, distinct MF PR spectra with additional features could be measured on various homo– or heteroepitaxial layer samples grown by molecular beam epitaxy (MBE) or metalorganic chemical vapour deposition (MOCVD) even if the layers are nominally undoped (n = $10^{15...16}$ cm^{-3}). This behaviour may be ascribed to the better crystal quality of the epilayers which has a more pronounced effect here than on GaAs.

The sequence of spectra in Fig. 2 demonstrates the evolution of the complex MF lineshape on InP epitaxial layers. The upper spectrum from an InP/Si heterostructure shows only a few damped FKOs. A value of F_s = 2.9 · 10^{6} V/m is obtained. The small number of oscillations might make the determination of the field more difficult but the fit shows that the experimental lineshape is regular and the graphical line plot method (see inset) [15] may be used to estimate F_s. Here, contributions from E_0 and $E_0 + \Delta_0$ are well separated, and the FKO component from the s–o split transition may be fitted in the same manner. The next spectrum has a more extended FKO structure where the change in the oscillation period due to the interference with the $E_0 + \Delta_0$ feature above 1.4 eV is clearly seen. A reasonable fit (F_s = 3.8 · 10^6 V/m) is achieved whereas the line plot shows deviations in the higher–energetic region. Furthermore, the kink below 1.35 eV indicates the superposition with the E_0 excitonic LF component [16] dominating the main peak region. The third example is again a spectrum from InP–on–Si chemically etched to a residual thickness of 700 nm. Here, the peak structure is further broadened, and the remaining FKOs fully interfere with the s–o transition. It is not easy to fit this kind of spectrum where field and broadening are both increased due to the higher carrier concentration in this sample presumably from indiffused Si. In this higher–doped sample, the vertical inhomogeneity of the field in the space charge region has to be taken into account in the modelling giving rise to the negative pre–structure below E_0 [17].

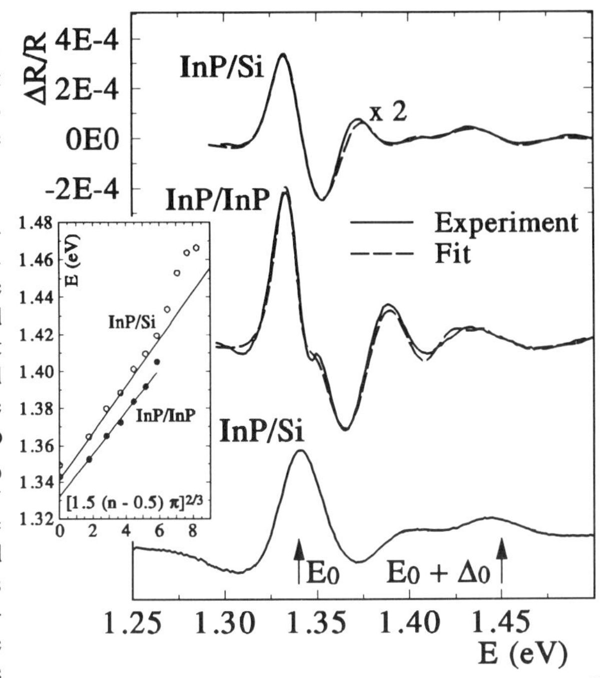

Fig. 2. MF spectra from n–InP epilayers. Top: InP/Si (n = 7.2 · 10^{17} cm^{-3}, d = 2.6 μm), middle: InP/SI–InP (n = 8.2 · 10^{16} cm^{-3}, d = 2.4 μm), bottom: InP/Si (d ~ 700 nm)

It has been shown that the PR lineshapes in InP are dominated by transitions with participation of the heavy hole (hh) valence band states [5,6]. The MF amplitude ratio of heavy and light hole (lh) contributions has been determined to $a_{lh}/a_{hh} \approx 0.2$ close to the theoretically expected figure. Many InP PR spectra may be well fitted using the hh component alone; however, taking $a_{lh}/a_{hh} > 0.25$ considerably worsens the fit quality. In GaAs, lh states have been shown to provide a larger contribution to the lineshape which, for example, gives rise to interference beats in the FKO region [18].

2.2. Fundamentals of the model and spectra analysis

Once the recorded experimental PR lineshapes have been identified with respect to their spectral components such as FKOs, sharp LF exciton–related features near E_0 as well as low–energetic interference oscillations (LEIO) in epitaxial layers (see 2.3.), the spectra may be analyzed quantitatively as shown above in the examples of Figs. 1 and 2. This is performed by means of extended simulation models based on the electroreflectance (ER) and PR theory from which multi–component non–linear regression routines have been derived.

The theoretical fundamentals of the numerical PR simulation have been given elsewhere [8]. There, the MF model starts from the broadened electrooptic functions G and F [19]. Their oscillation period depends on the electrooptic energy $\hbar\Omega = [e^2 \hbar^2 F_s^2/8 \mu]^{1/3}$ where μ is the reduced effective mass in field direction. The model may include both field inhomogeneity (by the use of a multilayer algorithm) and partial field modulation. Several components in superimposed spectra can be fitted simultaneously to the experimental data.

2.3. Application examples

All spectra shown in this paper were measured using 40 mW red HeNe laser modulation, unless otherwise indicated, and monochromatic illumination from a tungsten probe.

Since the PR signal reacts sensitive to changes in the electronic and optical properties, it can be conveniently used to investigate surface modifications and passivation treatments. In the present work, the effect of reactive ion etching (RIE) as a key technology for surface processing is studied on bulk p–InP. $CH_4/H_2/Ar$ is a well suited plasma system for ion etching of InP known as metalorganic RIE. Also, the separate influence of the three etch gas components as well as that of an O_2 plasma was investigated.

Fig. 3 illustrates the PR results before and after CH_4, H_2, Ar, $CH_4/H_2/Ar$, and O_2 plasma etching. The broad LF response of the untreated p–InP material transforms into MF spectra in the post–etch state. This is connected to a decrease in both the surface electric field and the broadening (see inset) which can be correlated to the carrier concentration given by C–V measurements. Whereas for O_2 and CH_4 etching the initial carrier concentration of $p = 1 \cdot 10^{18}$ cm^{-3} is not reduced, values of $2.4 \cdot 10^{17}$ cm^{-3} ($CH_4/H_2/Ar$), $1 \cdot 10^{16}$ cm^{-3} (H_2), and $(0.5 \ldots 1) \cdot 10^{17}$ cm^{-3} (Ar) were derived. Here, the hydrogen acceptor passivation [20] plays an important role together with the formation of sputter–related traps and material removal by the chemical component. Besides the free carriers, these traps also contribute to the broadening value. Rapid thermal annealing for 5 min at 300°C partly restores the free carriers ($p \approx 6 \cdot 10^{17}$ cm^{-3}) except for the Ar–etched sample ($1 \cdot 10^{16}$ cm^{-3}) but the MF condition is retained with increased fields due to a simultaneous rise in the surface potential. It shows that the carrier density degradation caused by Ar^+ is damage–related (also reflected by Γ) whereas the hydrogenation effect is not thermally stable. In this way, we are able to evaluate the complex electronic and optical material modifications during the optimum 100 W $CH_4/H_2/Ar$ RIE in more detail.

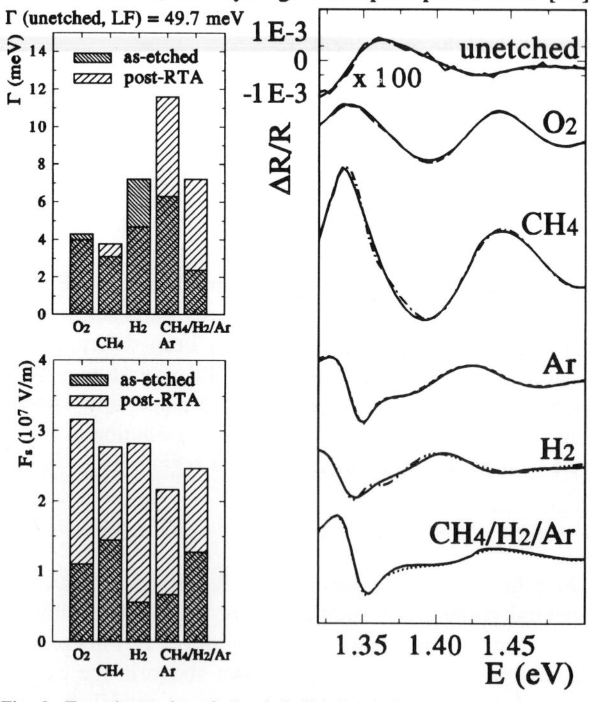

Fig. 3. Experimental and fitted (solid lines) PR spectra before and after O_2, $CH_4/H_2/Ar$, CH_4, H_2, and Ar ion etching (plasma power 100 W, pressure 10 Pa, etching time 5 min). Bar plots: fit parameters

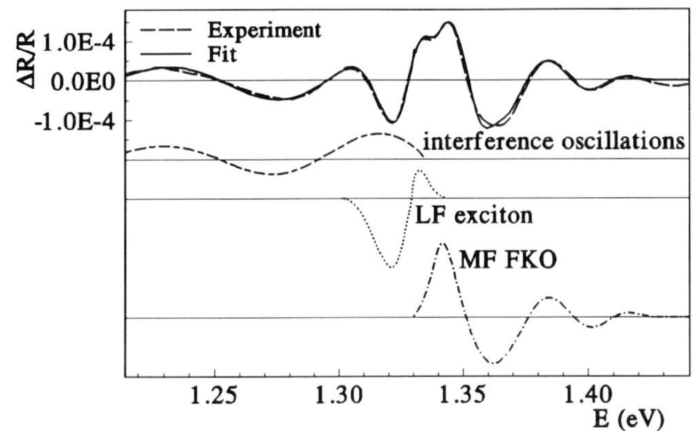

Fig. 4. PR from n–InP/SI–InP heterostructure. Experimental spectrum (2500 Hz HeNe laser modulation, λ = 633 nm) and fit with FKO (F_s = 3.12 · 10^6 V/m, Γ = 3 meV), LF (Γ = 9 meV) and LEIO component (d = 1728 nm)

These experimental results in connection with routine quantitative analysis methods demonstrate the potential of the optical PR spectroscopy also for on–line and in–situ monitoring of the InP surface region under realistic process conditions.

Further investigations were directed towards the characterization of homoepitaxial InP layers and strained–layer–heteroepitaxial structures of InP on Si. Weakly broadened MF spectra were recorded for n–type layers with carrier concentrations from 5 · 10^{15} cm^{-3} up to 7 · 10^{17} cm^{-3}. Being a main application of the PR spectroscopy in the MF range, the determination of the surface electric field from the oscillation period of the FKO could be used for the evaluation of the band bending eV_s at the surface. The present results yield values of eV_s < 0.1 eV for the air–exposed n–InP epilayer surfaces. Broadening parameters were found in the range 3 ... 9 meV for InP/InP and 5 ... 11 meV for InP/Si where a particular trend with the doping level was not discernible.

A multi–component spectral structure can be observed if the underlying InP substrate material has a carrier concentration which differs from that of the epilayer (Fig. 4). Due to the refractive index change, a signal modulated inside the layer or at the interface is back–reflected from the interface and interferes with the light reflected directly at the epilayer surface. These low–energetic interference oscillations (LEIO) have been investigated earlier on GaAs [8,21]. Here, they were detected on n–InP layers on SI substrate as n– or p–type substrate did not produce this kind of spectral structure.

From the PR measurements on homoepitaxial InP, the known gap energy of E_0 = (1.343 ± 0.002) eV is confirmed. The high spectral resolution even under room temperature conditions favours the use of this technique for the identification of deformation–induced spectral structures [6]. From a comparison of InP/InP PR spectra with (001) n–InP/Si samples (Fig. 5), a low–energetic shift of the hh–dominated PR lineshape to an average transition energy E_0(strained) = (1.334 ± 0.003) eV was derived. This corresponds to a residual thermally induced biaxial tensile stress of (0.74 ± 0.27) · 10^8 Pa. It should be noted that this interpretation is contrary to the lh–hh splitting model used earlier since no indication of peak splitting or stress–induced lineshape modifications were seen in the MF spectra except for increased fields caused by unintentional doping. LF spectra were also recorded on InP/Si layers with n < 5 · 10^{17} cm^{-3}.

3. Conclusions

The present work contains a compilation of PR results from n– and p–InP substrate, layers, and treated surfaces. The advantage of fitting procedures for analysing a number of relevant material parameters as well as of a suitable methodical approach assisted by synchronous phase and modulation power dependent investigations [8,9] is emphasized.

It is hoped that the present examples may contribute to a better theoretical understanding and experimental verification of PR spectral structures from InP.

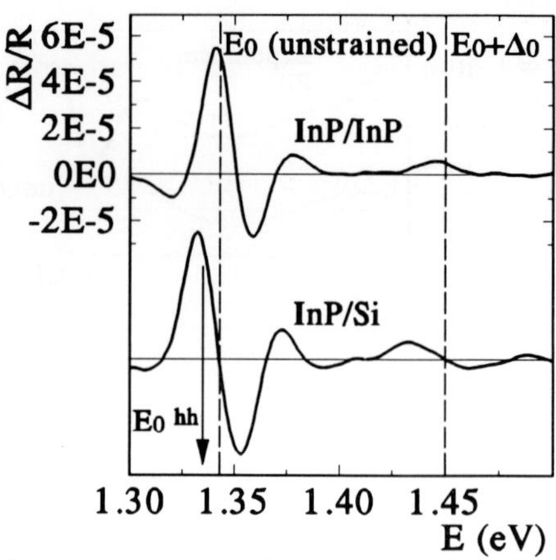

Fig. 5. MF PR spectra from InP/InP and strained InP/Si samples

Acknowledgements

The RIE investigations were performed at the National Microelectronic Research Centre Cork, Ireland, Materials Group, G M Crean, in cooperation with the Technische Hochschule Darmstadt, Germany, Institut für Hochfrequenztechnik. The authors would also like to thank A Schlachetzki from the Technische Universität Braunschweig, Germany, Institut für Halbleitertechnik, for supplying some of the InP epitaxial layers and for the joint work financially supported by the Deutsche Forschungsgemeinschaft (DFG).

References

[1] Pollak F H and Shen H 1993 *Mater. Sci. Eng. R* **10** 275

[2] Bhattacharya R N, Shen H, Parayanthal P, Pollak F H, Coutts T and Aharoni H 1988 *Phys. Rev. B* **37** 4044

[3] Berry A K, Gaskill D K, Stauf G T and Bottka N 1991 *Appl. Phys. Lett.* **58** 2824

[4] Zhou W, Dutta M, Shen H, Pamulapati J, Bennett B R, Perry C H and Weyburne D W 1993 *J. Appl. Phys.* **73** 1266

[5] Estrera J P, Duncan W M and Glosser R 1994 *Phys. Rev. B* **49** 7281

[6] Kuzmenko R, Gansha A, Schreiber J, Kircher W, Hildebrandt S, Mo S and Peiner E 1995 *phys. stat. sol. (a)* **152** 133

[7] see, e g 1995 *Proc. 17th Int. Conf. on Indium Phosphide and Related Materials* (New York: IEEE)

[8] Hildebrandt S, Murtagh M, Kuzmenko R, Kircher W and Schreiber J 1995 *phys. stat. sol. (a)* **152** 147

[9] Schreiber J, Kircher W, Hildebrandt S, Gansha A and Kuzmenko R 1995 *Proc. HOLSOS '95* (Frascati), to be published

[10] Aspnes D E 1973 *Surf. Sci.* **37** 418

[11] Shanabrook B V, Glembocki O J and Beard W T 1987 *Phys. Rev. B* **35** 2540

[12] Lu C R, Andersson J R, Stone D R, Beard W T, Wilson R A, Kuech T F and Wright S L 1991 *Phys. Rev. B* **43** 11791

[13] He L, Shi Z Q and Anderson W A 1991 *Proc. 3rd Int. Conf. on Indium Phosphide and Related Materials* (New York: IEEE) 531

[14] Spicer W E, Chye P W, Skeath P R, Su C Y and Landau I 1979 *J. Vac. Sci. Technol.* **16** 1422

[15] Bottka N, Gaskill D K, Griffiths R J M, Bradley R R, Joyce T B, Ito C and McIntyre D 1988 *J. Cryst. Growth* **93** 481

[16] Silberstein R P and Pollak F H 1980 *Solid State Commun.* **33** 1131

[17] Batchelor A, Brown A C and Hamnett A 1990 *Phys. Rev. B* **41** 1401

[18] Van Hoof C, Deneffe K, DeBoeck J, Arent D J and Borghs G 1989 *Appl. Phys. Lett.* **54** 608

[19] Seraphin B O and Bottka N 1966 *Phys. Rev.* **145** 628

[20] Pearton S J, Corbett J W and Shi T S 1987 *Appl. Phys. A* **43** 153

[21] Kallergi N, Roughani B, Aubel J and Sundaram S 1990 *J. Appl. Phys.* **48** 4656

Inst. Phys. Conf. Ser. No 149
Paper presented at DRIP 95

Semiconductor defect characterization in the Scanning Electron Microscope

H.-U. Habermeier

Max-Planck-Intitut für Festkörperforschung, Heisenbergstr. 1, D 70569 Stuttgart, Germany

Abstract. Defects in semiconductors usually act as sites of increased recombination and thus have a strong influence on the performance of microelectronic and optoelectronic devices. Among the variety of other techniques, both, the EBIC [electron beam induced current] as well as the CL [cathodoluminescence] have been proven to be powerful tools to localize the defects and characterize their electronic properties. In this paper these Scanning Electron Microscopy [SEM] are treated as quantitative research tools for defect investigation. Based on a review of the of the fundamental theoretical background for the quantitative analysis of the recombination properties the experimental requirements for a dedicated SEM are described. To demonstrate the capabilities of these techniques the investigation of gettering effects of dislocations in GaAs and the analysis of statistically distributed point-like defects in silicon are presented.

1. Introduction

Defect analysis in semiconductors plays a central role in device oriented research and consequently, a huge variety of different analytical tools have been developed and used for defect identification and the characterization of their electronic properties. Many of the standard techniques such as DLTS, Photoluminescence or the different resonance techniques like EPR, ENDOR, ODMR etc., however, probe large volumes of the semiconductor material and therefore their spatial resolution is rather limited. Individual defects e.g. dislocations or grain boundaries hardly can be studied by these methods, only ensemble averages are accessible.

From the point of view of device development, methods are required, which have a spatial resolution at least comparable to the minimum feature size of the relevant components in a device. Here the Scanning Electron Microscopy techniques with their spatially resolution of $\sim 1\mu m$ open attractive possibilities. Charge collection microscopy [EBIC] as well as cathodoluminescence [CL] are the methods to be considered. To use these techniques for defect characterization with high resolution they must not only fulfill the requirement of the localization of the defect position, the analysis of the electronic properties of the defect must be possible, too. It is obvious that these demands only can be met when a quantitative signal - and thus contrast - measurement is possible and a correlation of the signal strength with material properties is accessible. In this paper, the solutions for these fundamental requirements for the application of EBIC and CL are reviewed and examples for the application of the EBIC and CL technique. As examples for the application of the EBIC and CL technique as a quantitative research analytical tool the analysis of impurity gettering at dislocations in GaAs and investigations of etch-induced defects in silicon are presented.

2. Theoretical background for the quantitative contrast analysis

Both, EBIC and CL use the same basic principles to generate a signal [1]. We take a semiconductor whose top surface is covered with an optically transparent Schottky contact and an Ohmic contact at the back. The top surface is exposed to a focused scanning electron beam which produces a large amount of excess electron-hole pairs in a small volume close to the surface of the specimen. The total number N of generated carriers is

$$N = I_{beam} q (1-\eta_{BSE}) E_{beam}/E_{cr} \qquad (1)$$

where q is the electronic charge, I_{beam} and E_{beam} are the beam current and beam energy, respectively, and η_{BSE} is the backscattering coefficient. E_{cr} is the energy required to create an electron-hole pair which is roughly three times the gap energy of the material. The EBIC signal is generated by that part of the excess carriers which diffuse into the space charge region of the Schottky contact which acts as a collector. Electron-hole pairs reaching this region are se- separated and soaked away by the build in electrical field of the Schottky contact, giving rise to a measurable electrical current. In semiconductors with luminescent properties the part of the excess electron-hole pairs which recombine generate the CL signal. For homogeneous- , defect free semiconductors these signals can be calculated by solving a continuity equation with the corresponding boundary conditions [$\Delta n = 0$ at the interface of the Schottky barrier]:

$$D\nabla^2 \Delta n(r) - 1/\tau \Delta n(r) + g(r) = 0 \qquad (2)$$

Here, D denotes the diffusion coefficient of the charge carriers, Δn is the density of the excess carriers, τ the lifetime of the generated carriers and g(r) the generation function. The three contributions in Eq. (2) represent the diffusion, annihilation by recombination and generation of the excess charge carriers. Eq. (2) can be solved using the approximation of a point like generation function and the EBIC as well as the Cl signals are calculated.

$$I_{EBIC} \sim \int \frac{\partial \Delta n(r)}{\partial z} \Big|_{z=z_0} dF = 4\pi q D G \exp(-h/L) \qquad (3)$$

and similarly

$$I_{CL} \sim \int F(z) \Delta n(r) dV \qquad (4)$$

In these equations z is the depth of the collection barrier, L the diffusion length, G describes the generation conditions, h is the distance of the generation volume from the surface and F(z) represents the optical losses in the material .

In the presence of defects, which are regarded as sites of enhanced recombination, the situation becomes more complex. When the distance between the defect and the generation sphere is within the diffusion length, both signals are reduced due to the recombination properties of the defect. This shows, that the EBIC as well as the CL signals carry information about the defect position and the recombination properties and consequently a close relation between these two signals can be expected.

To quantitatively determine the signal strength in specimens with defects, several approaches have been published.
Donolato e.g. [2] introduced the defect as a volume of reduced excess carrier lifetime and calculated the signals within certain approximations. In this treatment, Eq. (2) is modified by the " defect function "

$$\gamma(r) = 1/D(1/\tau' - 1/\tau).e(r) \qquad (5)$$

with

$$e(r) = 1 \text{ for } r \in V_D \text{ and } e(r) = 0 \text{ elsewhere} \qquad (6)$$

This modified differential equation can no longer be solved analytically, Donolato has used the Born approximation, Hergert et al. [3] applied a perturbation approximation to solve the equation. In some special highly symmetric cases [e.g. a dislocation perpendicular to the surface, there are analytically accessible solutions as shown by Pasemann [4].
Jakubowicz [5] used a different approach to treat the semiconductor with defects. Here, the point like defect is assumed to act as a recombination site which reduces the minority carrier density to zero in a sphere with the radius a around the defect [i. e. the defect is characterized by an infinite surface recombination velocity]. Since different defects can have different recombination mechanisms, an effective defect radius γa is introduced, with $0 < \gamma < 1$. The effect of this approach is, that apart from the boundary condition $\Delta n = 0$ at z=0 only the condition $\Delta n = 0$ at $r = \gamma a$ has to be fulfilled. Using the technique of mirror images [7], the mathematical problem can be solved. Furthermore, with this approach, defects of any shape can be treated just by the superposition of the individual solutions for point-like defects.. Calculating the contrast, defined as

$$C = (I_0 - I)I_0 \qquad (7)$$

where I represents the signal close to the defect and I_0 that far away from it, it could be shown that the solutions given by Jakubowicz for the EBIC and the CL signal reveal a simple relation between the two

$$C_{CL} = \sigma C_{EBIC} \qquad (8)$$

where σ is a function of geometrical parameters, the diffusion length and the absorption coefficient, only [6].

3. Experimental Requirements

For the quantitative evaluation of the EBIC and CL data and the analysis of the physical recombination properties of defects any conventional SEM can be used [in our case a Cambridge Stereoscan 250 MK II] which is equipped with a cooling stage and a computer control for the main parameters. In our case we have installed a beam blanking system to allow the switching of the beam. Access to beam scanning, the SEM screen and the photographic system was accomplished by an additional external interface. The signals necessary to drive this interface are supplied by two 16 bit DAC converters and a 8 bit DDC converter directly hooked up to the control computer. Details of the set-up and the program structure are given in [7].

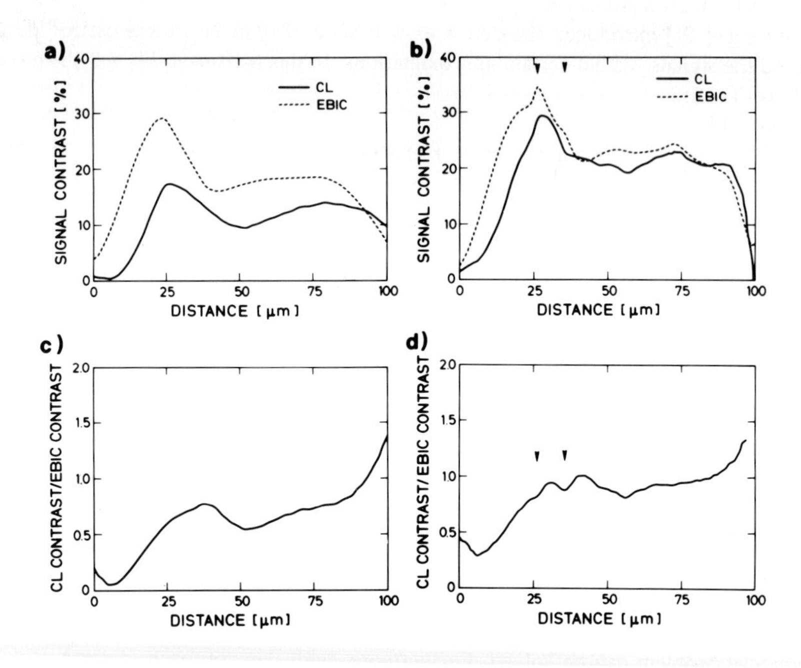

Fig. 1. Quantitative analysis of the EBIC and CL linescans along a dislocation in GaAs a, c before and b, d after copper diffusion Arrows indicate characteristic changes due to the formation of precipitates.

4. Analysis of Gettering Phenomena in GaAs

The interaction between dislocations and impurity atoms in a semiconductor single crystal results in an inhomogeneous distribution of point-like defects and thus in locally varying electrical and optical properties. As indicated above, simultaneous measurements of the EBIC and the CL signal allow a separation of the different contributions to the signal strength (defect structure, defect geometry, decoration etc.). Since the ratio of EBIC and CL contrast is only a function of geometrical parameters, the diffusion length of minority carriers and the optical absorption coefficient, an excellent reconstruction of the defect geometry as well as the analysis of the electronic properties of a defect is possible. Commercially available GaAs single crystals have been investigated before and after a copper diffusion which intentionally was used to decorate the grown induced dislocations. In both signal modes the dislocation is associated with a dark contrast due to an enhanced recombination via traps at and near the defect. A quantitative comparison of the signal contrasts taken along a dislocation line before and after Cu decoration is represented in Fig. 1.

together with the corresponding ratio C_{EBIC}/C_{CL}. The EBIC as well as the CL contrast profile shows an increase along the full length of the dislocation after Cu diffusion. Additionally, characteristic local changes can be observed. The curves representing the ratio of the contrasts look very similar before and after Cu diffusion, however, here small changes also can be observed [see arrows in Fig. 1 d]. The experimental results can be interpreted using computer simulations. The increase in the contrast (Fig. 1 a, b) is due to a homogeneous decoration of the defect, which results in an enhanced recombination rate. The local increase of the contrast [at around 25 μm in Fig, 1 b] is due to the formation of a precipitate after the annealing process. The slight shift of the contrast values observed after decoration are interpreted as a decrease in the bulk minority carrier diffusion length [8].

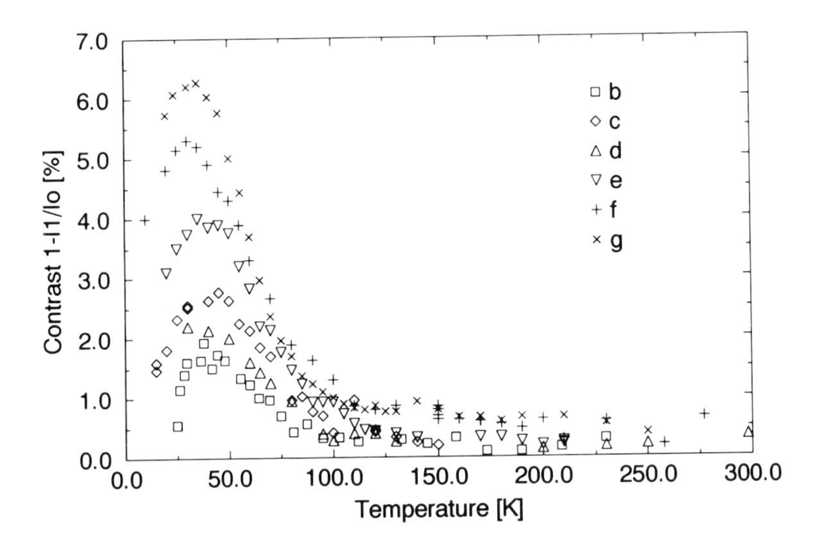

Fig. 2 EBIC contrast vs. temperature for different etch bias voltages (50 V steps; a: 300 V, g: 600 V).

5. Analysis of etch-induced defects in silicon

Dry chemical etching is one of the key technologies in state-of-the-art semiconductor device fabrication. Currently, the main-stream process lines are using the RIE [reactive ion etching] process for pattern definition. It is well known that the impact of reactive species to a semiconductor surface causes damages which are confined to a depth less than 100 nm [9]. In recent experiments we have shown that silicon wafers subjected to a RIE process generates electrically active defects whose investigation is accessible by the EBIC technique [10] in detail. Silicon wafers were patterned using SiO_2 masks and subjected to a RIE process with different processing conditions such as plasma composition, pressure and dc selfbias voltage. The EBIC analysis shows that the etched areas have a reduced signal due to the enhanced recombination at the etch-induced defects statistically distributed in the etched areas. In Fig. 2 the EBIC contrast is represented as a function of temperature. The different curves in Fig. 2

are measured using specimens with different dc selfbias voltages of the etching process which determine the velocity of the particles impinging the silicon surface. To determine whether the different bias voltages causes different types of defects or only affect their density, we analyzed the data given in Fig. 2 in an Arrhenius plot and found for all curves an activation energy of 17 meV. The increase of the contrast with decreasing temperature to the maximum at around 50 K is explained as due to compensation effects of the net impurity concentration by the defects which act similarly to thermal donors.

6. Conclusions

It is shown that quantitative EBIC and CL contrast calculations are possible for arbitrary defect geometries and especially the temperature dependent simultaneous EBIC and CL measurement paves a way to investigate the defect position in a semiconductor and the recombination properties separately. The temperature dependent contrast analysis allows the determination of the energy level of a defect in the band gap. The EBIC and CL techniques are not only applicable to isolated defects such as dislocations or grain boundaries, also statistically distributed defect of atomic dimension are accessible.

Acknowledgment

The author would like to thank A. Jakubowicz and his former PhD students M. Bode, M. Eckstein and G. Jäger-Waldau for their substantial contributions to develop the EBIC/CL technique which are summarized here. This work was partially supported by a grant from the European Union within the SUSTAIN project grant # CHRXCT -920060

References

[1] Holt D B and Joy D C 1989 *SEM Microcharacterization of Semiconductors* (London: Academic Press)

[2] Donolato C 1978 *Optik* **16** 19 - 36

[3] Hergert W Pasemann L an Hildebrandt S 1991 *Jornal de Physique IV Colloque C 6* **C6** 45 - 50

[4] Pasemann L 1981 *Ultramicroscopy* **6** 237 - 240

[5] Jakubowicz A 1985 *J. Appl. Phys.* **57** 1194 - 1198

[6] Bode M Jakubowicz A and Habermeier H.-U. 1987 Proceedings of *Conference on Defect Recognition and Image Processing in II-V Compounds* (Amsterdam: Elesevier) 115 - 121

[7] Bode M Jakubowicz A and Habermeier H.-U. 1988 *Scanning* **10** 169 - 176

[8] Eckstein M Jakubowicz A Bode M and Habermeier H.-U. 1989 *Appl. Phys. Lett.* **54** 2659 - 2661

[9] Oehrlein G 1989 *Mater. Sci. Eng. B* **4** 19 - 24

[10] Jäger-Waldau G Habermeier H.-U. Zwicker G Bucher E 1994 *J. Appl. Phys.* **75** 804 -808

Inst. Phys. Conf. Ser. No 149
Paper presented at DRIP 95

Fundamental Studies of Defect Imaging by SEM-CL/EBIC on Compound Semiconductor Materials

J. Schreiber, S. Hildebrandt, H. Uniewski, and V. Bechstein

Physics Department, Martin-Luther-University Halle-Wittenberg
Friedemann-Bach-Platz 6, 06108 Halle (Saale), Germany

Some fundamental aspects of the defect contrast formation at single dislocations in SEM-CL and SEM-EBIC micrographs are discussed from investigations on GaAsP, GaP and CdTe. The various features of contrast patterns from misfit dislocations in (001) epilayers GaAsP and GaP are analyzed in respect of geometric factors. At glide dislocations in CdTe intrinsic contrast behavior is studied. The temperature dependence of the defect contrast is investigated from 300 K to 5 K. The results obtained show thermally activated non-radiative recombination at the dislocations in GaP, but increasing radiative recombination at B(g) 60° dislocations in CdTe upon lowering the temperature.

1. Introduction

Cathodoluminescence (CL) and electron beam induced current (EBIC) mode are suitable tools of scanning electron microscopy (SEM) for recognition and identification of electronically active defects in many compound semiconductor materials. From the SEM-CL and EBIC micrographs important information on the density and geometric arrangement of the defects in the specimen crystal are obtained which are correlated with the formation, migration, and possible interaction of the observed defects. In particular, combined SEM-CL/EBIC investigations using high resolution defect imaging and quantitative contrast measurements provide appropriate insight into the contrast properties of single dislocations which are attributed to the electronic defect recombination activity and defect geometry. For detailed studies of the formation of dislocation contrasts in both modes it is necessary to take into account the intrinsic nature of dislocation recombination activity as well as effects due to dislocation - point defect interaction and the particular geometric contrast factors.

In this work the SEM-CL and EBIC contrast behavior of crystallographically defined dislocations is studied. The diversity of simple defect contrast patterns from the misfit dislocation arrangement appearing on (001) of the GaP and GaAsP epilayers is analyzed systematically. Temperature-dependent contrast examinations in the range 300 K - 5 K reveal significant changes of the defect recombination activity. The CL contrast measurements on GaP proved thermally activated non-radiative defect-bound recombination rates. CL contrast reversal is shown for 60° glide dislocation in CdTe, which is connected with a specific emission localized on single defects.

2. Experimental

SEM-CL/EBIC investigations were carried out using the experimental SEM-setup described in [1]. The SEM apparatus used is equipped with the CL system *CL302m* and liquid-helium cooling stage *CF302* (Oxford Instruments) allowing sample temperatures 400 K - 4 K. Specimens were n-type GaP ($n = 7\cdot10^{16}cm^{-3}$) and $GaAs_{1-x}P_x$ ($x = 0.375$, $n = 3\cdot10^{16}cm^{-3}$) epitaxial layers on (001) substrates, GaP and GaAs, respectively. Undoped CdTe ($n = 2\cdot10^{15}$ cm^{-3}) bulk crystal samples were used for the examination of glide dislocations. Microindentation performed at room temperature by a Vickers pyramid with 20 p load was employed to activate {111}<110> glide systems in (111)B. All CL/EBIC experiments were conducted under low injection condition (20 kV, 1 nA).

3. Fundamentals of CL and EBIC defect contrast

CL and EBIC contrasts of dislocations have to be described by the defects recombination activity. With respect to the experimental contrast measurements the *defect contrast profile* is defined:

$$C(\xi) = \frac{I(\xi) - I_0}{I_0}$$

Here, $C(\xi)$ is the variation of CL or EBIC signal with electron beam position ξ relative to the defect location ($\xi = 0$), and I_0 is the signal far away from the defect. $C(\xi=0)$ is the maximum contrast and is utilized as strength value for the *defect contrast*. The current theory of defect contrast [2] supplies a suitable quantitative description of CL and EBIC contrasts on surface-parallel dislocations. The contrast behavior can be calculated taking into account the electronic and optical bulk and defect properties and the geometrical factors as well. The defect contrasts are described by

$$C_{EBIC}(\xi,U_b) = \lambda_e \cdot C^*_{EBIC}(\xi,U_b,z_D,r_D,L)$$

for EBIC, and by

$$C_{CL}(\xi,U_b) = \lambda_e \cdot C^*_{CL,e}(\xi,U_b,z_D,r_D,L,\alpha) + \lambda_r \cdot C^*_{CL,r}(\xi,U_b,z_D,r^r_D,L,\alpha)$$

for CL; with U_b - beam voltage, z_D - defect depth, r_D - radius of defect induced carrier recombination, L - carrier diffusion length, α - optical absorption coefficient, r_D^r - radius of region with defect-induced radiative recombination. In these formulas the *defect recombination strength* and contrast pattern geometry may be separated.

The recombination activity of the defect is characterized by the quantities λ_e and λ_r, where λ_e corresponds to the change of *total recombination* rate at the defect, and λ_r describes the change in the *radiative recombination* at the defect and affects the CL contrast only. $C^*(\xi,U_b,z_D,r_D^r,L,\alpha)$ represents the specific contrast profile functions. The developed quantitative contrast theory allows detailed considerations of the fundamental contrast behavior of dislocations; in particular, the contrast behavior in dependence on U_b and z_D can be simulated.

4. Results and discussion

4.1 Analysis of single defect contrast patterns from misfit dislocation arrangement

Fig. 1 shows typical CL contrast patterns on individual single dislocations in the GaAsP epilayer samples. Various contrast features are displayed by the high resolution CL micrographs. The dot-, line- and coma-shaped contrast patterns may be considered as simple kinds of single defect contrasts attributable to various defect geometries. Assuming a homogenous line- and dot-shaped CL or EBIC pattern as basic dislocation contrast structures predicted in the framework of the carrier recombination defect model of [3], the specific features of the single defect contrast patterns can be correlated with either <001> or <112> threading dislocations or <110> surface-parallel misfit defect lines occurring on the (001) specimen surface.

The relationship between the various contrast pattern shapes and the corresponding defect geometry could be obtained by taking into consideration the information on the etch-pit structure and on the beam voltage dependent contrast behavior. Examples of characteristic contrast patterns are marked by d, e, f; their corresponding defect geometries are given below. In this way, the investigated part of the three-dimensional dislocation network could be reconstructed.

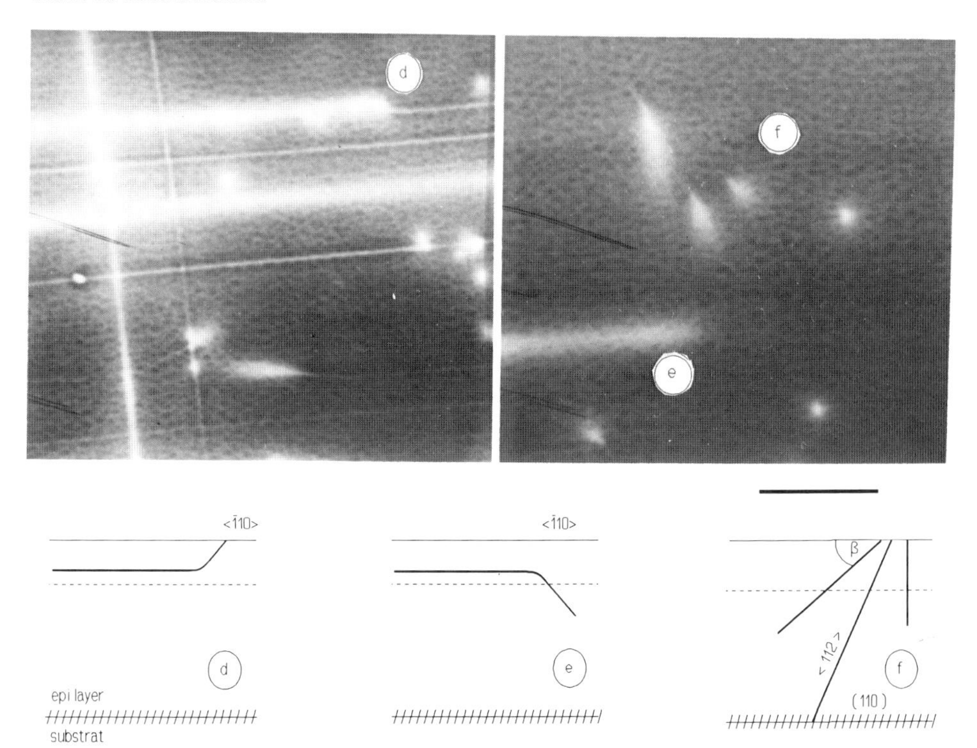

Fig. 1: GaAsP/GaAs sample. CL contrast patterns (inverted signal representation) shaped due to defect geometry. For the typical features (d, e, f) a schematic reconstruction of the misfit dislocation geometry is given in the lower part. The marker represents 5 μm.

By careful analysing the CL micrographs the elementary line- or dot-shaped contrast pattern of a single surface-parallel or surface-perpendicular dislocation, respectively, can be recognized and utilized to verify CL and EBIC basic dislocation contrast properties [2].

4.2 Temperature-dependent contrast behavior

4.2.1 Misfit dislocations in GaP

Fig. 2 (b) illustrates the results of CL contrast profile measurements performed on the GaP sample in the temperature range 300 K - 20 K [4]. They demonstrate a pronounced temperature dependence of the CL contrast of the misfit dislocations considered. Obviously, the surface-parallel (B,C) and surface-perpendicular (A) defects behave in like manner. The temperature dependence of the defect contrast from A, B, C is given in Fig. 2 (c). The different contrast values at 300 K mainly result from the different defect geometry, i.e. to different defect depth and surface-parallel or surface-perpendicular defect position, respectively. Except the maximum for the surface-perpendicular dislocation, the contrast behavior of the three defects follows the same trend characterized by a slow decrease of the contrast values upon lowering the temperature up to about 70 K. Below this temperature the contrasts drop steeper and vanish below about 30 K. The contrast-versus-temperature curves hint at two regions of distinct thermally activated defect-bound non-radiative recombination

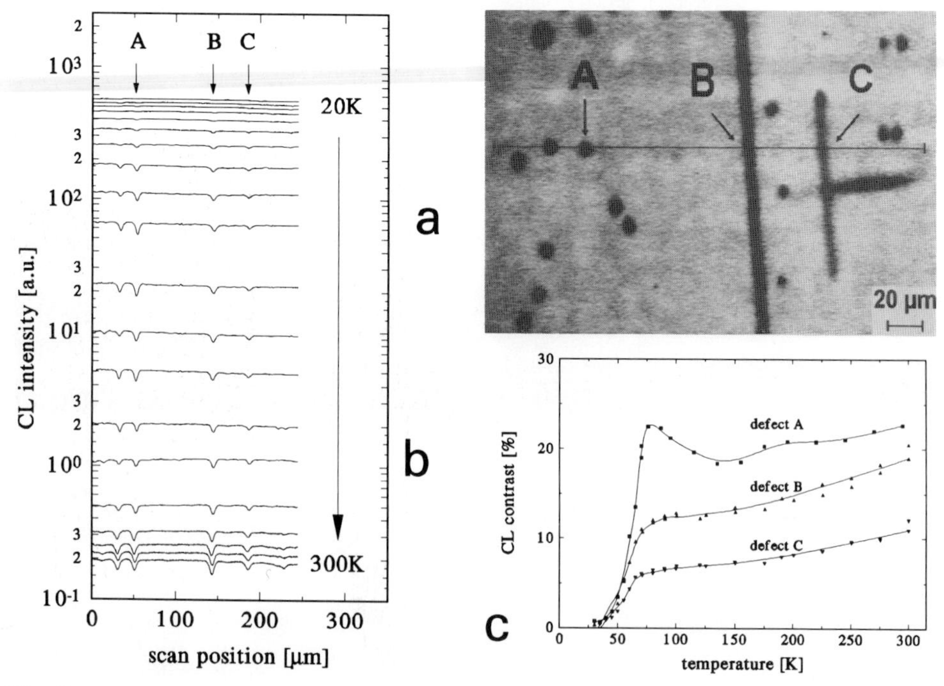

Fig. 2: Temperature dependence of CL contrast at misfit dislocations in GaP. CL profile measurements from 20 K to 300 K (b), CL contrast versus temperature (c) at defects A, B, C in (a), illustrating decreasing non-radiative rates at dislocations upon lowering the temperature.

rates. From these measurements, a lower activation energy of 18 meV and a higher one of 52 meV were estimated. A discussion of the dependence on temperature of the defect strength λ_e requires a quantitative analysis of the temperature-dependent contrast measurements. The CL contrast behavior found here on misfit dislocations in GaP is distinct from the nearly temperature independent dislocation contrasts in GaAs [5].

4.2.2 A(g) and B(g) glide dislocations in CdTe

CL defect contrasts on glide dislocations should reflect intrinsic recombination activity, thus different contrast behavior is expected for dislocations of various types. In particular, the polar A(g) and B(g) edge-type dislocations are accounted as differently active recombination centers. In [6], for GaAs a slightly higher contrast value at A(g) dislocations is reported. There are similar indications of type-dependent contrast behavior for InP. Our investigations on CdTe prove specific recombination behavior of the B(g), i.e. Te(g), dislocation at low temperature. Fig. 3 (b) shows the CL image (T = 5 K) in the region of the dislocation rosette at a microindent on (111)B. The micrograph displays CL bright contrasts which are attributed to the B(g) branch of each rosette arm. This is easily concluded from the surface glide geometry sketched in Fig. 3 (a). The defect configuration near the surface consists of 60° dislocations mainly. In the CL spectrum taken in a rosette arm a specific emission at 130 meV below the near band-gap peak is revealed. It could be localized at single defects and thus, may be identified as dislocation-bound radiative recombination. This dislocation luminescence [7] quenches above 120 K resulting in a rather weak dark defect contrast.

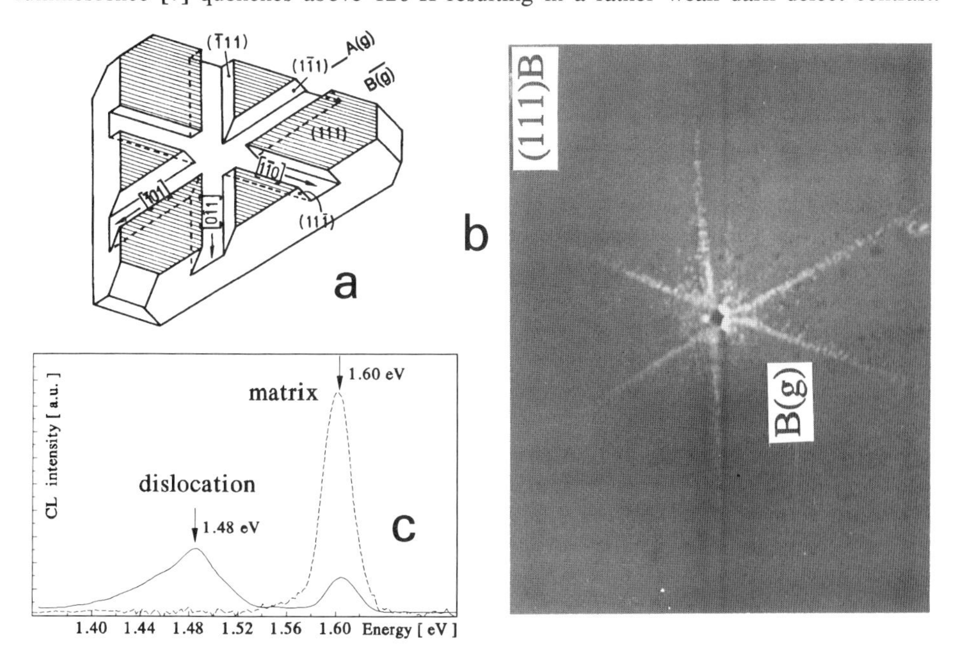

Fig. 3: CL bright contrast on B(g) 60° dislocations in CdTe at low temperature (b). The defect type is recognized by the surface glide geometry (a) in (111). Specific dislocation-bound luminescence emission peak in CL spectra (c).

Below 120 K, this dislocation luminescence increases strongly with decreasing temperature. From contrast profile measurements on a single surface-perpendicular dislocation an activation energy of 20 meV for the bright contrast is obtained.

Contrary to the B(g), the A(g) dislocation does not exhibit any defect contrast in the low temperature region. In this case, a CL dark contrast is found only at 300 K, which disappears with decreasing temperature. The low temperature luminescence at the B(g) dislocations is probably of excitonic nature as found for screw dislocation bound emission in CdS [8]. It should be interesting to observe CL bright contrasts at moving dislocations [9] which could support the suggestion of the intrinsic dislocation origin of the surrounding point defect induced emission.

5. Conclusions

SEM-CL and EBIC defect imaging allows comprehensive investigations of the electronically active defect configuration in the semiconductor samples. For dislocations, high resolution defect imaging reveals various contrast features which can be attributed to the intrinsic or extrinsic contrast behavior at the single defects in different geometric positions. The variety of the defect contrast patterns in particular case may be understood by means of basic image contrasts from surface-parallel and surface-perpendicular defect lines as derived in the framework of the defect contrast theory and/or taking into account decoration effects. The latter can screen the intrinsic defect contrast behavior.

Combined SEM-CL/EBIC studies [1,5] and the quantitative contrast analysis using an advanced defect contrast theory [2] can provide essential information on the recombination activity of the defects by supplying the defect strength value as an appropriate parameter for characterizing the carrier recombination at the dislocations. By studying the temperature dependence of the CL contrast on glide dislocations in CdTe radiative recombination could be established as the origin of the dislocation contrast. It was found, that preferentially the polar Te(g) 60° dislocations exhibit radiative contrast behavior. This finding manifests the distinct recombination nature of the polar A(g) and B(g) type defects.

References

[1] J. Schreiber, W. Hergert, Inst. Phys. Conf. Ser. **104** (1989) 97
[2] J. Schreiber, S. Hildebrandt, Mat. Sci. Eng. **B24** (1994) 115-120
[3] C. Donolato, Optik **52** (1978/79) 19
[4] V. Bechstein, Diploma Thesis, Halle 1995
[5] M. Eckstein, H.-U. Habermeier, J. Physique IV, **C6** (1) (1991) C6-23
[6] B. Sieber, Mat. Sci. Eng. **B24** (1994) 35-42
[7] S. A. Salkov, N. I. Tarbaev, G. A. Shepelski, J. Schreiber, Proc. XI Soviet Conf. on Semicond. Physics, Kishinev (1988) 59
[8] J. Gutowski, A. Hoffmann, Mat. Sci. Forms **38-41** (1989) 1391-6
[9] J. Schreiber, S. Hildebrandt, H. S. Leipner, phys.stat.sol.(a) **138** (1993) 705

Inst. Phys. Conf. Ser. No 149
Paper presented at DRIP 95

On the strain determination in cross-sectioned heterostructures by TEM/LACBED

A.Armigliato, A.Benedetti, R.Balboni and S.Frabboni†

CNR-Istituto LAMEL, Via P.Gobetti, 101, 40129 Bologna (Italy)

† Dipartimento di Fisica dell'Università, Via Campi 213/A, 41100 Modena, Italy

Abstract. The large angle convergent beam electron diffraction (LACBED) technique has been applied to the determination of the tetragonal distortion in coherent $Si_{1-x}Ge_x/Si$ heterostructures. This is related to the splitting of the Bragg contours, which is induced by the strain field in the TEM cross section.

1. Introduction

The recent progress of crystal growth techniques, such as molecular beam epitaxy or chemical vapour deposition has enabled the fabrication of coherent $Si_{1-x}Ge_x/Si$ heterostructures, which are presently employed in the fabrication of high frequency bipolar transistors (Crabbé *et al* 1993) or modulation doped FET (König and Dämbkes 1995). As the strain field due to such layers must be determined in localized volumes, TEM techniques become essential.
In this paper it will be shown how the large angle convergent beam electron diffraction (LACBED) technique can be applied to determine the tetragonal strain in $Si_{1-x}Ge_x/Si$ heterostructures, through the knowledge of the relaxation which takes place in cross-sectioned TEM specimens.

2. Experimental

The $Si_{1-x}Ge_x/Si$ heterostructures have been grown by molecular beam epitaxy (MBE). A thin (100 nm) Si buffer was grown at 650 C; the temperature was then decreased to 550 C and the Si-Ge layer has been grown. Finally a protective cap, 100 nm in thickness, has been grown at the same temperature on top of the heterostructures. The Ge concentration in the alloy film was 12.7 at.%, as determined by Rutherford backscattering spectrometry and double crystal X-ray diffractometry.

The preparation of the (100) TEM cross sections involved glueing, sawing, mechanical lapping down to 20 μm and ion beam milling to perforation. The LACBED patterns were taken at 300 kV by a Philips CM30 TEM. A Gatan liquid nitrogen cooled double tilting holder was used, in order to improve the pattern visibility.
A bright-field image of a cross sectioned heterostructure is reported in Fig.1. The dark, oblique piece of glue has been included to mark the imaged area in the LACBED patterns.

3. Surface relaxation of the TEM cross section

To describe the strain distribution in a two dimensional plate, it is commonly used the method proposed by Treacy and Gibson (1986), which assumes an ideal sinusoidal compositional modulation. However, this model cannot be used here, due to the aperiodical structure of the investigated heterostructures. Recently, Chou *et al* (1994) have proposed that the strain field due to a strained InGaAs layer, buried in GaAs, can be expressed as a superposition of the strain fields of an infinite number of sinusoidal compositional waves. Therefore, following their treatment and assuming the geometry and the coordinate system reported in Fig. 2, we consider a top-hat function $\phi(x)$ with height 1 and width w (i.e. equal to the thickness of the Si-Ge layer):

$$\phi(x) = \int_0^\infty F(\omega, w) cos(\omega x) d\omega \tag{1}$$

where $F(\omega, w) = (w/\pi) sin(\omega w/2)/(\omega w/2)$
The single strained layer is thus modeled by:

$$[a_0 - a(x)]/a_0 = \epsilon\phi(x) = \epsilon \int_0^\infty F(\omega, w) cos(\omega x) d\omega \tag{2}$$

Then the strain field of our heterostructure of constant composition is the superposition of strain fields of many compositional waves:

$$R_x(x, z, t, w) = \int_0^\infty p(\omega, z, t) F(\omega, w)[sin(\omega x)/\omega] d\omega \tag{3}$$

$$R_z(x, z, t, w) = \int_0^\infty q(\omega, z, t) F(\omega, w) cos(\omega x) d\omega$$

where p and q include the term $2\epsilon(1+\nu)/(1-\nu)$, ϵ being the tetragonal distortion and ν the Poisson ratio.
The strain field $\vec{R}$ is thus obtained for a cross section of the heterostructure of a given ϵ, w, t. For the specific case of the samples investigated in this paper, we have $\epsilon = 8.5 \cdot 10^{-3}$, w=100 nm and t=240 nm. This last value has been deduced from the analysis of the LACBED patterns (see next section).
The plot of the R_z component of the resulting strain field is reported in Fig.3.

4. Strain determination by LACBED

The LACBED technique reveals strains and surface relaxations through the shift or the splitting of Bragg contours which originate from planes belonging to the zero order Laue

Figure 1. Bright-field cross-sectional TEM image of the investigated heterostructure.

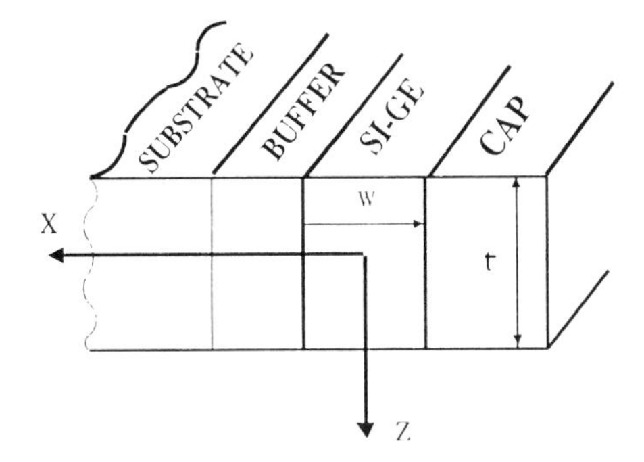

Figure 2. Sketch of the cross-section sample, with thickness t and a Si-Ge film of width w.

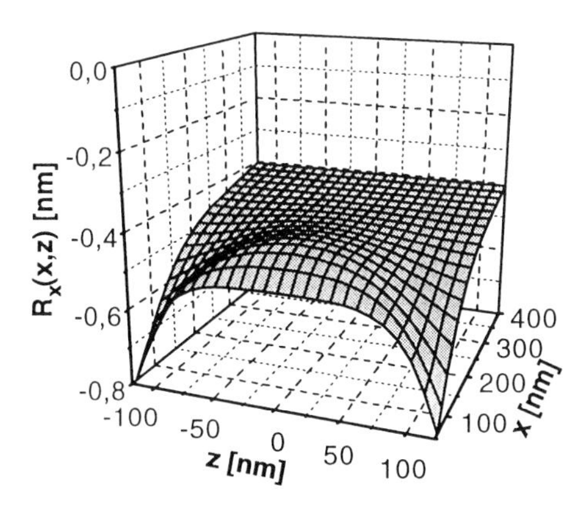

Figure 3. Computed distribution of R_x for t=240 nm, w=100 nm, $\epsilon = 8.5 \cdot 10^{-3}$.

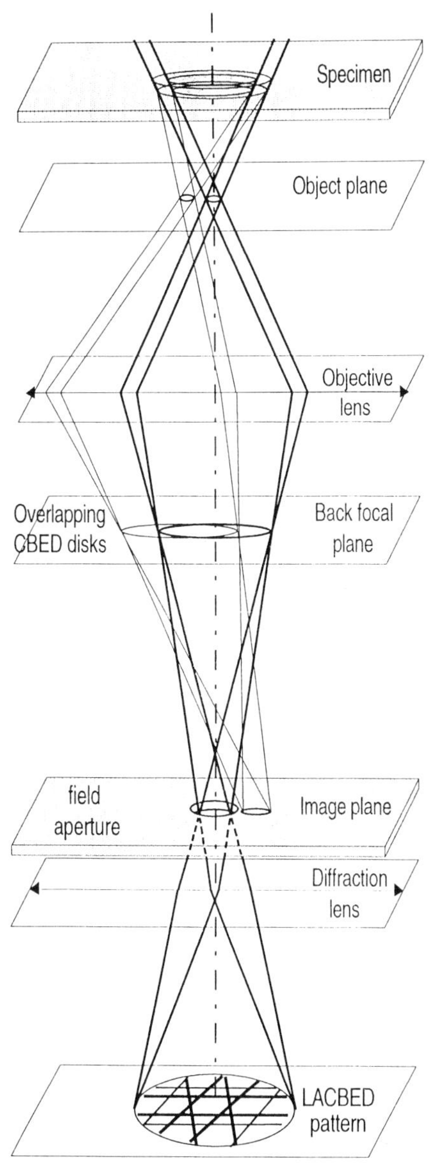

Figure 4. Ray path diagram to obtain LACBED patterns.

zone (Cherns and Preston 1989). These patterns are obtained by bringing an electron beam, with a convergence of a few degrees, i.e. higher than the Bragg angle of the excited zero order reflections, to a focus below the specimen and/or by raising the specimen from the eucentric position so that the focused spots due to the transmitted and the diffracted beams are spatially separated in the image plane of the objective lens. The ray path is reported in Fig.4, which shows that, inserting a selected area aperture to select a single beam and switching the microscope to diffraction mode, the transmitted beam with the Bragg contour deficiency lines or the rocking curve for a single diffracted beam can be recorded (Tanaka *et al* 1980). Due to the geometry depicted in Fig.4, the patterns include both real and reciprocal space information. This gives LACBED the unique advantage of yielding a diffraction pattern superimposed to a defocused image, with a correspondence between Bragg contours and image details limited by the probe size.

When applied to our cross-sectioned heterostructures, the LACBED patterns show that some Bragg contour is split into a number of subsidiary peaks; this is due to the surface relaxation, which modulates the scattering amplitude A_g for a column of crystal, according to the usual kinematical approach:

$$A_g(x,s,t,w) = \int_{-t/2}^{t/2} F_g \cdot \exp[-2\pi \mathrm{i}(\vec{\mathrm{g}} \cdot \vec{\mathrm{R}}(\mathrm{x},\mathrm{z},\mathrm{t},\mathrm{w}) + \mathrm{sz})]\mathrm{dz} \qquad (4)$$

where, apart from the symbols already defined in the previous equations, s gives the deviation from the Bragg position. In this way, if we choose for our analysis a vector $g = 800$, we can correlate the splitting of the corresponding Bragg contour, occurring at a certain distance from the center of the Si-Ge film (Fig.2), with the strain profile at the same position (Fig.2). Of course, $\vec{\mathrm{g}} \cdot \vec{\mathrm{R}}(\mathrm{x},\mathrm{z},\mathrm{t},\mathrm{w}) \neq 0$ only for the R_x component of the strain field. The applicability of the kinematical treatment is ensured by the fact that the extinction distance ξ_g of the 800 reflection is larger than the local thickness of the cross section (ξ_g=585 nm at 300 kV). A thickness of 240 nm has been deduced from the rocking curve of the 400 contour in the experimental patterns shown below.

5. Results

In Fig.5 are reported the experimental LACBED patterns (left) and the corresponding rocking curves of the 800 reflection (right), as computed according to eq.4. The top pattern is taken far from the $Si_{1-x}Ge_x$/buffer Si interface, i.e. in a region of undistorted silicon. As such, there is practically no variation of R_x through the sample thickness and no splitting of the Bragg contour is expected (as in a perfect crystal). The central LACBED pattern is taken in a region of the cross section closer to the interface (x=230 nm), where R_x varies continuously and significantly along z (see Fig.3); this results in a splitting of the contour, as revealed by both the experimental and the computed rocking curve. The calculated separation between adjacent peaks is $\Delta s = 5 \cdot 10^{-3}$, in fairly good agreement with the experimental one ($\Delta s = 7 \cdot 10^{-3}$). Finally, the bottom pattern corresponds to a position still closer to the $Si_{1-x}Ge_x$/buffer Si interface, where the variation of R_x with z is small, except at the two surface regions of the cross section (see Fig.3);

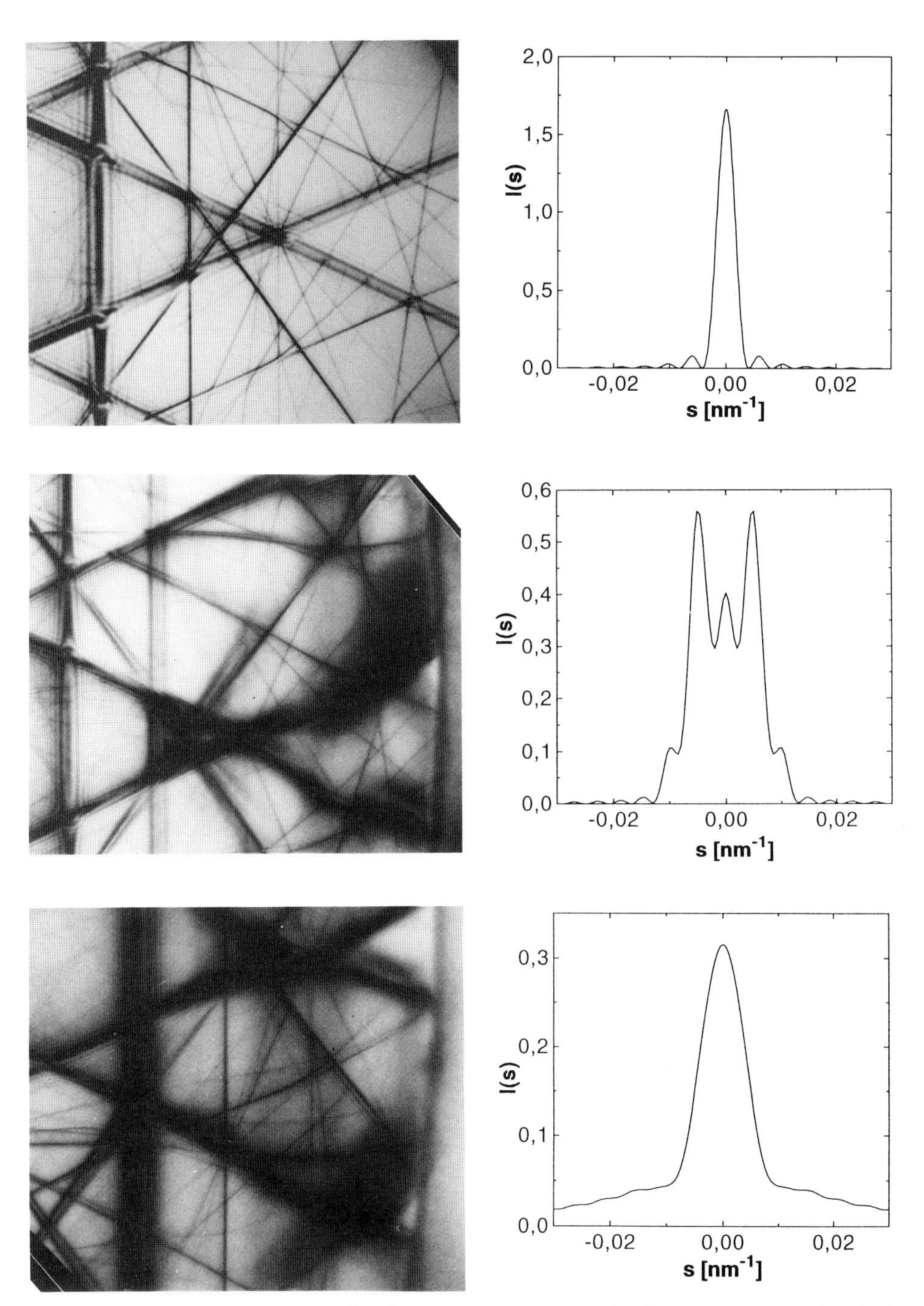

Figure 5. Experimental [012] LACBED patterns (left) and kinematical calculations of the 800 Bragg line profiles (right) for x=500 nm (top), 230 nm (middle) and 60 nm (bottom). The defocused image of the heterostructure is clearly visible in the middle and bottom patterns.

this is not enough to generate a splitting of the Bragg line, just a widening being expected. This is confirmed by a comparison between the top and the bottom patterns. The uncertainty in the position of the Bragg contour with respect to the center of the $Si_{1-x}Ge_x$ film, as deduced from the image visible in the LACBED pattern, is given by the spot size, which was in these experiments about 20 nm. Moreover, the strain sensitivity given by this technique is of the order of $5 \cdot 10^{-4}$; this means that a strain $\epsilon = 8.0 \cdot 10^{-3}$ (instead of $8.5 \cdot 10^{-3}$) would result in a significantly different splitting of the contour at the same position along the cross section.

6. Conclusions

In this preliminary work, it has been demonstrated that the LACBED technique can be employed to monitor the relaxation occurring in TEM cross sections of $Si_{1-x}Ge_x/Si$ heterostructures through the splitting of the Bragg contours. If this relaxation can be reliably modelled, the tetragonal distortion can be determined, by measuring the local sample thickness and the position of the contour with respect to the image of the heterostructure, which is superimposed to the diffraction in the LACBED pattern.

References

Armigliato A, Govoni D, Balboni R, Frabboni S, Berti M, Romanato F and Drigo A V 1994 *Mikrochim.Acta* **114/115** 175

Cherns D and Preston A R 1989 *J.Electron Microsc. Tech.* **13** 111

Crabbé E F, Meyerson B S, Stork M C and Harame D L 1993 *Proc. Int. El. Dev. Meeting (IEDM)* (Piscataway, NJ: The Institute of Electrical and Electronics Engineers) p 83

Chou C T, Anderson S C, Cockayne D J H, Sikorski A Z and Vaughan M R 1994 *Ultramicroscopy* **55** 334

König U and Dämbkes 1995 *Solid-State Electronics* **38** 1595

Tanaka M, Saito R, Ueno K and Harada Y 1980 *J. Electron Microsc.* **29** 408

Treacy M M J and Gibson J M 1986 *J. Vac. Sci., Technol.* **B4** 1458

Inst. Phys. Conf. Ser. No 149
Paper presented at DRIP 95

High resolution electron beam induced current measurements in an SEM-SPM hybrid system by tip induced barriers

R. Heiderhoff, R.M. Cramer, and L.J. Balk

Universität Wuppertal, Lehrstuhl für Elektronik, Fuhlrottstr. 10, 42097 Wuppertal, Germany
☎: ++49/ 202-439 2972 Fax: ++49/ 202-439 3040, e-mail: heiderho@wrcs1.uni-wuppertal.de

Abstract. High resolution electron beam induced current (EBIC) analyses were carried out in a scanning electron microscope (SEM) and scanning probe microscope (SPM) hybrid system by tip induced barriers. Therefore the tip of an SFM was evaporated with several materials to form either Schottky barriers or ohmic contacts with respect to the investigated semiconductor. EBIC analyses at the vicinity of these contacts were carried out by the integration of an SFM into the SEM. Because of the well known mechanism of signal generation, i.e. the creation of electron-hole pairs by the primary electrons, computer based simulation of the induced currents in conjunction with a device simulator allows estimation of the influence of microscopic electrical parameters of the devices on the induced current and on the device behaviour itself. The dependence of the electronic behaviour and the induced current on the electron beam parameters, e.g. frequency, could be demonstrated. It could be shown that using SFM tip induced barriers exhibits the opportunity of measuring spectroscopic microscopic electrical parameters with an obtainable spatial resolution in the nm region.

1. Introduction

Scanning electron microscopy (SEM) and scanning probe microscopy (SPM) are established tools for characterization and analysis of semiconducting materials and semiconductor devices. All SPMs exhibit an extremely high spatial resolution and a lot of efforts were made to substitute them for SEM techniques. Until now they are used as the electron source, which injects electrons into an extremely thin layer of a sample with no reduction of the lateral spatial resolution. As a consequence, several SEM measurement techniques are successfully applied to scanning tunnelling microscopes (STM), too, e.g., electron beam induced current (EBIC) measurements [1] or cathodoluminescence [2].

With the EBIC-microscopy as an established application of an SEM it is possible to obtain in-formations of electrical parameters, such as potential distributions within the sample or diffusion length. Because of the well known mechanism of signal generation, i.e. the creation of electron-hole pairs by the primary electrons, computer based simulation of the induced currents in conjunction with a device simulator allows estimation of the influence of microscopic electrical parameters of the devices on the induced current and on the device behaviour itself [3]. Recently, it could be demonstrated that an application of an STM to EBIC-investigations is possible [1], although an unavoidable influence of the tip potential on the potential distribution inside a device has to be considered, as shown by simulations.

To overcome this problem and to correlate directly the electrical properties with the morphology this paper shows that tip induced barriers as supplied by the tip of a scanning force microscope (SFM) in contact mode can be used to analyse the sample with no reduction of the lateral resolution. Finally a comparison of the CCM-SFM and EBIC results is given and an estimation of the possible applications of this method to characterize semiconductor substrates, epilayers, and nanostructures is illustrated.

2.Equipment

High resolution EBIC analyses were carried out using a method that allowed us to measure simultaneously morphological and electrical properties of semiconductors with no reduction of the lateral resolution. To achieve this, an SEM/SPM hybrid system was developed with the opportunity to analyse the sample by tip induced barriers supplied by the tip of an SFM in contact mode for the first time (see Fig. 1). These measurements were performed using an SFM (TopoMetrix prototype) equipped with a conducting tip, integrated into the analysis chamber of our SEM (Cambridge).

Therefore a pre-etched high-force silicon cantilever was evaporated with several materials to form either Schottky barriers (for example with a 100 nm thin Au film) or ohmic contacts with respect to the semiconductor. Due to the minimal contact areas, the measured current was of the order of several pA at an applied voltage of 20 V, the induced barriers were analysed with respect to their characteristics by the lock-in technique (Ithaco Dynatrac 3) as well as the use of a current amplifier (Ithaco 1212). Therefore a potential free voltage source was implanted in the circuit.

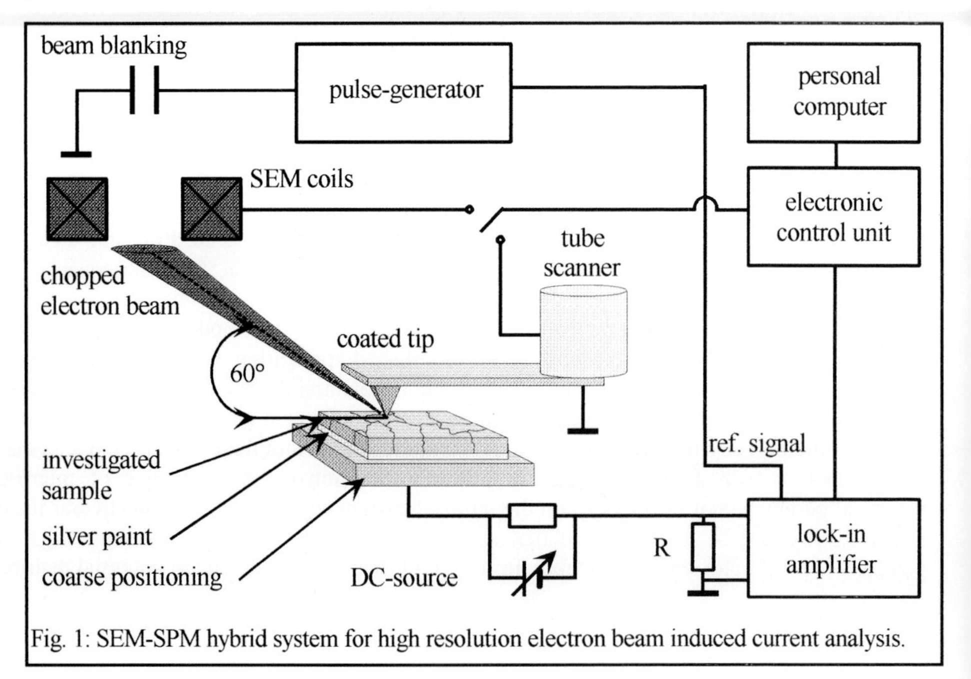

Fig. 1: SEM-SPM hybrid system for high resolution electron beam induced current analysis.

The induced current (excess carriers are generated in a small volumen of the sample by the chopped electron beam and are seperated by internal electrical fields) was determined by measuring the voltage drop across a serial resistor R. The lock-in amplified signal was AD converted, transferred to the computer, where the data were calculated and displayed. The modified computer program and the experimental set-up allowed simultaneous measurements of the electric and topographic structures.

The experimental method gives the opportunity to analyse these structures either by the electron beam while keeping the position of the tip and the induced barrier constant, as well as scanning the Schottky contact with the 1 μm tube scanner while maintaining a homogeneously electron irradiation. In the first case electronic parameters like diffusion length, potential distributions within the sample are detectable (see Fig. 2). The lateral resolution depends on the electron beam properties, diffusion lengths, as well as on the dispersion volume. Using Schottky contact displacement maintaining homogeneously electron irradiation on the other side, the lateral resolution only depends on the contact area and can therefore be used to analyse defects of the sample with no reduction of the position accuracy.

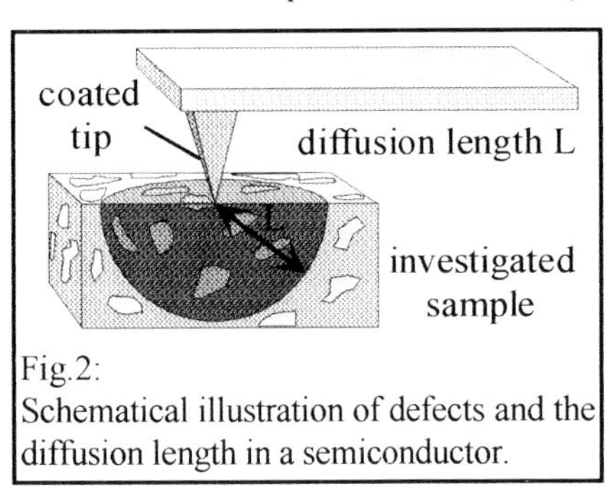

Fig.2:
Schematical illustration of defects and the diffusion length in a semiconductor.

First experiments were carried out on undoped CVD diamond films. The investigated representative sample in this study was grown heteroepitaxial by microwave plasma CVD [5] at a power-level of 4000 W and a reactor pressure of 170 mbar from standard C/H gas mixtures. The thickness of the polycrystalline film was 67 μm. The sample was cleaned in acetone and isopropanol by ultrasonics before it was mounted on a standard ceramic holder by silver paint following removal of carbon residues in argon plasma for 30 s at low energies.

The SEM-EBIC investigations were performed using an acceleration voltage of U_{acc}: 10 kV, a beam current of I_{beam}: 1 nA, and an applied frequency of f_{chop}: 100 kHz. A detailed description of SEM-EBIC can be found in [6, 7]. High resolution EBIC analyses maintaining homogeneously electron irradiation were carried out at U_{acc}: 20 kV, I_{beam}: 200 pA, and f_{chop}: 100 kHz to ensure a homogeneously distribution of electrons and holes in the sample.

3.Results

Schottky diodes on homoepitaxial and natural diamond show a barrier height of $\Phi_B > 1.5$ eV and a low ideality factor n = 1.1 (up to n = 65 for heteroepitaxial films [8]). High electron and hole mobilities of $\mu_- = 2000$ cm²/Vs and $\mu_+ = 1600$ cm²/Vs and a bandgap of $W_G = 5.45$ eV make them attractive for high temperature and high frequency applications. In this paper, we report for the first time, the utilization of the EBIC method in an SEM-SPM hybrid system to determine the leakage current generally observed in epitaxial diamond film diodes. Several possible origins for this excess leakage current were developed: a) field-induced barrier lowering and image force barrier lowering, b) carrier generation in the space charge region, c) fringing effects at the circumflex, d) lateral surface conductivity, e) tunnelling through defects, and f) homogeneously distribution defects, acting only on a small fraction of the diode surface area, bypassing the Schottky barrier contact [9].

High resolution EBIC analyses of a Schottky contact on CVD diamond produced with a Au tip induced barrier by keeping the position of the STM tip on a diamond grain constant are illustrated in Fig. 3. After the determination of the forward and reverse characteristics, p-type Schottky barriers were observed with a contact area of about 1 μm² (ideality factor n = 35, saturation current I_S = 4.16 nA), the induced current was measured, using an applied voltage between -8.5 V and 8.5 V. The image shifting is caused by the influence of the voltage on the electron beam.

At reverse bias (-8.5 V to 0 V) the space charge region is visible. The maximum induced current depends on the applied voltage and varies between 96 pA to 29 pA respectively. As can be seen from the signal at 0 V, the carrier separation is caused by the tip induced barrier. At an applied voltage of ≈2 V the barrier height was removed shown by no induced current. Contrary to the observed I-V characteristic also an EBIC signal can be detected at forward bias. Additional to the observable space charge region the grain boundary effect is visible [9].

This unexpected behaviour led us to carry out EBIC line scans by continuous variation of the applied voltage using an acceleration voltage of U_{acc} = 5 kV and a beam current of I_{beam} = 1 pA to ensure a high spatial resolution (see Fig. 4). From the vertical line scans of the

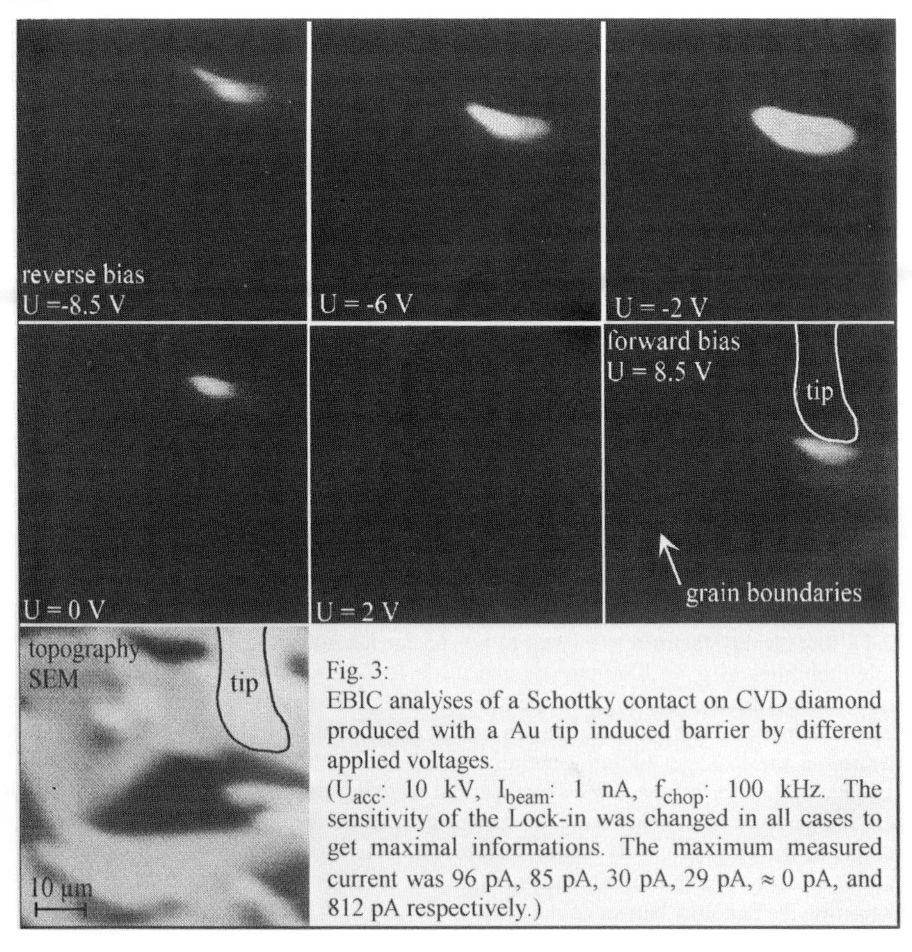

Fig. 3:
EBIC analyses of a Schottky contact on CVD diamond produced with a Au tip induced barrier by different applied voltages.
(U_{acc}: 10 kV, I_{beam}: 1 nA, f_{chop}: 100 kHz. The sensitivity of the Lock-in was changed in all cases to get maximal informations. The maximum measured current was 96 pA, 85 pA, 30 pA, 29 pA, ≈ 0 pA, and 812 pA respectively.)

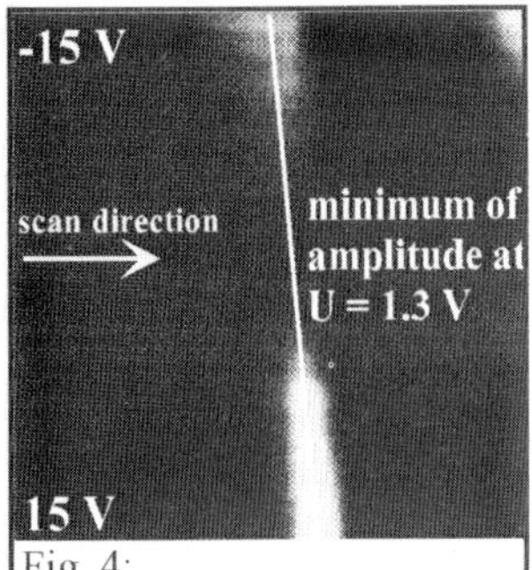

Fig. 4:
EBIC line scans by variation of the applied voltage (U_{acc}: 5 kV, I_{beam}: 1 pA, f_{chop}: 100 kHz).

amplitude and phase micrographs a barrier height of Φ_b = 1.3 V was found. The amplification varies by a factor 120. The relative phase (applied voltage const.) and phase signal (position const.) indicate a constant relationship at the space charge region and reversal of 180° between -15 and 15 V respectively. The diffusion length of the minorities could be determined by logarithmic illustration of the leading edges of the amplitude signals. Therefore a diffusion length of (1.8 ± 0.1) µm was found for the electrons. In the other case the diffusion length of the minorities was determined to be (1.6 ± 0.2) µm. These minorities are associated with holes. The ratio of these diffusion lengths correlates with μ_n/μ_p where μ_n and μ_p are the mobilities of the electrons respectively the holes.

Therefore, a new electronic model was developed which describes the results obtained from these EBIC analyses. In this circuit, two antiparallel diodes with different serial resistors are integrated. Capacitance and conductance analyses were carried out to confirm this hypothesis by using the lock-in technique. A saw-tooth signal superimposed with a sinus was applied between -25 V and 25 V, using the sinus as reference signal to determine the imaginary and real value. A degradation is observed at forward as well as at reverse bias by capacitance measurements indicating the presence of two anti parallel diodes. At reverse and forward bias a linear increasing of the conductance amplitude is detectable, supporting the equivalent circuit above. From these data the parameters of the diodes were determined and confirmed by PSPICE simulations.

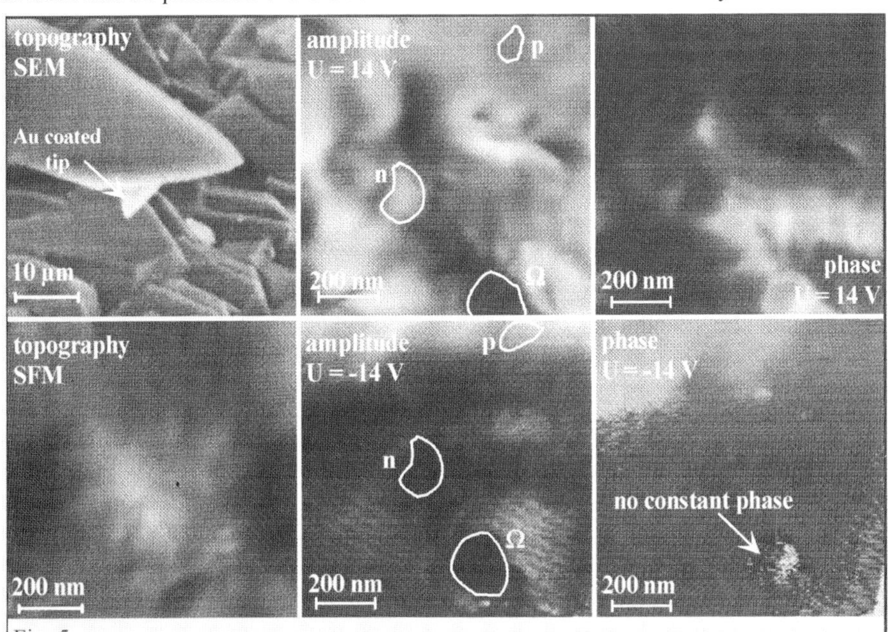

Fig. 5:
High resolution EBIC analyses by Schottky contact displacement in dependence on the applied tip voltage (U_{acc} = 20 kV, I_{beam} = 200 pA, f_{chop} = 100 kHz, scan speed = 1 line/s)

In addition of contacting single grains, an SEM-SFM hybrid system allows high resolution EBIC analysis by nano Schottky contact displacement. Therefore a Au tip was scanned with a constant loading force, while the sample was irradiated homogeneously by the electron beam. Fig. 5 illustrates the results in dependence on the applied tip voltage of a 1 μm^2 scan. The dark areas are indicating a low EBIC signal.

As can be seen from the micrographs, n-, p-, as well as ohmic areas can be separated with a lateral resolution down to 5 nm. Bright fields at 14 V (dark at -14 V) indicate n-type regions while bright fields at -14 V (dark at 14 V) indicate p- type regions. Additional there are areas visible which are dark at both applied voltages. These regions are ohmic contacts if there is no constant phase (indicating an electrical contact). These were confirmed by complementary I-V characteristics using the nano tip. In addition to the observed p-properties above the n-peculiarity showed high ideality factors and pronounced ohmic contacts at the grain boundaries.

4. Summary

High resolution EBIC analyses by tip induced barriers as well as by nano Schottky contact displacement were carried out in an SEM-SPM hybrid system for characterisation of polycrystalline CVD diamond films. The method used allowed us to develop an equal circuit, consisting of randomly spatially distributed antiparallel diodes, which describes the leakage current generally observed in epitaxial Schottky diodes. It could be demonstrated that tip induced barriers, as supplied by the tip of an SFM in contact mode, exhibit the opportunity of measuring microscopical electrical parameters with an obtainable spatial resolution in the nm region. In the investigated sample a barrier height of 1.3 V was observed for p-type Schottky barrier diode. Diffusion lengths of 1.8 μm and 1.6 μm were determined for the electrons, respectively for the holes.

Acknowledgements

The authors would like to thank Dr. P.K. Bachmann for supplying the diamond films. We also gratefully acknowledge financial support of the present work by the Deutsche Forschungsgemeinschaft (DFG)(Project No. Ba805/4-2) carried out under the auspices of the trinational "D-A-CH" German, Austrian, and Swiss cooperation on the "Synthesis of Superhard Materials".

References

[1] Koschinski P., Dworak V., Kaufmann K., and Balk L.J. 1993 *Inst. Phys. Conf. Ser.* **135** Chapt. 2 65-68

[2] Wenderoth M., Burandt C., Gregor M., Loidl G., and Ulbrich R.G. 1992 *Sol. State Comm.* **83** 536-537

[3] Kaufmann K. and Balk L.J.1992 *Microelectronic Engineering* **16** 513-520

[4] Maywald M., Pylkki R.J., and Balk L.J. 1994 *Scanning Microscopy* **8** 2 181-188

[5] Bachmann P.K., Leers D., and Lydtin H. 1991 *Diamond and Related Material* **1** 1-12

[6] Leamy H.J. 1982 *J. Appl. Phys.* **53** (6) R51-R80

[7] Balk L.J., Menzel E., and Kubalek E. 1980 *Proc. of the Eighth International Congress on X-Ray Optics and Microanalysis* Boston 18 - 24 August 1977 613-624

[8] Gomez-Yanez C: and Alam M. 1992 *J. Appl. Phys.* **71** (5) 2303-08

[9] Vescan A., Ebert W., Borst T., and Kohn E. 1995 *Diamond and Related Material* **4** 661-665

[10] Palm J. 1993 *J.Appl. Phys* **74** (2) 1169-78

Inst. Phys. Conf. Ser. No 149
Paper presented at DRIP 95

Recent improvements in NFO for semiconductor defect or device imaging

JP Fillard
LINCS-CEM Université de Montpellier 2
34095 - Montpellier cédex 05 - France

Abstract.
For some years increasing interest has been given to optical near field probe technologies. New sensors have been developped or improved which are able to collect or deliver photons with nanometric accuracy. This contributes to enlarging the field of the scanning probe microscopy techniques which are already well accepted in nanotechnology. Coupled investigations with optical means ans atomic force microscopy or shear force microscopy have been performed which lead to more consistent information. This also makes it possible either to computer assemble optical images or to make localized photo-excitation for the purpose of nano-lithography, quantum dot studies, optical connection or molecular luminescence, this list not being exhaustive. The aim of this communication is to review the present achievable means and the corresponding investigated applications especially devoted to the field of Integrated Circuits and opto-electronic device technology.

1. Introduction

In early semiconductor technology (*i.e.* 20 years ago), material qualification and device inspection was classically performed at a macroscopic scale by electrical or optical global tests. However the ever increasing shrinkage of the components has induced a compelling need for more detailed spatial analysis by optical microscopy imaging or scanning maps at a super-micron scale (SPM, OBIC, EBIC etc...). This has reveal a wealth of micron size grown-in or process induced defects. Now the time has come to enter into the mesoscopic range of dimensions (1 - 100 nm) for material analysis and also for the device technology itself because the Integrated Circuit as well as the OptoElectronic device critical dimensions have definitely gone beyond the 100 nm precision level.

Optical techniques in this area are of essential importance since they are generally easy to implement, they are not intrusive nor destructive and they do not require a special environment or preparation (vacuum, metallization etc...). Of course new methods of nanoscopy [1] have been flourishing and refered in the DRIP conference series such as scattering tomography (LST) or interferometry (PSM) which all belong to far field (FF) classical optics. However a new challenge has appeared with the optical near field (NF) techniques (NFO) associated to the scanning probe methods which have appeared following the Scanning Tunnel Microscope (STM) invention.

To do this, the optical probes can be obtained either from an optical fiber [2] which has been tapered to a tip end some 20 nm wide or from a standard atomic force (AFM) cantilever which is used for its transparency in the visible or near infra red range [3]. They all work as a high confinement funnel for the photons to match the macroscopic and the nanoscopic dimension worlds. Various kinds of experiments can then be performed using the probe as a nano light source or as a nano light detector: luminescence, topography, optical transmission or reflection photoelectrical injection etc... are obtained at a very high resolution well under the classical diffraction limit given by the Rayleigh criterion. This is of a particular interest for the semiconductor science and technology even if there has been to date only a very few investigations in this domain (see Betzig [4]).

In this paper which is devoted to a review of the NFO techniques we present in the first section the background of the NFO scanning probe methods and the emphasis is put on their ability to apply to semiconductor materials. The second part is devoted to the optical probe technology and the third section summarizes some recent results obtained on semiconductor materials, electronic circuits or optical integrated devices.

2. NFO Techniques

Any method in microscopy aims at giving two kinds of information: the first one relies on spectroscopy related features (*i.e.* chemical species) the second one is related to the topography morphology or metrology, both kinds are of interest for semiconductor materials or structures. The advantages of optical means with respect to others (mainly electrons) come from the large flexibility of the photon probe: polarization, wavelength, interaction coefficients can be changed or varied with a direct action on the result. Photons are widely used at a macroscopic scale in the bulk or at the surface of the samples in the diffraction limit by classical FF

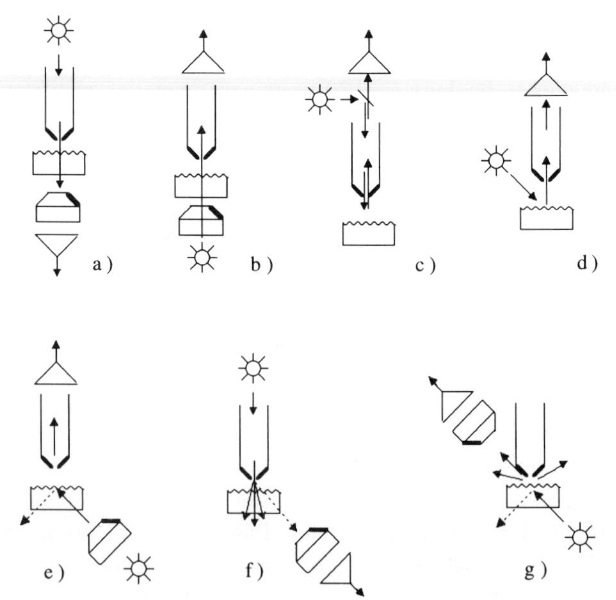

Fig.1 Some illumination techniques to achieve NF scanning microscopy:
SNOM : transmission (a,b), reflection (c), scattering (d) modes
PSTM: normal (e), inverted (f), scattering (g) modes

optical lens or mirror systems. However subwavelength illuminated objects are surrounded by an e.m. field which has a more complex structure including high frequency spatial frequencies;

the calculation becomes difficult for finding a solution to the Maxwell's equations since the usual approximations no longer hold: successively Helmholtz-Kirchhoff, aperture, paraxial, Fresnel and Fraunhofer. Several methods for discrete NF computations [5], [6], [7] have been proposed which are all complex to handle, given the physical limits of the objects. A qualitative but important point to be raised is that the intensity of the high frequency scattered NF decreases as the sixth power of the distance instead of the d^{-2} usual law of the FF, which explains why it is so confined to the close range of the objects. Then to pick-up these spatial frequencies requires a probe which also has a very small size. I have summarized in Fig.1 the various solutions to combine NF illumination and detection with an optical nano-probe. They are conventionally refered to as Scanning NF Optical Microscopy (SNOM) and Photon Scanning Tunneling Microscopy (PSTM) and they allow us to measure the transmitted or reflected or scattered light by a sample in the local range of the probe tip. Also it makes it possible to control the distance from the to the probe tip because of the very rapid increase of the intensity. The SNOM experiments relies on propagating photons (Fig.1 a, b, c, d) and the active region is limited to the NF range of scattered waves surrounding the tip. The PSTM method described in Fig.1(e, f g) is different in that it relies on the building of an evanescent wave which is obtained by total internal reflection; it is well known that the illumination intensity provided by such "virtual" photons decreases exponentially from the flat surface of the prism with a critical distance d_c given by:

$$I = I_0 \exp - \frac{d}{d_c} \quad ; \quad d_c = \frac{\lambda}{4\pi\sqrt{n^2 \sin^2\theta - 1}}$$

with n being the refractive index of the prism, θ the incidence angle ($\theta > \theta_c = \arcsin 1/n$) and λ the wavelength. Then the distance d can be controlled by a feed-back loop circuit to keep a constant intensity level which makes it possible to follow the relief evolution along a scan process.

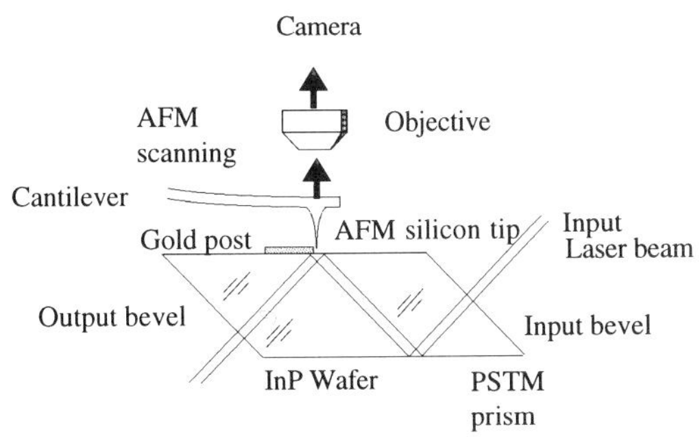

Fig.2 PSTM experiment.

The experiment is described in Fig.2 and looks very similar to the STM one excepted that d_c is in the range of 50-250 nm instead of some nanometers; in this case it is considered that photons can "tunnel" from the surface to the tip through the tip/surface gap where no propagating solution can be found. So PSTM is very convenient [8] for exploring mesoscopic objects such as those found in semiconductor technology; on top of that the especially high refractive index of these materials entails the possibility of obtaining an extended range of d_c = 10-150 nm. However it can be noted that transmission contrast and relief both reacts on the transmitted intensity and separating the respective contributions is not that easy. A well admitted conclusion now for any NFO experiment is to combine it with a simultaneous AFM or shear force (SFM) experiment in order for the latter to control the tip position close to the surface whereas the former provides the optical information. An example of such combined methods is given the poster session which shows an InP surface bearing semi-transparent metal layer, the AFM topographic image (contact mode) and the corresponding PSTM intensity have been simultaneously recorded. The wavelength is 1.06 μm and the probe is a silicon tip. It is observed that the definition of the optical image is sub-micronic. SNOM experiments can also be profitably coupled with an AFM control of the distance because the NF region extension at the tip end is still smaller than d_c . By the means it was made possible to obtain transmission nanometer resolution images [9] at solid surfaces or single molecule luminescence [10] . Extensive studies are currently being carried out to apply NFO methods to electronic related technologies such as nano-lithography [11], [12] or new techniques for data storage [13], [14].

3. NFO probes

At the beginning NFO was mainly orientated to transparent glass like materials and the corresponding probes were obviously obtained from optical fibers (OF) after a sharpening operation. Heating and pulling the OF makes it possible to obtain a very sharp end where the core and the cladding reduction keeps a constant ratio right up to the tip end. A peripheral metal coating also helps in confining more strictly the propagation inside the fiber. Such a system is very convenient since it is cheap and very easy to achieve. The OF tip is then directly glued onto the piezo drive of the scanning microscope keeping free the tapered end of the fiber. Then it becomes very easy to move and scan this probe close to the surface of the sample. Moreover it is possible to excite the piezo at the resonance frequency of the OF end in order to make it dither laterally over an amplitude of some nanometers. When the tip is brought close to the surface a friction force occurs which induces a damping of the oscillations [15]. This perturbation can be directly detected by optical means and will serve as a means for controling the tip/sample distance in the close range thus making possible a shear force microscopy imaging. More recent systems even use a "tuning fork" quartz oscillator [16] to directly detect the vibration by the reverse piezo-effect which makes the probe head more compact. However inspite of these unambiguous advantages the OF tip end is rather stiff, heavy and cumbersome at the nanometric scale; crashes are frequent which are quite destructive.

It has been proposed to use commercially available AFM cantilevers as optical sensors: in this case the tip is very small (some microns in length and diameter) and it is carried by a very soft arm (cantilever) which prevents the dramatic consequences of a crash. Two kinds of tips are available [3] : the more usual technology uses Si_3N_4 low stress films which are especially shaped as small and hollow pyramids the others are directly issued from the silicon technology and they are made of bulk silicon. It has been shown that these tips are able to conveniently capture and guide the light as an OF would do [17], [18] (Fig.2): Si_3N_4 is transparent all over the visible range whereas Si is limited to wavelengths larger than 1μm. The light which is captured and transmitted by such probes can be externally collected by a classical lens objective [3], [18] or by a large diameter optical fiber [19] (which requires a more compulsive positionning). A futuristic solution which has already been explored consists in integrating the photodetector (or an LED) directly on the tip which would sensibly enhance the signal to noise ratio of the output signal [20], [21]. We still are at the very beginning of a new era for such smart sensors which could be able to control at the same time several optical, mechanical [22] or electrical parameters... also possibly with some local artificial intelligence. In the special domain of semiconductor optical experiments such as photo-luminescence often requires low temperatures: a few works are reported [23] in the literature which refer to NFO experiments at temperature below LNT; the microscope is generally installed in a liquid helium container and the links to the external world are provided only by wires and OF.

4. Preliminary results in the semi-conductor field

SNOM configuration

Evanescent wave capture:
- guided mode structure [24] and losses [25]
- guide coupling field structure [26]

Propagating wave capture:
- optical guide output field [27]
- laser emission field [28]
- localised light emission
  - high resolution photoluminescence
  - individual quantum structure:
    - * MQW [30, 31]
    - * self organized Q dots

Super small light sources:
- local photoluminescence
- local photo-conductivity, NOBIC, minority carrier profiles [32, 33]
- transient absorption saturation [34]
- individual excitation of small devices

PSTM configuration:

Semi-conductor surface roughness [35] or features (etch pits or nano-indentations)
Epilayers and interface roughness [4]
Surface and sub-surface spectroscopy
Critical dimensions and metrology

Technological applications

Nanolithography [11], [12]
Data storage [13], [14]
Optical nano-connection

Table 1 NFO applications in the semiconductor domain.

NFO techniques can apply to the semi-conductor field with a tremendously large diversity of situations; we have tried to tidy up this mess and summarize in the Table 1 some of the main directions and corresponding references.

References

[1] Fillard "Near Field Optics and nanoscopy" - World Scientific Pub. (Singapore) to appear 1996
[2] Paessler MA, Buckland EL, Moyer PJ, Yakobson BI - 1992 Nato Asi series E**242** 287
[3] Fillard JP - 1995 Microscop. Microan. Microstruc.
[4] Hess HF, Betzig E, Yoo M, Harris TD, Pfeiffer LN, West KW - This conference
[5] Novotny L, Pohl DW, Hecht B, - 1995 Opt. Lett. **20** 970
[6] Dereux A, Vigneron JP, Lambin Ph, Lucas AA - 1991 Physica B **175** 65
[7] Girard C, Courjon D - 1990 Phys. Rev. B **42** 15 9340
[8] Fillard JP, Prioleau C, Lussert JM Castagné M, Bonnafé J - 1993 IOP DRIP V conf. **135** 249
[9] Zenhausen F, Martin Y, Wickramasinghe HK - 1995 Science **269** 1083
[10] Moerner WE, - 1994 Science **365** 46
[11] Ferrell TD, Sharp SL, Warmack RJ - 1992 Ultramicroscopy **42** 408
[12] Wegscheider S, Jirsh A, Bielefeld H, Meiners JC, Krausch G, Mlynek L - 1995 Proc. NFO 3 conf. Brno; to appear in Ultramicroscopy
[13] Shintani T, Nakamura K, Hosaka S, Imura R - 1995 Proc. NFO 3 conf. Brno; to appear in Ultramicroscopy
[14] Terris BD, Maurin HJ, Rugar D - 1995 SPIE Conf. San Diego **2535** Proc. to appear
[15] Betzig E, Finn PL, Weiner - 1992 Jour. Appl. Phys.**60** 2484
[16] Karraï K, Grober RD - 1995 Appl. Phys. Lett. **66** 1842
[17] van Hulst NF, Moers MHP, Noordman OFJ, Tack RG, Segerink, Bölger B, - 1993 Appl. Phys. Lett. **62** 5
[18] Castagné M, Fillard JP, Prioleau C - 1995 Appl. Opt. **34** 703
[19] Baïda F, Courjon D, Tribillon G - 1992 Nato Asi series E **242** 71
[20] Danzebrink HU, Ohlson O, Wilkening G - 1995 Proc. NFO 3 conf. Brno; to appear in Ultramicroscopy
[21] Akamine S, Kuwano H, Fukuzawa, Yamada H - 1995 MEMS'95 conf. Amsterdam IEEE proc. **145**
[22] Tortonese M, Barrett, CF Quate - 1993 Appl. Phys. Lett. **62** 8
[23] Ghaemi H, Cates C, Goldberg B - 1993 NFO 2 conf. Raleigh Proc. to appear in Ultramicroscopy
[24] Borgonjen E, Moers M, Ruiter T, van Hulst NL - 1995 Proc. NFO 3 conf. Brno; to appear in Ultramicroscopy
[25] Tsaï DP, Jackson HE, Reddick RC, Sharp SH, Warmack RJ - 1990 Appl. Phys. Lett. **56** 16 1515
[26] Jackson HE, Lindsay SM, Naghski DH, De Brabanter GN, Boyd JT - 1995 Proc. NFO 3 conf. Brno; to appear in Ultramicroscopy
[27] Obermüller Ch, Karraï K - 1995 Appl. Phys. Lett. to appear
[28] Isaacson M, Cline J, Barshatzky H, - 1991 "Scanned Probe Techniques" Wickramasinghe HK Ed. AIP Conf. Series **241** 23
[29] Pedarnig JD, Specht M, Hänsch TW - 1993 NFO 2 conf. Raleigh Proc. to appear in Ultramicroscopy
[30] Harris TD - 1995 Proc. NFO 3 conf. Brno; to appear in Ultramicroscopy
[31] Smith S, Orr BG, Kopelman R, Norris T - 1995 Proc. NFO 3 conf. Brno; to appear in Ultramicroscopy
[32] Obermüller C, Kolb G, Karraï K, Arbstreiter G - 1993 NFO 2 conf. Raleigh Proc. to appear in Ultramicroscopy
[33] Kolb G, Obermüller C, Karraï K, Arbstreiter G, Bohm G, Tränkle G, Weinman G - 1995 Proc. NFO 3 conf. Brno; to appear in Ultramicroscopy
[34] Smith S, Orr BG, Kopelman R, Norris T, - 1995 Proc. NFO 3 conf. Brno; to appear in Ultramicroscopy
[35] de Fornel F, Lesniewska E, Salomon L, Goudonnet JP - 1993 Opt. Com. **102** 1
[36] Betzig E, Trautman JK - 1992 Science **257** 189

Inst. Phys. Conf. Ser. No 149
Paper presented at DRIP 95

Cathodoluminescence and Electron Beam Induced Current techniques applied to the failure analysis of 980 nm pump lasers used in optical fiber amplification

R B Martins (1), G Salmini (2), and F Magistrali (2)

(1) CPQD TELEBRAS, PO BOX 1579, Campinas SP, 13.088-061 BRAZIL
(2) Pirelli Cavi S.p.A, Divisione Italia, Milan Italy

Abstract. In this work we analyze the possibilities of Electron Beam Induced Current and filtered Cathodoluminescence techniques as suitable tools for failure analysis of 980nm high power lasers. It is shown that a combination of both techniques together with a detailed band structure analysis allow to a much better understanding of the observed images and their signal formation.

1. Introduction

Among the Scanning Electron Microscope (SEM) techniques used for failure analysis of semiconductor devices, the Electron Beam Induced Current (EBIC) is the most common one, in comparison with Cathodoluminescence (CL), mainly due to a better understanding of EBIC signal formation and an effortless sample preparation and signal collection [1]. However, this is not the case in every sort of devices. For example, the major part of EBIC images used for the failure analysis of 980nm high power lasers, do not give direct knowledge about the degradation process of the samples. In this work, it is shown that, due to the laser structure complexity, the combination of EBIC and CL techniques give more helpful and reliable information than EBIC alone.

2. Experiment

The epilayers structure of the 980nm emitting laser is sketched in Fig.1. It consists of a GaAs/InGaAs/GaAs single quantum well (SQW) strained layer with graded index (GRIN) AlGaAs cladding layers [2]. For this work we analyzed laser chips bounded *p*-side up, allowing SEM observations without elaborated sample preparation. The damaged devices came from accelerated aging tests, carried out with very high current operation and case temperature (e.g.: ~500mA and ~100°C).

Figure 2 shows the experimental configuration utilized for collecting EBIC and CL signals simultaneously [3]. Also, the EBIC and CL images can be obtained by scanning either the top surface or the mirror facet of the samples. In the former case, the higher EBIC contrast was obtained with a primary electron energy of 30 keV, and for the mirror facet, it was used an electron energy of 10keV.

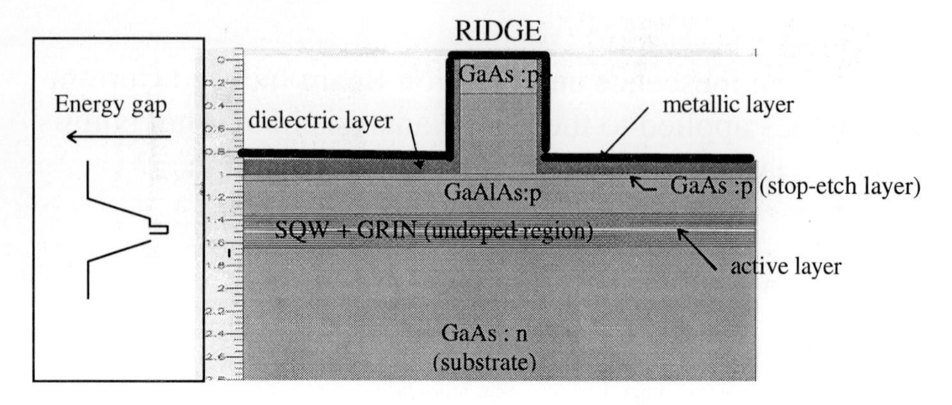

Fig. 1. Schematic description of the GRIN SQW laser structure. A schema of the conduction band energy gap is also indicated on the inset (left).

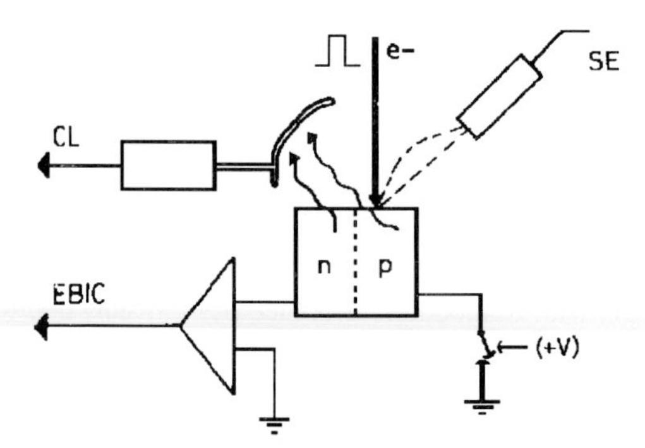

Fig. 2. Experimental SEM configuration allowing the simultaneous collection of EBIC and CL signals (an external bias V can also be applied to the device).

3. Preliminary EBIC observations

The top surface EBIC images collected from 980nm damaged lasers only show some "dark regions" if the EBIC current amplifier gain is overloaded. A typical EBIC image from a "dark region" is shown in Fig.3a. The same "dark region" when observed in normal gain conditions shows a very weak contrast as can be seen in Fig.3b. In this figure, an EBIC line profile is superimposed to a top surface secondary electron image of the ridge waveguide (in the same region shown in Fig.3a). Line profile measurements indicated that the EBIC contrast was not higher than 20% in the "darkest" regions observed - the EBIC contrast is set as $\frac{I_{max} - I_{def.}}{I_{max}}$, where $I_{def.}$ is the EBIC intensity at the defective region. Indeed, the majority of EBIC images collected from different samples do not show a clear correlation with the observed degradation.

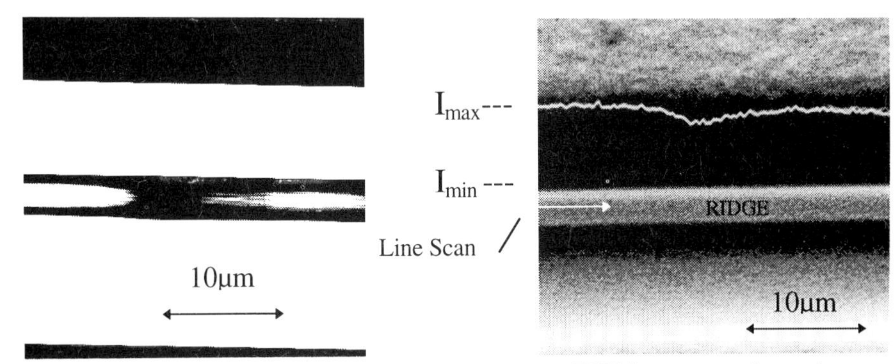

Fig. 3a. Top surface EBIC image obtained with overloaded gain conditions

Fig. 3b. EBIC line scan superimposed to a secondary electrons image of the top surface of the Ridge (same region as Fig. 3a) The white arrow indicates the line scan region.

4 Simultaneous observation of EBIC and CL images from mirror facets

Due to the lack of information carried by top surface EBIC images, we proposed the collection of EBIC images simultaneously with spectrally resolved CL ones, in order to investigate the EBIC signal formation. The same approach was already applied to analyze the influence of the gradual gap region on EBIC and CL signals collected from GRIN-SCH lasers[3]. Filtered CL images were collected at four different wavelengths: λ=740nm, due to the radiative emission from the GaAlAs confinement layers, λ=810nm which comes from the GaAs stop-etch layer emission, λ=860nm from the GaAs buffer layer and substrate, and finally, λ=980nm due to the radiative recombination from the SQW active layer.

Figure 4a shows EBIC image and Figs. 4b to 4d show CL images observed keeping the samples completely grounded (*p* and *n* sides short-circuited and grounded). It was possible to observe the EBIC signal (Fig. 4a) as well as the CL ones from the substrate (λ=860nm, Fig. 4b), from the confinement layers (λ=740nm, Fig. 4c), and from the GaAs stop-etch layer (λ=810nm, Fig. 4d). However, no CL signal was observed from the quantum well layer (λ=980nm) even at very high irradiation doses (electron beam current ~1μA and beam energy =35kV).

Applying a direct bias to the laser diode (voltage bias ranging from 0.8V to 1.4V) the 980nm CL signal starts to be observed, as can be seen in Figure 4e. The other CL filtered images are not influenced by the external bias. On the other hand, the EBIC signal almost disappeared. It was also observed that the EBIC signal (Fig. 4a) comes from the same region that the 980nm CL one (Fig. 4e). This is a clear evidence that, in the case of GRIN-SQW lasers, it exists a competition between EBIC and CL signal formation mechanisms. A reduction of the built-in electric field strength with a subsequent improvement of the excess carriers recombination in the SQW, can explain this phenomenon.

It was also observed that filtered CL images from the active layer (980nm) can be obtained by disconnecting one of the laser diode contacts (*p* or *n* sides). This effect could be due to the excess carriers generated by the electron beam which accumulate close to the depleted region, locally reducing the electric field strength, the same way as the external bias. This localized internal bias effect is named self-polarization.

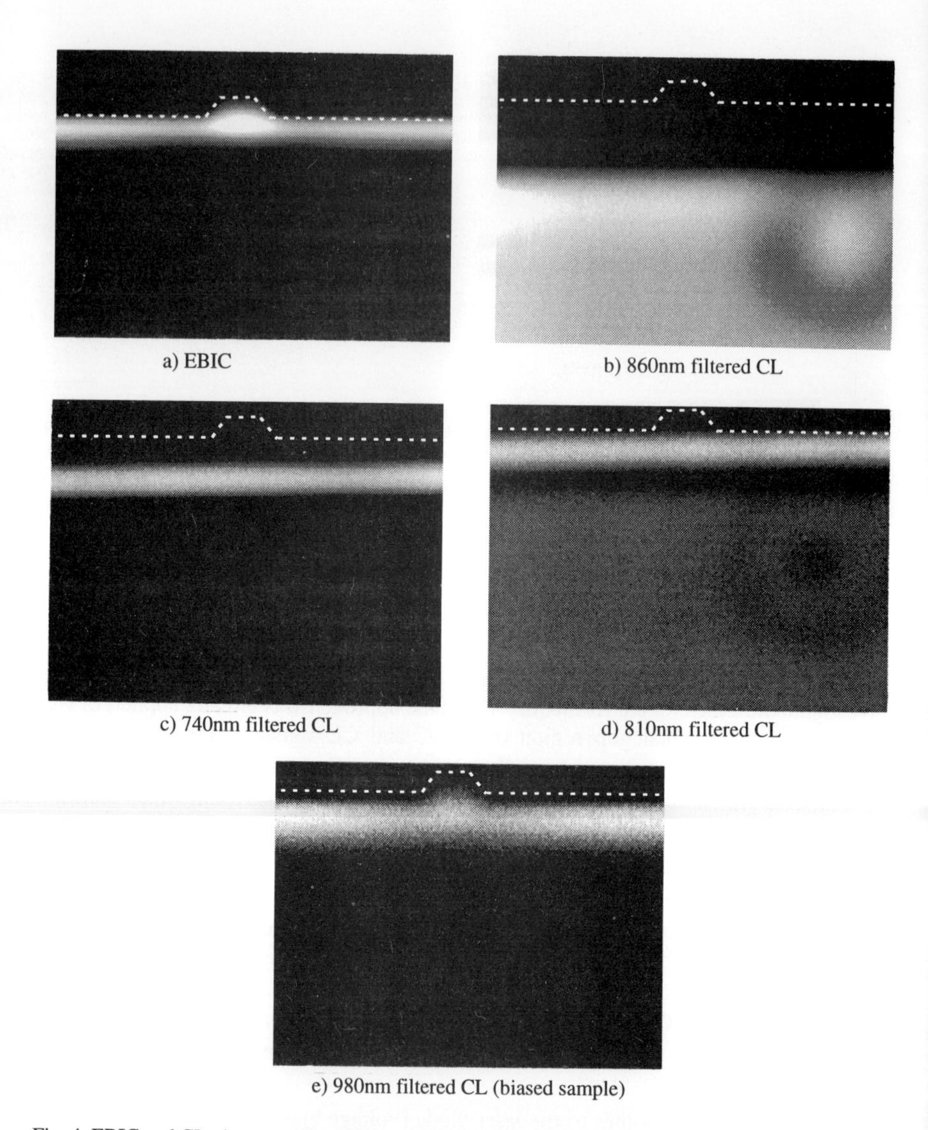

Fig. 4. EBIC and CL pictures collected from the mirror facet of a non damaged GRIN SQW device. The dashed white lines define the position of the 3μm width ridge waveguide. The EBIC and all CL images were collected with non biased samples, except picture 4e. The latter one was collected applying 1.4V bias to the laser diode.

In order to investigate the bias effect on the 980nm CL emission, as well as on the EBIC signal, we estimated the influence of the external bias on the GRIN-SQW band structure. The valence and conduction band diagrams were obtained by solving numerically the Poisson and Continuity equations and taking into account the detailed laser structure (see Fig.1). It was assumed a residual doping on the order of $10^{16}cm^{-3}$ in the GRIN region. Figure 5 shows the energy band diagrams of the GRIN-SQW structure at zero bias (Fig. 5a) and at 1.4 Volts bias (Fig. 5b). At zero bias (Fig. 5a), the SQW is inside a high strength electric field. In this condition, the excess electrons accelerate towards the *n*-side of the junction instead of recombine in the SQW, avoiding 980nm CL emission. This effect also explains the lack of information about the actual active layer condition, carried by EBIC images. Since the built-in electric field avoid the excess carriers recombination in the SQW, a local modification of the SQW carrier lifetime is not able to change the EBIC signal intensity.

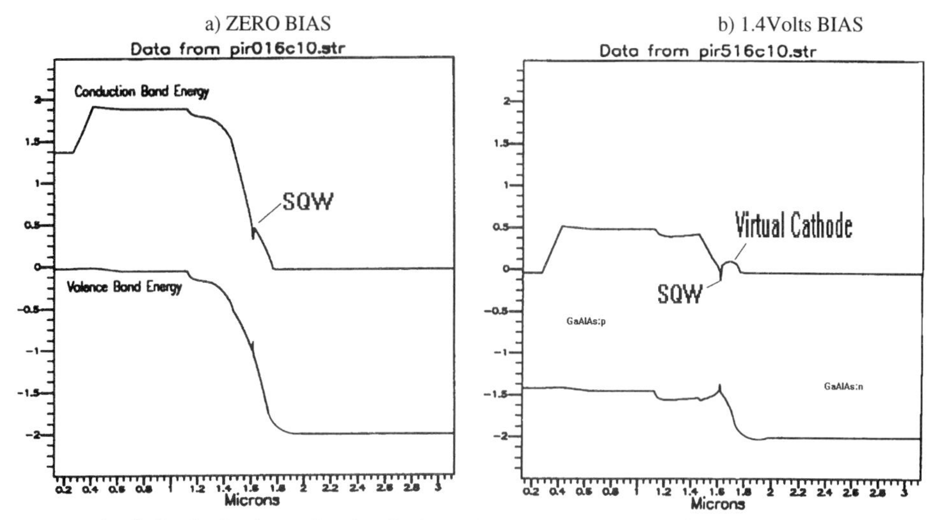

Fig. 5. Conduction (upper lines) and valence (lower lines) energy as a function of the vertical coordinate (see Fig.1) calculated for a) V=0, and b) V=1.4V biased GRIN SQW laser device.

Applying a direct bias to the GRIN SQW laser, the strength of the built-in electric field vanishes and a competition between the *p-n* junction electric field and the increasing gradual gap takes place. This competition gives rise to a virtual cathode[5] (region where the electric field is null, although being inside a depleted region), which works as an energy barrier for the excess carriers. In this condition, the CL signal from the SQW can be observed, instead of the EBIC image, because the excess carriers can now recombine in the SQW and no longer travel towards the n-side of the junction, as confirmed by our observations (cf. Fig.4).

5. CL images of damaged regions

Finally, after removing the *p*-side contact metallic layer, the CL signal could be collected from the top surface of some damaged lasers. CL images were collected by selecting the same four wavelengths as before, and at the same regions previously analyzed by EBIC. The main differences observed are the following:

1) some regions, that did not display contrast in the EBIC mode, exhibited strong dark region contrast in the 980nm filtered CL images;

2) some EBIC "dark regions" (Fig. 6a) also showed contrast on the filtered CL images. In these regions, the 810nm CL images showed the same contrast as the overloaded EBIC ones (Fig. 6b). On the other hand, the 980nm CL images (Fig. 6c) exhibited higher contrast and larger dark regions than those observed in the EBIC or 810nm CL images. Both observations indicate that the EBIC signal can only display a meaningful (although weak, cf. Fig.3b) contrast if the GaAs stop etch layer was damaged.

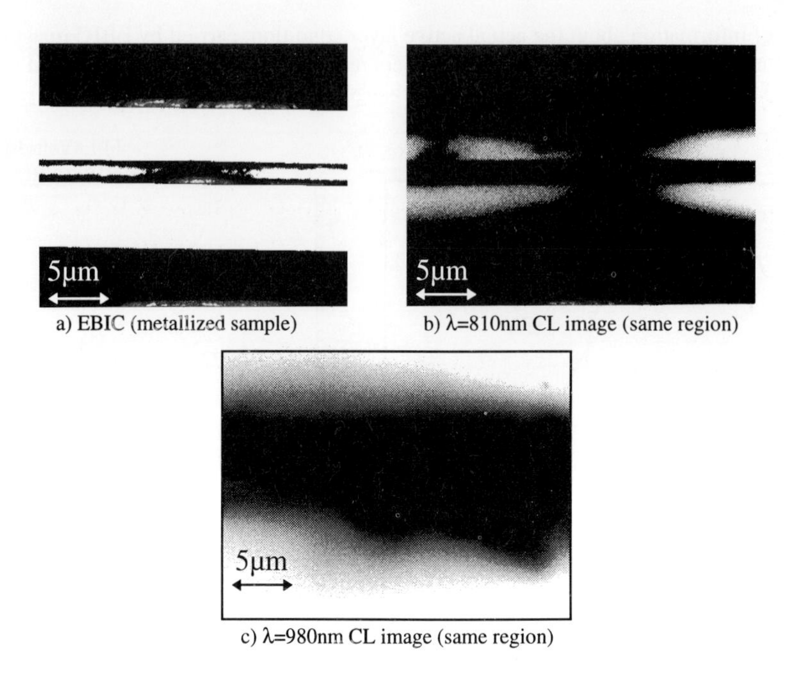

a) EBIC (metallized sample)

b) λ=810nm CL image (same region)

c) λ=980nm CL image (same region)

Fig. 6. EBIC and CL micrographs of a failed GRIN-SQW laser device. These images were collected at the same region and with the same magnification, before (EBIC) and after (CL) removing the metallic contact layer.

6.Conclusion

The GRIN SQW structure concept is spreading out in many high power laser diodes used in Telecommunications, laser printers, solid state pumping and so on, mainly due to the improvement in the laser efficiency and light output power. On the other hand, the well known EBIC and CL techniques must be used carefully for the failure analysis of these structures, avoiding misleading interpretations.. For the 980nm pumping lasers analyzed in this work, it was demonstrated that only in the case where the p-n junction collection probability was locally altered by some damage in the stop-etch layer, the EBIC images can display a significant contrast. On the other hand, filtered CL revealed to be a more suitable SEM technique to observe damaged regions present in the active layer of 980nm GRIN-SQW lasers, providing that a self-polarization effect takes place, or an external bias was applied to the sample, allowing a detectable radiative recombination in the SQW. It is also pointed out that the simultaneous collection of EBIC and CL signals together with a band structure analysis allow a much better understanding of the signal formation mechanisms.

References

[1] Dale E Newbury, David C Joy, Patrick Echlin, Charles E Fiori and Joseph Goldstein in "Advanced Scanning Electron Microscopy and X-Ray Microanalysis" 1986: Plenun Press, NY.

[2] see for example, P Blood in "Physics and Technology of heterojunction devices" ed. by D. Vernon Morgan and Robin Williams, 1991: Peter Peregrinus Ltd. London.

[3] R B Martins, P Henoc, B Akamatsu, G Bartelian, and M N Charasse 1990 J. Appl. Phys. **68** 937-942

[4] A A Grinberg, S Luryi, M R Pinto, and N L Schryer 1989 IEEE Trans. Electron. Devices **36** 1162- 1166

Inst. Phys. Conf. Ser. No 149
Paper presented at DRIP 95

Degradation and defects in electron-beam-pumped $Zn_{1-x}Cd_xSe$/ ZnSe GRINSCH blue-green lasers

J-M Bonard, D Hervé[1], J-D Ganière, L Vanzetti[2], J J Paggel[2], L Sorba[2], E Molva[1] and A Franciosi[2]
Institut de Micro- et Optoélectronique, Département de Physique, Ecole Polytechnique Fédérale de Lausanne, CH-1015 Lausanne, Switzerland
[1] LETI (CEA-Technologies Avancées) Département Optronique, 17 rue des Martyrs, F-38054 Grenoble Cédex 9, France
[2] Laboratorio Nationale TASC-INFM, Area di Ricerca, Padriciano 99, I-34012 Trieste, Italy

ABSTRACT: We explore the defect distribution and the degradation in electron beam pumped $Zn_{1-x}Cd_xSe$/ ZnSe laser structures by combining cathodoluminescence measurements in a scanning electron microscope with transmission electron microscopy. We found that degradation occurs via the formation of <100>-oriented dark line defects, and that it involves the formation of a characteristic type of defect, namely dislocation loops within the ZnCdSe quantum well, which grow to form a characteristic network as degradation progresses.

1. INTRODUCTION

An intense effort has been devoted in the last few years to the realisation of incoherent and coherent light emitters based on II-VI semiconductors and operating in the blue-green range of the visible spectrum. Although important progress has been made, II-VI blue-green laser diodes operating at room temperature and in cw mode are still far from showing reliability and lifetime comparable to those of their III-V infrared counterparts, since the longest reported lifetime amounts to ~2 hours for cw operation at room temperature [1]. Doping and ohmic contact limitations, as well as high defect densities, increase dissipation in wide gap II-VI laser diodes, and contribute to the rapid degradation of the structures under lasing action.

Recently, microgun-pumped blue and blue-green lasers have been demonstrated up to 223K [2]. Such electron beam (e-beam) pumped devices use $Zn_{1-x}Cd_xSe/ZnSe$ graded-index, separate confinement heterostructures (GRINSCH) grown by molecular beam epitaxy (MBE), in combination with a lithographically patterned microtip cathode, which acts as a miniaturized electron source for injection. Because e-beam pumping does not require doping or contact fabrication, this approach allows one to circumvent some of the major limitations of current II-VI laser diodes, and evaluate the remaining intrinsic material limitations.

We study here the degradation (i.e. the decrease of the luminescence yield) of e-beam pumped $Zn_{1-x}Cd_xSe/ZnSe$ GRINSCH laser structures by combining cathodoluminescence (CL) measurements in a scanning electron microscope (SEM) with transmission electron microscopy (TEM) studies of selected structures. In the former type of measurements, the SEM directly supplies e-beam pumping for electron-hole pair excitation, while the CL detection mode allows us to monitor the degradation in real time with submicron spatial resolution. If this type of degradation studies are performed on thin-foil samples prepared for TEM measurements, a direct correlation between the degradation behaviour and the formation of defects is possible.

2. EXPERIMENTAL DETAILS

All laser structures are grown by molecular beam epitaxy on $In_{0.01}Ga_{0.99}As$ (001) wafers. Two buffer layers, comprised of $In_{0.04}Ga_{0.96}As$ (1μm thick, lattice matched to ZnSe) and ZnSe (1μm thick), are grown sequentially on the substrate prior to the deposition of the GRINSCH. The GRINSCH itself includes a

$Zn_{1-x}Cd_xSe$ quantum well (QW) embedded between two 500nm thick $Zn_{1-x}Cd_xSe$ graded layers, where x varies continuously from x=0.05 (at the well boundaries) to x=0. Typical well thicknesses examined are in the 50-100Å range, and typical Cd concentration in the well are in the $0.15<x<0.25$ range.

Plan-views, as well as cross-sections, are prepared from all structures by mechanical thinning down to a thickness of 50μm followed by 5keV-Ar^+ ion bombardment in a Gatan PIPS (Precision Ion Polishing System). The TEM observations are carried out on a Philips EM430ST at 300keV.

The CL measurements are done on a Cambridge S-360 SEM, with a modified stage (Oxford Instruments 402) allowing measurements between 10K and room temperature. All observations are performed by scanning the beam across the (001) growth surface. The sample CL is collected by an ellipsoidal mirror and focused either on a Si-photodiode for polychromatic imaging and measurements or through an optical fiber on the entrance slit of a Jobin-Yvon HR250 monochromator equipped with a Si-CCD camera for spectral acquisition. The theoretical spectral resolution is better than 1Å with 0.2mm entrance slits and a 1200 lines/mm holographic grating blazed at 500nm.

3. TRANSMISSION ELECTRON MICROSCOPY

TEM observation of cross-sections of the non-degraded structures reveal a complex defect distribution, as exemplified in Fig. 1. The different layers are easily visible, and no defects appear in the III-V buffer. Strong contrasts are present at the III-V/II-VI interface, and threading dislocations initiating at the interface rise in the ZnSe buffer layer. Only a small number of them persist in the GRINSCH layers (none on Fig. 1), but a high density of defects is present in the lower graded-index (GRIN) layer.

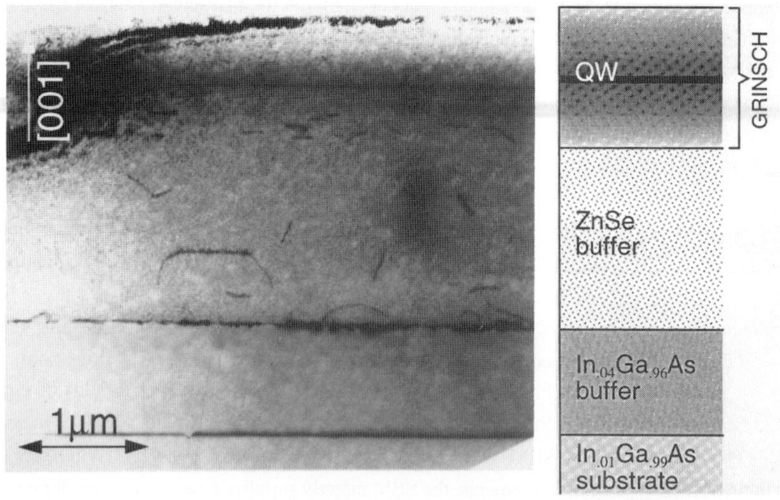

Fig. 1 Cross-sectional view of sample 372 in bright field near the [110] zone axis, with a schematic representation of the structure.

An analysis on planviews allows us to identify the different defects (not shown here). Only threading dislocations, and on some samples stacking faults, are present in the upper GRIN layer. An array of dislocations running approximately along the [100] and [010] directions is located below the quantum well, and can be seen in the cross-sectional view of Fig. 1. Their origin is misfit relaxation, since the GRINSCH material (ZnCdSe) and the ZnSe are not lattice-matched. High densities of stacking faults and threading dislocations are found in the ZnSe buffer, and another array of misfit dislocations is visible at the III-V/II-VI interface. The ZnSe layer is therefore relaxed, although it should be lattice-matched to the underlying $In_{0.04}Ga_{0.96}As$. X-Ray diffractions studies [3] show that the ZnSe layer is actually under a slight biaxial tensile strain, as a result of the presence of the GRINSCH structure. Finally, misfit dislocations at the substrate-buffer interface indicate radation profiles in the scan mode for four "lasing" samples at E_b=20keV and I_b=10nA are displayed. Samples #372 and #296, with the lowest resp. highest densities show corresponding decay times (on Fig. 2) of >20000s (~5.5h) resp. 590s. (~0.15h). Since there is a factor 10 between the CL yields of the two samples (see section 4.), this difference could be due only to the variation of the emitted CL intensity . However, the comparison of samples #367 and #292, showing nearly identical CL intensities, proves that the degradation rate is clearly influenced by the defect density. #367, with a ~100x smaller defect density than #292, shows a 2x higher CL decay time (2150s against 1110s). Samples with identical QWs but different defect densities show the same behaviour.

Degradation experiments in spot mode show that they are two types of degradation behaviour, one typical of the majority of bright spots, the other typical of DSD but also encountered on some bright spots. The result of a degradation experiment on such a bright spot (i.e. showing the same behaviour as a DSD) in the spot mode is illustrated in Fig. 3 with the help of CL maps. The CL intensity profile of Fig. 3 shows actually two abrupt decreases that can be correlated to the extension of the DSD in the CL maps. The first decrease corresponds to the rapid creation of a DSD that joins three DSD with the bombarded one. The second decrease coincides with the further extension of the newly created DSD, about 4.5μm from the impact point on the sample surface. The extensions grow clearly in [100] and [010] directions.

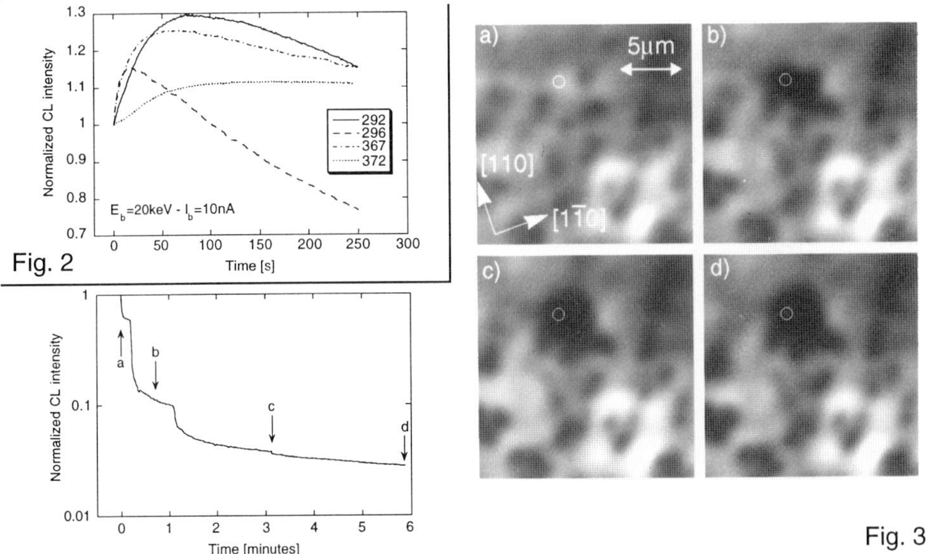

Fig. 2 Polychromatic CL intensity as a function of time in the scan mode at 5000X, I_b=10nA, E_b=20keV at 300K for samples #292, #296, #367, #372.

Fig. 3 CL polychromatic maps as a function of time, with corresponding graph displaying CL intensity versus time in spot mode on a DSD-like bright spot of sample #367 (T=300K, E_b=20keV, I_b=10nA). The position of the beam is indicated by a circle on the CL images.

Fig. 4b displays a CL intensity proflle representing the second type of behaviour under bombardment (typical of the majority of bright spots). The intensity increases at first (or decreases slowly), prior to a catastrophic degradation after a certain lapse of time. CL maps acquired before and after this rapid decrease show that it coincides with the creation of a DSD.

CL maps indicate that degradation experiments in the scan mode, on samples with high defect densities (#292, #296), result in the random extension of existing DSD and the creation of new DSD. On samples with low defect densities (#367, #372) however, the DSD grow and extend in [100] or [010] directions, for a part of them actually along existing DLDs.

that the $In_{0.04}Ga_{0.96}As$ buffer is slightly relaxed, but no additional threading dislocations or stacking faults are present in the buffer.

4. CATHODOLUMINESCENCE PROPERTIES

All examined structures show strong cathodoluminescence (with typical beam powers of 2×10^{-5} W) up to 300K. Polychromatic maps show non-uniform CL; the defects affecting the CL are visible mainly as dark spot defects (DSD), with a few dark-line defects (DLD) oriented along [100] and [010].

The following table resumes the measured defect densities before degradation, by CL and by TEM (for the GRINSCH layers only), along with lasing power threshold density at 83K under e-beam pumping:

Sample Nr.	threshold [$kWcm^{-2}$]	TEM misfit	[cm^{-2}] threading	CL DLD	[cm^{-2}] DSD
292	11	9.2×10^7	1.6×10^8	not visible	1×10^7
296	8.3	2.6×10^7	3.8×10^7	not visible	2×10^7
367	3.8	4.2×10^7	5×10^6	1×10^6	9×10^6
372	4.3	3.5×10^7	3×10^6	2×10^6	4×10^6

We see that the defects observed in CL do not readily correspond to defects observed in TEM, since neither for DLDs nor for DSD the densities match. The CL contrast is probably caused by dislocations crossing the well, but other non-radiative recombination centres, such as clusters of point defects, may also play a part.

5. ELECTRON-BEAM INDUCED DEGRADATION

Besides lasing power threshold density and wavelength, the device lifetime is an important characteristic of a laser. The best actual devices are still far from meeting industry standards, and it is therefore crucial to study the mechanisms that lead to device failure. It is known that the lifetime is directly related to the resistance to degradation of a device, and since our structures are designed for operation under electron-beam pumping, CL is certainly the best tool available for the investigation of the degradation.

Two kinds of degradation experiments are undertaken in the SEM. Scan mode degradation studies are performed by scanning the beam over a limited area of the sample surface at a fixed magnification of 5000X with TV scan rate. This scan mode emulates the operating mode of the laser under electron-beam pumping, in which typical regions of typically 150μm x 600μm in size are excited by the electron beam. In contrast to the lasing experiments, we cannot excite the samples with a pulsed beam and the beam powers that may be achieved with our SEM at this magnification (up to 50 W cm^{-2}) are two orders of magnitude lower than the actual lasing threshold powers at 83K (typ. 5 kW cm^{-2}). Spot mode degradation studies are performed by placing the beam at a point of the sample surface, and record the behaviour of bright and dark spots under bombardment. The spot mode allows us to reach beam powers comparable to the threshold powers (up to 25 kW cm^{-2} if we consider a generation volume 1μm in diameter).

The degradation in the scan mode involves actually three distinct phases (phases one and two are visible on the CL intensity versus time curves of Fig. 2). During the first phase, the CL intensity actually increases, with the duration and the amplitude of the increase depending strongly on the injection conditions. The second phase is visible as a near exponential decrease of the CL intensity, where we can define a CL decay time, corresponding to the time constant of the exponential fitting the intensity profile in a given time range. The third phase shows as a gradual increase of the CL decay time, followed by a very slow degradation, with CL intensities that are typically less than 2% of the initial intensity. Degradation occurs at 100K as well, but at a slower rate than at room temperature: the decay times are about 3 times longer at 100K than at 300K.

The influence of the threading dislocation density on the degradation is shown on Fig. 2, where CL deg-

6. CL AND TEM ON THIN-FOIL SAMPLES

CL and TEM observations performed on the same areas of thin-foils samples prepared for TEM measurements (planviews and cross-sections) permit a direct correlation between the CL contrast and the presence of defects, as well as between the degradation behaviour and the formation of defects.

In past studies, the DSD observed in CL or in electroluminescence were very often attributed to threading dislocations and/or stacking faults crossing the QW [4][5], because of the similarity between defect densities measured in CL and TEM. As for DLDs running along [100] directions, they are interpreted as alignments of point defect clusters [6][7], since no corresponding dislocations could be observed in TEM after degradation. It is important to verify these assumptions before undertaking both the study of the degradation defects and the interpretation of the degradation behaviour.

The comparisons of CL and TEM images on non-degraded samples show that, in most cases, DSD are caused by threading dislocations crossing the well. However, not every threading dislocation acts as a non radiative recombination center, since dislocations can be present at a CL bright spot. This fact is actually confirmed by the different values for the defect densities measured by TEM and CL (section 4.). We can assess from our comparisons that 88% of the DSD are due to threading dislocations. The other DSD are caused either by point defects in the well or by fluctuations of the thickness or of the chemical composition of the QW. As for stacking faults, only a minority of them gives raise to a faint CL contrast in CL, and are therefore not a dominant issue in our structures.

The observations of TEM thin foils before and after degradation reveals that the degradation induces the creation of a new type of defect. The TEM image of Fig. 4a is taken after a degradation in spot mode on a CL bright spot, and the corresponding CL intensity versus time profile (Fig. 4b) shows that the intensity decreased slowly in the first instants of the degradation, and then dropped rapidly with a loss of 85% of the intensity in less than 5s.

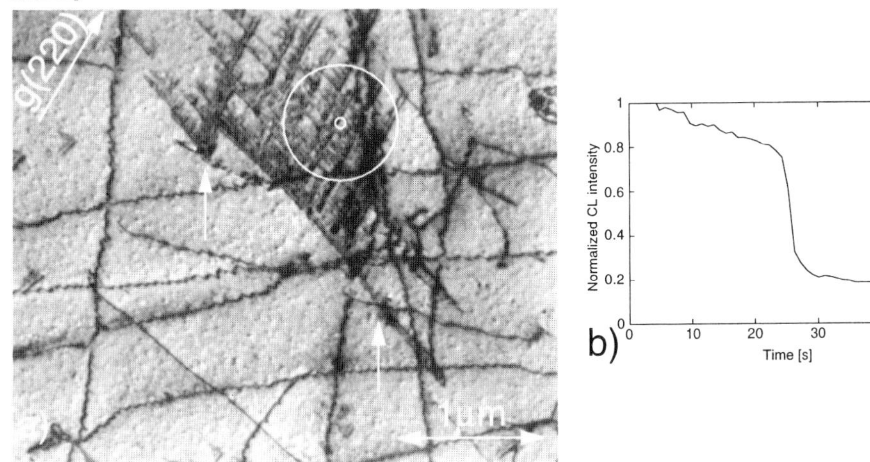

Fig. 4 (a) Bright field TEM image (in two beam (000)-(220) diffraction conditions) of a degraded bright spot, 20keV, 10nA. The large white circle indicates the extension of the generation volume. (b) Corresponding CL intensity versus time curve.

The TEM image allows an interpretation of the CL intensity profile: in the first phase, degradation defects are created at two threading dislocations (indicated by arrows) that are situated about 1µm from the point of impact of the beam (indicated by a little circle, while the large circle corresponds to the extension of the generation volume). Both defects grow with time, with the CL intensity slowly decreasing since an increasing number of carriers recombine non-radiatively at the defects after diffusion in the well. When the defects reach the region where electron-hole pairs are generated by the beam, they extend very rapidly to the whole generation region, causing the catastrophic decrease of the CL intensity. The defects continue thereafter to propagate, which is consistent with the observations on the CL maps taken during degradation experiments (see section 5.) where we saw that the DSD induced by the degradation continues its extension after the rapid CL intensity decrease. After this catastrophic event, the CL intensity decreases further, but at a very slow rate since nearly all generated carriers recombine non-radiatively at the degradation defects.

The defects visible on Fig. 4 have also been observed on II-VI laser structures designed for optical and/ or electrical pumping [4][6][8]. They are comprised of small dislocation loops, each approximately 20-60nm in length, that are arranged in V-shaped treelike structures. The "V" opens in a [100] or equivalent direction, and the branches of the "V" are nearly symmetric. The direction of the dislocation line varies from defect to defect and show an angle of 29-38° with the [100] or equivalent direction. We performed stereomicroscopy analysis on planviews in order to determine where the defects lie in the structure. It appears that the degradation defects lie in a plane parallel to the interfaces, higher in the structure than the misfit dislocation network. Further observations on cross-sections assert clearly that they are located in the QW or at the QW boundaries.

Some conclusions can be drawn from the observations of sections 5 and 6. Measurements performed at different beam energies with a constant beam power show that the CL decay time increases when the emitted CL intensity decreases. Furthermore, degradation defects can extend or be created as far as 5μm from the point of impact of the beam. These facts suggest that part of the processes governing the degradation are photoactivated. Carriers pairs, created by the beam and/or by the absorption of an emitted photon, can recombine non radiatively at a dislocation (or another recombination center). Such a non radiative recombination process transfers enough energy to the lattice to cause the diffusion of point defects, and hence the formation of dislocations [6].

7. CONCLUSION

We found that degradation in microgun-pumped blue-green GRINSCH lasers occurs via the formation of DSD and/or <100>-oriented dark line defects, in a manner compellingly similar to what has been observed in blue-green diode lasers. Our combined CL-TEM studies clearly indicate that the initial nonradiative regions visible in the CL images do not always correspond to extended defects visible in TEM such as threading dislocations or stacking faults, as usually believed. We observe furthermore that the degradation involves the formation of a characteristic type of defect, namely small dislocation loops within the ZnCdSe quantum well, which grow to form a characteristic network as degradation progresses.

We would like to thank G.Peter, B. Garoni and B.Senior at the Centre Interdépartemental de Microscopie Électronique (CIME) for the expert technical support with the microscopes.

REFERENCES

[1] Itoh S and Ishibashi A 1994 Proc. SPIE **2346** 2 and private communication
[2] Hervé D, Molva E, Vanzetti L, Sorba L and Franciosi A, 1995 Electron. Lett. **31** 459
Hervé D, Accomo R, Molva E, Vanzetti L, Sorba L and Franciosi A 1995 Appl. Phys. Lett. **67** 2144
[3] Li J H, Bauer G, Vanzetti L, Sorba L and Franciosi A 1995 J. Appl. Phys. (in press)
[4] Guha S, DePuydt J M, Haase M A, Qui J and Cheng H 1993 Appl. Phys. Lett. **63** 3107
[5] Tomiya S, Morita E, Ukita M, Okuyama H, Itoh S , Nakano K and Ishibashi A, Appl. Phys. Lett. 1995 **66** 1209
[6] Hovinen M, Ding J, Salokatve A, Nurmikko A V, Hua G C, Grillo D C, Li He, Han J, Ringle M and Gunshor R L 1995 Appl. Phys. Lett. 66, 2013
[7] Guha S, Cheng H, DePuydt J M, Haase M A and Qui J 1994 Mat. Sci. Eng. **B28** 29
[8] Hua G C, Otsuka N, Grillo D C, Fan Y, Han J, Ringle M D, Gunshor R L, Hovinen M and Nurmikko A V 1994 Appl. Phys. Lett. **65** 1331
[9] Bonard J-M, Ganière J-D, Akamatsu B and Araùjo D 1995 J. Appl. Phys (to be published)
[10] Bonard J-M, Ganière J-D, Vanzetti L, Sorba L, Franciosi A, Hervé D and Molva E 1995 Inst Phys Conf Ser (in press)

Inst. Phys. Conf. Ser. No 149
Paper presented at DRIP 95

Imaging of defects in InGaAs/InP avalanche photodetectors created during electrostatic discharge stress

H.C.Neitzert, R.Crovato, G.A.Azzini, P.Montangero and L.Serra

Centro Studi e Laboratori Telecomunicazioni (CSELT),
Via G. Reiss Romoli 274, 10148 Torino, Italy

Abstract. InGaAs/InP avalanche photodiodes failed during electrostatic discharge tests at voltages between 700V and 1000V. A failure analysis using photoluminescence imaging revealed the formation of star-like defects on the active area of the devices having an inhomogeneous photoresponse before the test. Differential phase contrast images of the damaged device showed metal migration from the guard ring towards the defect location. The reverse bias current after degradation increased for several orders of magnitude. For the degraded photodiodes with defects on the active layer a negative differential resistance has been observed in the forward current-voltage characteristics.

1. Introduction

Electrostatic discharges (ESD) pose a severe problem to the reliability of active components in fiber-optic data transmission systems. With increasing data rates and hence decreasing device capacitances the ESD sensitivity of optoelectronic components becomes more and more critical. Several studies regarding the impact of electrostatic discharge on lasers can be found [1,2], but few data have been published concerning the ESD sensitivity of photoreceivers. In particular a study of the electrostatic discharge sensitivity of avalanche photodiodes with a detailed failure analysis does not exist. Recently we performed a comparative study of the ESD sensitivity of four different families of InGaAs/InP based avalanche photodiodes (APDs), all from the separate absorption grading multiplication type [3]. In this study [4] for three families the ESD testing resulted in the device breakdown at the perimeter of the diodes. The position of the ESD related defects for the APDs from family four, however, was dependent on the homogeneity of the electric field across the active area of the devices. Only for APDs that have large inhomogeneities in the photoresponse, the ESD damage has been observed on the active area. Here we will present detailed results of electrical measurements and a defect analysis of one of these latter APDs, where the created defects have been observed on the active photodiode area.

2. Device structure and test conditions

The investigated devices were commercially available top illuminated separate absorption grading multiplication (SAGM) avalanche photodiodes, grown by liquid phase epitaxy. On a <111> n^+ - InP substrate covered by a n - InP buffer layer is firstly grown the n - InGaAs absorption layer, followed by an InGaAsP grading layer and a n - InP multiplication layer. The multiplication region is defined to a diameter of 50μm laterally by the mesa formation via chemically etching of this n - InP multiplication layer, which is successively covered by a n^- - InP layer. This slightly doped n^- - layer and a p^- - InP guard ring are used to reduce electrical breakdown at the junction edges. A p^+ - InP layer is finally formed by Cd diffusion and a SiN layer is deposited on top, which serves as surface passivation layer and antireflective coating. Ti/Pt/Au metallic contacts are deposited as a ring structure on the top of the guard ring region. More details about the device structure can be found in **[5]** and a detailed electro-optical characterization of these devices before ESD testing has been given in **[6]**.

The electrostatic discharge tests have been performed using an IMCS System 700 ESD tester. One positive and subsequently one negative ESD pulse - conforming to the human body model (HBM) - with successively increasing amplitudes have been applied to the devices, starting from 100V until 400V in steps of 50V and subsequently in steps of 100V. Positive pulse means in this case the application of the voltage in the forward direction of the photodiode. The dark current-voltage characteristics of the APDs has been monitored after each ESD pulse.

3. Results and Discussion

3.1 Electrical Characterization

Six APDs of the family described above have been tested. All devices failed within a rather narrow voltage interval between 700V and 1000V and in all but one case this occurred during the application of the negative ESD pulse. All measurements subsequently shown have been performed on a particular device of these six APDs, that failed during a negative pulse with an amplitude of 800V.

The current transients during the application of the ESD pulses have been measured using a wideband current probe. Typical current transients for the above mentioned device are reported in Fig.1. The transient, labelled A, has been taken during the application of a positive pulse with 800V amplitude, corresponding to a maximum current value slightly above 500mA. This value is consistent with the HBM conditions, where a 100pF capacitor is discharged through the device under test in series with a 1.5kΩ resistor. The forward biased APD can be approximated to a small additional series resistance and the current pulse decay is consequently purely exponential. Negative pulses with pulse amplitudes below 800V showed a similar exponential decay, but with a minor distortion of the leading edge due to the impedance change of the device under test (DUT), when the APD changes from the breakdown regime to voltages below breakdown and vice-versa. The negative pulse with a voltage amplitude of 800V (see Fig.1B), however, exhibits a sharp distortion of the current transient a few ns after pulse application. The reverse dark current of the APD after this negative 800V pulse increased for several orders of magnitude compared to the dark current before. A comparison of the current transients during ESD testing of all six APDs confirmed that this type of pulse distortion is always and exclusively seen for negative ESD pulses leading to the failure of the device. This distortion of the current pulse can be explained by a fast and dramatic change of the DUT impedance during the damaging of the photodiode. It is the consequence and not the origin of the defect creation. In the case of the particular device presented in this work, a second negative ESD pulse with an amplitude of 800V has been applied after the first destructive pulse with the same amplitude of 800V. As can be seen in Fig.1C the disturbance in the current transient is not observed during this second negative pulse. The failed device has already a very low resistivity also in reverse bias regime before the second pulse at -800V. Therefore the circuit resistance is dominated in the whole course of the transient by the 1.5kΩ resistor like in the case of the positive pulse and no pulse distortion is observed.

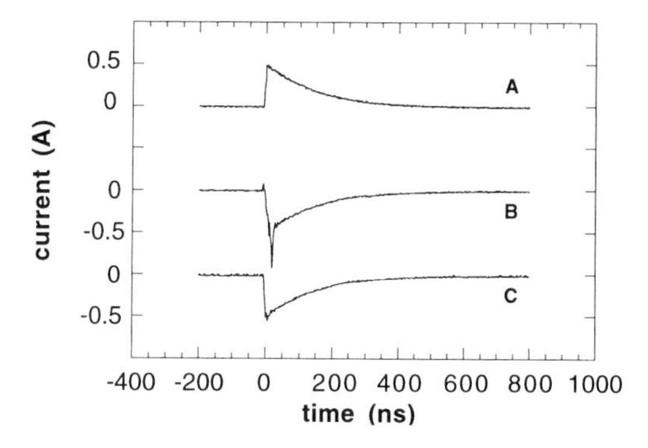

Fig.1 Current transients measured: A) during a positive ESD pulse of +800V, B) during a negative ESD pulse of -800V, C) during a second ESD pulse of -800V

In Fig.2A the current-voltage (I-V) characteristics of the reverse biased APD before ESD testing and after the application of pulses up to -700V are displayed. It can be seen that for this device no changes in the dark I-V are observed after the application of ESD pulses with amplitudes slightly below the voltage, where the APD failed. After the -800V pulse, however, the current-voltage characteristics is completely changed, as seen in Fig.2B. In reverse direction the current increased now for more than seven orders of magnitude and also in forward direction a slight current increase can be seen. The first measurement after ESD-testing (full line) revealed another interesting feature. At about 0.6V the forward current reaches already a value of 400μA, but decreases then instantly for further increasing voltages to about 250μA and increases finally again. After some cycles of this type of measurements the phenomena of negative differential resistance (NDR) vanished and the final stable I-V characteristics (dotted line in Fig.2B) has been found. This NDR feature in the forward I-V-characteristics after ESD testing has been found only for an increasing voltage ramp in forward bias direction and under the condition that the device has been polarized in reverse before. A possible explanation of this phenomena is related to the trapping of electronic charges during the application of a negative bias voltage. This trapped charge lowers the height of a barrier ruling the transport in the forward bias regime. For a given current value in forward bias direction, however, this charge is suddenly released and consequently the barrier height increases and the current drops instantly.

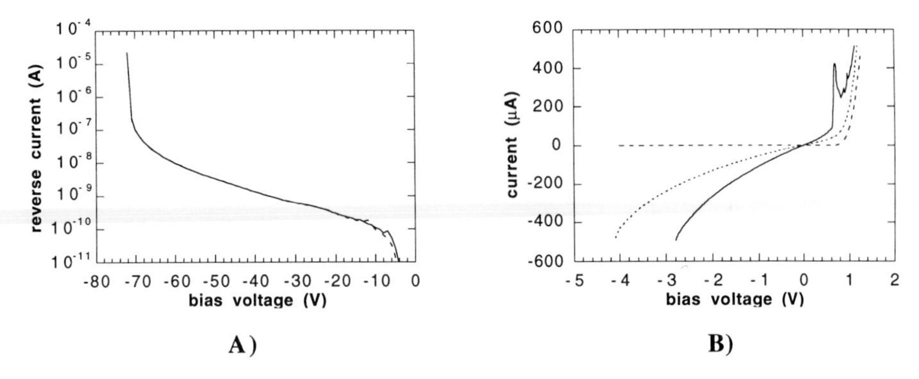

Fig.2 Dark current-voltage characteristic of an APD measured at 25°C: A) In the reverse bias regime before ESD test (full line) and after an ESD pulse with -700V (dotted line)
B) Before ESD test (broken line), first measurement after damage by an ESD pulse of -800V (full line) and final stable measurement (dotted line). The measurements have been performed, starting from the most negative bias voltage.

3.2 *Failure analysis*

In order to identify the position and type of the created defects, photoluminescence and reflected light images of the APDs have been taken before and after ESD testing. The scanning optical microscope (SOM), which has been used to obtain these images has been described in detail elsewhere **[7]**. The photoluminescence signal with an excitating beam at a wavelength of 1064nm has been measured using an InGaAs detector and the incident light beam has been suppressed by suitable notch filters.

Photoluminescence images with excitation at different wavelengths, taken before the ESD test, did not show any defect feature. Information about the homogeneity of the electric field across the active area can be obtained by imaging of the optical beam induced current (OBIC). Fig.3 shows an OBIC image with an applied reverse bias voltage of 71V using an excitation wavelength of 1320nm, which is mainly absorbed in the InGaAs layer.

Fig.3 OBIC image of an APD measured at 25°C with illumination at 1320nm (I_0=5W/cm^2) for a bias voltages of 71V. Bright areas correpond to regions of enhanced photocurrent.

The OBIC image shows large inhomogeneities of the photoresponse across the whole active area of the APD. A regular pattern of parallel stripes is observed with alternating photocurrent minima and maxima. At the bias voltage of 71V, which is close to the reverse bias breakdown, the multiplication factor of the photocurrent is very sensitive to variations of the local electric field. The observed stripes may therefore reflect variations in the local dopant concentration or layer thickness.

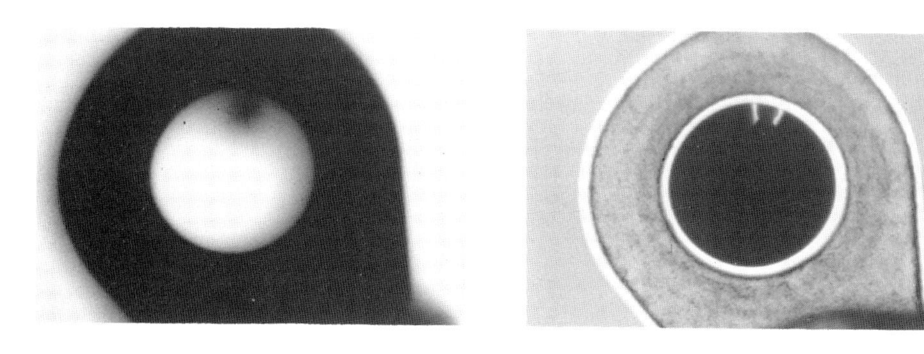

A) B)

Fig.4 A) Photoluminescence image (excitation at 1064nm) and B) reflected light image (illumination at 488nm) of an APD after ESD damage

The photoluminescence measurement (excitation at 1064nm) after ESD damage of the device is shown in Fig.4A. A star-like dark structure with decreased photoluminescence efficiency can be seen on the active area of the APD. This local defect touches in three points the nearby guard ring metallization. The reflected light image (Fig.4B) reveals also two of these connections with an increased reflectivity. This structure has also been observed in the differential phase contrast mode (illuminated with 488nm light) of the SOM (see Fig.5), where the beginning of metal lift-off at the edge of the guard ring and the groove formation with metal migration towards the defect can be seen. This must be caused by local heating due to the high current density towards the short-circuit in the center of the star-like feature.

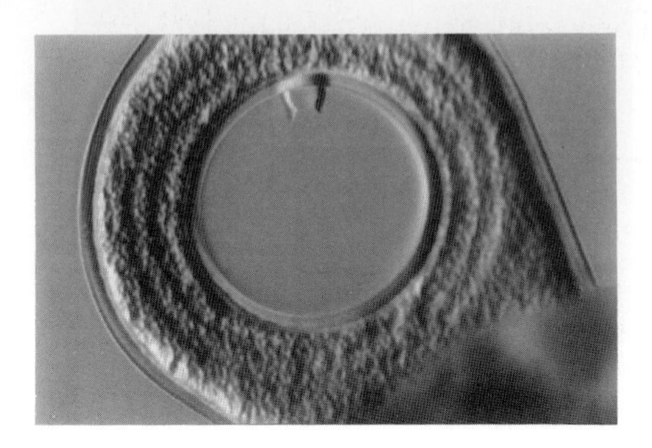

Fig.5 Differential phase contrast image (illumination at 488nm) of an APD after ESD damage

It should be noted that the negative differential resistance seen in Fig.2B has been found for all devices from the described APD family, which during ESD testing developed the above described star-like defects.

4. Conclusions

InGaAs/InP based avalanche photodiodes failed during electrostatic sensitivity tests for pulse amplitudes between 700V and 1000V. For devices with an inhomogeneous electric field distribution, photoluminescence imaging revealed star-like defects on the active photodiode area after the degradation. Differential phase contrast imaging showed metal migration from the guard ring towards these defects.

5. References

[1] Twu Y, Cheng L S, Chu S N G, Nash F R, Wang, K W and Parayanthal P 1993 *J. Appl. Phys.* **74** 1510-1520

[2] Jeong J, Park K H and Park H M 1995 *IEEE J. Lightwave Technol.* **13** 186-190

[3] Pütz N 1991 *J. Crystal Growth* **107** 806-821

[4] Neitzert H C, Crovato R, Montangero P, Azzini G A and Serra L, unpublished results

[5] Kobayashi M, Yamazaki S and Kaneda T 1984 *Appl.Phys. Lett.* **45**, 759-761

[6] Montangero P, Azzini G A, Neitzert H C, Ricci G and Serra L, *Microelectron. Reliab.*, to be published

[7] Azzini G A, Arman G and Montangero P 1992 *Microelectron. Reliab.* **32** 1559-1604

Inst. Phys. Conf. Ser. No 149
Paper presented at DRIP 95

On the nature of large-scale defect accumulations in Czochralski-grown silicon

V P Kalinushkin[1], A N Buzynin, V A Yuryev[2], O V Astafiev and D I Murin

General Physics Institute of the Russian Academy of Sciences, 38, Vavilov Street, Moscow, GSP-1, 117942, Russia

Abstract. Czochralski-grown boron-doped silicon crystals were studied by the techniques of the low-angle mid-IR-light scattering and electron-beam-induced current. The large-scale accumulations of electrically-active impurities detected in this material were found to be different in their nature and formation mechanisms from the well-known impurity clouds in a float zone-grown silicon. A classification of the large-scale impurity accumulations in CZ Si:B is made and point centers constituting them are analyzed in this paper. A model of the large-scale impurity accumulations in CZ-grown Si:B is also proposed.

1. Introduction

The detection of the large-scale impurity accumulations (LSIAs) with the sizes ranged from several to several tens μm in CZ Si by means of the low-angle light scatter (LALS) [1] was reported for the first time in Ref. [2]. It was supposed in that work that LSIAs are analogous in their nature to the oxygen and carbon clouds observed in FZ Si in Ref. [3]. It was shown, however, as a result of the research of Si crystals grown at variable growth rate done by LALS and EBIC that most of LSIAs in CZ Si have a shape close to cylindrical [4] which contradict the cloud model [2]. In the present work, an attempt is made to select different in their nature types of LSIAs in CZ Si and an information about their parameters as well as the influence of different thermal treatments on them is given.

2. Experimental details

Industrial 76 and 100-nm substrates of CZ Si:B grown in the <100> and <111> directions — $\varrho \sim$ several Ω cm — were studied in this work. The oxygen concentration in them

[1] E-mail: VKALIN@KAPELLA.GPI.RU.
[2] E-mail: VYURYEV@LDPM.GPI.RU.

ranged from 6×10^{17} to 10^{18} cm^{-3}, the carbon concentration was less than 10^{16} cm^{-3}. The thickness of 76-mm substrates was 380 μm, that of 100-mm ones was 500 μm.

The investigation was carried out by LALS and EBIC. CO- and CO_2-lasers oscillating at the wavelength of 5.4 and 10.6 μm, respectively, were used in LALS to select the scattering by free carrier accumulations [5]. To determine the activation energies (ΔE) of the centers constituting LSIAs, the temperature dependances of LALS were investigated in the range from 85 to 300 K [6]. A shape of LSIAs was determined from the dependances of the LALS diagrams on the sample orientation with respect to the detection plane. The plasma etching of the sample surface in a special regime before the Schottky barrier creation greatly increased the sensitivity of EBIC to electrically-active defects in crystals [7].

In the experiments on annealings, wafers were cut into four sections. One of them was not treated, the others were subjected to either isothermal processes at 600 or 800°C for 24, 48 and 120 h, respectively, or high-temperature treatments at 965, 1100, 1150, 1200 and 1250°C for several tens minutes. The treatments at $T > 1200$°C resulted in the formation of a large amount of defects of structure which were revealed by the selective etching (SE). The substrates grown in the <100> direction were subjected to both the former and the latter treatments, while those grown in the <111> direction were treated only in the latter way.

3. Results

3.1. Initial samples

Fig. 1 demonstrates the EBIC images of defects. The samples contain many non-uniformities with the sizes from several to several tens μm — mainly cylindrical. In addition, spherical defects are also seen.

Cylindrical defects (CDs). There are sections of LALS diagrams, the shape of which is dependent on the sample orientation to the detection plane (Fig. 2, $\theta < 4.5°$). These sections are well fitted with the curves of scattering by cylinders [1] with the diameters from 3–4 to 8–10 μm and length from 15 to 40 μm depending on a sample. They predominantly oriented along the <110> direction. It is seen from the EBIC patterns that CDs have rather elliptical or curved-cylindrical shape (Fig. 1 (a, b)). We assume the CDs revealed by EBIC and LALS to be the same defects similar to CDs observed in Ref. [4].

The concentration of CDs — the most usual defects in CZ Si:B — estimated from EBIC ranged from 10^6 to 10^7 cm^{-3}. We could not find a dependance of the CD concentration on oxygen concentration, growth direction, ingot diameter or location on a wafer. Nonetheless we found their concentration to vary within a wafer as well as in different wafers.

LALS at 10.6 and 5.4-μm wavelength showed CDs to be domains of the enhanced free carrier concentration [5]. Using the CD concentrations from EBIC, the variations of the dielectric function ($\Delta\varepsilon$) and the maximum free carrier concentration in them (Δn_{max}) were evaluated [1]: they are (1–4) $\times 10^{-4}$ and (3–10) $\times 10^{15}$ cm^{-3}, respectively. CDs occupy less than 3 % of crystal volume and the total amount of impurities contained in them (N_i) is not greater than 3×10^{14} cm^{-3}.

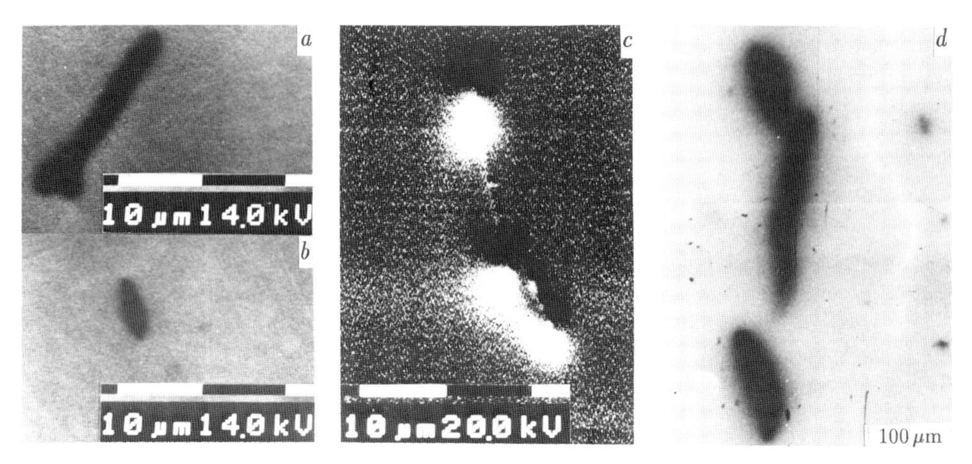

Figure 1. EBIC microphotographs of as-grown CZ Si:B: cylindrical (a, b), spherical (c) and superlarge (d) defects.

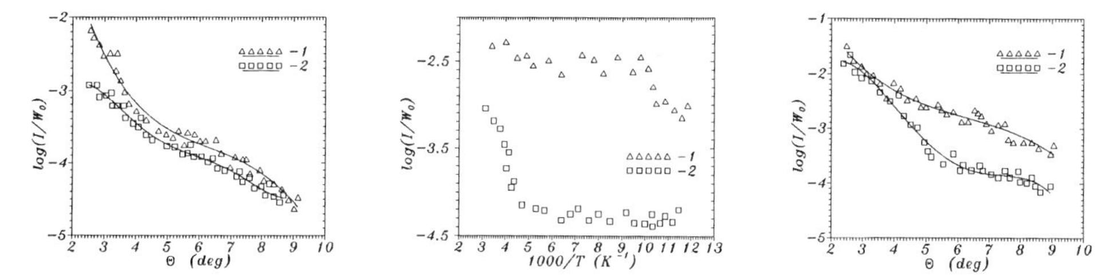

Figure 2. LALS diagrams for initial CZ Si:B, orientation with respect to detection plane (deg): 0 (1), 90 (2).

Figure 3. LALS temperature dependances for cylindrical (1) and spherical (2) defects in initial CZ Si:B.

Figure 4. LALS diagrams at different temperatures (K): 300 (1), 110 (2).

LALS temperature dependances showed for CDs a small (2–3 times) drop of the scattering intensity (I_{sc}) at about 90 K (Fig. 3, curve 1). This allows us to state that LALS by CDs at 300 K at 10.6 μm is controlled by the centers with $\Delta E \approx$ 40–60 meV containing in CDs. Naturally another defects, such as deep or compensating centers as well as precipitates, inclusions and structural imperfections can also be contained in CDs.

Spherical defects (SDs). Besides the above sections of LALS diagrams, those independent of the sample orientation were also observed (Fig. 2, $\theta > 4.5°$). These sections are well fitted with the curves of light scattering by spherical defects with the Gaussian profile of ε [1] and sizes from 5–8 to 20 μm. Such defects are also seen in the EBIC pictures (Fig. 1 (c)). Their concentration is usually about 10^5 cm^{-3}, $\Delta\varepsilon \approx (1\text{–}3) \times 10^{-3}$, $\Delta n_{max} \approx (3\text{–}9) \times 10^{16}$ cm^{-3} [1]. SDs occupy less than 0.04 % of the crystal volume, $N_i \lesssim 4 \times 10^{13}$ cm^{-3}.

LALS temperature dependances showed SDs, like CDs, to be domains with enhanced concentration of ionized at 300 K impurities with $\Delta E \approx 120$–160 meV. These impurities are "frozen out" at about 250 K, so SD-related scattering is smothered in the range from 90 to 250 K which enables the accurate selection of CD-related scattering in LALS diagrams (Fig. 4).

Superlarge (SLDs) and *small* (SmDs) *defects.* The sections of LALS diagrams well approximated with the curves of light scattering by defects with the sizes greater than 50 μm were sometimes observed. These defects appeared to have an asymmetrical shape. Sometimes SLDs were seen in the EBIC photographs (Fig. 1 (*d*)).

The sections of LALS diagrams independent of the scattering angle ("plateaux") were often observed at 5.4 μm (and sometimes at 10.6 μm) which correspond to defects with the sizes less than 4–5 μm. I_{sc} for SmDs was also independent of the probe wavelength, so they are domains with the enhanced free carrier concentration. SmDs were sometimes observed in the EBIC picture as well. Although we could not unambiguously determine their shape, SmDs seem to be very small CDs and SDs rather than a separate class of defects. This was verified by LALS temperature dependances: SmDs were "frozen out" at 90 K when CDs predominated and at 250 K if SDs predominated.

3.2. Annealed samples

The LALS diagrams and EBIC pictures for the annealed crystals did not differ in general features from those for the as-grown samples. The following peculiarities may be emphasized.

1. In crystals annealed in the temperature range of 600–1100°C, the light scatter by SDs was greatly (but not completely) suppressed, and CDs and SmDs predominated. EBIC showed mainly CDs and SmDs too.

2. Annealing at $T > 1100$°C resulted in predominance of SD-related scattering and general growth of I_{sc}. A great number of SDs was observed by EBIC (Fig. 5).

3. After annealing at $T > 1200$°C, the centers with the same ΔE as in the as-grown samples composed LSIAs.

4. Annealing at 800°C and short (up to 48 h) treatment at 600°C did not change ΔE of the centers composing CDs and SDs. Longer treatment at 600°C resulted in prevailing of the centers with $\Delta E \approx 70$–90 meV in CDs and SDs.

5. After 120-h annealing at 600 and 800°C, SLDs became more habitual than in the as-grown samples. The centers with $\Delta E \approx 130$–170 mev were contained in SLDs.

4. Discussion

It is difficult now to determine the nature of LSIAs in CZ Si:B unambiguously. Ii is clear, however, that CDs are domains with the enhanced free carrier concentration caused by point centers with $\Delta E \approx 40$–60 meV. CDs look like the defects observed by EBIC after oxidizing annealings [8]. Our research gave an evidence to the presence of these defects in initial crystals, moreover the CD-related sections of LALS diagrams do not change after high-temperature annealings. In our opinion, however, CDs are the impurity atmospheres (IAs) around defects-precursors, e.g. stacking faults (SFs). Remark that some authors connect the contrast of EBIC patterns with the formation of precipitate colonies around

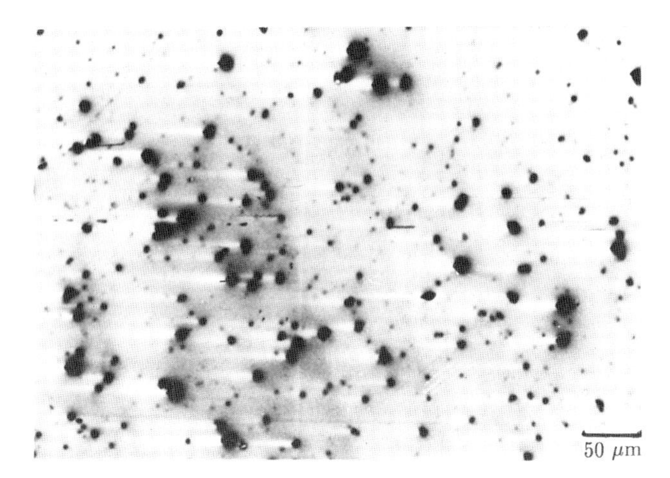

Figure 5. Typical EBIC microphotograph of defects in CZ Si:B annealed at $T > 1100°C$.

SFs [8]. As these colonies may have no influence on the free carrier concentration in IAs, the following scenario might be proposed.

In initial wafers, the precipitate concentration in IAs is low. At the same time, the dissolved ionized impurity concentration is high enough but insufficient for the precipitate formation, hence I_{sc} is high and the recombination contrast in EBIC is low. During high-temperature annealing, precipitate colonies arise in CDs but the dissolved impurity concentration and Δn change weakly (e.g. because the impurity concentration in CDs in the as-grown samples was close to the saturation limit or due to the growth of the compensation degree). The recombination contrast will have grown and I_{sc} will have changed weakly and randomly.

From the other hand, the enhanced EBIC contrast may be caused by the specificity of the sample preparation. Two variants are possible. The centers enhancing the EBIC contrast may arise as a result of the plasma etching applied. Alternatively, an "exhaustion" of CDs may be a result of the chemical etching usually applied for sample preparation for EBIC.

Thus, the hypothesis according to which CDs are IAs around SFs does not meet contradictions. As to the point defects composing CDs, we suppose them to be the "new" thermal donors [9] whose ΔE is close to the estimates made [3]. The influence of 600°C annealing on ΔE of the centers composing CDs indirectly verifies the assumption [4]. New experiments are required to obtain more evidences to the model proposed, though.

SDs also are domains with the enhanced dissolved impurity and free carrier concentrations. We assume SDs to be IAs around defects of structure (e.g. precipitates). This hypothesis is confirmed by the growth of the SD concentration during the high-temperature annealings and correlation with the appearance of structural defects revealed

[3] Although, some alternatives exist [10], and B and [Cu–O] are among them.

[4] The growth of ionization energy of "new" thermal donors as a result of long-term annealing at 650°C was reported in [11].

by SE. This assumption also have no sufficient evidences [5] and require an additional research, however.

Some more definite assumption can be made for the centers composing SDs. It was supposed in [12] that SDs likely contain the centers with the negative correlation energy. Regarding annealing at 600°C resulting in predominance of the centers with changed ΔE in SDs, we assume that, like in CDs, double thermal donors are contained in SDs [6].

SmDs are likely very small CDs and SDs. The nature of SLDs is hard to be discussed now. It is evident only that they are domains with the enhanced concentration of the free carrier and dissolved impurities.

5. Conclusion

On the basis of the above we can summarize in the conclusion that at least two types of LSIAs different in their nature and composition exist in CZ Si:B. Their parameters determined for the investigated in this work group of crystals are rather typical for the industrial Si:B with the specific resistivity of several Ω cm. Some additional details of this research can be found in Ref. [10].

References

[1] Voronkov V V, Voronkova G I, Zubov B V *et al* 1981 *Sov. Phys.-Solid State* **23** (1) 65–75
Kalinushkin V P 1988 *Proc. Inst. Gen. Phys. Acad. Sci. USSR* vol 4 (New York: Nova) pp 1–79

[2] Voronkov V V, Voronkova G I, Murina T M *et al* 1983 *Sov. Phys.-Semicond.* **17** (12) 2137–42

[3] Voronkov V V, Voronkova G I, Zubov B V *et al* 1979 *Sov. Phys.-Semicond.* **13** (5) 846–54

[4] Buzynin A N, Zabolotskiy S E, Kalinushkin V P *et al* 1990 *Sov. Phys.-Semicond.* **24** (2) 264–70

[5] Kalinushkin V P, Masychev V I, Murina T M *et al* 1986 *Journ. Tech. Phys. Letters* **12** (3) 129–33

[6] Zabolotskiy S E, Kalinushkin V P, Murin D I *et al* 1987 *Sov. Phys.-Semicond.* **21** (8) 1364–8

[7] Buzynin A N, Butylkina N A, Lukyanov A E *et al* 1988 *Bul. Acad. Sci. USSR. Phys. Ser.* **52** (7) 1387–90

[8] Schmalz K, Kirscht F-G, Niese S *et al* 1985 *Phys. Status Solidi* (a) **89** 389–95

[9] Bouret A 1985 *Proc. 13 Int. Conf. on Defects in Semiconductors* ed L C Kimerling 129–34

[10] Astafiev O V, Buzunin A N, Buvaltsev A I *et al* 1994 *Sov. Phys.-Semicond.* **28** (3) 407–15

[11] Kanamori A and Manamori M 1979 *J. Appl. Phys.* **50** 80–6

[12] Valiev K A, Velikov L B, Kalinushkin V P *et al* 1990 *Sov. Phys.-Microelectronics* **19** (5) 453–9
Abdurahimov D E, Bochikashvily P M, Vereschagin V L *et al* 1992 *Microelectronics* **21** (1) 21–7

[5] Some alternatives to the assumption — the cloud models — are discussed in [10].
[6] Possible alternatives to these centers are discussed in [10], they are B_i, [O–V], [C_i–C_s], *etc.*

Inst. Phys. Conf. Ser. No 149
Paper presented at DRIP 95

Recognition and Characterization of Precipitates in Annealed Cz and FZ Silicon by S.I.R.M. and L.B.I.C.

C. VEVE, N. GAY, J. GERVAIS and S. MARTINUZZI

Laboratoire de Photoélectricité des Semi-Conducteurs E.A. 882 "D.S.O."
Faculté des sciences et techniques de Marseille-St. Jérôme
13397 MARSEILLE Cedex 20 - FRANCE

Abstract: A Scanning InfraRed Microscope, a Light Beam Induced Current mapping equipement, a Fourier Transform infrared spectroscope, and a minority carrier diffusion length measurement tool have been associated to detect precipitates in annealed Czochralski silicon wafers and to evaluate their recombination strenght with or whithout metallic contamination. The influence of a phosphorus diffusion was also investigated. A comparison is made whith Float Zone silicon which contains less oxygen.

1. Introduction.

In Czochralski grown (Cz) silicon oxygen is present under the form of a supersaturated interstitial solid solution in a large temperature range. When the temperature is sufficiently high in order that oxygen atoms become mobiles, they tend to agglomerate and to form precipitates of silicon oxyde with associated crystallographic defects. These precipitates and defects could be electrically active because they can recombine minority carriers or they can also induce leakage currents in the space charge region of p-n junctions.

The oxygen precipitates have been largely studied and several review papers were recently written, describing their nucleation, their morphology and chemical composition [1-3]. Despite this scientific production knowledge is still missing in order to predict and control the behaviour of oxygen in wafers used for device processing. The role of metallic precipitates is also a subject of controversy, as the morphology, the density and the electrical activity of these precipitates is depending on the presence of oxygen precipitates.

One of the reasons of this imperfect understanding has been the lack of non destructive techniques which can detect and recognize the precipitates and which can evaluate their recombination strength.Conversely, in Float Zone grown silicon wafers (FZ) the interstitial oxygen concentration $[O_i]$ is sufficiently low in order to avoid the formation of precipitates and it could be easier to investigate the properties of metallic precipitates.

In the present work, we have associated a Scanning Infrared Microscope (S.I.R.M.), the Light Beam Induced Current (L.B.I.C.) mapping tool or the Microwave Detected Photoconductivity Decay technique and the Fourier Transform Infrared Spectroscopy (F.T.I.R.), to investigate the electrical influence of precipitates and to follow their evolution during different treatments.

2. Experimental.

We used two types of (111) oriented silicon wafers. The Cz ones were P-type, boron doped ($10^{15}cm^{-3}$), both sides polished and $[O_i]$ is about $10^{18}cm^{-3}$. The FZ ones were N-type phosphorus doped ($8.10^{13}cm^{-3}$) by neutron transmutation doping (N.T.D.) and $[O_i]$ is

Table 1 : Nature and purpose of the different treatments

Purpose		Process	Temperature & time of treatment	Label
Oxygen precipitation	Nucleation	Anneal in pure Ar	750°C for 16h	**N**
	Growth	Anneal in pure Ar	900°C for 24h	**G**
Injection of Si_i	Phosphorus diffusion	Anneal under $POCl_3$ flow	900°C for 4h	**P**
Cu contamination		Electron gun evaporation & anneal in pure Ar	950°C for 1h	**Cu**
Ni contamination		Cathodic pulverization & anneal in pure Ar	950°C for 1h	**Ni**

about $2.10^{17}cm^{-3}$. Different thermal treatments were applied, like annealing in pure argon at 750°C for 16h (treatment N) and then at 900°C for 24h (treatment G), in order to create nucleation sites and to grow oxygen precipitates, respectively. Some samples were also submitted to metallic contamination by copper (treatment Cu) or nickel (treatment Ni). Annealed and contaminated samples were also submitted to a phosphorus diffusion from a $POCl_3$ source at 900°C for 4h (treatment P), in order to inject an excess of self-interstitials (Si_i) in the material. The table 1 summarizes the nature and the goal of the different treatments applied to the investigated samples.

$[O_i]$ was determined at room temperature by F.T.I.R. spectroscopy by means of the $1107cm^{-1}$ absorption band and using the conversion coefficient $3.03.10^{17}cm^{-3}$ [4].

Minority carriers diffusion lengths (L_n) were determined from the spectral variations of the photocurrent and of the light absorption coefficient in the near infrared ($860 < \lambda < 980$ nm). The current collecting structures were semitransparent Al-Si diodes or p-n junctions (phosphorus diffusion) for the Cz samples, and Au-Si diodes or n-p junctions (boron implantation) for the FZ ones. In FZ samples life time of minority carriers was also measured by means of microwave detection photoconductivity, the surfaces being passivated by immersion of the wafer in an iodine solution.

The L.B.I.C. tool with the extension to the diffusion length mapping has been described already [5] and we recall only that the light spot diameter is below 10μm when the wavelength varies in the range between 800 and 1000nm.

The S.I.R.M. realized in our laboratory following the idea of Booker et al. [6] has also been already described [7]. Let us recall that a three dimentional picture of the sample can be built and precipitates can be detected even with a size as small as 0.1μm and distinguished if separated by more than 2-3μm. The contrast obtained when precipitates are metallic in nature is higher compared to that given by a dielectric one [8].

3. Results and discussion.

3.1.Thermal treatment

Table 2 Results obtained after the different oxygen precipitation annealling treatments on Cz samples

Label	F.T.I.R.	S.I.R.M.	Ln	L.B.I.C.
N	1080 cm^{-1}	no precipitates	15μm	ring like distribution
N+G	1080 cm^{-1}&1225 cm^{-1}	precipitates (different size & contrast)	2μm	ring like distribution
N+G+P	1120 cm^{-1}&1225 cm^{-1}	no precipitates	7μm	ring like distribution

3.1.1 Cz samples. The as-received Cz samples are found to have value of L_n higher than 200μm homogeneously distributed over the entire wafer (as verified by the L.B.I.C.maps), and no precipitates are revealed by S.I.R.M.. The table 2 displays the results concerning oxygen precipitation after N and/or G treatments applied to the Cz samples.

Treatment N leads to a degradation of L_n, correlated with a ring like distribution of photocurrent in L.B.I.C. maps and with an absorption band located at 1080 cm^{-1} in F.T.I.R spectroscopy, which could be due to nucleation centers. No precipitates are revealed by S.I.R.M.. After two steps annealing (N+G) the S.I.R.M. detects the presence of dark spots identified as oxygen related precipitates. Indeed, two absorption bands located at 1080 cm^{-1} and 1225 cm^{-1} appear in F.T.I.R. spectra, which are attributed to SiO_x amorphous precipitates and platelets respectively [9;10]. The L.B.I.C. map reveals a ring like distribution of recombination centers, and L_n collapses to 2 μm.

When phosphorus diffusion is applied after the two step annealings (treatments N+G+P) no precipitates can be revealed by S.I.R.M. certainly because of their shrinkage by Si_i injection. The absorption band at 1080 cm^{-1} is replaced by another one at 1120 cm^{-1} often associated to that at 1225 cm^{-1} [10] which is always present, and associated to the ring like distribution in the L.B.I.C. map. The mean value of L_n increases to only 7 μm. This can be due to the strong recombination strength of defects related to the 1225 cm^{-1} absorption band and the ring like distribution of L_n. Indeed, the Si_i emitted by the formation of amorphous precipitates can contribute to the formation of oxygen stacking faults (O.S.F.) and platelets, which follows the well known oxygen striations of the as-grown material [11].
Note that a blank for treatment P has been done, i.e. 900°C for 4h in argon flow. In as-grown samples, it leads to the decrease of $[O_i]$ to about $5.10^{17} cm^{-3}$, to the formation of precipitates detected by S.I.R.M. and to the appearence of the ring like distribution in L.B.I.C. maps, while L_n decreases to about 40μm. In the two step annealed wafers there is not a marked difference after this additional treatment. This let us conclude that phosphorus shrinks and inhibates the formation of large precipitates revealed by S.I.R.M., but has no effects on the defects associated to L_n ring like distribution.

3.1.2. FZ samples. On the as-grown FZ samples the diffusion length is around 250μm (certainly due to the damages related to N.T.D.[12]) The thermal treatments (N+G) give no comparable results excepted a decrease of L_n to 100μm, confirming that the S.I.R.M.-detected precipitates present in Cz material are oxygen related.

3.2. Metallic contamination.

Table 3: Results concerning the copper and nickel contamination of Cz samples

Label	F.T.I.R.	S.I.R.M.	Ln	L.B.I.C.
Cu	1107 cm^{-1} $[Oi]\#1.0\ 10^{18}$ at.cm^{-3}	large starshaped precipitates	51μm	starshaped precipitates
N+G+Cu	1080 cm^{-1} & 1225 cm^{-1}	small precipitates	2μm	ring like+stacking faults
Cu+P	1107 cm^{-1} $[Oi]\#1.0\ 10^{18}$ at.cm^{-3}	no precipitates	12.2μm	——
Ni	1107 cm^{-1} $[Oi]\#1.0\ 10^{18}$ at.cm^{-3}	spot + rod like precipitates	40μm	spot-like precipitates
N+G+Ni	1080 cm^{-1}	spot + rod-like	2μm	ring like distrib+spots

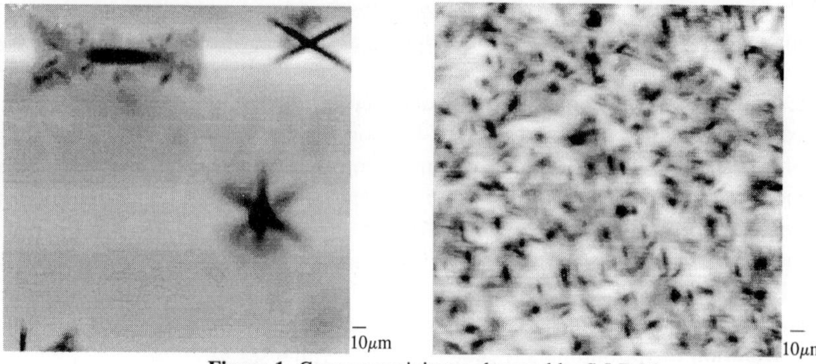

Figure 1: Copper precipitates observed by S.I.R.M
a) contamination of as-grown sample b) contamination of two-steps annealed sample

3.2.1. Cz samples. Results concerning influence of metallic contaminations of Cz samples containing or not oxygen precipitates are given in the table 3.

The introduction of copper in the as received samples (treatment Cu) gives starshaped colonies of precipitates visible in the S.I.R.M. images (fig.1a), as already observed by Laczik et al.[13]. In the L.B.I.C. maps (fig.2a) they appear as regions of very poor photocurrent intensity. This well known particular precipitation is due to the high rate of silicon self-interstitial emission when the $SiCu_3$ phases are formed during the "slow" cooling down. The S.I.R.M. indicates also that the starshaped precipitate density is higher near the free surfaces of the wafers than in the bulk. This is due to the absorption by these surfaces of the self-interstitials emitted during the formation of copper precipitates. The oxygen concentration does not vary.

In the contaminated and two step annealed samples (N+G+Cu), precipitates are revealed by S.I.R.M. as shown in figure 1b. The comparison of figures 1a and 1b suggests that when oxygen precipitates are present, copper atoms precipitates around them: the size of the metallic precipitates is smaller, since copper is precipitated on a larger number of nucleation sites. The L.B.I.C. detects also circular stacking faults which the dislocation loops are decorated by copper (fig.2b), and L_n # 2 μm like after treatments (N+G).

The use of slow cooling down could inhibate the recombination activity of copper atoms, due to their very fast precipitation. This is verified for non contaminated samples in which the same values of L_n are found after a thermal treatment at 950°C for 1h without copper (blank for Cu). The starshaped Cu precipitates could behave only as light absorbing particles, without any recombining activity for minority carriers.

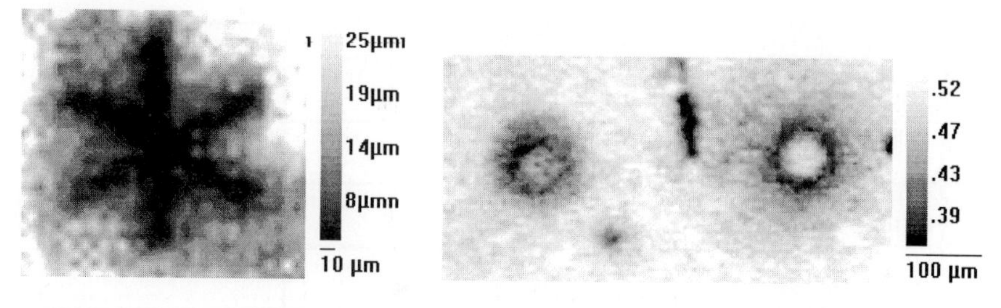

Figure 2: L.B.I.C. maps of copper contaminated Cz sample
a) starshaped colony of Cu precipitates in as-grown sample (diffusion length)
b) Cu decorated stacking fault after two-step annealing (photocurrent)

Table 4 : Results concerning nickel contamination on FZ samples

Label treatment	F.T.I.R.	S.I.R.M.	Ln	L.B.I.C.
Ni	1120 cm^{-1}	rod like defects+spots	50μm	few spot-like precipitates
N+G+Ni	1120 cm^{-1}	spot precipitates & rod like defects on surface	40μm	lot of spot-like precipitates
Ni+P	1120 cm^{-1}	nothing	400μm	few spot-like precipitates

The diffusion of phosphorus leads to the disappearance of copper precipitates from the S.I.R.M. images, but there is no improvement of L_n value.

Nickel contamination gives approximately the same results concerning L_n and L.B.I.C. maps, and precipitates are also detected by S.I.R.M. as shown by fig.3a. Rod-like defects [14;15] are detected near the surface and spot-like precipitates are revealed near the surface and in the bulk of Ni contaminated as grown samples (treatment Ni). The same rod-like defects and precipitates are revealed near the surface but not in the bulk for two step annealed and contaminated samples (treatment N+G+Ni), probably because the formation of oxygen related precipitates is accompagnied by the emission of self-interstitials which inhibate the formation of $NiSi_2$ precipitates [15]. Notice that the S.I.R.M. technique has the advantage towards X-ray topography to be able to detect and recognize Ni precipitates althought the lattice constant of $NiSi_2$ is very close to that of silicon crystal.

3.2.2. FZ samples. The results obtained after nickel contamination of FZ samples are given in the table 4.

After contamination of as-grown samples there are a lot of spot-like precipitates and only a few rod-like defects revealed near the surfaces of the wafer as shown in the figure 3b. After a longer thermal treatment (1050°C;16h), a larger number rod-like defects are observed as shown by the figure 4.

Some precipitates are detected in the volume of the two step annealed and contaminated samples, which did not appear in the non contaminated wafers. These results confirm that the nickel which diffuses in the bulk can form by itself precipitates in FZ samples as it does not require the presence of foreign nucleation centers [13].

The diffusion length decrease is less important than in Cz material, due to the smaller quantity of oxygen precipitates and related crystallographic defects.

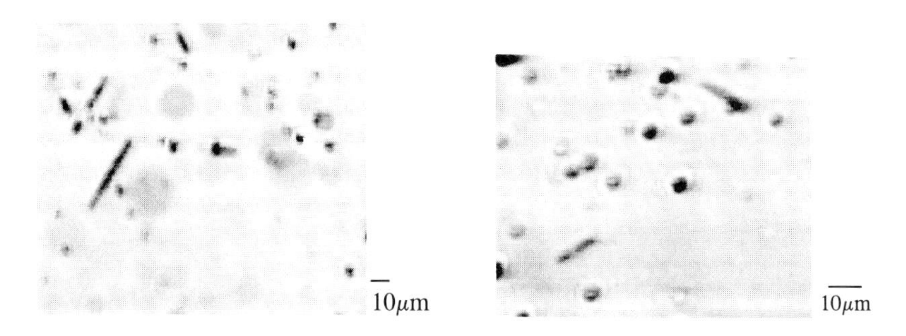

Figure 3: S.I.R.M. image of Ni precipitates on as-grown contaminated samples
a) in Cz sample b) in FZ sample

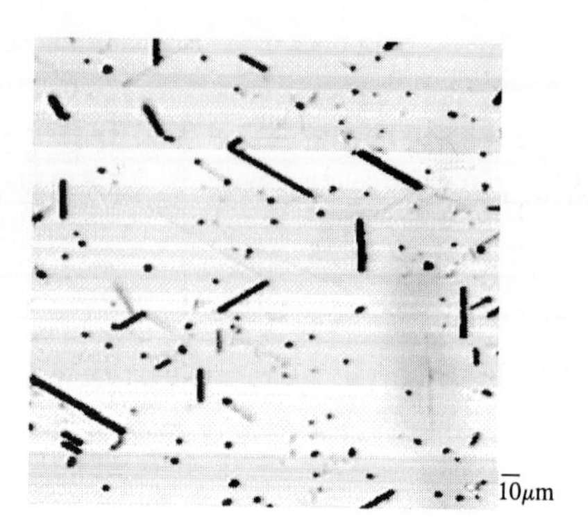

Figure 4: S.I.R.M. image of Ni precipitates after long thermal treatment (1050°C-16h)

4. Conclusion

Non destructive techniques like S.I.R.M., L.B.I.C. mapping and F.T.I.R. appear to be able to detect and recognize oxygen and metallic atoms related precipitates in silicon. In addition the recombination strength of these defects can be evaluated. Copper precipitates are easily recognized due to their starshaped colonies. Nickel precipitates are also identified and the techniques used in the present work appear to be more efficient than X-ray topography when the lattice parameters of the precipitate is closed to that of the host crystal, as it is the case for $NiSi_2$.

The techniques are also able to describe the behaviour of the precipitates after phosphorus diffusion near the surface: they disappear in the S.I.R.M. images and the minority carrier diffusion lengths are restored in oxygen poor silicon wafers.

References

[1] Hu S.M 1986 Mat. Res. Soc. Symp. Proc. **59** 249-267
[2] Endrös A.L 1993 Solid State Phenomena **32,33** 143-154
[3] Bender H. and Vanhellemont J. 1992, Mat. Res. Soc. Symp. Proc. **262** 15-29
[4] Yatsurugi Y., Akiyama N., Endo Y., Nozaki T 1973. J. Electrochem. Soc. **120** 976-81
[5] Stemmer M 1993. Appl. Surface Science **63** 213-216
[6] Booker G.R., Laczik Z., Kidd P 1992 Semicond.Sci. Technol. **7** A110-A121
[7] Vève C., Gay N., Stemmer M. and Martinuzzi S. 1995 J. Phys.III 1353-1363
[8] Stemmer M. : Thesis of Doctorate, University of Marseilles Sept.1994
[9] Tempelhoff K., Spiegelberg F.,Gleichmann R., Wruck D 1979 Phys.Stat. Sol. (a) **56** 213 214-223
[10] Sun Q., Yao K.H., Gatos H.C. and Lagowski J 1992 J. Appl. Phys. **71** 3760-65
[11] Iino E., Takano K. Fusegawa I. and Yamagishi H. 1994 Semi cond Silicon 148-155
[12] Hartung J. and Weber J., 1995 J.Appl.Phys., **77** 118
[13] Laczik Z., Booker G.R., Falster R. 1989 Solid State Phenomena **6 &7** 395-402
[14] Xu D.L. Zhou F.S. Guo R.F. 1987 GADEST'87 ed. by H. Richter p175
[15] K. Graff 1995 « Métal Impurities in Silicon Device Fabrication » Springer Series in Material Science 24, Berlin, p57-59

Acknowledgments: This work was supported by CNRS-ECOTECH-France and by European Union-DG XII-Joule Programme-Multichess contract.

Inst. Phys. Conf. Ser. No 149
Paper presented at DRIP 95

Identification of the mechanism of stacking fault nucleation in $In_xGa_{1-x}As$ and $In_yAl_{1-y}As$ layers grown by MBE on InP substrates

F Peiró, A Cornet and J R Morante.

EME, Enginyeria i Materials Electrònics, Dept. Física Aplicada i Electrònica
Universitat de Barcelona. Avda. Diagonal 645. 08028 Barcelona (Spain).
e-mail: paqui@iris1.fae.ub.es

Abstract. This work presents three different origins of stacking faults in $In_xGa_{1-x}As/In_yAl_{1-y}As/InP$ matched and mismatched systems, depending on the growth conditions: as a stacking error during the growth, as a consequence of the dissociation of misfit dislocations, and finally as a proper strain relieving mechanism. The occurrence of these different mechanisms are discussed in both homogeneous samples and samples with composition modulation.

1. Introduction

The origin of bidimensional defects as stacking faults (SF's) in III-V (In/Ga/As/Al/P) semiconductor epilayers has been a matter of discussion during several years. Different authors related them to the existence of contamination at the epilayer-substrate interface[1]. Others argued that SF's may arise from a thermal mismatch because of the different thermal expansion coefficients between the layer and the substrate[2]. Some reports also exist illustrating the origin of SF's during the coalescence of islands if the growth is tridimensional due to the higher mismatch of the system[3]. However, the main point of controversy was always centered in whether stacking faults contribute or not to the relaxation of stress in strained layers. Nowadays, it seems well assumed that partials dislocations and thus, the associate SF, are the primary defects on the relaxation of tensile strained layers[4],[5].

In the recent years our effort has been devoted to the characterization of $In_xGa_{1-x}As/In_yAl_{1-y}As$ matched and compressive mismatched layers grown by Molecular Beam Epitaxy (MBE) on (100) InP substrates, for the fabrication of high electron mobility field effect transistors (HEMT). Depending on the growth temperature and system mismatch, some of these epilayers exhibited coarse contrast modulations along the $\langle 010 \rangle$ directions, related to composition variations driven by the existence of a miscibility gap for the growth of III-V compounds. Within the frame of the analysis of these systems depending on the In molar fraction (x_{In}), growth temperature (T_g) and epilayer thickness (t), we have been able to distinguish different mechanisms for SF's nucleation. A summary of the most significative results is presented here.

2. Experimental details

The characterization of the samples was carried out by Transmission Electron Microscopy (TEM) using both conventional (CTEM) and high resolution imaging modes (HRTEM). The specimen preparation for TEM analysis consisted in mechanical flat grinding and polishing up to a specimen thickness of 25-30μm, and final Ar^+ beam thinning in a cooled stage. All the samples were examined in plan view (PV) and cross-sectioned (XTEM) views. For the former, an accurate control of the milling rates allowed us to locate thin foil regions at different distances from the interface. Thus, the

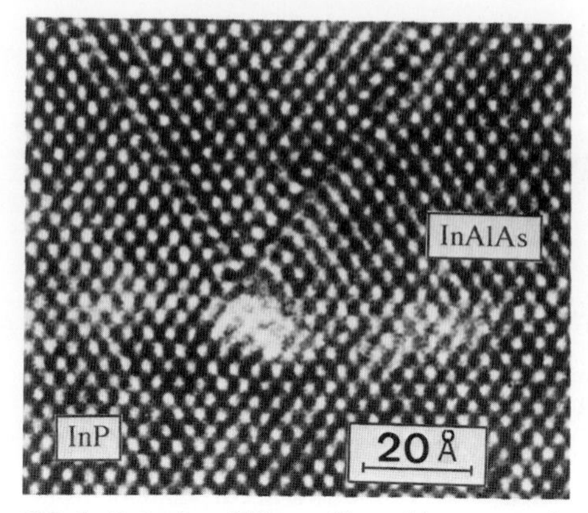

FIG. 1. Nucleation of SF at oxide nuclei present at the interface due to deficient oxide desorption at $T_h < 530°C$.

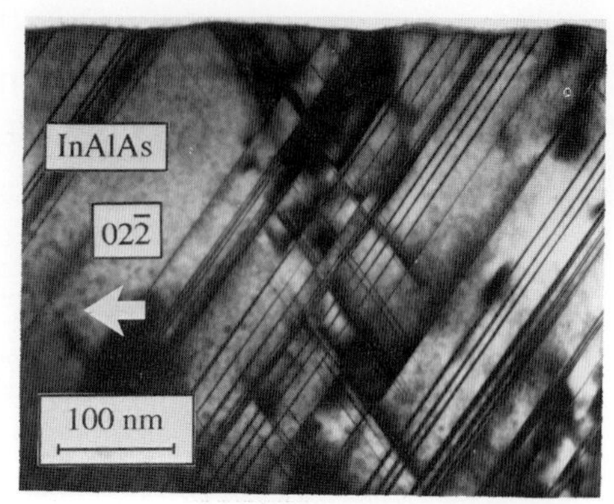

FIG. 2. [011] Cross-section showing the preferential formation of twins on the planes $(\bar{1}\bar{1}1)$-$(11\bar{1})$.

evolution of the sample morphology along the (100) zone axis, was analyzed in detail across the whole epilayer. For the latter, the observations along both the (011) and $(01\bar{1})$ zone axis, revealed, in some cases, a clear anisotropy between the $\langle 110 \rangle$ directions.

3. Homogeneous samples

3.1 $In_y Al_{1-y} As$

Driven by the application of this material as buffer layer or wide gap barrier in HEMT devices, our attention was focused in the optimization of the growth of $In_{0.52}Al_{0.48}As$. Two different aspects were studied: the influence of the preheating treatment temperature (T_h) for InP cleaning, and the substrate temperature during InAlAs growth (T_g). The analysis of different cleaning process at T_h in the range 480°C-530°C, revealed that low T_h lead to a rough growth front and that the rests of some oxide nuclei due to the deficient desorption temperature acted as preferential sites for SF nucleation (Fig. 1). The optimum T_h was found to be 530°C. A far as the InAlAs T_g was concerned, we could establish a clear differentiation in three ranges.

3.1.1 Low growth temperature. The InAlAs layers grown at $T_g \simeq 300°C$ presented a twinned structure extremely anisotropic: most of the twins formed on the $(1\bar{1}1)/(11\bar{1})$-[011] slip system (Fig. 2), whereas, threading dislocations (with densities $\simeq 10^{10}$ cm^{-2}) were the preferent type of defects in the $(111)/(1\bar{1}\bar{1})$-$[01\bar{1}]$ slip system. Taking into account the thermal expansions coefficients of the InAlAs (5.2×10^{-6} K^{-1}) and InP (4.5×10^{-6} K^{-1}), the relatively low T_g, and that the twins did not appear if T_g was slightly increased, the twinned morphology should not had been developed as a relaxation of the thermal mismatch (just 0.05%). The origin of those bidimensional defects must be found on the stacking errors during the first stages of growth because of the limited MBE kinetics at such a low T_g, specially due to the low diffusion of the species of the group III. The low surface mobility of III atoms, particularly Al, does not allow the ad-atoms to migrate up to the preferential incorporation sites inducing a rough growth with Al vacancies that favours a faulted growth giving rise to the observed twinned structure. The anisotropy between the $\langle 011 \rangle$ directions, is explained on the basis of asymmetric III/V free bond distribution during the growth: whereas the surface step propagation in the $[01\bar{1}]$ direction takes place by the saturation of an As free bond, the propagation along [011] is accomplished by the saturation of a III bond. Besides, the migration of III species at low T_g, is limited due to the As overpressure in the MBE chamber needed to reach an As-stabilized surface. As a consequence, the propagation of growth steps along $[01\overline{1}]$ is faster than along [011], and then, the probability of faulted incorporation for III atoms is increased, giving rise to the nucleation of twins preferentially on the planes $(1\overline{1}1)$ and $(11\overline{1})$.

3.2.2 Intermediate T_g. As T_g rised in the range 440°C-530°C, the twinned morphology disappeared giving rise to TD and SF also preferentially on $(1\overline{1}1)$ and $(11\overline{1})$, with a monotonically reduction of the defects density (ρ_{SF} and ρ_{TD}) in all the T_g range. Besides, most of the SF's were found to nucleate at the buffer-substrate interface, despite the fact that, after the optimization of the preheating InP treatment, the HRTEM images demonstrated that there were no rest of contamination over the substrate that could induce the nucleation of these planar defects. The analysis of the evolution of ρ_{SF} with T_g and the evolution of the SF anisotropy, [higher ρ_{SF} on the planes $(1\overline{1}1)/(11\overline{1})$-[011] ($\rho_{02\overline{2}}$), than on the planes $(\overline{1}\overline{1}\overline{1})/(111)$-$[0\overline{1}\overline{1}]$ (ρ_{022}) (Fig. 3), where the subindex $02\overline{2}$ and 022 refer to the g reflection for which the respective SF are visible in TEM] reveal that the SF's arised from the dissociation of dislocations near the interface, where there exist a tensile stress field due to the presence of a InAs interfacial layer because of the P-As exchange during the thermal cleaning under an As-overpressure. In effect, the correlation of χ vs T_g by an Arrhenius plot $\chi \simeq \exp(-E_a/kT)$ leads to an $E_a \simeq 1.9$ eV, which is in the range of 1-2 eV commonly reported for the activation energy required for the generation and movement of 60° dislocations[6]. The different mobilities of the 90°+30° partials resultant depending on their α or β character would explain the anisotropycal rate of partial recombination, it is SF annihilation, between the $\langle 110 \rangle$ directions.

3.2.3 High T_g. As T_g was increased in the range 530°C$<T_g<$590°C, we observed a dramatic increment of ρ_{SF} and precipitate nucleation at the buffer substrate interface, also with higher density (ρ_p) as T_g increases. The reason for precipitates nucleation was found in the $In_{0.52}Al_{0.48}As$ alloy clustering favoured at high T_g. The precipitates formed preferentially at the interface, because previous InAs clusters existed there due to the As-P exchange favoured at higher T_g. These centres of located stress have two main effects: to activate the glide process on the {111} planes, leading to the formation of stacking faults, and to induce an inhomogeneous growth around these complex defect sites, giving rise to the square shaped inhomogeneities observable in Fig. 4.

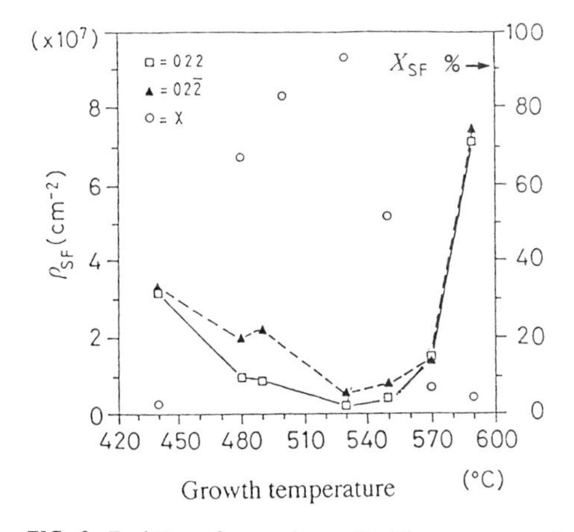

FIG. 3. Evolution of ρ_{SF} and χ vs T_g. The parameter of anisotropy χ is defined as $(\rho_{02\overline{2}} - \rho_{022})/[(\rho_{02\overline{2}} + \rho_{022})/2]$.

FIG. 4. SF nucleation on precipitates located at the InAlAs/InP interface of a sample grown at T_g=590°C.

1. Praseuth J P, Goldstein L, Hénoc P, Primot J and Danan G 1987 *J. Appl. Phys.* **61** 215

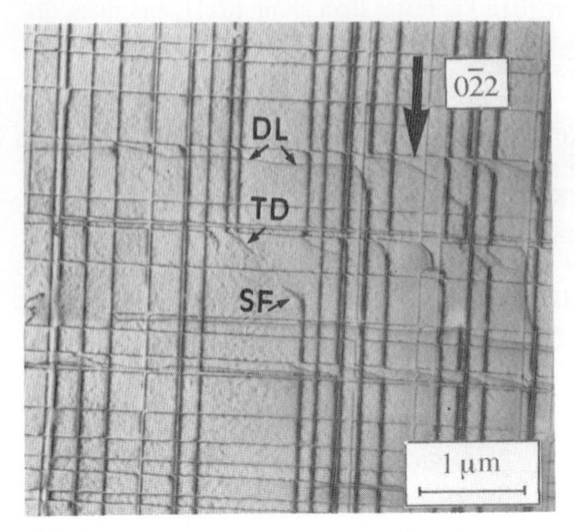

FIG. 5. $In_{0.63}Ga_{0.37}As$ layer (t=0.5μm). Some dislocations turn up by threading segments (TD), or are pinned at SF's. Dislocation loops on {111} planes (DL) are also observed.

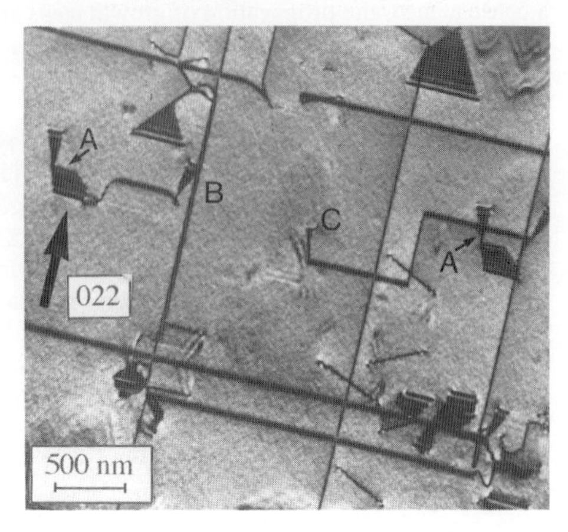

FIG. 6. Onset of stress relaxation in a $In_{0.6}Ga_{0.4}As$ layer t=20nm: dislocations segments at the interface limited by threading segments or stacking faults (SF).

3.2 $In_x Ga_{1-x} As$

A preliminary study was done on four $In_xGa_{1-x}As$ layers with the same thickness (t=0.5μm) but different x_{In} ranging from 54.1% to 62.5%, grown at 515°C on (100)-InP, with an initial lattice mismatch (f) between $9x10^{-4}$ and $6.5x10^{-3}$. Samples having $x_{In}>60\%$ exhibited a network of 60° misfit dislocations with asymmetric distribution along the two ⟨011⟩ directions.

Once established that the relaxation process started at a certain thickness below 0.5 μm in layers with In composition of $x_{In}\simeq0.60$, we went further into the study of the onset of stress relaxation by the analysis of the morphology of compressive $In_{0.6}Ga_{0.4}As/In_{0.52}Al_{0.48}As$ layers grown by MBE on (100)-InP, with t in the range 5-25 nm. Whereas in the layer with t=20nm there were no evidence of strain relaxation, the InGaAs well with t=25nm exhibited a network of 60° dislocations irregularly distributed on the layer surface, coexisting with SF's and partial dislocations (PD) (Fig. 6). Thus, the critical thickness (t_c) for plastic relaxation was found $20nm<t_c<25nm$. According to the model of Dodson et al.[7] the excess stress σ_{ex} (which is the driving force for the nucleation and motion of dislocations) is determined by a balance between the stress due to the elastic strain, the stress of the dislocation line itself and the stacking fault energy if the dislocations concerned are partial. Assuming a value γ_{SF}= 25mJ/m^2, the critical thickness for 90°, 30° and 60° dislocations at the given composition resulted $t_c^{90°}$=210 Å, $t_c^{60°}$=228 Å and $t_c^{30°}$=234 Å, in good agreement with our experimental t_c. In fact recent reports have shown that the energy required for the 90° partial is much lower than that of 60° or 30° dislocations[8]. However, it is well established that in compressive stress fields, the nucleation of the 90° partial to relieve the strain is not formed before the 30° partial[5]. It has been also suggested that partial dislocations may appear as the border of a growth error propagating the stacking fault during further growth, up to the critical thickness of the second partial is overcomed. Taking into account that in the layer 20nm thick we did not observed neither SF nor dislocations, those SF may have appear as a strain relieving mechanism favoured by the existence of some irregularities. We have found that there exist some SF nucleated at the InAlAs buffer/substrate interface that may act as nucleation centers of further SF at the interface between the buffer and the InGaAs well, giving rise to the configurations labelled A in figure 6. As growth continues, all the t_c are overcome and some of the partials would recombine. This point would correspond to the configuration frozen in figure 6. In fact, by converging the electron beam over the sample surface, we induced the annihilation of SF by recombination of partials dislocations as marked by B and C labels in Fig. 6.

4. Samples with composition modulation

The III-V compounds often present composition modulations when grown at T_g below the critical temperature for spinodal decomposition. The main TEM features is then the coarse quasiperiodic contrast modulation (CM) observed with g=022 type reflections, with contrast bands oriented along the ⟨010⟩ directions and separated a distance of hundreds of nm (modulation wavelength Λ). In the following we will deal with the strain relaxation mechanism in those inhomogeneous layers.

4.1 $In_x Ga_{1-x} As$

Five slightly compressive $In_{0.54}Ga_{0.46}As$ layers (f=0.09%), with t=0.29, 0.49, 0.74, 0.98 and 1.96 μm were grown at T_g=515°C, and all of them exhibited coarse CM. We noticed a monotonical reduction of the modulation wavelength Λ from $\simeq$400nm to $\simeq$230nm as layer thickness increased. The coarse modulation however, extended just up to 0.5μm above the interface. Misfit dislocations were not observed in any case, but stacking faults appeared in samples with t>0.5μm (Fig. 7). The remaing strain were also measured by DCXRD. The results revealed that samples with t=0.3μm and t=0.5μm were totally strained, whereas the strain started to relax for t<0.5μm (the limit beyond which SF appeared). Finally, there was the same residual strain ($\varepsilon \simeq 10^{-4}$) and also similar values of Λ ($\simeq$230nm), in the layers t=1μm and t=2μm thick, having the latter higher ρ_{SF} than the former.

4.2 $In_y Al_{1-y} As$

Compressive $In_yAl_{1-y}As$ layers with y_{In}=55% and 58.9% were grown at 580°C and 570°C. These samples exhibited also CM. Again, misfit dislocations were not observed, being SF the preferent type of defects. In this case, however, we could classify the SF in two groups according to the configuration of the defect on different planes of the {111} family and the location of their origin with respect to the interface. Hence, SF_{SI} refers to SF nucleated near the layer-substrate interface, which frequently affected two or even four of the {111} planes; SF_{SS}, refers to SF on isolated {111} planes and nucleated above the interface (Fig. 8). Although we could relate the observed SF to the high T_g used, a detailed comparison of these layers with the InAlAs matched samples described above revealed that the SF_{SS} were not due to such alloy clustering related effects[9]. In view of these results on $In_xGa_{1-x}As$ and $In_yAl_{1-y}As$ layers let us know explain the possible origin of the SF in those composition modulated samples. According to the model of Glas[10] a composition modulation would have an associated elastic energy W_{MOD} depending on t and Λ, given by the equation (1):

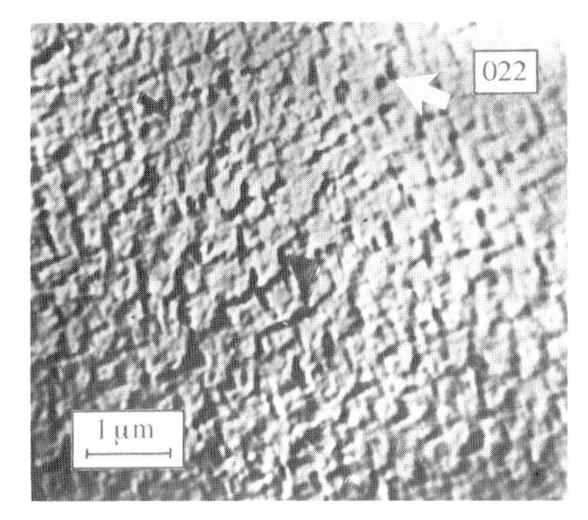

FIG. 7. $In_{0.543}$ $Ga_{0.437}$ As layer, (t=0.74μm), exhibiting ⟨010⟩ contrast modulation and the onset of SF nucleation.

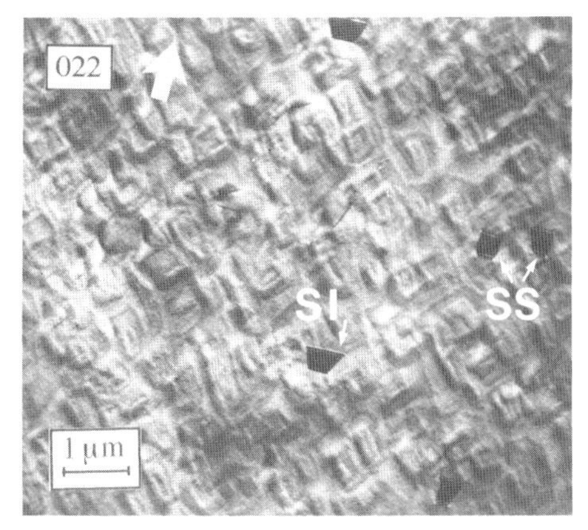

FIG. 8. $In_{.589}$ $Al_{0.411}$ As layer (t=2μm), with coarse modulation, and stacking faults of type SS and SI.

$$W_{mod} = \frac{1}{2}\left[\frac{Y}{(1-\nu)}\right] \varepsilon_o^2 t \left[1-(1+\nu)\frac{(1-e^{-\alpha t})^2}{\alpha t}\right] \quad (1) \qquad W_f = \frac{dE}{dS} = 2\mu \left(\frac{1\nu}{1+\nu}\right) \varepsilon^2 t \quad (2)$$

being Y the Young modulus, ν Poisson coefficient and $\alpha=2\pi/\Lambda$. The calculation of this energy for the $In_xGa_{1-x}As$ layers with modulation amplitude $\varepsilon_o=10^{-3}$ leads to the values $W^A_{mod}=15.4$, $W^B_{mod}=30.9$ and $W^C_{mod}=50.9$ mJ/m^2 for layers with $t_A=0.29$, $t_B=0.49$ and $t_C=0.74$ respectively. If those layers were homogeneously strained (taking $\varepsilon \simeq f$) instead of presenting CM, they should have an elastic energy $W^A_f=22.1$, $W^B_f=37.4$ and $W^C_f=56.4$ mJ/m^2 respectively given by (2). If we compare the results, we can notice that $W^B_{mod}-W^A_{mod} \simeq W^B_f-W^A_f$ and $W^C_{mod}-W^B_{mod} \simeq W^C_f-W^B_f$, it is, the reduction of the modulation wavelength seems to store the same elastic energy as if the layer was homogeneously strained. At this point we would like to remember that the CM just arrived up to 0.5μm above the interface, and that beyond this thickness SF appeared. Looking at W^B_{Mod} values, we noticed that at this specific thickness $W_{MOD} \simeq \gamma_{SF}$. Hence, we suggest that the composition modulation accommodates the elastic energy by Λ reduction up to an apparent critical thickness t_{ca}, beyond which SF start to appear because W_{MOD} equals γ_{SF}. As layer thickness increases beyond this value SF are favoured more than further reduction of Λ and CM vanishes. Nevertheless, the calculated energy values are much lower than the energy required for partial dislocation nucleation ($\simeq 10^4$ mJ/m^2), and thus, these defects could not appear as strain relieving defects, but just as a stacking errors during the growth. This should be also the origin of the SF observed in the $In_{0.55}Al_{0.45}As$ layer ($t_E=1\mu$m), for which $W^E_{Mod} \simeq 78$ mJ/m^2, being W^E_f still lower than the energy required to overcome the limit of plastic relaxation by dislocation nucleation. Conversely, for the layer with $y_{In}=58.9\%$, ($t_F=2\mu$m), at a thickness of 0.5μm we would already have a $W^F_f=1.5\times10^3$mJ/m^2, and hence, SF would appear as a result of partial dislocations nucleation as a strain relieving mechanism. This could be the origin of the SF_{SS} observed in the higher mismatched $In_yAl_{1-y}As$ layer. Further interaction of partials to annihilate the SF and give 60° dislocations is avoided by the stress field related to the CM.

5. Conclusions

TEM has been used as the main tool for the characterization of $In_xGa_{1-x}As$ and $In_yAl_{1-y}As$, homogeneous and composition modulated layers. The observations along the (100), (011) and $(01\bar{1})$ zone axis, has allowed us to analyze in detail the observed SF and distinguish three different origins: as a stacking error when the layer is grown at low T_g in homogenous samples or when the energy of the composition modulation equals that of the SF; as a result of 60° dislocations dissociation activated at high T_g or as a strain relieving mechanism in higher mismatched inhomogeneous layers.

Acknowledgments

This work was funded by the Spanish CICYT program MAT95-0966. We acknowledge A. Georgakilas from FORTH (Crete) and S. Clark from Cardiff, for the growth of the samples.

References

1. Praseuth J P, Goldstein L, Hénoc P, Primot J and Danan G 1987 *J. Appl. Phys.* **61** 215
2. Gerthsen D, Ponce F A, Anderson G B and Chung H F 1988 *J. Vac. Sci. Technol.* **B6** 1310
3. Ernst F and Pirouz P 1988 *J. Appl. Phys.* **64** 4526
4. Hwang D M, Bhat R, Schwarz S A and Chen C Y 1992 *Mat. Res. Soc. Symp. Proc.* Vol. **263** 421
5. Min D and Hwang D 1995 *Mat. Chem. and Phys.* **40** 291
6. Gerthsen D, Biegelsen D K, Ponce F A and Tramontana J C 1990 *J. Cryst. Growth* **106**, 157
7. Tsao J Y, Dodson B W, Picraux S T and Cornelison D M 1987 *Phys. Rev. Lett.* **59**, 2455
8. Chen Y, Liliental-Weber Z and Washburn J 1995 *Appl. Phys. Lett.* **66**, 499.
9. Peiró F, Cornet A and Morante J R 1994 *J. of Electron. Materials*, **23**, 969
10. Glas F 1987 *J. Appl. Phys.* **62**, 3201

Inst. Phys. Conf. Ser. No 149
Paper presented at DRIP 95

Defect-Induced Oxidation of TiN in Ion-Beam-Assisted Deposition

H.Kubota*, M.Easterbrook, M.Tokunaga*, M. Nagata***, I.Sakata* and M.A.Nicolet****

* Dep. of EE & Computer Science, Kumamoto University, Kumamoto 860, Japan
** California Institute of Technology, 116-81 Pasadena, California 91125, USA.
***Kumamoto Industrial Research Institute, Higashi-machi, Kumamoto-shi, Kumamoto 862

Abstract. We focus on oxygen contamination of TiN during ion-assisted deposition, where enough oxygen is well introduced such that defects show up at the surface under competition between thin film growth and ion assisted etching. The analysis for the phenomena gives not only a standard of TiN oxidation in ion process, but also a chance of development of the defect-induced oxidation process.

1. Introduction

Although oxidation and oxygen contamination during thin film growth always degrade LSI processes seriously, the phenomena could inherently be effective to formation of the useful oxide layers, if we can accurately control the deposition of the oxides and their mother elements with no chemical or high temperature treatment. Highly dielectric materials, for instance Ta_2O_5 and TiO_2 are intensively investigated for ULSI and could be generally applied to GBit generation. In this paper, TiN depositions are demonstrated in cases of no oxidation and with Ti oxides, by only adjusting the beam current in Ion-Beam-Assisted processes. The successful results are described due to surface defects which show up under competition between film growth, ion assisted etching, and reacted captured oxygen.

Meanwhile the optimum deposition conditions for TiN are not established because of difficulties which are originated to that the resistivity, the composition and the color of TiN, its defects, surface roughness, grain size and grain boundaries all depend strongly on the formation conditions[1,2]. The oxidation of TiN is a matter of primarily importance for the diffusion barrier performance in LSI medullization process. Then we discuss a standard of TiN oxidation which occurrs in any ion assisted processes.

2.Experimental

The deposition system that we used is schematically shown in Fig. 1(a) [2]. A 3-inch reactive magnetron sputtering target and nitrogen ion beam from a Kaufman source (Millatron of Commonwealth Co., Inc.) are located on opposite sides along the circumference of the top plate of the same deposition chamber and share the source nitrogen gas mixture. The parameters are described in detail in ref. [1]. The openings can simultaneously be positioned below the Ti target and the ion source. The shutter openings are smaller than the ion source and the Ti target, which ensures lateral uniformity of the grown film. The shutter openings have the shape of a 40° section of a ring to ensure that all points of the sample are exposed to the target and the beam for the same duration in each cycle (see Fig. 1(b)). The sample table was rotated with the period of ΔT=35.4 s (about 100 [turn/h]). Under these conditions and without the use of the ion beam, the TiN film grew at a time-averaged rate of about 20 [Å/min].

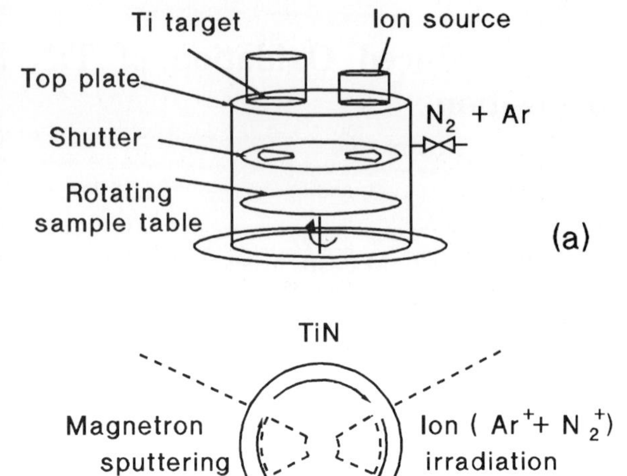

Figure 1 (a) Sequential ion-beam-assisted deposition system. (b) The magnetron sputtering targets deposit a film of thickness ΔD.

The direction of incidence of the broad unfocused ion beam was almost perpendicular to the substrate surface. Two beam parameters, the beam voltage, V_B, and the beam current, I_B were set independently. The ion plasma parameters (cathode heater current and DC discharge current) were automatically adjusted to keep V_B and I_B constant as the gas pressure changed. The parameter values we used range from 0.1 to 1 keV for V_B and from 0 to 45 mA (equivalent to ~2.5×10^{16} ions/cm^{2}cycle) for I_B.

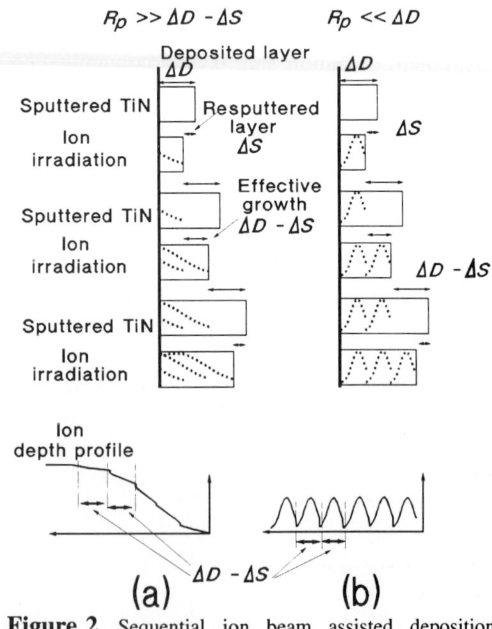

Figure 2 Sequential ion beam assisted deposition process.

The fabricated specimens were analyzed by backscattering spectrometry for thickness (in units of Ti atoms/cm^2), atomic composition and contamination of oxygen. We also measured the thickness in Å with a Dectac profilometer. The result is roughly consistent with that of backscattering analysis. Because the backscattering data are the more reliable set of the two, we use the number of Ti atoms/cm^2 as the unit of thickness and then Ti atoms/cm^{2}sec as the unit of growth rate in this paper.

We model the sequence of processing steps encountered in sequential ion-beam-assisted deposition as follows (Fig. 1(b)). First, in the reactive magnetron sputtering section of the system, TiN is deposited on the substrate in the stable B1 structure, together with other ions. Second, the sample is moved by the carousel to the section under the ion source. The major effects there will be 1) resputtering of TiN, 2) implantation of gaseous species, and 3) surface migration, dissociation, and nucleation effects. Figure 2(a) shows this sequential ion-beam-assisted deposition process schematically. R_p is the projected range of ions implanted in TiN, ΔD is the thickness of the layers deposited during one pass under the magnetron sputtering section, and ΔS is the thickness of the layers resputtered through one pass under the ion beam under steady-state conditions. The resulting growth observed over the whole cycle is ΔD-ΔS.

We define the growth rate, G, as

$$G \equiv \frac{(\Delta D - \Delta S)}{\Delta t}, \tag{1}$$

referred to in the following as the effective growth rate. Δt is the time per cycle for which a point on the rotating table (carousel) lies under the opening of the shutter below the magnetron sputtering section or the ion beam section, hence $\Delta t = \Delta T \cdot 40°/360°$ (see Fig.2).

3. Results and Discussion

The ratios of the {220} to {111} X-ray signal heights for layers with various beam voltages V_B are plotted in Fig. 3 with the effective growth rate G as the abscissa. The arrow indicates the effective rate corresponding to the deposition of two (111) Ti layers in one cycle of magnetron sputter deposition and ion beam irradiation. Above that critical growth rate the intensity ratio changes little. Below that critical growth rate, the structure changes suddenly and shows preferred [220] orientation. Since the deposited layer ΔD is kept almost constant at a value of 0.94x10^{15} Ti atoms/cm^{2}cycle which corresponds to 3.2 (111) Ti layers/cycle, a thickness ΔS of 1.2 (111) Ti layers/cycle is removed by the ion beam at the critical effective growth rate as shown by the arrow. Assume that a TiN grain grows layer-by-layer along the [111] direction in the Frank-Van der Merwe fashion. When a (111) nitrogen layer is exposed on the growth surface, atomic substitutions, interstitial occupancy by other atoms, and vacancy formation are likely, because nitrogen-nitrogen

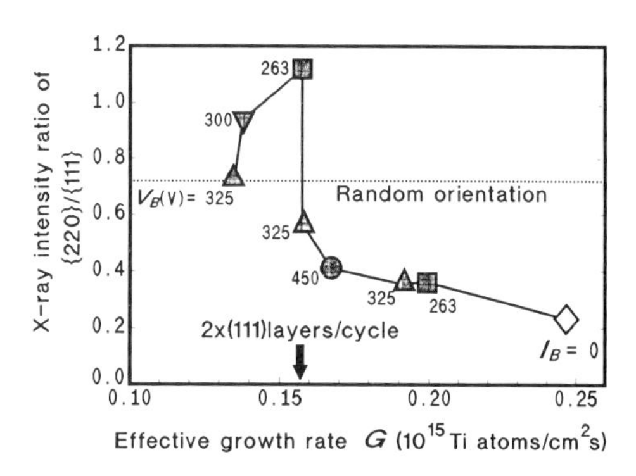

Figure 3 Dependence of the Cu-Kα X-ray intensity ratio of TiN {220}/{111} on the effective growth rate.

bonding in TiN is weak according to theoretical work based on X-ray photoemission spectra [5]. It would be desirable to have an equally simple atomistic model to explain the preferential orientation that is observed for effective growth rates of less than two (111) Ti layers/cycle.

The existence of a critical growth rate of two (111) bilayers of TiN per deposition-irradiation cycle could be understood in consideration of stabilization of crystal energy through packing nitrogen atoms of <111> surface. Below that rate, the columnar growth microstructure changes into a granular one with azimuthal texture, while growth morphology of the TiN thin film does not alter in any ion assisting in the range above. This experimental result suggests that the nitrogen <111> layer sandwiched by the Ti <111> bilayer is secured by strong guard of the Ti atomic layer's networks against oxygen atoms from the outer space.

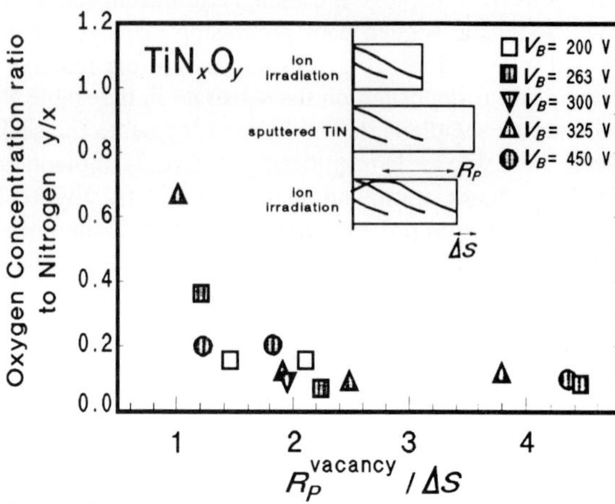

Figure 4 Dependence of oxygen concentration, y/x on a parameter indicating location of vacancies, $R_p^{vacancy}/\Delta S$.

Figure 4 shows the dependence of oxygen-N ratio, y/x on the value of $R_p^{vacancy}/\Delta S$, where $R_p^{vacancy}$ is an average value of the projected range of vacancies for Ti and N by injected ions, and ΔS is the thickness of the layers resputtered through one pass under the ion beam. The values of $R_p^{vacancy}$ were evaluated by Monte Carlo simulations (e.g. TRIM) and ΔS obtained by resultant film thickness. Generally, one can image that the decrease of the growth rate will increase oxygen contamination in the deposition system related to Ti. Titanium, Ti has strong tendency to be in favor of oxidation rather than nitridation, because the free energy for Ti and oxygen reaction process is much lower than that for Ti and nitrogen[4]. The result of Fig. 4 is roughly in agreement with the above consideration, but does not describe the particular increase of y/x at $R_p^{vacancy}/\Delta S = 1$ [6]. $R_p^{vacancy}/\Delta S = 1$ means that the vacancies after the irradiation turn almost all distribute on the surface of the sample during the growth process. Therefore, the abrupt increase of oxygen below $R_p^{vacancy}/\Delta S=1$ in Fig. 4 is a result of surface vacancies and introduced oxygen contamination in ion-assisted process.

Bonding in TiN consists of a complex combination of localized metal-metal and metal-nonmetal interactions resembling both metallic and covalent bonding with a small amount of ionic bonding. There is no apparent nonmetal-nonmetal interactions[5]. Because of the absence of N-N bonding the electrical properties are dominated by the metallic sublattices, and hence TiN possesses metallic properties not too different from those of Ti metal[5]. The Ti and N (111)--planes stack alternatively along <111> direction. An N atom which has three Ti atoms above itself is stabilized against oxidation, but with an opening in the top (111) plane of Ti, there can be (1) substitution of O atoms for the N atoms and/or (2) take O atoms moving into interstitial sites on the (111) plane of N.

With the two main results, 1)nitrogen layer sandwiched by the Ti <111> bilayer is stable,

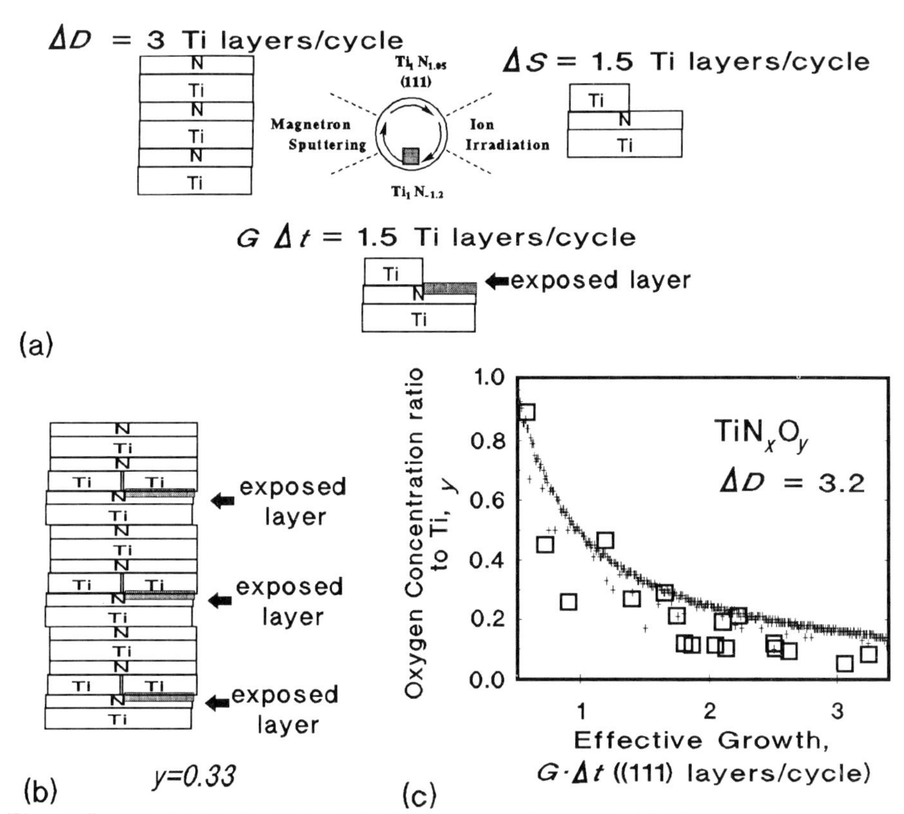

Figure 5 A model of oxidation during thin film process under sequential ion-beam assistance.

2)surface vacancies precede the oxygen contamination in ion-assisted process, one can reach the simple model in which openings in the top Ti layer introduce oxygen atoms into nitrogen layer. Figure 5(a) shows atomic layers' construction at the first turn (cycle) where ΔD=3 (Ti layers/cycle) and ΔS=1.5 $G\Delta t$=1.5 (Ti layers/cycle), and the value of the window opening results in 0.5 (Ti layer) and leaves 1/2 of the nitrogen atoms exposed to oxygen and moisture. The process sequence is shown in Fig. 5(b) and then finally leads to 0.33 of Ti openings. When the openings perfectly substitute nitrogen atoms with oxygen, the value of 0.33 gives a maximum of oxygen concentration (y=0.33). The calculated values for various $G\Delta t$ are shown as small crosses in Fig. 5(c) with experimental results indicated by squares. Though our experiments were not under the atomic layer epitaxy, the results still prefer the simple model, because titanium openings at the growing surface is important for the oxidation.

According to Figure 4, the dependencies of the TiN oxidation on assisted ion beam voltage and ion doses, which are real parameters used at thin film growth, are shown in Fig. 6. Curves on the bottom plane indicate boundaries where the equations mentioned above equals zero, in another words, higher voltages and ion doses over the curve do not increase film growth. In conclusion, the assisted ions should be set on biases for cases of obtaining

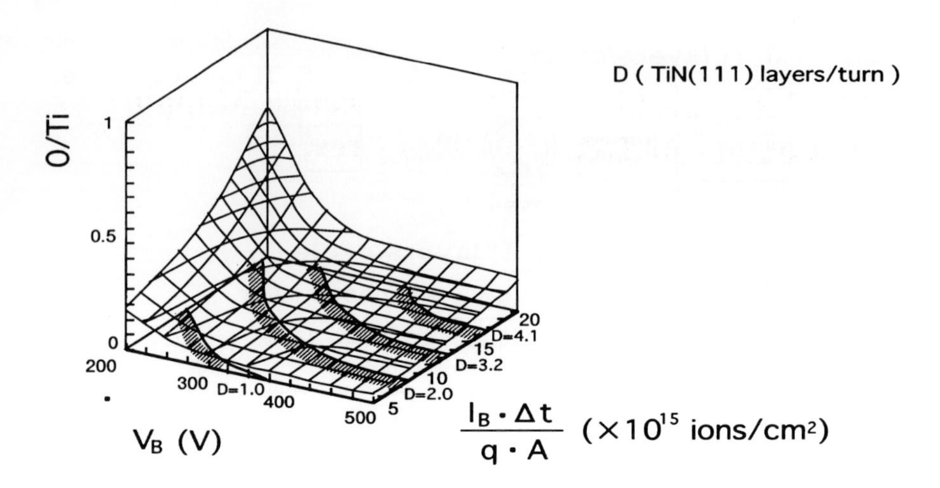

Figure 6 The dependencies of the TiN oxidation on assisted ion beam voltage and ion does. Curves on the bottom plane indicate film growth boundaries for ΔD.

oxygen or reducing oxygen as referred in Fig. 6.

We have treated here the sequential process of oxidation. Dynamic (simultaneous) process can be considered likely, that is, the low level limitation of the ion doses in Fig. 6 is the dynamic mode, because the sequential mode at Δt =0 should be equivalent to the dynamic mode at continuous ion assistance..

Acknowledgments

All authors wish to express thanks to Rob Gorris and Elzbieta Kolawa of Caltech for their technical assistance, to Hironori Moto of Kumamoto University for preparation of ion-beam, and to Takaaki Fukumoto and Hiroshi Kanno of Mitsubishi Electric Corporation for their exciting discussions.

References

[1] H. G. Tompkins, J. Appl. Phys. **70**(7) (1991) 3876-3880

[2] H. Kubota, M. Nagata, M. A. Nicolet et.al., Jpn. J. Appl. Phys. Vol. **32** (1993) 3414-3419.

[3] K. Hashimoto and H. Onoda, Appl. Phys. Lett. **54** (1989) 120.

[4] S.R. Nishitani and S. Yoshimura, J. Mater. Res. **7** (1992) 754-764.

[5] N. Savvides and B. Window, J. Appl. Phys. **64** (1988) 225.

[6] H. Kubota and M. A. Nicolet, Applied Surface Science **82/83** (1994) 565-568.

Inst. Phys. Conf. Ser. No 149
Paper presented at DRIP 95

Infrared Microscopic Photoluminescence Mapping on Semiconductors at Low Temperatures

M. Tajima [a)]

Institute of Space and Astronautical Science, Sagamihara 229, Japan

Abstract. Highly spatially resolved mapping of photoluminescence (PL) in the infrared region at low temperatures has been carried out for the analysis of deep-levels in semiconductors. A unique scanning-laser-beam type of apparatus was developed with an excitation beam of 10 μm diam, a scanning area of 1 mm x 1 mm, a wavelength region between 600 and 1800 nm, and a temperature range between 15 and 300 K. The microscopic mapping of deep-level PL from an annealed Czochralski-grown Si crystal with slip dislocations was analyzed for the first time. An opposite intensity contrast between the 0.77 eV band (D_b band) and dislocation-related D-lines gives strong evidence for the idea that the D_b band is due not to dislocations but to oxygen precipitates.

1. Introduction

Microscopic distributions of impurities and microdefects provide us with quite useful information for understanding their interaction and the mechanism of defect formation. Photoluminescence (PL) mapping with a high spatial resolution is known as one of the most effective methods by which to visualize the microscopic distribution of impurities and microdefects.

From a technical point of view, measurement at room temperature is easier than that at low temperature, similarly measurement in the visible region is easier than that in the longer wavelength region. Therefore, microscopic PL measurement was first reported for the visible band-edge emission from $GaAs_{1-x}P_x$ and GaP at room temperature. Extensive works were then performed for the band-edge emission from GaAs (*e.g.* Hovel and Guidotti 1985), where the wavelength region was around 830 nm, just outside the visible region. The author (1990) extended the wavelength region to 1.8 μm and improved the optical throughput of the apparatus, which enabled him to measure the deep-level emission as well as band-edge emission from Si and GaAs. For low temperature measurement, Heinke and Queisser (1974) first reported the microscopic mapping of the band-edge emission in GaAs at 10K. Although several methods for high-resolution and low-temperature mapping were recorded following this, all of them were limited in the wavelength region up to 1.0 μm, and very sophisticated cryostats were required.

The purpose of the present paper is to document microscopic mapping in the infrared region at low temperatures. A unique PL mapping apparatus of a scanning-laser-beam type

---

a) Electronic mail: tajima@newslan.isas.ac.jp

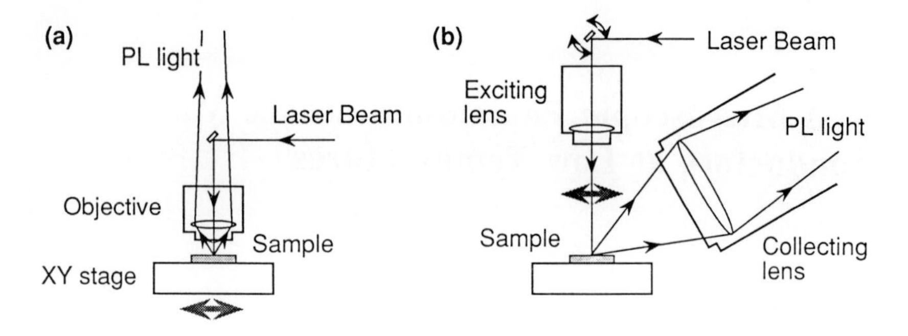

Figure 1. Principle of microscopic PL mapping. (a) Scanning sample type (confocal microscopy). (b) Scanning beam type.

was developed which is more advantageous than conventional systems of a scanning-sample type. To the best of the author's knowledge the microscopic mapping of weak PL in the infrared region at low temperatures is successfully carried out here for the first time using this apparatus.

2. Approach

A conventional method for microscopic PL mapping is shown in figure 1(a). The apparatus used in this method is simple: the basic design is essentially the same as an optical microscope. A laser beam for the excitation is introduced through the sample illumination optical path. An objective lens serves both as a focusing lens for the excitation beam and a collecting lens for the emitted light (confocal microscopy). Mapping of PL is obtained by scanning the sample. One of the problems of the technique, however, is a loss of light. Since conventional objectives are designed for visible light, the measurement of deep-level PL in the infrared region involves a serious loss of light. On the contrary, an infrared objective is advantageous for infrared light, but loss of the excitation light cannot be ignored.

Another problem is the short working distance of the objective, which is on the order of 10 mm, thus preventing the use of a conventional cryostat. Although a long working distance objective is available, its efficiency for collecting the emitted light is poor.

It is appropriate to point out the effect of the deviation from the focal point of the objective on the PL mapping data. To make the explanation simple we assume that the size of the radiative area is equal to the excitation area, S. The excitation intensity I_{EX} is inversely proportional to the excited area: $I_{EX} \propto S^{-1}$. In general, the PL intensity (I_{PL}) has an n-th power dependence on I_{EX}. The total amount of the PL light (I_{total}) is then expressed by

$$I_{total} \propto I_{PL} \times S \propto (I_{EX})^n \times S \propto (S^{-1})^n \times S \propto S^{1-n}.$$

The n value for the band-edge emission is close to 2, since the recombination process is ambipolar. In contrast, the n value for the trap-related emission is unity. The n value for the commonly observed deep-level emission is less than unity because of the saturation effect in the trap-related recombination. The value of S increases with increase in the deviation. Therefore, the PL signal decreases with increasing deviation for the band-edge emission, while it increases for the deep-level emission.

This was experimentally demonstrated as shown in figure 2: the intensities of the band-edge emission and the deep-level emission involving the EL2 defect in semi-insulating GaAs are plotted against the deviation. The measurement of the excitation intensity dependence

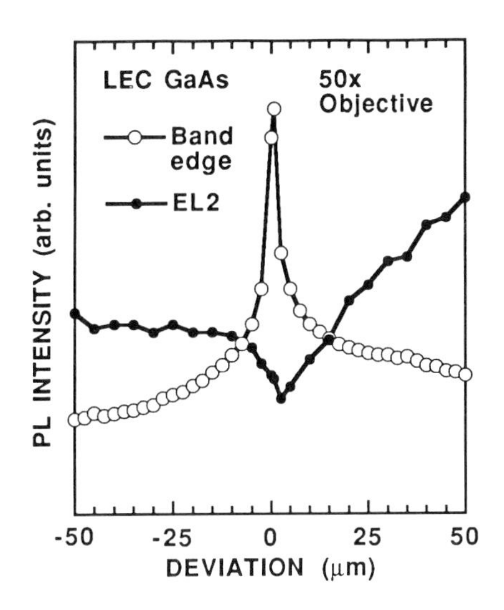

Figure 2. PL intensity variation caused by deviation from focal point of objective. Band-edge emission (○) and EL2-related band (●) in undoped semi-insulating GaAs at room temperature.

showed that the n values for the band-edge and EL2-related emissions are 2 and 0.8, respectively. If the scanning plane of the sample is not perpendicular to the laser beam, the deviation increases in a certain direction. The band-edge emission then decreases and the EL2-related emission increases along that direction. Obviously, the observed anti-correlated pattern is not inherent in the sample but is caused by a misalignment.

To solve these drawbacks we adopted the scanning beam type method with a long working distance shown in figure 1 (b). Independent lens systems are used for the excitation and collection of the PL light. A laser beam is deflected by mirrors and focused on the sample surface with a long working distance lens. The scanning area is limited so as not to increase aberration. The PL light is collected by a lens system with small *F* number. Since the scanning area is extremely small compared with the diameter of the lens, the light is collected by the lens efficiently regardless of the scanning position.

The independent lens systems enables us to design the respective systems optimally. The working distance of the lens systems is designed to be long so that a conventional cryostat can be fitted in. At the same time the long working distance is beneficial for the effect caused by the deviation from the focal point (figure 2). The disadvantage of its inferior focusing ability is not serious for our purpose.

3. Description of the instrument

Small beam size of the excitation, long working distance, and large scanning area are preferable; however, these are not independent in the apparatus shown in figure 1(b). First, the beam size was set at 10 μm, from our experience of deep-level PL mapping at room temperature, and from the fact that the excited carriers diffuse beyond a few tens of microns in Si at low temperatures. The working distance was then set at 80 mm so that our cryostat could be installed. These conditions determined the size of the scanning area to be 1 mm x 1 mm, which was appropriate for our experiments. Our cryostat was a temperature-variable, He closed-cycle type with two optical windows for the excitation and collection of PL at an angle of 60° with a field of view of 30°. The transfer of mechanical vibration from the piston to the

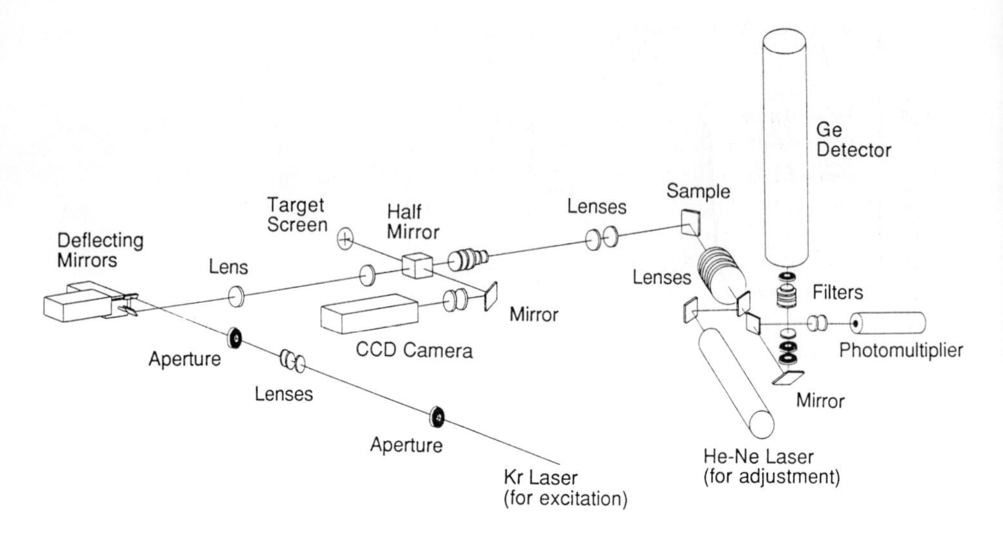

Figure 3. Schematic configuration of scanning-beam microscopic PL mapping system.

sample was reduced to ≤ 2 μm, using He gas isolation between the cold finger and the sample holder.

The schematic configuration of our apparatus is shown in figure 3. The 488 nm line of an Ar laser or the 647 nm line of a Kr laser was introduced on the deflection mirrors with a beam size of 1 mm, being guided properly with apertures and a target screen. The beam was deflected by the mirrors and focused on the sample surface with a diameter of 10 μm. The excitation beam spot on the sample surface was monitored by a charged-coupled-device camera. The PL from the sample was collected with an F=1.5 lens system. The collected light was passed through narrow bandpass filters to extract a specific spectral component, and then transferred to a Ge detector or a photomultiplier.

Careful consideration was necessary for image transfer from the sample to the detector. The image size on the photosensitive element in the detector had to be magnified large enough so that it was not influenced by a nonuniformity of its photosensitivity. At the same time this magnification should be done only of the image but not of the scanning area, otherwise the image would be easily off the element. This image transfer was realized by setting the optical path of the collection lens system to be essentially the reverse of that of the excitation lens system. In fact, the image size on the optical element in our apparatus was 3.5 mm, which was comparable with the size of the element (5 mm). The position of the image on the element remained nearly unchanged regardless of the scanning position.

The apparatus enabled us to measure PL with an excitation beam of 10 μm diam, with a scanning area of 1 mm x 1 mm, in the wavelength region between 600 and 1800 nm, and at temperatures between 15 and 300 K.

4. Experiment

The author and his colleagues used the present apparatus to analyze the oxygen precipitation around dislocations in annealed Czochralski-grown Si crystals (Tajima *et al* 1995a) and to investigate the behavior of impurities and point defects around dislocations in Si-doped GaAs grown by the vertical boat method. The former analysis is reviewed in the following part, and details of the latter will be given in a separate paper (Tajima *et al* 1995b).

Deep-level PL lines labeled D1-D4 are known to appear at low temperatures in deformed float-zone (FZ) Si crystals containing dislocations (Drozdov *et al* 1976). Similar D lines have also been observed in annealed Czochralski(CZ)-grown Si crystals, where small dislocation loops are generated from oxygen precipitates (Tajima and Matsushita 1983). The deep-level emission remains at room temperature as the 0.77 eV band (Tajima *et al* 1992), and the intensity of this band correlates positively with the precipitated oxygen concentration (Kitagawara 1992). The 0.77 eV band lies in the same photon energy region as the D1 and D2 lines.

A question arises here whether or not the 0.77 eV band and the D1/D2 lines have the same origin. If the two emissions do have the same origin, the effect of the oxygen precipitation on the appearance of the 0.77 eV band will be secondary: this band will be due to the dislocation loops generated from the oxygen precipitates. On the contrary, if the 0.77 eV band has a different origin from the D1/D2 lines, the concept that the 0.77 eV band is due to oxygen precipitates will be strongly supported. To solve this point we analyzed the 0.77 eV band and the D1/D2 lines in annealed CZ crystals with grown-in dislocations. The present microscopic PL mapping apparatus allowed us to measure the distribution of the two emissions around dislocations for the first time.

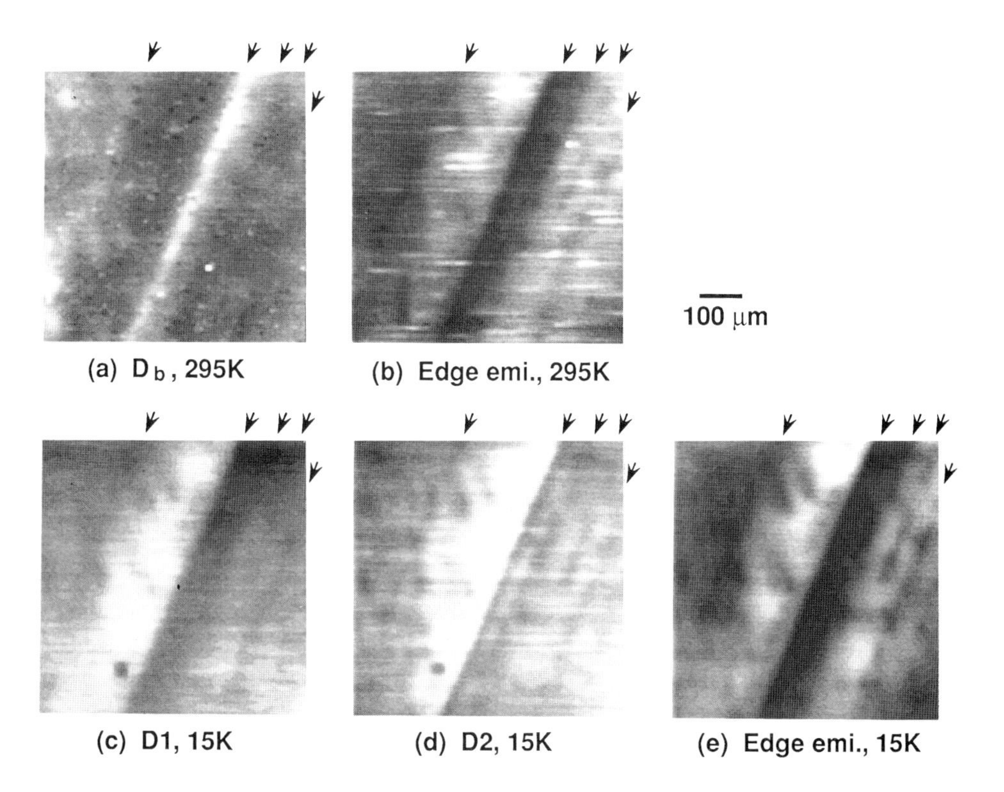

Figure 4. Microscopic PL mapping around dislocations in annealed Czochralski-grown Si crystals. (a) D_b band (0.77 eV band), and (b) band-edge emission at room temperature. (c) D1 line, (d) D2 line, and (c) band-edge emission at 15 K. Whiter gradation indicates higher intensity level. Positions of dislocations are marked with arrows.

The sample used for this study was a rapidly cooled and dislocated CZ Si wafer with an oxygen concentration of about 1 x 10^{18} cm^{-3} (conversion factor in infrared analysis: 3.14 x 10^{17} cm^{-2}). The distribution of dislocations was obtained by X-ray topography. A wafer was sliced from the ingot parallel to the growth direction, and a small wafer chip with a dislocated area was cut for the measurement. The sample was annealed at 1000°C for 16 h with pre-annealing at 450°C for 64 h, so that a considerable number of oxygen atoms were precipitated.

The PL spectra of the sample were measured at temperatures from 11 to 295 K. The D1 and D2 lines at 0.81 and 0.84 eV, respectively, dominate the spectrum at 11 K. The intensities of the D1/D2 lines decrease with temperature. Above 150 K a broad band with a peak around 0.8 eV becomes dominant. The peak position of the broad band shifts parallel to the band gap, and its peak position at room temperature is 0.77 eV. The broad band, known as the 0.77 eV band at room temperature, is renamed the D_b band because of this peak shift.

The intensity mappings of the D_b band (the 0.77 eV band) and the band-edge emission at room temperature are shown in figure 4 (a) and (b), respectively. Whiter gradation indicates higher intensity level. The X-ray topography on the same sample revealed that slant line patterns correspond with bundles of slip dislocation lines on {111} glide planes. The intensity of the D_b band is raised along the dislocation lines. This region is surrounded by a low-intensity region (denuded zone). The intensity is raised again in the outer region of the denuded zone, although this is not obvious in figure 4 (a) because of the overlapping of neighboring denuded zones. A reverse intensity pattern was observed for the band-edge emission.

The intensity mappings of the D1, D2 and band-edge emissions at 15K on this same area are shown in figure 4 (c) - (e). The intensity patterns for the three emissions are essentially the same as that in figure 4 (b): the intensity is low along the dislocation lines, high in the surrounding area, and low again in the area further away.

The opposite intensity contrast between the D1/D2 lines and the D_b band is strong evidence for the idea that the two emissions are of different origin. It should be pointed out that the decrease of the D1/D2 line intensity along the dislocation lines was also reported in low-temperature cathodoluminescence images of a plastically deformed FZ Si crystal (Higgs *et al* 1992). The optical transition was interpreted to be associated with the point defects trapped within the strain field of dislocations. It is the author's opinion that the D_b band is associated with the oxygen precipitation; this is based on the fact that the intensity of this band correlates positively with the precipitated oxygen concentration (Tajima *et al* 1992, Kitagawara 1992). The increase of the D_b band intensity along the dislocation lines indicates the enhancement of the oxygen precipitation on these lines. This is due to the preferential precipitation of oxygen atoms at the dislocations.

5. Summary

A unique PL mapping apparatus was developed with an excitation beam of 10 μm diam, a scanning area of 1 mm x 1 mm, a wavelength region between 600 and 1800 nm, and temperature range between 15 and 300 K. The apparatus enabled us to analyze the deep-level PL from an annealed Czochralski-grown Si crystal with slip dislocations. The D_b band (the 0.77 eV band at room temperature) showed an opposite intensity contrast with respect to the D1/D2 lines, suggesting that the D_b band is not associated with dislocations but with oxygen precipitates.

Acknowledgments

The authors would like to thank R. Shimizu for setting up of the apparatus, M. Warashina, Y. Kawate and M. Tokita for the measurement and discussion, H. Takeno for the Si sample preparation, and R. Toba for the GaAs sample preparation. This work was partly supported by a Grant-in-Aid for Developmental Scientific Research from the Ministry of Education, Science, and Culture of Japan.

References

Drozdov N A, Partin A A and Tkachev V D 1976 *Sov. Phys.-JETP Lett.* **23** 597-599

Heinke W and Queisser H J 1974 *Phys. Rev. Lett.* **33** 1082-1084

Higgs V, Lightowlers E C, Tajbakhsh S and Wright P J 1992 *Appl. Phys. Lett.* **61**, 1087-1089

Hovel H J and Guidotti D 1985 *IEEE Trans. Electron Devices* **ED-32** 2331-2338

Kitagawara Y, Hoshi R and Takenaka T 1992 *J. Electrochem. Soc.* **139** 2277-2281

Tajima M and Matsushita M 1983 *Jpn. J. Appl. Phys.* **22** L589-L591

Tajima M 1990 *J. Cryst. Growth* **103** 1-7

Tajima M, Takeno H and Abe T 1992 *Defects in Semiconductors 16* ed G Davies, G G DeLeo and M Stavola, *Materials Science Forum* **83-87** (Trans Tech Publications: Switzerland) pp 1327-1332

Tajima M, Tokita M and Warashina M 1995a *Defects in Semiconductors 18, Sendai, 1995* (to be published)

Tajima M, Kawate Y, Toba R, Warashina M and Nakamura A 1995b *This Conference*

Inst. Phys. Conf. Ser. No 149
Paper presented at DRIP 95

Evaluation of localized area epitaxy by spectrally resolved scanning photoluminescence

M F Nuban S K Krawczyk M Buchheit R C Blanchet

Laboratoire d'Electronique, URA CNRS 848, Ecole Centrale de Lyon,
BP 163, 69131 Ecully, FRANCE

S C Nagy B J Robinson D A Thompson J G Simmons

Center for Electrophonic Materials and Devices, McMaster University,
Hamilton, Ontario, L8S 4L7, CANADA

Abstract. In this contribution, room temperature spectrally resolved scanning photoluminescence technique with high spatial resolution ($<1\mu$m) is introduced and applied to control the uniformity of the composition and of the thickness of quantum well (Q.W.) structures obtained by localized area epitaxy. Furthermore, this technique is applied here to study lateral uniformity of Q.W. InGaAs/InP heterostructures grown by localized area Gas Source Molecular Beam Epitaxy (GSMBE) at various conditions (temperature, Arsine flow rate) and as a function of stripe width and spacing.

1. Introduction

Localized and selective area epitaxy (SAE), processes in which the growth occurs in the areas opened in a dielectric mask, offer a possibility of varying the thickness and composition of the grown layers by controlling the width and/or spacing of the open areas [1]. Thus, in a single epitaxy run it is possible to obtain localized heterostructures with different electro-optical properties. These technologies are attractive for fabricating advanced optoelectronic devices, such as photonic integrated circuits [2, 3, 4, 5]. As an example, let us mention a twelve-channel strained-layer InGaAs-GaAs-AlGaAs buried heterostructure quantum well laser array for wavelength division multiplexing, which was fabricated by selective-area MOCVD [5].

All commonly used epitaxial techniques potentially allow the growth of good quality layers in localized unmasked regions. Both, MOCVD and MOMBE, carried out at standard conditions, are capable of highly selective growth, which means that no material is deposited on the dielectric mask. However, in the case of MBE and GSMBE, the temperature necessary to obtain this selectivity is normally too high (over 700°C) to grow high quality crystals in the opened windows but at lower temperatures, a polycrystalline material is deposited on the masked regions [6].

In the case of each growth technique, there is a competition between a large number of physico-chemical processes [7], including absorption, desorption, and diffusion of the reactants on the topographically varying surface (masked and unmasked areas). Thus, the size and spacing of lithographically defined regions affect not only the thickness, the composition and the morphological quality but evidently also the lateral uniformity of the localized-area layers.

In such a complex situation, it seems extremely important to be able to check easily the lateral uniformity of the composition, of the thickness and of the electrical quality of heterostructures grown by localized area epitaxy.

The purpose of this contribution is to introduce room temperature spectrally resolved scanning photoluminescence with high spatial resolution ($< 1\mu m$) as a powerful technique to control the uniformity of the above mentioned parameters of Q.W. structures obtained by localized area epitaxy. Also, this technique is applied here to study lateral uniformity of Q.W. InGaAs/InP heterostructures grown by localized area GSMBE at various conditions (temperature, Arsine flow rate) and as a function of stripe width and spacing.

2. Experimental

2.1. Samples

To form the patterned substrate for growth, a 100nm thick SiO_2 layer was deposited by PECVD on n-type (100) InP substrates. Stripe patterns along the [011] direction with various widths (ranging from $5\mu m$ to $50\mu m$) and spacings (ranging from $10\mu m$ to $50\mu m$) were opened by standard photolithography. The crystal growths were carried out by GSMBE at various temperatures (ranging from 460°C to 510°C) and Arsine flow rates (ranging from 2.2sccm to 6.0sccm). The grown multilayer structures consisted of an InP buffer layer, a bulk InGaAs layer (50nm), an InP barrier (50nm), an InGaAs Q.W. (nominally 5nm thick) and an InP cap layer (50nm). All InGaAs layers were nominally lattice matched to InP. Fig. 1 shows a typical cross-sectional SEM image of the border between a masked and unmasked regions after the growth of a full structure.

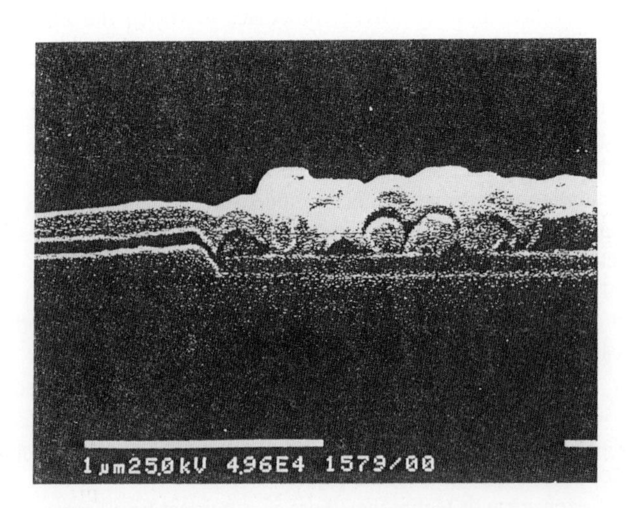

Fig.1. Cross-sectional SEM image of a border between masked and unmasked regions after the growth of a full structure.

2.2. Measurement technique

Spectrally resolved scanning photoluminescence (SPL) measurements were carried out at room temperature using a SCAT-SPEC IMAGEUR of SCANTEK (France). The photoluminescence signal was excited with a He-Ne laser (632.8nm) and measured with a LN_2 InAs photodetector. The diameter of the laser spot on the sample was $1\mu m$ and the scanning step was $0.5\mu m$.

No PL signal was detected on the polycrystalline material deposited on the dielectric mask, which indicates a very high density of non-radiative recombination centers. Room temperature PL spectra measured in unmasked regions consists of 2 partially overlapping peaks (see Fig.2 as an example). The low energy and high energy peaks are due to the emission from the bulk and from the Q.W. InGaAs layers, respectively.

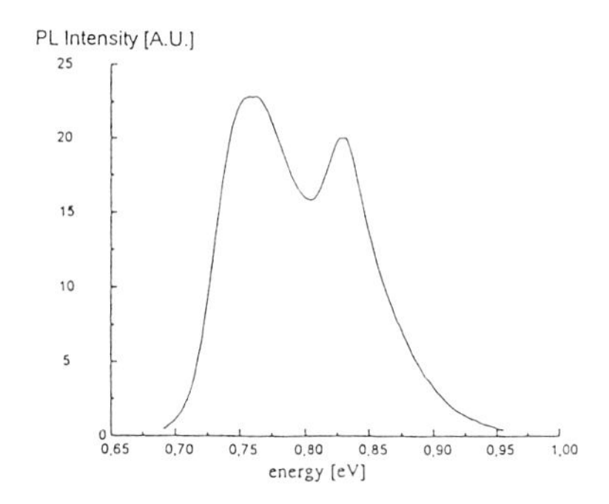

Fig.2. Typical shape of the PL spectra measured in unmasked regions

Using a dedicated software, we performed the deconvolution (decomposition into 2 constituents) of the above spectra. The shape of each constituent was approximated by a theoretical expression given in Ref. [8], which takes into account the radiative transition energy fluctuations as an adjustable parameter.

From the position of the lower energy PL peak we determined the band gap of the thick InGaAs layer. For compositional fluctuations in the range of few percents around the composition of the lattice matched InGaAs, a 50nm thick InGaAs layer can be supposed to be totally elastically strained [9]. Thus, the local composition of this layer was deduced taking into account elastic strain [10]. The difference of energy between the first quantum levels of electrons and holes (E1) in the InGaAs Q.W. was deduced after deconvolution from the higher energy PL peak. Local thickness of the Q.W. is calculated using Schrödinger equation. Thus, both the local composition and the thickness of the Q.W. can be mapped in a fast and non destructive way with about $1\mu m$ lateral resolution.

3. Results and discussions

As an example, Fig. 3 shows typical high resolution ($100\mu m$x$100\mu m$) SPL images at wavelengths corresponding to the emission from the InGaAs Q.W. and from underlying thick InGaAs layer.

The bright stripes correspond to the PL emission from the epitaxial layers in the opened areas in the SiO_2 mask. Note the presence of dark spots (few μm in diameter)

attributed to the dislocations emerging from the InP substrate and reproduced in both InGaAs layers. In addition, the PL peak intensity from both layers decreases close to the edges with the SiO_2 mask.

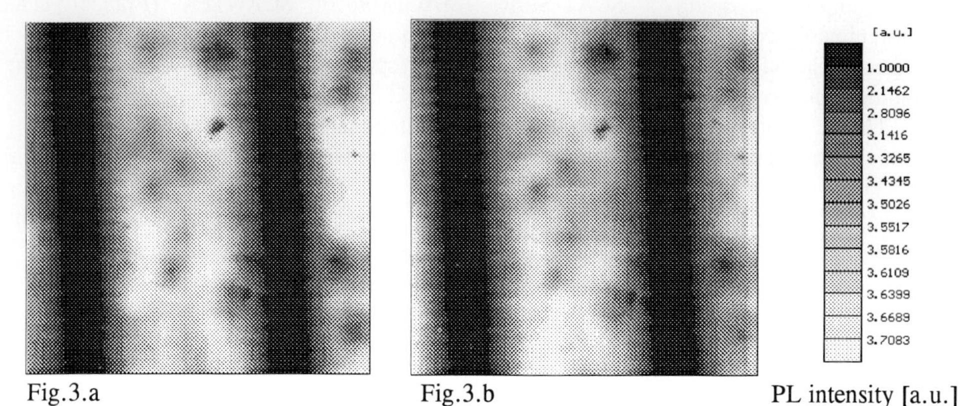

Fig. 3 High resolution (100μmx100μm) SPL images at wavelength corresponding to the PL emission from (a) the InGaAs Q.W. and (b) from underlying InGaAs layer. The growth of this structure was carried out on 50μm stripes with 10μm spacing at 460°C with 2.2sccm Arsine flow rate.

Our principal aim here is to study the uniformity of the deposited layers across the stripes. Thus, spectrally resolved SPL line scans were carried out across the stripes. Then, according to the procedure described in §2 we calculated and plotted :
- Q.W. thickness,
- composition,
- first quantum levels of electrons and holes in the Q.W. layer (E1)
- peak PL intensity from the Q.W. layer, PL(E1).

The Fig. 4 shows the lateral distributions of the above parameters across the stripes of different width (10μm, 20μm and 50μm) for a sample ("A") grown at 460°C with an Arsine flow rate of 2.2sccm. We found that the spacing between the stripes ranging from 10μm to 50μm doesn't affect these results.

It can be noticed in Fig. 4a that the Q.W. thickness increases near the mask edges. In the case of 20μm and 50μm stripes, the growth planarizes beyond a 5μm distance from the edges. Note, in particular, that the Q.W. thickness in the middle of the stripe increases considerably with decreasing stripe width. In the case of 10μm stripes, the increase of the Q.W. thickness in the center exceeds 1nm (20% above the nominal thickness). The increase of the Q.W. thickness in the stripes is explained by the migration of In and Ga atoms from the SiO_2 edges [6].

Fig. 4.b shows the compositional variations across the 10μm, 20μm and 50μm stripes. In all the cases, an increase of the In concentration is observed near the stripe edges. However, in the case of narrow stripes (10μm), a significant increase of the In concentration (2% above the nominal value) is observed in the middle of the stripes. The increase in the In to Ga ratio near the stripe edges, can be attributed to the difference in the migration length of In and Ga atoms from the edges [11,12].

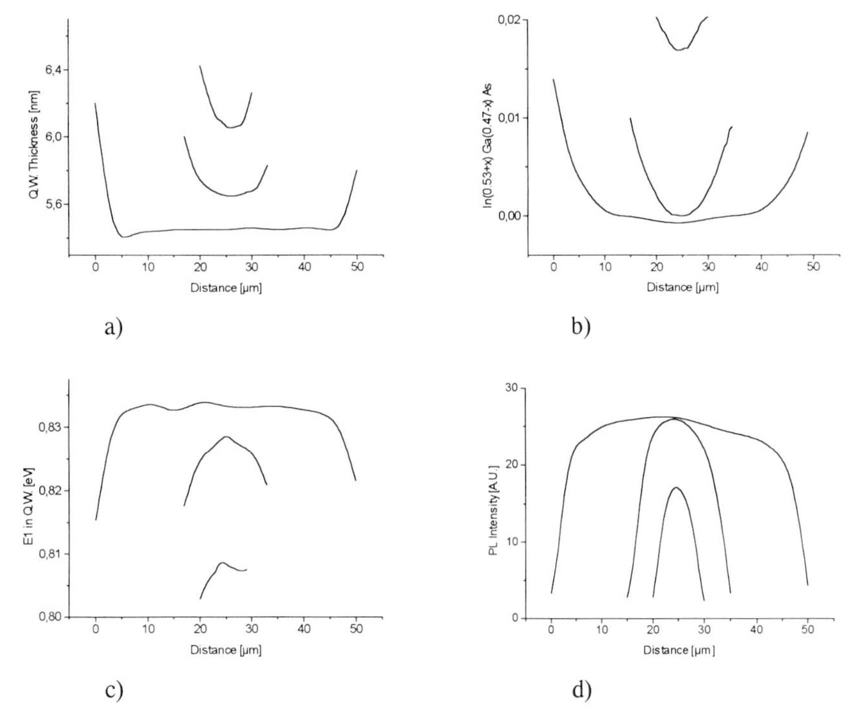

Fig. 4. Distribution across the stripes of different widths (10μm, 20μm and 50μm) of : a) Q.W. thickness, b) composition, c) energy E1, d) PL intensity from Q.W.. This sample was grown at 460°C with an Arsine flow rate of 2.2 sccm.

Fig.4.c, shows lateral variations of the energy E1 in the Q.W. across the stripes. For all stripe widths the energy E1 decreases near the edges. For narrow stripes, a significant decrease of E1 is also observed in the stripe center. The difference between E1 values in the middle of 10μm and 50μm stripes exceed 20meV. Both the increase of the growth rate and the increase of In mole in Q.W. fraction contribute to the red shift of the PL(E1) peak : an increase of the Q.W. thickness reduces the quantum size effects and an excess in In reduces the InGaAs band gap energy.

Fig. 4.d shows the lateral variations of the PL(E1) peak intensity from the Q.W. across the stripes. It appears that the PL peak intensity decreases near the SiO_2 edges. When the stripe width decreases from 50μm or 20μm to 10μm, the PL intensity in the middle of the stripe decreases. In the proximity of stripe edges, the density of structural defects is expected to be high [7]. These defects introduce non-radiative recombination centers which are responsible for the reduction of the PL signal. It has been demonstrated by our research group that the PL intensity is inversely correlated to photodiode leakage current [13]. In addition, Wang et al [14] have shown that the removal of edge material reduces the leakage current of an InGaAs/InP p-n junction made by SAE. So, the uniformity of the PL(E1) peak intensity is an additional indication of the useful parts of a stripe.

The above described analysis of the uniformity of Q.W. properties were carried out also on three other samples grown at the following conditions:

- sample B : growth temperature = 460°C Arsine flow rate = 6.0sccm.
- sample C : growth temperature = 510°C Arsine flow rate = 3.9sccm.
- sample D : growth temperature = 510°C Arsine flow rate = 2.8sccm.

The important parameters to be compared are the following :

- the shift of the difference between the quantum levels of electrons and holes (E1) in the middle of the stripes.
- the lateral uniformity of the stripe, which can be expressed as a width of the stripe where the uniformity of E1 is better than 2meV.

The results obtained on the stripes of different widths for different growth conditions can be summarized as follows :

- in the case of large stripes (50μm) at all the growth conditions used in this work, there is no shift of E1 in the middle of the stripe. This may be explained by the fact that the diffusion length of both Ga and In are much lower than a half of the stripe width (i.e. 25μm), since elements are supposed to diffuse from the stripe edges.
- concerning the narrow stripes (10μm), the most pronounced E1 red shift (25meV) is observed in the case of the samples A and D. This shift is 5meV and 8meV in the case of 10μm stripes on the samples B and C, respectively. This indicates that Arsine flow rate and growth temperature play a particularly important role in controlling the diffusion processes. It is also interesting to notice that for all growth conditions, the width of the "uniform" part of the 10μm stripe is roughly the same (about 8μm). Thus, it appears that by reducing the Arsine flow rate or by increasing the growth temperature, it is possible to increase the E1 red shift, without reducing the uniformity width.

4. Conclusion

We have introduced here spectrally resolved scanning photoluminescence, carried out at the room temperature with high spatial resolution, as a fast and non-destructive way to control spatial uniformity of heterostructures grown by selective area epitaxy.

References

[1] Temkin H, Hamm R A, Feygenson A, Cotta M A, Harriott L R, Ritter D and Wang Y L 1993 *Mat. Res. Soc. Symp. Proc.* vol. 300

[2] Lammert R M, Mena P V, Forbes D V, Osowski M L, Kang S M and Coleman J J 1995 *IEEE Photonics Technology Letters*, **7** 247-250

[3] Lammert R M, Cockerill T M, Forbes D V and Coleman J J 1994 *IEEE Photonics Technology Letters*, **6** 1167-1169

[4] Joyner C H, Zirngibl M and Meester J P 1994 *IEEE Photonics Technology Letters*, **6** 1277-1279

[5] Cockerill T M, Lammert R M, Forbes D V, Osowski M L and Coleman J J 1994 *IEEE Photonics Technology Letters*, **6** 786-788

[6] Kuroda N, Sugou S, Sasaki T and Kitamura M 1993 *Jpn. J. Appl. Phys.* **32** 1627-1630

[7] Mallard R E, Thrush E J, Gibbon M A and Booker G R 1993 *Mat.Sci. and Eng.* **B20** 48-52

[8] Herman M A, Bimberg D and Christen J 1991 *J. Appl. Phys.* **70** R1-R52

[9] Wang T Y and Stringfellow G B 1990 *J. Appl. Phys.* **67** 344-352

[10] Bhattachyra P *Properties of InGaAs lattice matched and strained* (EMIS data review N°8. INSPEC publication)

[11] Nishida T, Sugiura H, Notomi M and Tamamura T 1993 *J. Cryst. Growth* **132** 91-98

[12] Arent D J, Nilsson S, Galeuchet Y D, Meier H P and Walter W 1989 *Appl. Phys. lett* **55** 2611-2613

[13] Klingelhoffer C Ph.D. Thesis no 95-24, Ecole Centrale Lyon, 69130 Ecully, France

[14] Wang Y L, Feygenson A, Hamm R A, Ritter D, Weiner J S, Temkin H and Panish M B 1991 *Appl. Phys. Lett.* **59** 443-445

Inst. Phys. Conf. Ser. No 149
Paper presented at DRIP 95

Microscopic Photoluminescence Mapping of Si-Doped GaAs around Dislocations at Low Temperatures

M. Tajima[1], Y. Kawate†, R. Toba‡, M. Warashina and A. Nakamura†

Institute of Space and Astronautical Science, 3-1-1 Yoshinodai, Sagamihara 229, Japan

† Faculty of Science, Science University of Tokyo, 1-3 Kagurazaka, Shinjyuku 162, Japan

‡ Dowa Semiconductor Co., Ltd., 1 Sunada, Iijima 011, Japan

Abstract. Microscopic intensity variations of photoluminescence(PL) bands around dislocations were studied on Si-doped, liquid-encapsulated vertical boat grown GaAs at low temperatures. We measured the PL mappings for four emission bands: the 1.49eV band of free-to-acceptor and donor-acceptor transitions involving Si_{As}, the 1.33eV band associated with B_{As}, the 1.15eV band due to V_{Ga} complex, and the 0.95eV band which appears commonly in n-type GaAs but has not yet been identified definitely. The PL intensity patterns of each emission band are explained based on the concept that defects have been gettered by the dislocations and that the area near the dislocations is more As-rich than that farther away.

1. Introduction

Si-doped GaAs is essential for optoelectronic devices. Defects in semiconductors, such as dislocations, impurities and vacancies, influence electrical and optical properties which affect device performances. Nonuniform distribution of these defects can be the cause of inhomogeneity in electrical and optical properties of semiconductor wafers. Therefore, the understanding of defect behavior and correlation of point defects and dislocations is very important. A highly spatially resolved photoluminescence(PL) measurement is one of the most successful methods to reveal the inhomogeneity of defect distribution around dislocations. Measurements of microscopic PL mappings at room temperature[1] and those of near band edge emissions alone at low temperatures[2] have been reported.

In this study, we have performed microscopic mapping measurements of deep-level as well as near band edge PL using a unique PL mapping system[3]. The distribution of impurities and defects and the variation of stoichiometry around dislocations are discussed on the basis of intensity variation of various PL bands.

[1] E-mail tajima@newslan.isas.ac.jp

2. Experiment

The investigated samples were n-type Si-doped GaAs crystals with carrier concentrations of $(3 \sim 40) \times 10^{17}/cm^3$ and were prepared by the liquid-encapsulated vertical boat growth method which is a combination of the vertical Bridgeman and the vertical gradient freeze methods. These crystals were grown along the [001] direction and had diameters of 2-inches. Wafers were sliced in the plane perpendicular to the growth direction. The average dislocation densities were less than $300/cm^2$ and depended on the crystal growth condition, the Si concentration and the position of the ingots at which the wafers were sliced. The dislocations, revealed as etch pit patterns by KOH solution, were located primarily in the central area and four radial areas along the <100> direction(figure 1(a)) [1, 4]. In radial areas along the <100> direction, microscopic etching features by modified super oxidizer (MSO) photoetching[5] show dislocation cores and traces moved by thermal stress during the cooling period after solidification (figure 1(b))[6]. Microscopic PL mapping measurements were carried out on the band edge emission and the 0.95eV at room temperature in various wafers sliced from various ingots[1] and two wafers which showed characteristic PL intensity patterns were chosen. Their average dislocation densities of the two wafers were 300 and $70/cm^2$ and carrier concentrations were 3 and $8 \times 10^{17}/cm^3$,respectively. PL mapping measurements were made on the radial and central dislocated areas of each wafers and the results of the two measurements showed the same tendency as described below. The results are shown here in the radial dislocated peripheral area of the wafer with $n = 8 \times 10^{17}/cm^3$.

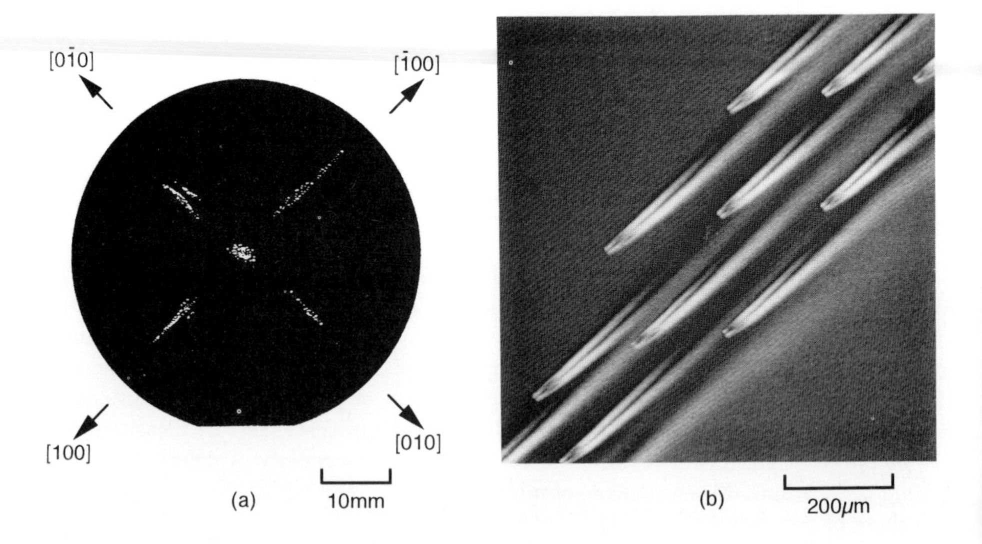

Figure 1. (a) Macroscopic dislocation pattern on Si-doped GaAs wafer revealed by KOH etching. (b) Microscopic etching feature in radial dislocated area along the [100] direction revealed by MSO photoetching.

Microscopic PL mappings at 16K were measured using a unique PL mapping system[3]. Excitation laser was focused on a sample surface with 10μm diameter with scanning region of 1mm×1mm. Each PL component was extracted with band pass filters. This apparatus enabled us to take mappings of weak PL emission in the infrared region with a spatial resolution of 10μm at temperatures from 16K to 300K. Details of the apparatus are described in [3]. The 647nm line from a Kr-ion laser was used as an excitation source and the power densities on the sample surfaces varied from 10^{-2} to 10^{3}W/cm^{2} depending on the PL intensity.

3. Results and Discussion

The PL spectrum of the sample at 4.2K is shown in figure 2. In this spectrum, four PL emission bands are observed. The 1.49eV band is due to donor-acceptor pair and free-to-acceptor recombination involving Si_{As}. The 1.33eV band is related to B_{As} or its complex which acts as an acceptor.[7] This assignment was supported by the fact that the band did not appear in Si-doped crystal grown by the horizontal Bridgeman method in which the B_2O_3 encapsulant was not used.[1] The 1.15eV band is associated with the $V_{Ga} - Si_{Ga}$ complex.[8] The 0.95eV PL band is often observed in n-type GaAs. Although the origin of this band has not yet been identified, we have shown that the defect responsible for the band acts as an acceptor from the macroscopic correlation between the distribution of carrier concentration and the intensity variation of the 0.95eV band[1]. Difference in the spectral position of the four bands is not very significant between 16K and 4.2K. Measurements of PL mapping were performed on these four bands at 16K.

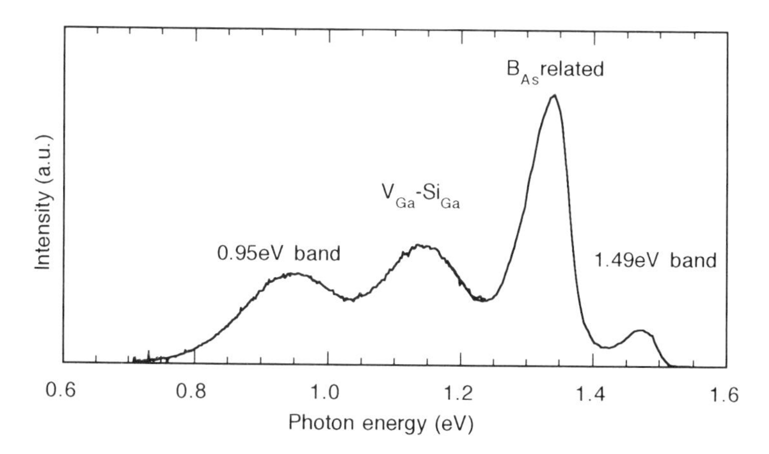

Figure 2. PL spectrum of Si-doped GaAs at 4.2K.

Intensity variations of each PL emission band at 16K are shown in figures 3-5. These measurements were done on the peripheral side of the dislocated area along the [010] direction with the spatial resolution of 10μm. Whiter gradation in these figures indicates higher intensity. Figure 3 shows the intensity pattern of the 1.49eV band. The peculiar pattern corresponding to etch pit pattern(figure 1(b)) can be seen. The intensity is lower in the core region of dislocations, higher in the surrounding area and much lower in the area far from dislocations. Since this band is due to donor-acceptor and free-to-acceptor transition, one might assume that the cause of higher intensity was the higher concentration of Si_{As} acceptor. However, it was shown that the carrier density in the dislocated area is higher than the dislocation-free area.[1] This tendency of carrier density is inconsistent with the assumption connected with Si_{As} concentration. We speculate that this intensity pattern is caused by the distribution of defects acting as killer centers of carriers, and that these defects are gettered by dislocations in the cooling period after solidification of crystal.

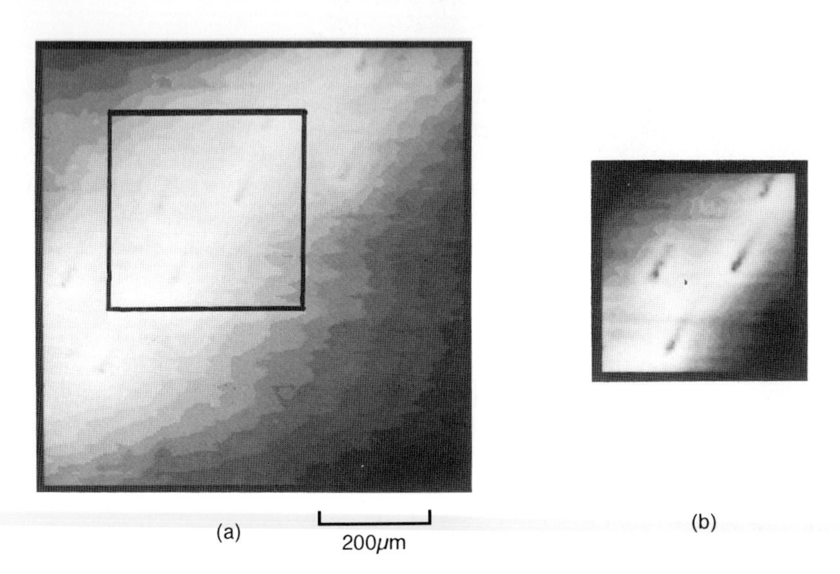

Figure 3. (a)Microscopic PL mapping of the 1.49eV band on the dislocated area. (b)Mapping under more enhanced contrast in the area outlined by the square in (a). Whiter gradation indicates higher level of intensity.

Figure 4 shows the intensity patterns of the bands related to B_{As} and $V_{Ga} - Si_{Ga}$ complex in the same area as figure 3. The feature corresponding to moving traces as well as core regions of dislocations is observed in these PL patterns. In these regions, the intensity of the B_{As} band is lower and that of $V_{Ga} - Si_{Ga}$ band is higher. Concentrations of the defects responsible for the two bands depend on stoichiometry. It is expected that the concentration of B_{As} is higher in the Ga-rich condition[9] and that of $V_{Ga} - Si_{Ga}$ is higher in the As-rich condition[8]. Therefore, the opposite intensity tendency between the two PL mappings is explained by the concept that the core regions of the dislocations and their moving traces are more As-rich than the areas farther away.

Macroscopically, intensity tendency of the band due to $V_{Ga} - Si_{Ga}$(figure 4(b)) and that of the 0.95eV band(figure 5) are similar, and this tendency is reverse to that of the band-edge 1.49eV band. The intensities of these two deep-level PL emissions are lower in the dislocated areas than in the areas far from dislocated areas. This macroscopic tendency shows that $V_{Ga} - Si_{Ga}$ and defects responsible for the 0.95eV band are killer centers for the 1.49eV band and that these defects or the constituents of the defects are gettered by dislocations, resulting in the formation of the denuded zone around dislocations.

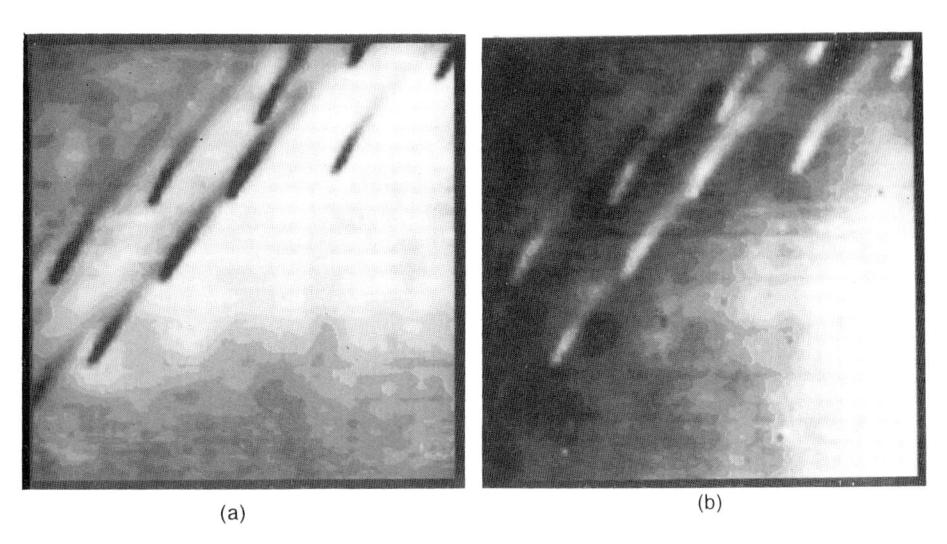

Figure 4. Microscopic PL mappings of (a)the 1.33eV band related to B_{As} and (b)the 1.15eV band related to $V_{Ga} - Si_{Ga}$ on the same area as Fig. 3(a).

It was found that microscopic intensity variation of the 0.95eV band has excitation power dependence around dislocations:A reduction in intensity in the core region was observed under an excitation power density higher than $5W/cm^2$ in this sample(figure

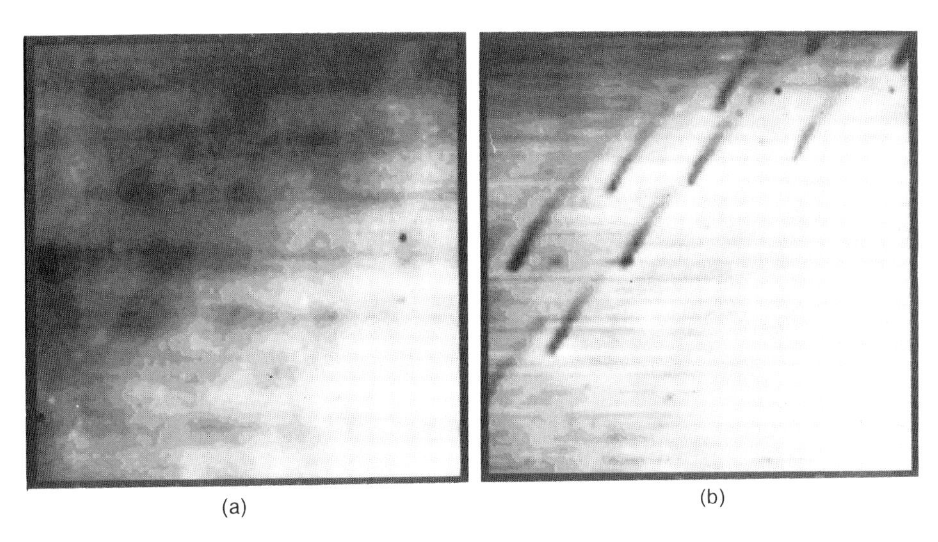

Figure 5. Microscopic PL mappings of the 0.95eV band. (a)weak excitation($3 \times 10^{-2}W/cm^2$). (b)strong excitation($5 \times 10W/cm^2$).

5). Dependence of the microscopic intensity pattern reflects the fact that the excitation power dependence of PL intensity in the core regions and the moving traces of dislocations differ from that in the surrounding area. The excitation power dependence of PL intensity is influenced by the concentration of defects capturing carriers. We speculate that the cause of this excitation power dependence is the higher concentration of various defects in the dislocation core regions and moving traces than in the surrounding area. The macroscopic tendency, however, in which intensity of a dislocated area is lower than that of a distant area, is independent of excitation power. This type of excitation power dependence was not observed for the 1.49, 1.33 and 1.15eV bands under our experimental conditions. Further investigation is necessary to understand the mechanism.

4. Conclusion

We have shown the intensity variations of four of the PL emission bands in Si-doped GaAs at low temperatures. Each PL pattern shows different characteristics. The pattern of the 1.49eV band reflects the distribution of killer centers. PL mappings of the bands related to B_{As} and $V_{Ga} - Si_{Ga}$ have revealed that the core regions and moving traces of dislocations are more As-rich than the surrounding areas. Macroscopic negative intensity patterns of the deep-level 1.15 and 0.95eV bands with respect to the band-edge 1.49eV band lead us to suggest that the defects causing the deep-level PL bands act as killer centers for the 1.49eV band, and that the defects or the constituents of the defects are gettered by dislocations.

Acknowledgment

This work was partly supported by a Grant-in-Aid for Developmental Scientific Research from the Ministry of Education, Science, and Culture of Japan.

References

[1] Tajima M, Toba R, Ishida N and Warashina M to be published *Proc.1st Int. Conf. Materials for Microelectronics, Barcelona, 1994* (Mater. Sci. Technol.)

[2] Heinke W and Quiesser H J 1974 *J. Appl. Phys.* **33** 1082–1084

[3] Tajima M to be presented at this conference

[4] Toba R, Tajima M and Warashina M to be published *Proc. 18th Int. Conf. on Defects in Semiconductors, Sendai, 1995*

[5] Nishizawa J, Oyama Y, Tadano H, Inokuchi K and Okuno Y 1979 *J. Crystal Growth* **47** 434–436

[6] Frigeri C and Weyher J L 1990 *J. Crystal Growth* **103** 268–274

[7] Brierly S K, Hendriks H T, Hoke W E, Lemonias P J and Weir D G 1993 *Appl. Phys. Lett.* **63** 812–814

[8] Williams E M and Bebb H B 1972 *Semiconductors and Semimetals* Vol 8 ed Willardson R K and Beer A C (New York: Academic Press) p 359

[9] Elliot K R 1984 *J. Appl. Phys.* **55** 3856–3858

Inst. Phys. Conf. Ser. No 149
Paper presented at DRIP 95

Influence of photoexcitation depth on luminescence spectra of bulk GaAs single crystals: application to defect structure characterization

V A Yuryev†[1], V P Kalinushkin†, A V Zayats‡[2], Yu A Repeyev‡ and V G Fedoseyev‡

† General Physics Institute of the Russian Academy of Sciences, 38, Vavilov Street, Moscow, GSP–1, 117942, Russia

‡ Institute of Spectroscopy of the Russian Academy of Sciences, Troitsk, Moscow region, 142092, Russia

Abstract. The results of investigation of bulk GaAs photoluminescence are presented taken from near-surface layers of different thicknesses using for excitation the light with the wavelengths which are close but some greater than the excitonic absorption resonances (so-called "bulk" photoexcitation). Only the excitonic and band-edge luminescence is seen under the interband excitation, while under the "bulk" excitation, the spectra are much more informative. The interband excited spectra of all the samples investigated in the present work are practically identical, whereas the bulk excited PL spectra are different for different samples and excitation depths and provide the information on the deep-level point defect composition of the bulk materials.

1. Surface and quasi-bulk photoexcitation: two approaches to luminescence characterization of semiconductors

Photoluminescence (PL) technique is a powerful tool for the investigation of defects in semiconducting materials. In most cases, the luminescence of semiconductors is investigated under the interband excitation, so that electrons are excited from the valence band to the conduction band. After the excitation, the electrons and holes thermalize to the thermoequilibrium distribution in the conduction and valence bands due to electron–phonon and electron–electron (hole–hole) interactions in a time less than 0.1 ps for the holes in the valence band and of about 1 ps for the electrons in the conduction band [1, 2], and then recombine. Main channels of the radiative recombination in semiconductors are the following: free and bound exciton luminescence, band-edge PL involving

[1] E-mail: VYURYEV@KAPELLA.GPI.RU.
[2] E-mail: AZAYATS@UALG.PT. Present address: University of Algrave, P–8000, Faro, Portugal.

shallow acceptor-like defects, and luminescence via the states of the deep-level defects or the defect associations. All three types of the PL yield the information on the defect structure of semiconductors.

1.1. Do PL measurements show real deep-level defect structure at band-to-band excitation?

In case of the interband excitation ($\hbar\omega > E_g$), the exciting light is absorbed in the region less than 0.1 μm close to the surface. For this reason, the spectra of the observed luminescence reflect mainly the defect structure of the near-surface layer but not of the sample bulk, whereas the information about the defect structure of the crystal bulk is of primary interest for solving many fundamental and applied issues. The case is that the procedures of preparation of experimental samples or technological wafers — such as cutting, abrading, and lapping — enter a great amount of defects in the near-surface layer, and despite the following etching removes this damaged layer, a considerable concentrations of non-uniformly distributed defects entered by these procedures remains in the near-surface layers [3]. Moreover, even if the damaged layer is etched off up to the depth where the influence of the former treatments seems to be practically absent, the etching itself changes the defect structure of the near-surface layer [4], e.g. by removing the impurities, intrinsic defects, precipitates and their agglomerations, which are present in the crystal bulk, or by entering some new defects. Besides, some sample treatments, such as annealing (or those involving annealing), *etc*, lead to creation and/or redistribution of the defects between the crystal bulk and its near-surface layer, so the defect structure of the near-surface region becomes not identical to the bulk one and nonuniform. *Therefore, interband excited PL reflects neither the real defect structure of the crystal bulk nor its changes after the exposures given to the samples.*

Unfortunately, in the PL measurements and particularly in the PL mapping of semiconducting materials neither the above considerations nor the circumstance, that the near-surface layer modified with special treatments — e.g. with etching — is investigated rather than the material bulk, are often taken into account. (It should be noted that we do not mean etching simulating one of the technological steps of a device fabrication — in this case the procedure used would be correct, but only etching aimed to reduce the non-radiative recombination beneath the surface and enlarge the PL efficiency[3].) The inferences valid for the particular way of surface preparation are often spread to whole the material bulk or near-surface regions subjected to other kinds of pre-experimental treaments.

1.2. Photoluminescence at band-tail excitation: a way to obtain true information on subsurface defect structure

Nonetheless some modifications of the routine PL measurements might easily be done, which enable obtaining more correct information about the defects in the bulk of crystals. To study the photoluminescence of the crystal bulk, the light with a wavelength close but some greater than the excitonic absorption resonances for the PL excitation can be used. At these wavelengths, the absorption is expected to be still effective for PL excitation due to the band tails. At the same time, the absorption coefficient for this light is not

[3] Note that this remark is valid for some other methods of investigations of semiconductors such as EBIC, OBIC, DLTS, *etc*, as well.

so large as for the interband excitation hence the absorption length is large enough to excite a bulk of a sample (a layer with effective thickness up to 100 μm might easily be studied).

All three types of the luminescence mentioned above for the interband photoexcitation can be excited under the "bulk" excitation. Nevertheless, the excitonic and band-edge PL generated in the sample bulk could hardly be seen in the PL spectra due to absorption and re-radiation by the near-surface layer because of the large absorption coefficient at these wavelengths. It means that the surface generated excitonic and band-edge PL is always seen in the spectra irrespectively to the "surface" or "bulk" excitation is used. Another situation takes place for the deep-level luminescence. The wavelengths of this luminescence are far from the band gap and the absorption is weak at these wavelengths, so the spectrum of the bulk-excited deep-center luminescence is less affected by the surface layer. It gives one the possibility to investigate the defect structure in a bulk of samples by means of the deep-level-center PL excited by the bulk-absorbed light.

In the present paper, the authors strive to demonstrate how the band-tail excited PL might be used for the investigation of the effect of sample processing on the deep-level structure of the crystal bulk, and the vapour phase epitaxy (VPE) of GaAs as well as the VPE-simulating annealing are taken as an example.

2. Experimental details and basic result

2.1. PL setup

The photoluminescence spectra were recorded using 20-ps pulses from a tunable parametric oscillator consisting of two DKDP crystals pumped by the second harmonic of the YAG:Nd^{3+}–laser radiation [5]. The output radiation wavelength could be continuously tuned from 370 to 1890 nm. The emission with the wavelengths of 580, 810, 835 and 845 nm was used for PL excitation. The maximum energy of the pulses was of 0.1 and 0.5 mJ at 580-nm and 810-nm excitation wavelengths, respectively, and 1 mJ at 835-nm and 845-nm excitation wavelengths. The unfocused beam was used to avoid too high excitation densities at which nonlinear effects could predominate. Relaxation processes in GaAs single crystals were expected to be fast enough in order to consider 20 ps-pulse excitation as quasistationary one [1]. Taking into account the difference between an absorbance at the excitation wavelengths and between the energies of pump pulses, the concentration of the excited electrons was estimated to be ranged from 10^{15} to 10^{17} cm^{-3} under all the excitation wavelengths. In average, the effective thicknesses of the layers excited in our experiments were estimated to be of around 0.1, 0.5, 1, and 10 μm at 580, 810, 835, and 845-nm excitation wavelengths, respectively.

The spectra were taken at the temperature of 80 K using a liquid nitrogen cooled cryostat. For the spectral analysis, a MDR–3 diffraction monochromator (LOMO) and photomultiplier were used. The PL signal was averaged over 20 pump pulses whose intensity was in the preset range.

2.2. Samples

To investigate the effect of the vapour phase epitaxy (VPE) on the defect structure of the bulk substrate, undoped GaAs epitaxial layers were grown using the trichloride VPE

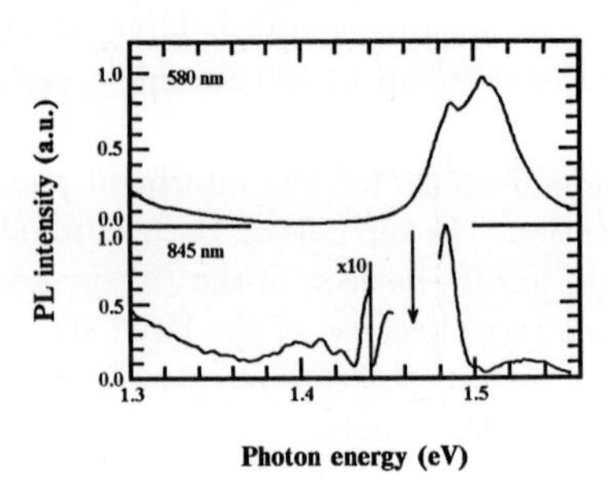

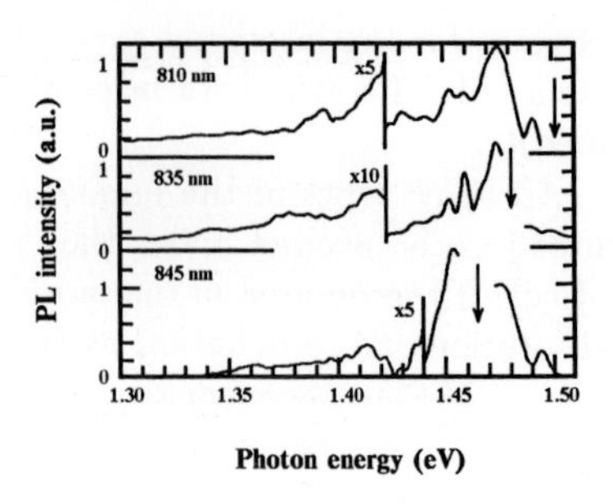

Figure 1. Photoluminescence spectra of the as-grown GaAs:Cr wafer measured at "surface" ($\lambda_{ex} = 580\,\mathrm{nm}$) and "bulk" ($\lambda_{ex} = 845\,\mathrm{nm}$) excitations.

Figure 2. PL spectra of the initial GaAs wafer obtained at different excitation depths (excitation wavelengths are indicated in the plot).

process on substrates of chromium-doped LEC gallium arsenide. The wafers intended for epitaxial growth were cut from the same ingot. The dislocation density in the substrates was 10^4–$10^5\,\mathrm{cm}^{-2}$ and their resistivity was in excess of $10^8\,\Omega\,\mathrm{cm}$. The substrate surfaces were oriented at angles of $2-6°$ with respect to the (100) plane. The thicknesses of the substrates were $300\,\mu\mathrm{m}$. The thicknesses of the films grown were about $6\,\mu\mathrm{m}$, the film growing rate was $0.1\,\mu\mathrm{m/min}$, the growth temperature was 720°C. The carrier concentration in the epilayers did not exceed $10^{14}\,\mathrm{cm}^{-3}$ at room temperature.

Some of the samples were subjected to a thermal treatment alone at 720°C in a hydrogen atmosphere, which simulates the VPE procedure. The annealing lasted 1 h. Reference samples were cut from the same ingot being usually adjacent wafers in the ingot to the processed ones.

The samples were etched in the liquid etchant before the PL measurements, so as the epilayers in the samples subjected to VPE as well as about $10\,\mu\mathrm{m}$ thick near-surface layers in the samples, which were not subjected to the VPE process, were etched off.

The defects in the analogous samples after the same exposures were previously investigated by means of the low-angle mid-IR-light scattering technique (LALS) [6].

2.3. Basic result

As it is clearly seen from Fig. 1 where the spectra for the as-grown GaAs sample are presented under the interband (580 nm) and band-tail (845 nm) excitations, only the excitonic and band-edge luminescence is observed under the interband excitation, while under the bulk excitation the spectrum is much more informative. The interband excited spectra of all the samples investigated are practically identical, whereas the influence of the sample treatment is clearly seen under the band-tail excitation (Figs. 2–4). Three groups of bands are observed in the spectra of the band-tail excited PL: excitonic and hot PL ($\hbar\omega \gtrsim 1.49\,\mathrm{eV}$), band edge luminescence ($1.46 \lesssim \hbar\omega \lesssim 1.49\,\mathrm{eV}$), and deep centers related luminescence ($\hbar\omega \lesssim 1.46\,\mathrm{eV}$). The relative intensities and spectral positions of the PL bands depend on the excitation wavelength as well as the wafer treatment. Excitonic-related and hot photoluminescence depends weakly on the excitation wavelength, that reflects its surface nature, but this PL is very sensitive to the sample treatment.

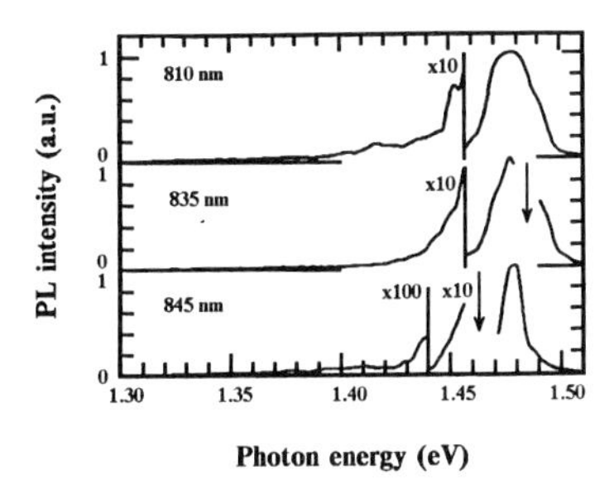

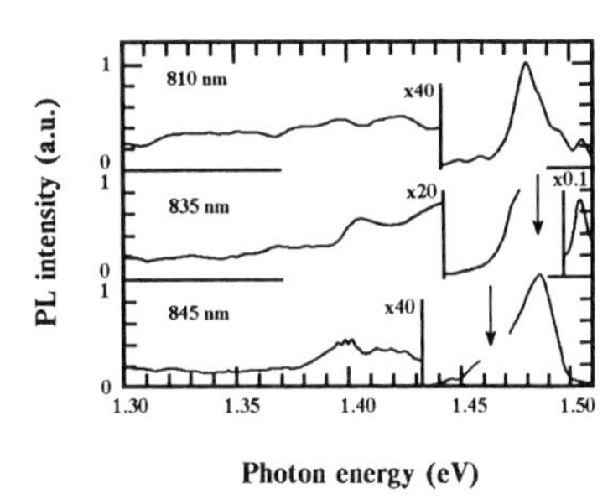

Figure 3. PL spectra of the GaAs wafer after VPE obtained at different excitation depths (excitation wavelengths are indicated in the plot).

Figure 4. PL spectra of the annealed GaAs wafer obtained at different excitation depths (excitation wavelengths are indicated in the plot).

This is a basic result of the current work from which a technique for correct analysis of the subsurface layer deep-level defect structure can be easily derived in application to bulk semiconductors and thick films.

The intensity of the deep-level related PL is not always a straightforward measure of the center concentration. The ratio of the extrinsic to intrinsic luminescence components is to be used to compare the center concentrations from the luminescence intensity [7]. For this reason, all the presented spectra are normalized on the band edge luminescence intensity. The excitation intensities varied from sample to sample and from one excitation wavelength to another by less than 8% during the measurements, therefore the difference between the monomolecular and bimolecular types of the deep-level and interband types of recombination is expected to be small enough. So we can use the normalized PL intensities from the spectra of different samples to compare the defect concentrations in the samples.

Not only the intensities of the corresponding luminescence bands but also the PL-bands wavelengths reflect the changes of the composition and/or concentration of the defects in the bulk and in the near-surface region because of the formation of different complexes and defect associations in the bulk as well as close to the surface that leads to the shift and/or splitting of the PL bands. Annealing or etching of the wafers can result in the changes of the intensities and the spectral positions of the PL bands because the treatments change the concentrations and compositions of defects as well as the defect associations.

3. Example of defect structure characterization: defect redistribution as a result of vapor phase epitaxy and simulating annealing of GaAs

After the treatments, dramatic changes of the PL intensities were observed. Relative intensity of the excitonic PL increased compared to the band-edge PL, while the deep-level related PL decreased, and the degree of this decrease depended on the treatment. The defect-related PL from the crystal bulk decreased by about 6 times after the annealing, and more than 10 times after the epitaxy (Figs. 2–4), i.e. during the treatments either the defect diffused to the surface, either the compensation of the sample changed which led to the radiative recombination centers became nonradiative, or the defects gathered in the precipitates, where they could not participate in the radiative recombination. The epitaxial growth resulted in the decrease of the defect-related luminescence. This decrease was stronger for PL at 845-nm excitation (about 100 times) and weaker for that at 810-nm excitation (about 10 times). The 810-nm excited PL from the annealed wafer and the epitaxy subjected one had comparable intensities for the bands at 1.39 and 1.41 eV, while the concentration of more deep centres close to the surface was much lower after the epitaxy than after the annealing.

Deep centers related PL had a strong dependence on the excitation wavelength that reflected the nonuniform distribution of the defects between the bulk and the near-surface layer rather than the resonance excitation of the luminescence, since all the excitation wavelengths were far from the locations of direct deep-level-to-band transitions and this PL was excited via the interband or band-tail absorption. When the excitation changed from "surface" one to "bulk" one, the increase of the deep centres related luminescence intensity with respect to the band-edge PL intensity could be clearly seen for all the investigated samples (Figs. 1 – 4).

4. Summary

Photoluminescence at band-tail excitation is a fruitful way to obtain a true information on a real subsurface deep-level defect structure of bulk semiconductors and thick layers. It enables the investigation of the defect distribution deep inside the crystal bulk as well as the redistribution of defects in subsurface layers resulting from sample treatments.

References

[1] Scholz R, Stahl A, Zhou X-Q, Leo K and Kurz H 1992 *IEEE J. Quantum Electron.* **28** 2473–85

[2] Nintunze N and Osman M A 1994 *Proc. SPIE* vol 2142 *Ultrafast Phenomena in Semiconductors* ed D K Ferry and H M van Driel pp 286–97

[3] Kalinuhkin V P, Yuryev V A and Astafiev O V 1995 *Mater. Sci. Technol.* **11** in press
Astafiev O V, Kalinushkin V P, Yuryev V A *et al* 1995 *Proc. 1995 MRS Spring Meeting* vol 378 *Defect- and Impurity-Engineered Semiconductors and Devices* ed S Ashok, J Chevallier *et al* (Pittsburgh: MRS) in press

[4] Baude P F, Tamagawa T and Polla D L 1990 *Appl. Phys. Lett.* **57** 2579–81

[5] Zayats A V, Repeyev Yu A, Nikogosyan D N and Vinogradov E A 1992 *Journ. Luminesc.* **52** 335–43

[6] Yuryev V A and Kalinushkin V P 1995 *Mater. Sci. Eng.* B **33** 103–14

[7] Bryskevich T, Bugajski M, Lagowski J and Gatos H C 1987 *J. Cryst. Growth* **85** 136–41
Dean P J 1982 *Progress in Crystal Growth and Characterization* vol 5 ed B Pampin (Oxford: Pergamon) p 89

Inst. Phys. Conf. Ser. No 149
Paper presented at DRIP 95

Photocurrent Mapping of Fe-doped Semi-insulating InP

J.Jimenez[1], M. Avella[1], A.Alvarez[1], M. Gonzalez[1], R. Fornari[2]
1) Fisica de la Materia Condensada, ETS Ingenieros Industriales, 47011 Valladolid, Spain..
2) MASPEC-CNR Institute, Via Chiavari 18/A, 43100 Parma, Italy

Abstract. Using a physical model which accounts for the photocurrent contrast on the basis of the electronic transitions associated with electrically active Fe, we have investigated numerous InP wafers, which were grown under different conditions and submitted to different thermal treatments. It is observed that the photocurrent fluctuations are mainly due to non-uniform distribution of the neutral Fe (Fe^{3+}) atoms. Different samples were studied in order to determine their homogeneity of different growth and treatment conditions.

1. INTRODUCTION

The semi-insulating (SI) InP wafers are used as substrates for the fabrication of optoelectronic integrated circuits and high-frequency devices. The high resistivity (> 10^7 Ωcm) of InP is generally obtained by doping with iron. The Fe atoms are substitutional to indium and give deep acceptor levels (at about 0.64 eV from CB) suitable for the compensation of the residual donors, mainly S and Si. There are however a few drawbacks connected with the Fe doping: i) longitudinal concentration profiles due to the very small distribution coefficient (top and tail wafers from the same InP crystal have very different Fe contents), ii) lateral inhomogeneities (such as growth striations for instance) which are very detrimental to device manufacturing, iii) Fe diffusion takes place during thermal treatments, giving rise to a new source of inhomogeneity and contamination of the epilayers. It is thus desirable to provide quick and reliable means for the study of the non-uniformities typical of Fe-doped semi-insulating InP.

Recently we have found that photocurrent (PC) is a very sensitive technique for the characterization of Fe-related optical transitions in InP. An experimental setup for mapping the photocurrent response has been developed while a physical model suitable to describe the complex relationships between the photocurrent intensity and the carrier photoionization (at different wavelengths) and recombination rates has been proposed [1].

In this paper we report the principles of this technique and discuss a physical model which can account for the photocurrent contrast on the basis of the electronic transitions associated with a 1.32 μm excitation . As an application of this mapping method, we show and discuss different photocurrent maps obtained on several as-grown and thermally-treated semi-insulating InP wafers.

2. EXPERIMENTAL

The SI InP samples investigated in this study were grown by the LEC (Liquid Encapsulated Czochralski) technique with addition of pure iron to the melt. One sample was initially undoped semiconducting and was made SI by Fe-deposition and diffusion at high temperature. The characteristics of the samples are in table I.

Table I:

Sample #	Treatment	Properties before annealing		Properties after annealing	
		ρ (Ωcm)	μ (cm²/Vs)	ρ (Ωcm)	μ (cm²/Vs)
155-70	as-grown (8 mm/h)	$1.47 * 10^7$	3.058		
155-70	48h @ 750°C in P ambient			$4.47 * 10^6$	1.650
161-50	as-grown (3 mm/h)	$2.7 * 10^7$	2.845		
159-75	as-grown (8 mm/h)	$2.3 * 10^7$	2.218		
R34 (+)	60 h @ 900°C in P ambient			$3.7 * 10^7$	3.830
R35-h3 (++)	45 h @ 900°C in P ambient			$8.1 * 10^6$	3.240

(+) This sample is from the top a lightly Fe-doped ingot and was semiconducting before the thermal treatment.
(++) The SI properties were obtained by diffusion of Fe into undoped InP: pure Fe was deposited onto the sample surfaces prior to thermal treatment.

PC measurements were performed according to the experimental set up shown on fig.1. The electrodes necessary for PC measurements are obtained by evaporating a thin layer of gold on both wafer surfaces. The sample is mounted on the cold finger of a cryostat especially designed for the X-Y stage of the microscope. The dark current at low temperature (Liquid Nitrogen) is below the detection limit of the electrometer used for measurements. Optical excitation is done with a YAG laser (λ=1.32 μm) focused through the objective of the microscope. The probe beam is nearly cylindrical with a beam diameter of about 3 μm. The volume excited by the laser beam increases its conductivity by several orders of magnitude, while the rest of the sample maintains its high resistivity. The measured photocurrent thus corresponds to the volume probed by the laser beam. The X-Y stage is controlled by two high precision stepping motors. The system is programmed in order to perform successive scans and to draw a photocurrent map. The measurements are carried out under low bias, in the linear part of the Iph vs. V plot in order to avoid non linear electrical effects.

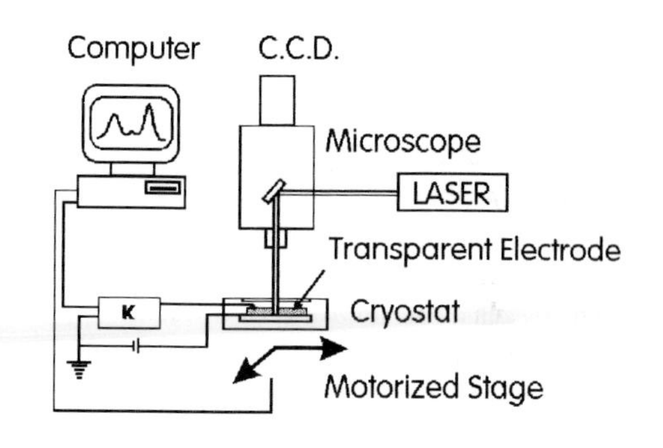

Fig.1.- Experimental setup for photocurrent mapping.

3. NATURE OF THE PHOTOCURRENT CONTRAST

Iron is a substitutional impurity in InP and gives a deep acceptor level (about 0.64 eV from the conduction band). The substitutional Fe can have two charge states: neutral (Fe^{3+}) or singly ionized (Fe^{2+}). The total electrically active iron [Fe_{act}] is thus given by the sum [Fe^{2+}]+[Fe^{3+}]. It should be noted that, in semi-insulating material, the concentration [Fe^{2+}] is nothing but the concentration of the net shallow donors: [Fe^{2+}] = (N_d-N_a).

The physical properties of the deep states associated with Fe in InP have been reported and discussed in a number of reports [2-6] whereas the PC spectra of SI InP have been reported in [7-9]. The main electronic transitions for 1.32 μm excitation are:

$$Fe^{3+} + h\nu(0.92\ eV) \longrightarrow Fe^{2+}\ (^5E) + 1h^+$$
$$Fe^{2+} + h\nu(0.92\ eV) \longrightarrow Fe^{3+} + 1e^-$$

Recombination of electrons is achieved at Fe^{3+}, while the photogenerated holes recombine at Fe^{2+}. According to these photogeneration-recombination processes a system of rate equations was proposed in order to obtain quantitative information about the origin of the photocurrent fluctuations. Trapping of carriers by other levels was neglected, since no changes in the shape and intensity of PC profiles were observed after subsequent scans on the same line. This proves that the relative population of $[Fe^{2+}]$ and $[Fe^{3+}]$ is restored after the excitation with 1.32 μm light. The set of rate equations describing these transitions are:

$$\frac{dp}{dt} = \sigma_p^o\left[\mathrm{Fe}^{3+}\right]\Phi - pC_{pFe}\left[\mathrm{Fe}^{2+}\right]$$

$$\frac{dn}{dt} = \sigma_n^o\left[\mathrm{Fe}^{2+}\right]\Phi - nC_{nFe}\left[\mathrm{Fe}^{3+}\right]$$

$$\frac{d\left[\mathrm{Fe}^{3+}\right]}{dt} = -\sigma_p^o\left[\mathrm{Fe}^{3+}\right]\Phi + pC_{pFe}\left[\mathrm{Fe}^{2+}\right] + \sigma_n^o\left[\mathrm{Fe}^{2+}\right]\Phi - nC_{nFe}\left[\mathrm{Fe}^{3+}\right]$$

$$\frac{d\left[\mathrm{Fe}^{2+}\right]}{dt} = \sigma_p^o\left[\mathrm{Fe}^{3+}\right]\Phi - pC_{pFe}\left[\mathrm{Fe}^{2+}\right] - \sigma_n^o\left[\mathrm{Fe}^{2+}\right]\Phi + nC_{nFe}\left[\mathrm{Fe}^{3+}\right]$$

$$I_{ph} = (n\mu_e + p\mu_p)eES$$

Average carrier mobility was determined for each specimen by Hall effect. The data were fitted to the experimental photocurrent transients, which allowed the determination of the electron and hole capture coefficients. The ratio between hole and electron capture coefficients was found to be nearly 0.1, which agrees with previous reports [2, 10].

The results of the calculation are summarized in Fig. 2, where we show the type of influence that $[Fe^{2+}]$ and $[Fe^{3+}]$ can have on the photocurrent transient. According to these results the initial slope of the photocurrent is dependent on $[Fe^{3+}]$ fig.2a, remaining practically unchanged for different $[Fe^{2+}]$. The photocurrent saturation level depends on both $[Fe^{2+}]$ and $[Fe^{3+}]$, fig.2b. When a photocurrent scan is carried out, the measured PC should depend on the scanning speed across the sample. The numerical

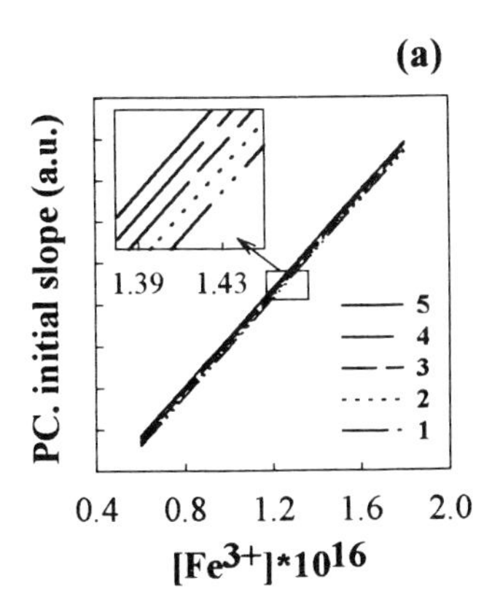

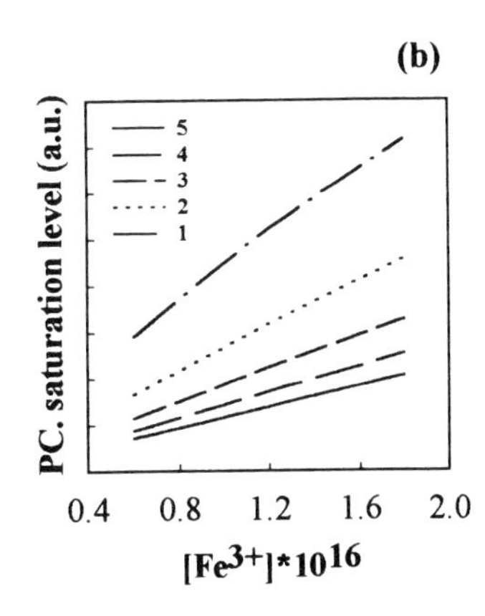

Fig.2.- a) Initial slope of the photocurrent transient vs. $[Fe^{3+}]$ for different $[Fe^{2+}]$:
1) $0.1*10^{16}$ cm^{-3} 2) $0.2*10^{16}$ cm^{-3}, 3) $0.3*10^{16}$ cm^{-3}, 4) $0.4*10^{16}$ cm^{-3}, 5) $0.5*10^{16}$ cm^{-3}
b) Saturation level of the photocurrent transient vs. $[Fe^{3+}]$ for different $[Fe^{2+}]$:
1) $0.1*10^{16}$ cm^{-3} 2) $0.2*10^{16}$ cm^{-3}, 3) $0.3*10^{16}$ cm^{-3}, 4) $0.4*10^{16}$ cm^{-3}, 5) $0.5*10^{16}$ cm^{-3}.

solution represented in fig.2 revealed that the main source of photocurrent fluctuation is the non uniform distribution of [Fe^{3+}] when the PC scan shape is independent of the scanning speed. .

4. RESULTS AND DISCUSSION

We obtained several photocurrent maps by scanning the laser beam across the sample surface as explained in the experimental section. All the maps presented herein were obtained under the same experimental conditions (T=77 K, λ=1.32 μm, Φ= 1019 cm^{-2} s^{-1} and scanning speed 125 μm /s).

The magnitude of the photocurrent fluctuations over as cut samples can be quite different from sample to sample. This can be a consequence of the history of the studied wafer: total dopant amount, position in the ingot, pulling rate... . Generally, a photocurrent scan presents both long and short range fluctuations; the central part of the wafer and the edges normally present an enhanced photocurrent in relation to the other parts of the wafer. Strangely, the two-dimensional maps never revealed growth striations, although these dopant inhomogeneities were observed by chemical etching. Probably the absence of sharp striations has to be ascribed to the fact that the photocurrent is representative of a cylindrical volume with a section corresponding to the diffraction spot and a height equal to the wafer thickness. Since the striations are tilted with respect to the wafer planes the probe beam will actually be probing several striations simultaneously thus giving an averaged signal. This can justify the lack of periodicity in the short range fluctuations, which normally appear as peaks in the photocurrent scans.

Fig.3a.- Photocurrent map of sample 155-70 as-grown. (Scale bar=1mm; White line sample contour)

Let us examine now the features of the different samples, starting from a comparison between as-grown and annealed InP (compare Figures 3a and 3b). In the as-grown sample of Fig.3a we observed the long range fluctuation mentioned above. After a prolonged annealing at 750 °C in P ambient, the sample (which is cut from the same wafer of Fig.3a, in symmetric position with respect to the wafer diameter) exhibits an improved homogeneity. It should be noted that the central area of the specimen is much more homogeneous that before the thermal treatment, although some Fe accumulation at the sample edges can be observed. Some suggestion that outdiffusion takes place also comes from the electrical measurements reported in Table I:

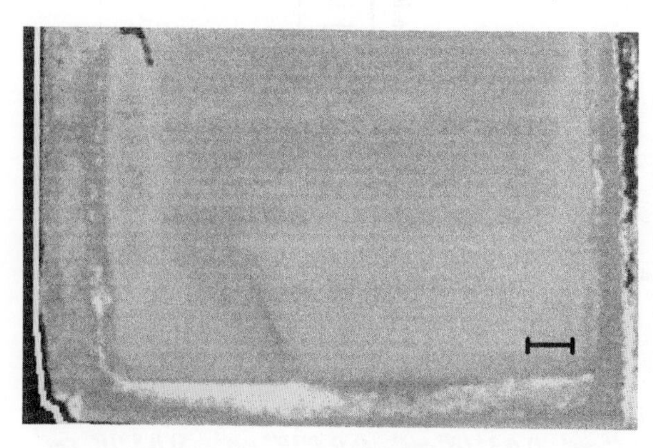

Fig.3b.- Photocurrent map of sample 155-70 thermally annealed.(Scale bar=1mm; White line sample contour)

the heavily-doped wafer 155-70 shows a significant reduction of the resistivity and mobility, consequent to annealing, which reveals that the compensation is now just sufficient.

We observed a limited area depleted of iron close to the Fe-rich edges in other cases as well. This is particularly conspicuous in Fig. 4, where we report the PC map of a lightly Fe-doped wafer (R34). This was originally semiconducting and converted to semi-insulating after thermal annealing at 900 °C for 60 hours under P pressure. Again the central part of the sample is quite uniform but a large depression in the PC profile can be observed close to the edge. Our interpretation of these data is that we have somehow reactivated the inactive Fe atoms spread in the matrix but, since at the periphery they tend to outmigrate, there is a drop of resistivity because the compensation is insufficient. The central part of the sample presents a significant iron activation, as deduced of the photocurrent response, which has a value typical of the as grown semi-insulating InP. It should be noted that the high photocurrent measured suggests that a significant [Fe^{3+}] has been activated, since [Fe^{2+}] should give very low photocurrent intensity for this excitation wavelength. The doping gradients between these regions appear very sharp. From these observation it can be concluded that a proper annealing procedure is suitable for improving the homogeneity of Fe doped InP, though some outdiffusion of iron (and corresponding Fe accumulation at the edges) is likely to take place, which avoids the usefulness of those parts of the wafer.

Fig.4.- Photocurrent map of an undoped sample diffused with iron. (Scale bar = 500 μm; White line sample contour).

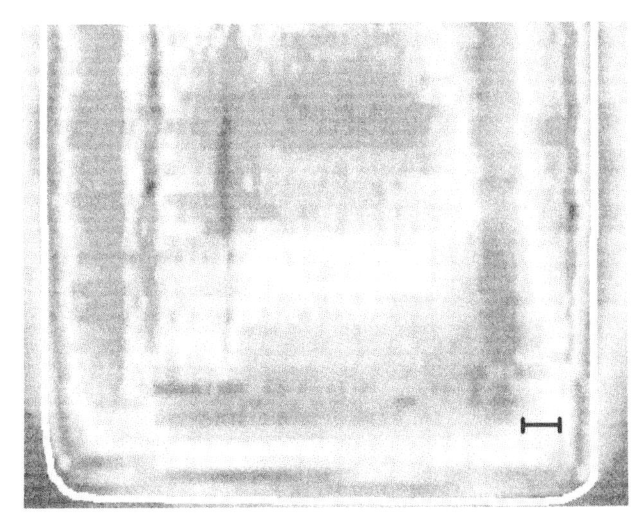

Fig.5.- Photocurrent map of an undoped sample diffused with iron. (Scale bar = 500 μm; White line sample contour).

We studied as well an undoped InP sample after it was made SI by iron diffusion at high temperature. In this case the photocurrent map (Fig.5) shows some new features: i) some vertical lines with very high PC intensity, probably connected with scratches which gettered Fe during the diffusion, ii) a denuded zone close to the edges as observed in the other annealed samples. Furthermore, a diagonal gradient of Fe concentration can be recognized that was not seen in the other specimens. This might be related to temperature gradients in the diffusion furnace but we are still investigating this data.

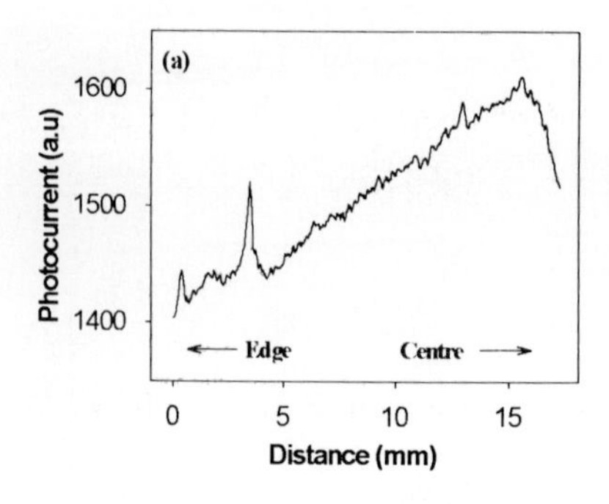

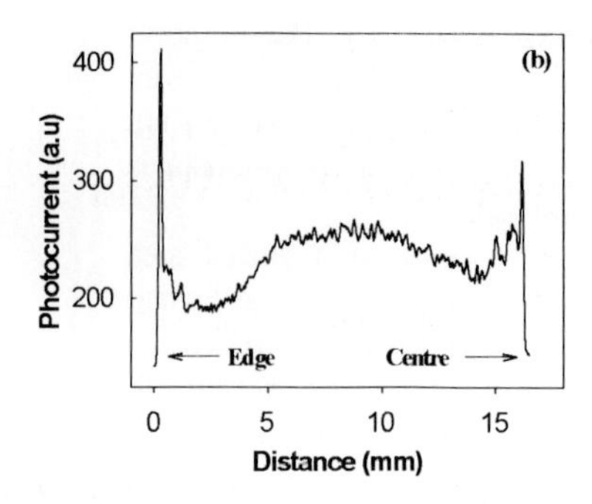

Fig.6.- Photocurrent scans ; a) Sample 161-50 , b) Sample 159-75 . (see table I)

The last example that we want to report here concerns the comparison of the Fe uniformity in LEC crystals pulled at different rates: 8mm/h (standard) and 3 mm/h (slow). It is expected that the latter crystal, which remained at high temperature for a longer time, is more uniform.

The long range distribution of iron could be associated with an enhanced iron activation for sample pulled at 3 mm/h, fig.6, specially at the centre of the wafer where photocurrent is enhanced. The short range fluctuations appear improved for slower pulling rate.

5. CONCLUSIONS

Photocurrent mapping provides a meaningful technique for observing inhomogeneities related to electrically active iron in InP wafers. The fluctuations of [Fe^{3+}] appear as the main causes for the inhomogeneity. Thermal annealing homogenizes large surface of the samples and produces electrical activation of iron.

ACKOWLEDGEMENTS
This work is supported by DIGICYT Mat 94.0042.

REFERENCES

[1] A. Alvarez, M. Avella, J. Jiménez, M.A. Gonzàlez and R. Fornari, Submitted to Semicond. Sci. Technol.
[2] P.B. Klein, J.E. Furneaux and R.L. Henry, Phys. Rev. B 29 (1984) 1947
[3] P. Leyral, G. Bremond, A. Nouailhat and G. Guillot, J. of Luminescence 24/25 (1981) 245
[4] C. Backhouse and L. Young, J. Electrochem. Soc. 138 (1991) 3759
[5] W.H. Koschel, U. Kaufmann and S.G. Bishop, Solid St. Comm. 21 (1977) 1069
[6] T. Takanoashi and K. Nakajima, J. Appl. Phys. 65 (1989) 3933
[7]. L. Eaves, A.W. Smith, P.J. Williams, B. Cockayne and W.R. Mc Ewan, J. Phys. C 14 (1981) 5063
[8] S. Fung, R.J. Nicholas and R.A. Stradling, J. Phys. C 12 (1979) 5145
[9] J. Jiménez, M. A. Gonzàlez, V. Carbayo and J. Bonnafé, Phys. Stat. sol a77 (1983) k69
[10]. A.Evan Iverson, D.L.Smith, N.G. Paultier and R.B.Hammond, J. Appl. Phys. 23 (1987) 234

Inst. Phys. Conf. Ser. No 149
Paper presented at DRIP 95

A New Method for Measuring Mobile Charge in SiO_2 on Si; The First Real-Time Wafer Mapping Capability

P. Edelman

Semiconductor Diagnostics, Inc., 6604 Harney Road, Suite F, Tampa, FL 33610

J. Lagowski, L. Jastrzebski, A.M. Hoff, A. Savchuk

Center for Microelectronics Research, University of South Florida, Tampa, FL 33620

Abstract. We demonstrate, for the first time, whole-wafer mapping of the mobile charge in SiO_2 on silicon. The technique uses no test structure. The measurement is done in non-contact mode in a total time of about 20 minutes per wafer. Determination of the mobile charge is based on monitoring the ion drift by measuring corresponding changes in the contact potential difference. The drift is induced by corona charge on the entire SiO_2 surface.

1. Introduction

Mobile charge in SiO_2, especially Na^+ and K^+, is a critical contaminant from oxidation furnaces during silicon IC manufacturing. It deteriorates the performance and manufacturing yield of the most advanced MOS devices and memory chips. All traditional methods for measuring the mobile charge require fabrication of MOS capacitor test structures. This is followed by characterization of each individual capacitor combined with a bias temperature stress, causing ion drift across the oxide. The traditional methods of furnace qualification give the results days after completion of an oxidation process and are clearly unsuited for wafer mapping in a reasonable time.

We have recently proposed a different approach [1] which eliminates two most time consuming steps: a preparation of MOS test capacitors and a point-by-point characterization of individual capacitors. In this paper, we present principles of the approach and its application to wafer-scale mapping of the mobile charge.

2. Principle of the Approach and Scenario of Mobile Charge Mapping

The surface energy band diagram of silicon with an SiO_2 film in contact with air is shown in Fig. 1. Positive sodium ions in SiO_2 contribute to two potential barriers: the surface barrier qV_s and the oxide barrier qV_{ox}. Of the two, only the latter depends upon the distribution of ions in the oxide. The surface barrier is a function of the total surface charge, Q_T, determined by the sum of: the interface trapped charge, Q_{it}; the fixed oxide charge, Q_{Fox}; the oxide trapped charge, Q_{tox}; and the mobile ion charge, Q_m. It does not, however, depend on the exact distribution of charges across the oxide.

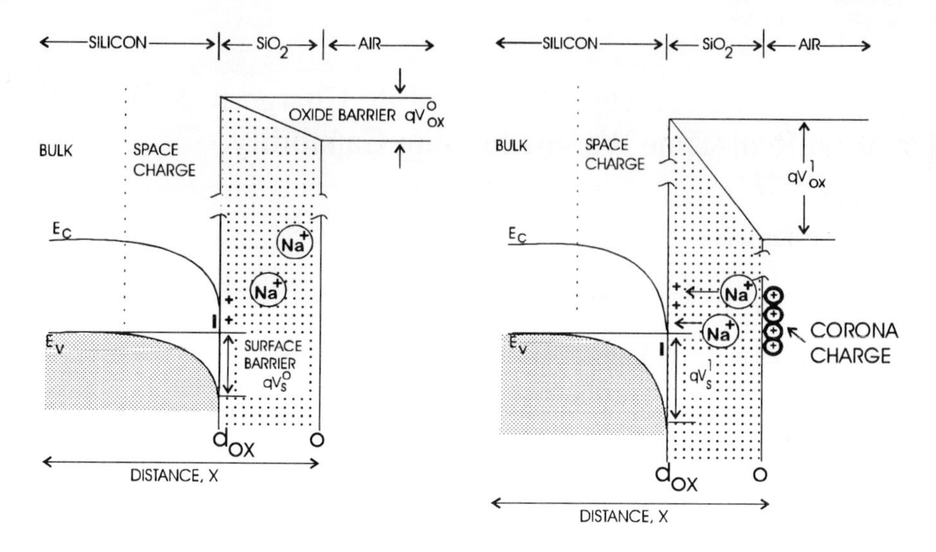

Figure 1. Surface energy band diagram of Si-SiO_2-Air.

The diagram on the left side in Fig. 1 represents the initial state in our consideration. The diagram on the right side represents the same system, however after a placement of an additional charge on the SiO_2 surface. The charge, Q_c, is deposited instantaneously and very uniformly over the entire wafer with a corona charging station similar to the devices used in Xerox-copiers. The charge is deposited at room temperature, when the ions in SiO_2 are not mobile. There are three important consequences of Q_c. First, the net surface charge changed, producing a new surface barrier, qV^1_s. Second, the electrical field in the oxide changed, leading to a new oxide barrier, qV^1_{ox}. Finally, the field in the oxide created a force acting on ionic charges. Ionic charges, however, cannot move because of a too low temperature.

At this stage, a whole wafer mapping of $q(V^1_s + V^1_{ox})$ is performed. The mapping is done with a surface potential sensor placed 0.5 mm above the moving wafer. To be precise, the sensor measures the contact potential difference, $V_{cpd} = V_s + V_{ox} + \text{const}$ [1]. High density mapping of a 200 mm wafer is completed in about 5 minutes. Next, the wafer is placed on a hot-plate at a temperature of 200°C, sufficient to make the ions mobile. This causes a drift of ions across the oxide. In an analogy to a bias temperature stress (BTS) in the MOS method [2], this drift is referred to as corona temperature stress (CTS). Ion re-distribution is completed in 5 to 10 minutes. As a result, the oxide barrier changes to a new value, qV^2_{ox}, while the surface barrier remains qV^1.

The wafer is cooled to room temperature and a whole wafer mapping of the surface potential is performed again. After about 5 minutes, mapping is completed. The new map is subtracted from the original, producing a map of the oxide barrier shift, $\Delta V_{ox} = \Delta V_{cpd}$, caused by the mobile ion drift. This map is converted into mobile oxide charge. Calculation is similar to that of the mobile charge from the flat band voltage, V_{FB}, shift in MOS capacitors [2]. Computer calculation is practically instantaneous. Therefore, the scenario described above produces a mobile charge map of the whole wafer in about 15 to 20 minutes. In practice, an additional, preparatory step can be added to the sequence in order to obtain a well-defined, mobile charge distribution prior to the first mapping of V_{cpd}. Such a complete sequence of steps is outlined in Fig. 2. Also included in Fig. 2 is a corresponding sequence used in conventional MOS/C-V measurements based on the flat-band voltage shift caused by a drift of mobile charge.

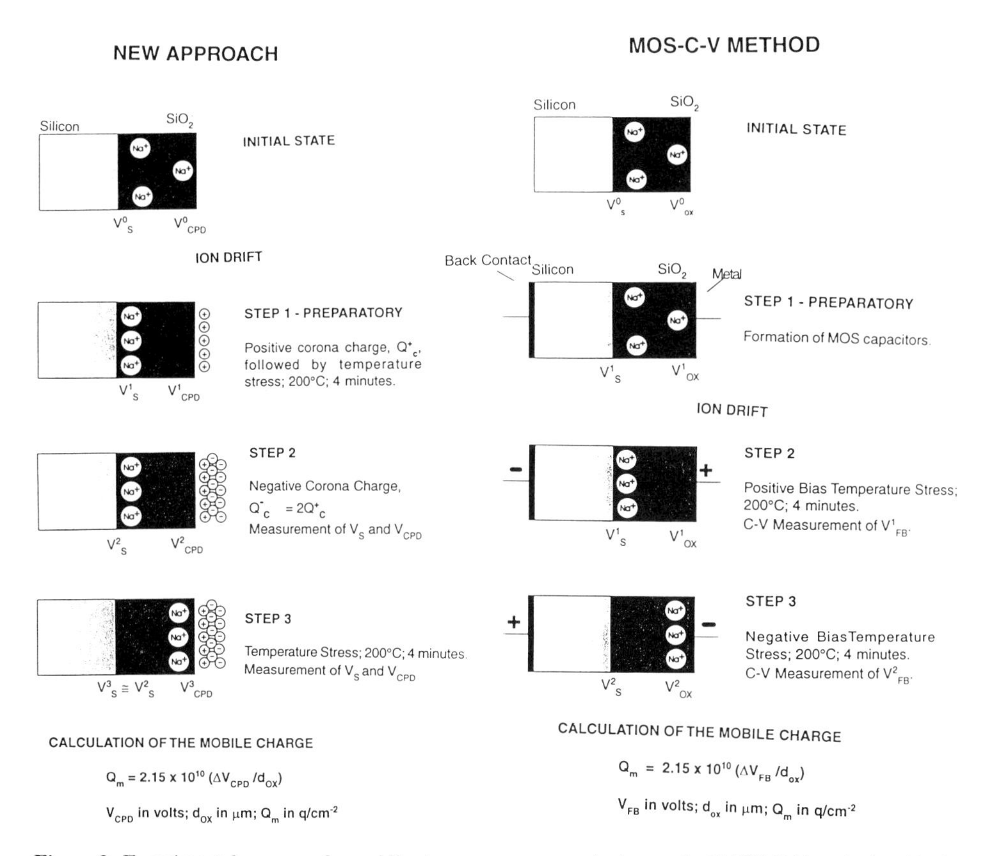

Figure 2. Experimental sequence for mobile charge measurement in the standard MOS-C-V method and in the new approach.

3. Experimental

Two types of commercial surface potential sensors (or contact potential difference sensors) were used in this study with basically similar results: the Kelvin Probe and the Monroe Probe [3]. In both of them, the measurement relies on an AC signal generated between the reference electrode and the wafer. The other contact completing the circuit was provided by a large wafer-chuck capacitance. This is essential for non-contact measurements. In the Kelvin Probe, a reference electrode vibrates perpendicularly to the surface and generates an AC signal by changing the electrode-wafer capacitance. In a Monroe probe, the reference electrode is stationary and an AC signal is generated by a horizontal vibration of a grounded fork which shields the probe from the wafer.

The sensors are typically used in a compensating mode in which an AC signal is zeroed by a DC bias equal to the contact potential difference, V_{cpd}, between the probe and the wafer. This mode is slow, but very precise. For fast mapping, we, therefore, have used a null-off measuring mode based on measuring the magnitude of the AC signal. With a vibration frequency of about 1 kHz, this mode facilitates V_{cpd} measurements with a speed of 20 measurements/ second and 1 mV sensitivity. The mapping and scanning was performed by moving the wafer under the probe.

The surface barrier, V_s, was measured with a commercial SPV apparatus which uses the photovoltage saturation method described in Ref. [4]. The surface potential probe and the SPV probe were mounted on the same head, next to each other, enabling simultaneous measurements of V_{cpd} and V_s. All measurements were done using computerized data acquisition and processing.

4. Tests of the Approach

Initial feasibility tests have been reported in our recent publication [1]. Key results can be summarized as follows: 1. Ion drift is very effective in corona fields in the mid-10^5 V/cm range at temperatures of 180-200°C similar to TBS drift in MOS capacitors; 2. No corona charge loss from the surface was detected during CTS (note that such loss would be detected in SPV measurements); and 3. Ion drift changes only the oxide barrier, but not the surface barrier.

The present results confirm these findings and also provide quantitative comparison with other methods. In Fig. 3, a comparison is given between the results of corona temperature

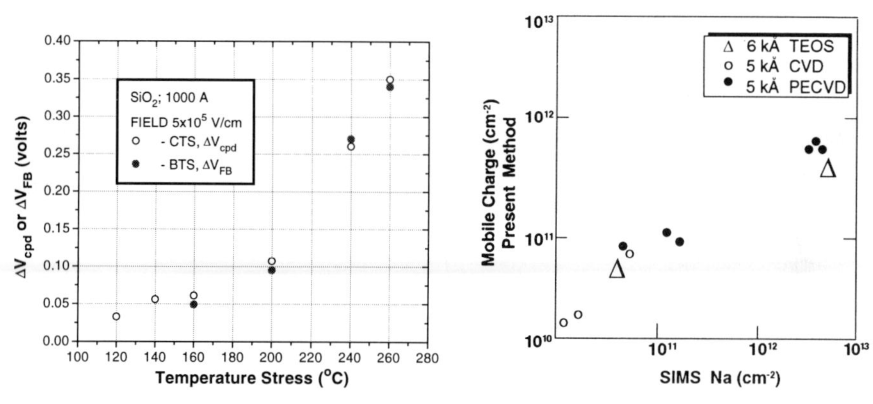

Figure 3. Comparison between the effects of corona temperature stress and bias temperature stress during isochronal annealing.

Figure 4. Correlation between sodium concnetratoins in SiO_2 determined with the present approach and with SIMS.

stress and bias temperature stress. Both results were obtained on the same 1000Å, thermally-grown oxide, using the sequence in Fig. 2. BTS was performed on MOS capacitors. CTS and CPD measurements were done in regions between the capacitors. In both cases, the same drift field of 5 x 10^5 V/cm was used. Isochronal temperature stress refers to Step 3 in Fig. 2, i.e. to negative bias and corona charge. The agreement between the two methods is very good. For positive bias (corona charge), BTS often gives flat-band shifts lower than Δ_{cpd} shift in CTS. We observed differences as large as a factor of 1.5 to 2.5. This difference may originate from ion interaction with a metal electrode suggested as a reason for often observed bias asymmetry in BTS [2, 5].

In Fig. 4, the mobile charge measured by the present method is compared with the results of SIMS analysis for oxides prepared using three different methods. Considering different probing characteristics of the two methods, the correlation seems to be satisfactory, especially for low sodium concentration below 5 x 10^{11} cm^{-2}. For high Na concentration, SIMS gives larger values than the present approach. This is not surprising since CTS loses effectiveness when the mobile charge becomes comparable to corona charge.

5. Mobile Charge Maps

In mapping experiments, we used 190°C, 5 minute annealing with a corona field of about 7 x 10^5 V/cm. This is sufficient for Na^+ drift, but not for a drift of K^+ ions [2, 5]. Therefore, the maps presented below reflect primarily Na^+ contamination. We start with Fig. 5, in which three maps are shown to illustrate the sequence in Fig. 2. Measurements were done on a 6-inch diameter wafer with 1000Å thick thermal SiO_2 contaminated with sodium in the top right corner. V_{cpd} map in Fig. 5a was taken after positive corona charge, but prior to CTS. No contamination pattern is evident. After CTS, the contaminated region becomes visible (see Fig. 5b). Positive corona charge repels Na^+ ions toward the Si/SiO_2 interface and decreases V_{cpd}. The sodium contaminated section is indeed manifested, in Fig. 5 & 6, by lower V_{cpd} values. After negative CTS (step 3 in the sequence), the contrast between the contaminated region and the rest of the wafer reverses because a negative corona attracts Na^+ and increases V_{cpd}. The sodium concentration map in Fig. 5c was calculated from ΔV_{cpd}, caused by step 3 of the sequence.

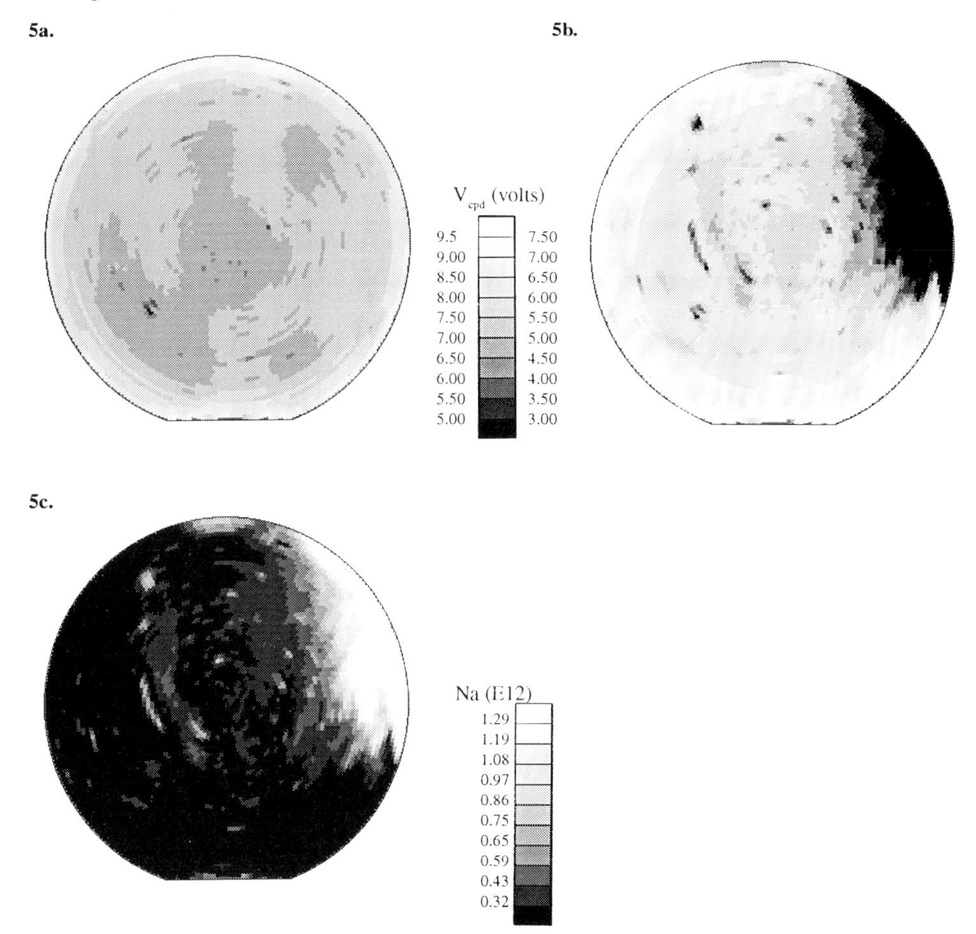

Figure 5. The fast, high resolution, 6000-point maps of a 6-inch oxidized wafer: a.) V_{cpd} after positive corona but no CTS; b.) V_{cpd} after positive corona and CTS; c.) sodium concentration (ions/cm^2).

An example of a different Na^+ pattern is shown in Fig. 6. Again, the contamination is barely resolvable in the map before CTS (Fig. 6a). On the other hand, the vertical sodium contamination trail became clearly visible after CTS, leading to the Na^+ map in Fig. 6b.

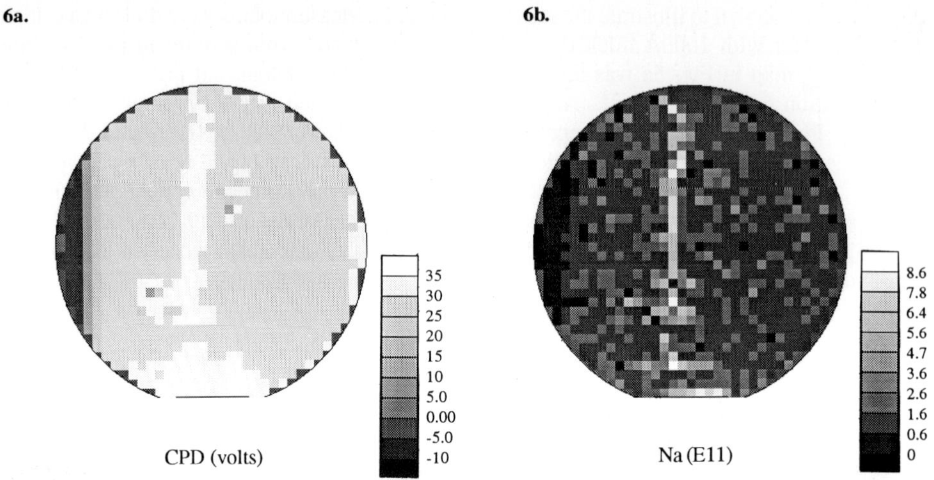

Figure 6. 35 x 35 matrix (about 800 points) maps of a 6 inch oxidized wafer: a.) V_{cpd} after positive corona but no CTS; b.) sodium concentration (ions/cm^2).

In summary, we have demonstrated a new technique for mapping mobile ions in SiO_2 on processed silicon wafers in a realistic time. We believe that this approach opens new possibilities for real-time sodium contamination diagnostics and also for the identification of Na contamination sources from the contamination patterns.

The authors are indebted to AT&T Microelectronics, Orlando, Florida for providing the oxidized silicon wafers used in this study and for performing SIMS analysis.

References

[1] Edelman, P.; *New Approach to Measuring Oxide Charge and Mobile Ion Concentration*; SPIE Vol. 2337; October 20, 1994.

[2] Schroder, D.K.; *Semiconductor Material and Device Characterization*; John Wiley & Sons, Inc.; New York, 1990; Chapter 6.2.5 and references therein.

[3] Tada, Y., Tomizawa, Y.; 1993 *Jpn. J. Appl. Phys. Lett.*; Vol. 34 (1995), p. 643-648.

[4] Edelman, P. et al; *MRS Symposium Proceedings*; 261, p. 223 (1992).

[5] R. Williams; *J. Vac. Sci.* **14** p. 1106 (1977).

Inst. Phys. Conf. Ser. No 149
Paper presented at DRIP 95

Surface Photovoltage and Contact Potential Difference Imaging of Defects Induced by Plasma Processing of IC Devices.

K. Nauka

Hewlett-Packard Company, Palo Alto, CA 94304.

J.Lagowski, and P.Edelman

Semiconductor Diagnostics, Inc., 6604 Harney Road, Tampa, FL 33620.

Abstract. Contact Potential Difference and Surface Photovoltage have been employed to investigate materials defects introduced by plasma processing during the manufacturing of IC devices. Employed analytical techniques facilitated detection of defects formed by different interactions between the Si wafer and plasma environment offering a comprehensive picture of plasma induced damage. Correlation between the materials defects introduced by plasma processing and degradation of electrical devices indicate that the proposed techniques could be used for plasma damage monitoring in the IC processing.

1. Introduction

Integrated circuit (IC) manufacturing involves a large number of plasma processes ranging from anisotropic etching, isotropic stripping and cleaning, to plasma enhanced deposition and implantation. It frequently occurs that undesirable reactions between the partially completed IC devices and plasma environment can lead to detrimental and permanent changes of device properties. These reactions include: charge exchange between the plasma sheath and processed wafer, UV and particle radiation, unwanted chemical reactions, and introduction of impurities into the wafer. They can be determined by: plasma properties, plasma generation conditions, and type of surface or material that is exposed to the plasma. Electrical charging of dielectric surfaces and damage caused by radiation are among the most commonly encountered sources of device damage (Figure 1). The first process results from a local lack of balance between the positive and negative charge fluxes exchanged between the surface of a wafer covered with dielectric layer and the plasma sheath. It can cause gradual surface charge build-up and subsequent damage of a dielectric. This effect can be particularly severe in the case of thin gate dielectrics when electrical connections between the gate and large conductive areas acting as charge collectors can accelerate plasma induced damage of gate dielectric (antennae effect). Highly energetic particles and UV radiation penetrating into a device can cause chemical bond breakage, formation of structural and electrical defects, and generation of excessive amounts of electron - hole pairs.

Most of the analytical techniques employed to detect plasma damage rely on fabrication of devices acting as plasma damage sensors. They can include memory cells sensitive to a speficic types of plasma damage [1], or circuits designed to accentuate interactions between different parts of an IC device when exposed to plasma [2]. Present work demonstrates application of relatively simple materials analytical techniques, namely, Contact Potential Difference (CPD) and Surface Photovoltage (SPV) to predict damage caused by charging of an oxide surface (CPD), and UV and particle radiation damage (SPV).

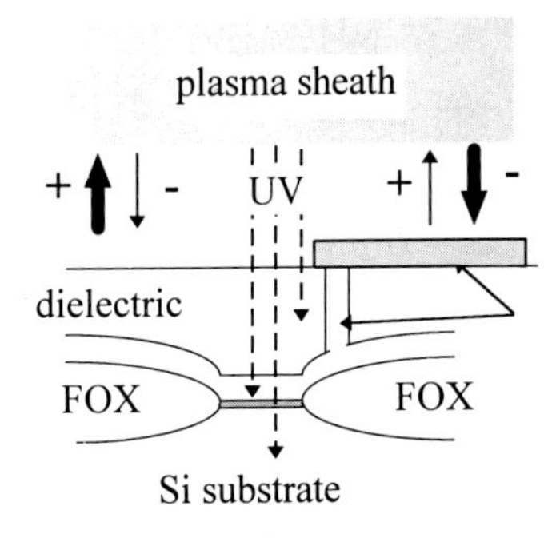

Figure 1. Schematic representation of antennae damage and UV radiation damage in MOS devices. (+) and (-) represent positive and negative charge fluxes between the Si wafer and plasma sheath.

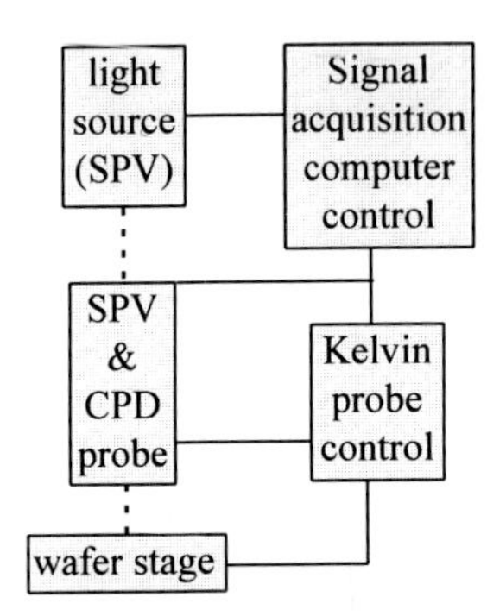

Figure 2. Experimental set-up used for plasma damage monitoring.

2. Experimental

The SPV signal is obtained by illuminating one side of a Si wafer with photons having energies higher that the Si bandgap, while the other side remaines in darkness. Nonequilibrium carriers generated near the illuminated surface can cause a partial collapse of the surface potential barrier and introduce a potential difference V_{SPV} between the illuminated front and dark back surfaces. At low light intensities V_{SPV} can be directly related to the minority carrier diffusion length in Si substrate, and to the Si surface recombination rate. At high illumination intensities surface potential barrier disappears and V_{SPV} becomes saturated. Thus, the SPV value obtained under saturation conditions can provide information about the magnitude of surface potential barrier and related surface charges. In the case of an oxidized Si wafer this mode of SPV can be employed to obtain information about the net oxide charges, and in particular about the interfacial charges. This technique has been employed to detect oxide charges and damage in the underlaying Si caused by UV and particle radiation during the plasma processing.

CPD measurement employes a vibrating Kelvin probe to measure a difference (V_{cpd}) between the known work function of a reference metal electrode and the work function of measured sample. In the case of an oxidized Si wafer, the last quantity is primarily determined by the oxide's surface charges. CPD has been employed to investigate distribution and density of charges deposited on the oxide's surface by the plasma.

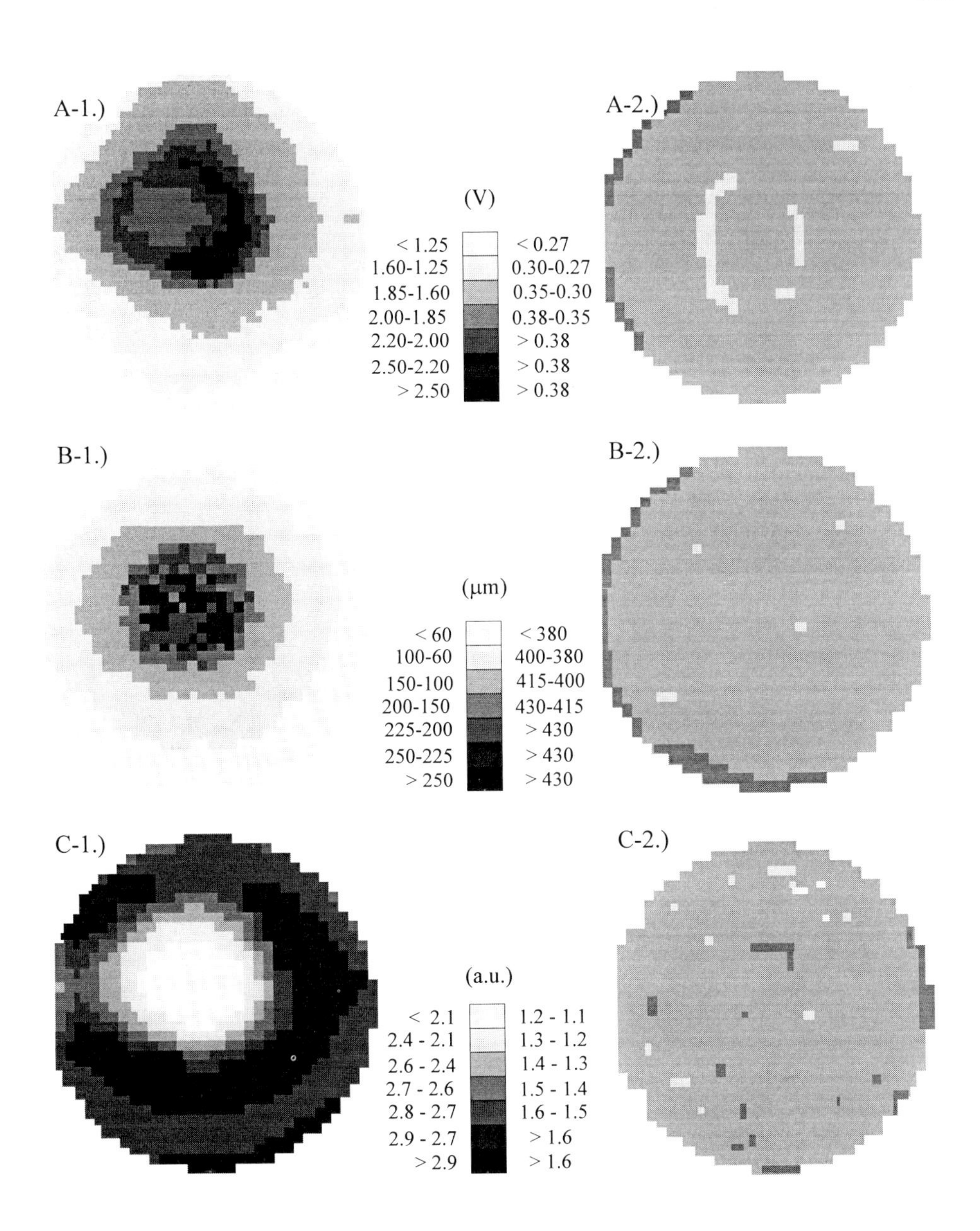

Fig.3. Si wafer after Ar plasma clean; A-1.) CPD map of the oxide surface charges, B-1.) map of the minority carrier diffusion length, C-1.) map of the net bulk oxide charges; A-2.), B-2.), and C-2.) - respective maps obtained for the reference wafers (wafers before plasma treatment).

Figure 2 shows experimental set-up used in the present experiment. It facilitated fast, non-contact SPV and CPD measurements not requiring any wafer processing. In most cases results were obtained in form of defect distribution maps, as shown in Figure 3. All measurements were conducted using 150 mm diameter p-type Si wafers with resistivities from the range 10 Ω cm - 25 Ω cm. Wafers were RCA cleaned and thermal oxides with thicknesses either 100 nm or 40 nm were grown. Blank oxidized wafer were subjected to plasma treatments analogous to plasma processes used in IC manufacturing. Following plasma processes were investigated: Ar plasma clean preceding metal sputtering, Ar plasma strip used in PVD cluster metal tool, O plasma resist ashing following dielectric etch, and plasma enhanced dielectric deposition. All plasma processing was conducted using state-of-the-art IC manufacturing equipment that was operated within the range of processing conditions allowed by the equipment vendor. Additionally, wafers with as-grown oxides (no plasma processing) were measured. DI rinse of plasma processed wafers caused removal of defect patterns observed with CPD, thus demonstrating that the CPD measured charges resided on the oxide's surface.

In some cases aformentioned plasma processes were employed to fabricate test devices. They included antennae devices (devices with abnormally large antennae pads), standart MOSFETs (channel length equal 0.35 μm and gate oxide thickness from the range between 7 nm and 10 nm) and MOS gate capacitors. Device were tested for the gate oxide leakage at 3.3 V, and for the transistor threshold shift after the Fowler-Nordheim stress. Purpose of these experiments was to compare the device data with the corresponding materials characteristics obtained by the CPD and SPV, and to evaluate usefulness of these techniques for detecting plasma processing condictions having deletorious effect on IC devices [3].

3. Examples of applications.

Figure 3 shows examples of plasma damage imaging using SPV and CPD. They were obtained for the wafers exposed to a plasma strip in the plasma enhanced CVD dielectric reactor. No dielectric was deposited in this case, plasma strip caused removal of approx. 20 % of the initial thermal oxide. Quasi-circular pattern observed in Figure 3 is due to the wafer rotation. Similarly complex defect patterns were seen for all the investigated plasma processes.

Presented data demonstrate complexity of defect generation during the plasma processing. Plasma exposure caused changes of all three measured material properties, i.e. oxide surface charges, Si substrate diffusion length, and net bulk oxide charges. In addition, in each case plasma generated defects that were nonuniformly distributed across the wafer.

It was also observed that defect formation was strongly dependent on the plasma generation conditions, such as gas pressure, distribution and strength of electrical and magnetic fields, wafer rotation, and shape of plasma sheath. Even small changes in the plasma conditions had large impact on the appearance of CPD and SPV generated defect maps. Example showing sensitivity of defect generation processes to plasma conditions is shown in Figure 4. It summarizes data obtained for the wafers subjected to O_2 plasma ashing. Purpose of this experiment was to investigate uniformity of defect distribution across the wafer as a function of the O_2 pressure. Variability of defect distribution is expressed in terms of the standard deviation of measured CPD signal.

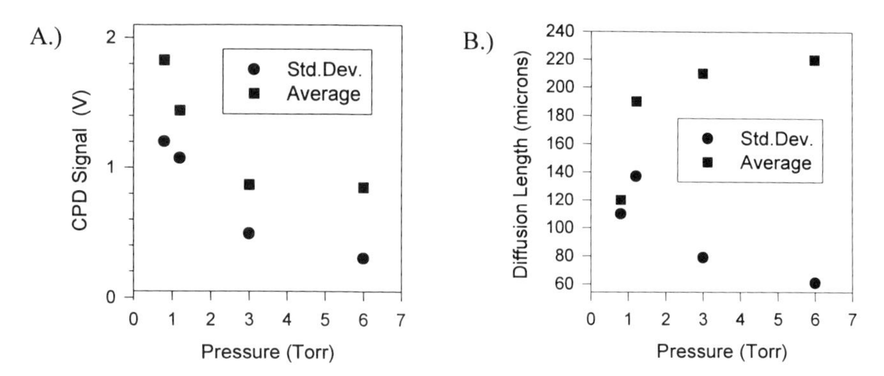

Fig.4. Surface oxide charges (CPD signal) (A.), and Si minority carrier lifetime (B.) after O_2 plasma ashing.

4.Correlation with device properties.

Usefulness of CPD and SPV as tools for predicting damage induced by plasma processing in IC devices can only be tested by correlating the CPD and SPV measured defect maps with the IC device parameters. The most important difference between the CPD and SPV materials analysis and the IC device testing is due to the fact that, in the first case one measures material properties of a simple system consisting of an oxidized Si wafer that was subjected to a single plasma process, while in the later case measured device has a very complex structure that was created by a large number of processing steps. Thus, damage created by a given plasma process can be masked by defects formed during the subsequent processing steps, or it can even be eliminated by the subsequent processing. Sensitivity of IC devices to plasma induced defects depends on device design and size; smaller devices could be more sensitive to plasma induced damage.

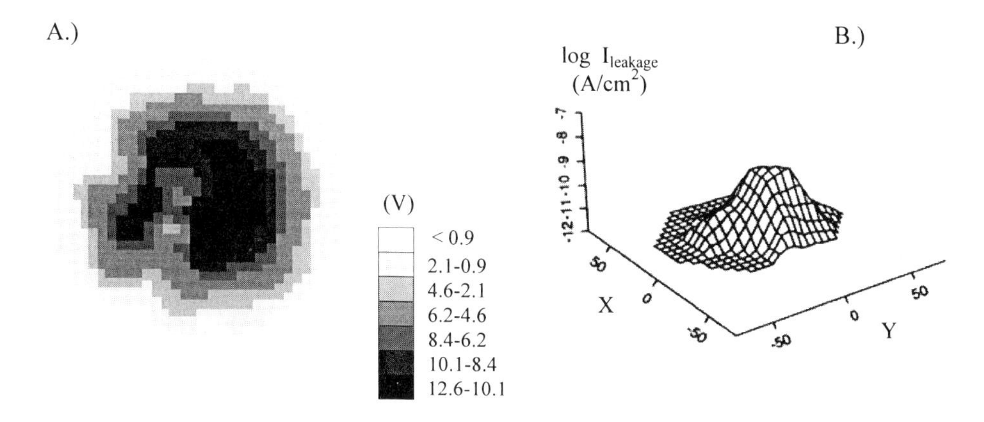

Fig.5. A.) Oxide surface charge maps (CPD units); B.) corresponding maps of MOS capacitor leakage for the 0.35 μm devices.

Fig.5 presents correlation between the CPD measured surface charges and submicron MOS capacitor leakage resulting from pre-metal Ar plasma sputtering. An excellent correlation between the high oxide surface charge density and capacitor leakage can be seen for these devices. Minority carrier lifetime and bulk oxide charges weren't changed by the sputtering. Figure 6 summarizes correlation between the submicron device electrical properties and variations of the oxide surface charge obtained for this plasma process.

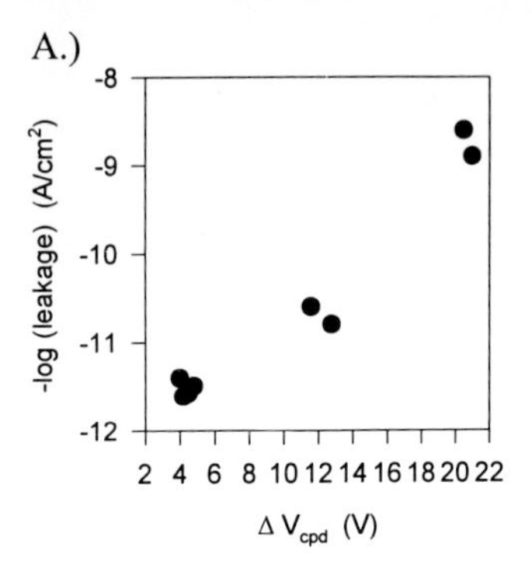

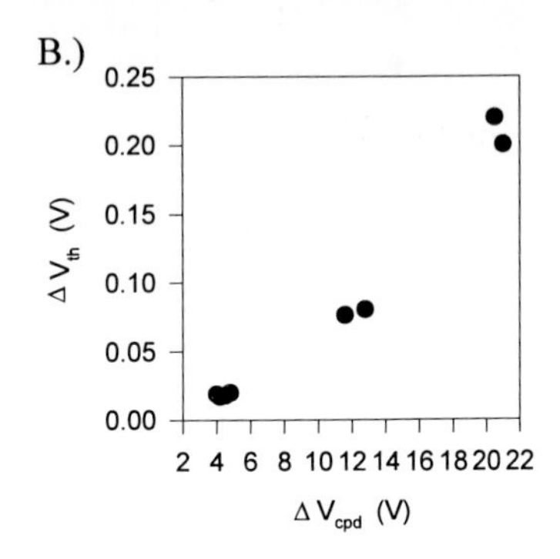

Fig.6. Correlation between the submicron device parameters (A. - capacitor leakage current, B. - transistor threshold voltage (V_{th}) shift) and variation of the oxide surface charges expressed as standard deviation of the CPD signal (ΔV_{cpd}).

4. Conclusions

Contact Potential Difference and Surface Photovoltage have been employed to monitor defects introduced by plasma processing of IC devices. CPD was used to monitor plasma deposited oxide charges responsible for "antennae"damage in IC devices, while SPV was employed to detect oxide and Si substrate damage caused by UV radiation. It was found that density and distribution of defects induced by plasma processing is strongly dependent on the plasma generation conditions. Correlation between the materials defects induced by plasma processing and IC device degradation has been observed for the selected plasma processes. CPD and SPV could be used for routine monitoring of plasma IC processes after the detailed correspondence between the CPD and SPV measured materials defects and the IC device elecrtical characteristics is established for each of these processes.

Acknowledgements

Collaboration of J.B.Kruger, W.W.Dixon, W.E.Greene and B.W.Langley (Hewlett-Packard Company) in monitoring of the plasma induced damage is acknowledged.

References:

[1] Lukasek W., et al., 1992 IEEE/SEMI Adv. Manufact. Conf. Proc., IEEE Publ. p.148.
[2] Fonash S.J., et al., Sol.St.Technol, July 1994, p.99.
[3] Nauka K. et al., 3rd Internat.Workshop on Adv.Plasma Tools Proc., AVS Publ., 1995.

Inst. Phys. Conf. Ser. No 149
Paper presented at DRIP 95

Scanning internal-photoemission microscopy: an imaging technique to reveal microscopic inhomogeneity at metal-semiconductor interfaces

Tsugunori Okumura

Dept. of Electronics & Inform. Eng., Tokyo Metropolitan University
1.1, Minami-ohsawa, Hachiohji, Tokyo 192-03, Japan

Abstract. We have developed scanning internal-photoemission microscopy (SIPM) which is capable of imaging Schottky-barrier distribution at "buried" metal-semiconductor interfaces. By using this technique, inhomogeneous reaction at annealed interfaces of Ti/Pt/Au/GaAs and epitaxial-Al/Si(111) systems has been studied in relation to their microscopic as well as macroscopic electrical properties.

1. Introduction

The relation between Schottky barrier height (SBH) and microscopic structure of metal-semiconductor interfaces has been of fundamental as well as practical interest in the semiconductor science and technology over several decades. From the practical point of view, we shall have to meet the challenge of establishing the process technology of routinely making a large number of homogeneous contacts in order to achieve a very large scale integration of circuits. The homogeneity of contacts should be guaranteed not only metallurgically but also electrically over a large area of semiconductor wafers. The interfacial structure is generally modified by reaction and diffusion upon heat treatment used during device fabrication processes (e.g. Poate et al.1978). Such modification might occur inhomogeneously and thus the SBH at the real interfaces is not guaranteed to be homogeneous. While structural as well as compositional analyses of the interfaces are now available in a microscopic or atomic scale, the SBH is usually characterized by using "macroscopic techniques", such as current-voltage (*I-V*) and capacitance-voltage (*C-V*) methods. It is often observed that the SBH's determined by different experimental methods are not always in good agreement with each other (Okumura et al. 1983). Inhomogeneity and/or non-abruptness at the real metal-semiconductor interfaces are considered to be one reason for the above discrepancy among the measured SBH's (Ohdomari et al. 1980, Tung 1992).

Internal photoemission measurements have been used to characterize various semiconductor interfaces. This is usually called as a "photoresponse" technique and an accurate and direct method of determining the SBH because of its spectroscopic nature (e.g. Sze 1981). When the light beam used for the photoresponse measurements is focused at the interface, local

SBH's can be measured (DiStefano 1971). This paper presents that scanning internal-photoemission microscopy (SIPM) is a powerful technique to reveal microscopic inhomogeneity in SBH's at buried metal-semiconductor interfaces, which has not been disclosed by the conventional *I-V* and *C-V* techniques (Okumura et al. 1989).

2. Internal photoemission and its microscopy

When ultraviolet (UV) light or x-ray is incident upon a metal, the excited electrons in the metal can emit into vacuum. This effect is well-known as the photoemission effect. A similar effect takes place at interfaces between two solids. This is called as the internal photoemission effect. The basic model for the internal photoemission process is shown in Fig. 1(a), where we show a metal layer deposited on an *n*-type semiconductor. When a light with a photon energy of $h\nu > q\phi_{Bn}$ is incident upon a metal, the excited electrons in the metal can surmount the Schottky barrier and then photocurrent is generated due to the built-in field in the depletion region. In general, the value of SBH's is smaller than that of semiconductor energy gaps E_G. Therefore, we can excite electrons in a metal layer from the semiconductor side, even when the metal thickness is extremely large. This configuration is the so-called "back illumination". This is the reason why the internal photoemission effect is applied to a non-destructive evaluation for "buried interfaces". To determine the SBH from the internal photoemission measurements, we usually apply an approximate Fowler's formula

Two dimensional mappings of photoemission current are obtained by translating (scanning) a specimen with respect to the axis of the focused laser beam, or vice versa as shown in Fig. 1(b). Thus, the technique is a kind of scanning microscopy. It should be noted that this current is due to majority carriers in contrast to minority carriers in the optical-beam induced current (OBIC) as well as the electron-beam induced current (EBIC) techniques. Therefore, the diffusion length of minority carriers does not limit the spatial resolution in SIPM.

We used a 1.3 μm semiconductor laser as a light source; the corresponding photon energy from the laser diode is 0.95 eV and less than the energy gaps of both Si (1.12 eV) and GaAs (1.42 eV). Therefore, the

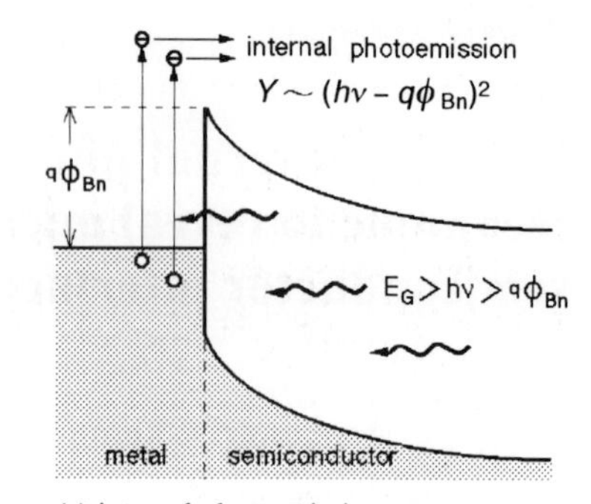

(a) internal photoemission process at a meta--semiconductor interface.

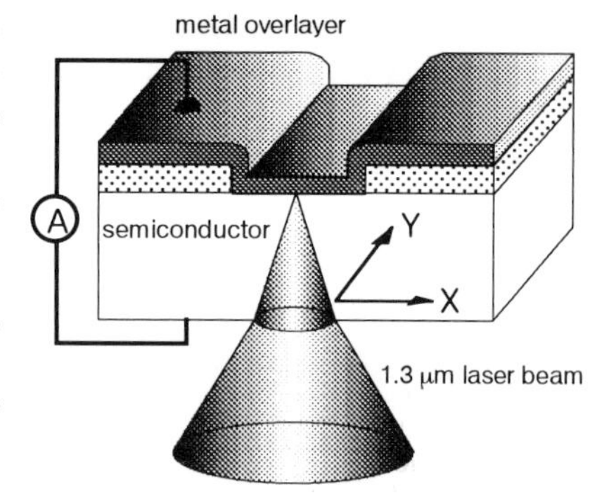

(b) schematic illustration of a sample illuminated by focused laser beam.

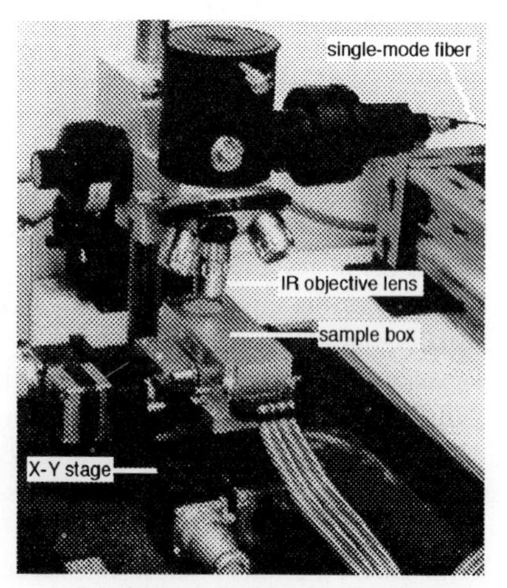

(c) photograph of our developed microscope (optics).

Fig. 1. scanning internal photoemission microscopy (SIPM).

laser beam is able to impinge through substrates directly on metal-Si and metal-GaAs interfaces. On the other hand, the interface is probed through metal overlayers in ballistic electron emission microscopy (BEEM), which is an attractive technique to characterize Schottky contacts in a nanometer scale (Kaiser et al. 1988). However, it was pointed that an apparent inhomogeneity in the BEEM images often stems from the thickness variation of the metal overlayer, and that the attenuation length of hot electrons in metals restricts the thickness of the metal overlayer (several tens of nanometers).

Figure 1(c) shows a photograph of the SIPM developed by our group (Shiojima et al. 1991a). A beam diameter of less than 2 μm was achieved in free space by using a "pigtail-type" 1.3 μm laser diode (LD) as a light source, whose emission wavelengths are scanned by controlling the operational temperature from 0 to 60 °C. The energy accuracy of SBH's depends on the stability of the total system as well as quality of a sample Schottky diode. For a typical value of $\Delta Y/Y = 2\times10^{-3}$, the resultant accuracies are estimated as 4 and 15 meV for the SBH's of 0.8 and 0.7 eV, respectively. Thus, even such a narrow span of photon energy would be sufficient for the evaluation of most metal-Si and metal-GaAs systems.

3. Thermal degradation of Au/Pt/Au contacts to GaAs

The inhomogeneous degradation of Ti/Pt/Au Schottky contacts to n-GaAs upon annealing were evaluated by SIPM (Shiojima et al. 1991b). A good correlation was found between the local formation of a low-SBH phase and the degradation of macroscopic *I-V* characteristics.

The substrate was *n*-GaAs(100) with a carrier density of 1×10^{17} cm^{-3}. Electron beam evaporation was used to deposit 50 nm of Ti, then 50 nm of Pt, and finally 200 nm of Au. Standard photolithographic techniques were used to define an array of 50 μm by 200 μm rectangular diodes. The periphery of each diode was covered with PE-CVD silicon dioxide. Isochronal and isothermal annealing were conducted at temperatures between 200 and 500 °C. Anneals were carried out in a furnace that was purged with argon.

SIPM measurements revealed that diodes annealed at temperatures up to 400 °C gave a uniform internal-photoemission yield, within ± 2%, across the area of a diode. Upon annealing at 400 °C, all diodes had SBH's of about 0.72 eV and exhibited good *I-V* characteristics with an *n*-value of 1.2. For diodes annealed above 450 °C, the internal-photoemission yield, however, varied significantly across the area of a diode, and rectifying characteristics in *I-V* measurements disappeared. Isothermal annealing was performed to investigate the effects of anneals at temperatures above 450 °C. It was found that the SBH decreased and the *n*-value increased as the annealing time increased. This trend was particularly noticeable for annealing time between 15 and 20 min. The data also showed that there was considerable variation in values measured for different diodes. In the *I-V* curves for the diodes with a high *n*-value, an excess current was superimposed on the thermionic emission current. The excess current was approximately proportional to the applied voltage, *i.e.*, quasi-ohmic.

SIPM measurements revealed that the regions of high photoemission yield , which appear bright in the maps, were formed for the sample annealed at 480 °C longer than 15 min, where the *I-V* characteristics were degraded. Frequently, the bright regions started in the vicinity of the contact edge and moved in toward the center of the electrode pattern. Figure 2(a) shows representative maps of the high internal-photoemission yield (high-*Y*) region detected in the sample annealed for 40 min at 480 °C. The measured SBH at the center was lower than that for the entire interface by 0.06 eV; the SHB in normal regions was around 0.82 eV determined with wavelength dependence of photoyield in SIPM measurements. Figure 2 also shows scan-

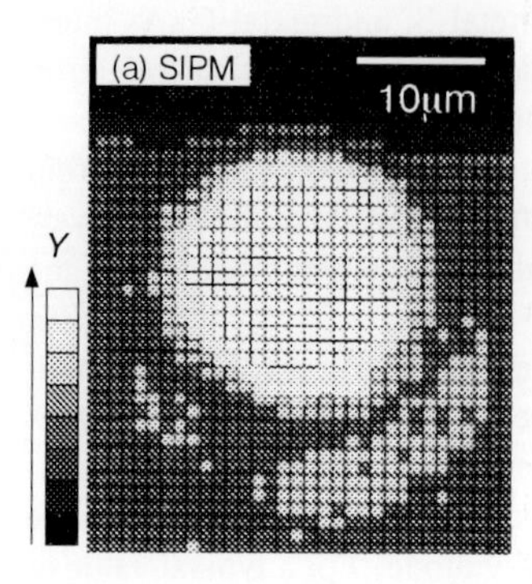

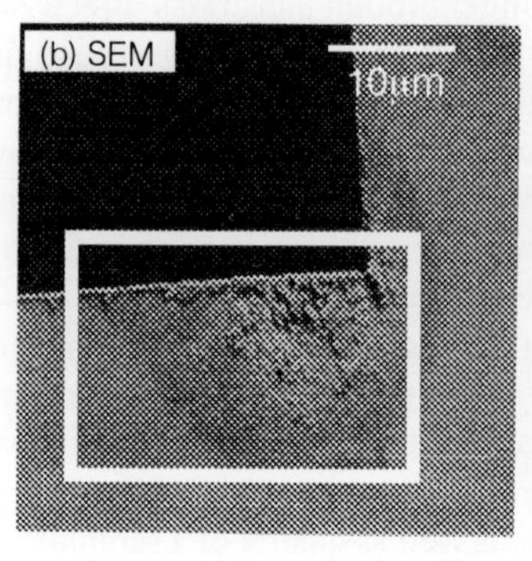

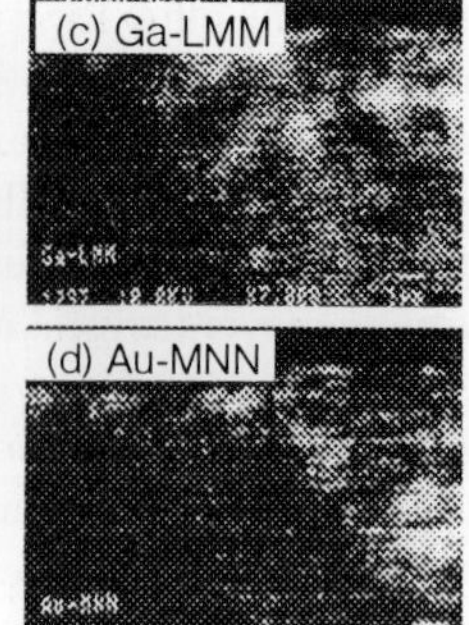

Fig. 2. Interface images for Ti/Pt/Pt-GaAs annealed at 480 °C taken by (a) scanning internal-photoemission microscopy (SIPM) and scanning Auger microscopy with (b) secondary electron (SEM), (c) Ga-LMM and (d) Au-MNN signals. SEM and Auger singals were taken in a different but equivalent place from the area as shown in (a) after removal of the GaAs substrate.

ning images with (b) secondary electron, (c) Ga-LMM and (d) Au-MNN Auger signals by using samples whose GaAs-substrate was selectively removes.

Tests showed that the total area of the high-Y regions tended to increase as annealing time increased. The excess current measured at a forward bias of 0.1 V and the number of the high-Y pixels were correlated with each other. It has been reported that the characteristics of gold-to-GaAs contacts change from rectifying to quasi-ohmic after anneals at temperatures above 400 °C, where the β-phase gold-gallium intermetallic phase formed (Beam III et al. 1985). Gold probably reached the interface by migration around the edges of the contact or by diffusion along grain boundaries in the platinum film. It is postulated that the degradation in rectifying characteristics observed after anneals at 480 °C arose because regions of β-AuGa with quasi-ohmic I-V characteristics shorted out regions that were still rectifying.

4. SBH inhomogeneity at epitaxial Al/(111)Si interfaces

It is often observed that the SBH's determined by different methods are not always in good agreement with each other. Al-Si interfaces are known as one of such cases (e.g. Miura et al. 1992). Since SIPM can reveal the local distribution of SBH, we apply this technique to annealed Al/(111)Si interfaces (Miyazaki et al. 1994a, b). Epitaxial Al films with a thickness of 100 nm were grown at room temperature (sample I) and 220 °C (sample II) on n-Si(111) substrates with resistivity of 10-20 Ωcm . The samples were annealed under N_2 flow for 1 hours at 400-550 °C and then quenched in water. Schottky diodes, 500 μm in diameter, were fabricated with the as-deposited and annealed samples. In the as-deposited state, Al/Si(111) interfaces consisted exclusively of type B orientation and mainly of type-A with partial type-B grains, for sample I and sample II, respectively.

For sample I, the SBH's determined by I-V and C-V, ϕ_{IV} and ϕ_{CV}, were agree well with each other, after annealing at 400 °C, while ϕ_{CV} was slightly larger than ϕ_{IV} in the as-deposited state. This discrepancy in the as-deposited state could be attributed to radiation damage in a subsurface region during e-gun evaporation of Al. Like sample I, the Al films deposited at 250-300 °C also gave a single domain with a type B orientation entire through the interface. Note that such interfaces were thermally stable and that both ϕ_{CV} and ϕ_{IV} did not change up to 550 °C (slightly below the eutectic pint of 577 °C) for such single domain interfaces.

After annealing at 400 °C, however, SIPM image revealed microscopic inhomogeneity of SBH's as shown in Fig. 4 (a). In this image, a 12-step gray scale was used to express the photoyield value in reverse order; darker pixels correspond to the region with low SBH. The most interesting feature of the image is that dark spots align almost along [110] direction. Figure 4(b) shows a line profile of the SBH around a particular dark spot determined with photoemission yields at two different wavelengths. The dashed line shows the simulated profile, which was obtained by assuming an SBH and a dark-spot diameter of 0.65 eV and 3 μm. Local reaction between Al and Si cannot account for the formation of low SBH regions, since the formation of an Al-doped layer enhances the effective SBH. On the other hand, a local strain, which is induced by a difference in thermal expansion coefficients for Al and Si, might modify the band structure of the Si surface, and hence the SBH. It is noted that the density of such an anomalous spot was much lower on average than this particular image as shown in Fig. 4 (a). Therefore, we think that macroscopic values of ϕ_{CV} and ϕ_{IV} were not hardly affected.

For sample II, on the other hand, the discrepancy between ϕ_{IV} and ϕ_{CV} appeared upon annealing at temperatures above 400 °C; ϕ_{CV} became larger than ϕ_{IV}. The value of ϕ_{CV} took the maximum of 0.78 eV around 500 °C and then decreased to 0.73 eV, while ϕ_{IV} slightly decreased by 0.01 eV upon annealing at 400 °C, and then remains almost constant. The discrepancy between ϕ_{CV} and ϕ_{IV} is often attributed to SBH inhomogeneity at the interface (Ohdomari et al. 1980). Figure 4 shows typical images of the internal-photoemission yield (Y) for the Al/n-Si(111) interfaces annealed at (a) 500 °C and (b) 550 °C. Both images show the dark (low SBH) spots as well as tiny bright (high SBH) spots; their origins are discussed in a separate paper at this conference (Miyazaki et al. 1995). Note that the local SBH at the dark spot in Fig. 4 is comparable to that in the average region at the 400 °C-annealed sample shown in Fig. 3(a). Here, we compare the number of anomalous spots between Fig. (a) and (b). The density of the dark spot is almost the same for both interfaces. This fact can account for the fact that ϕ_{IV} is independent on the annealing temperature above 400 °C, since ϕ_{IV} is mostly determined by lower SBH regions because of $\exp(-\phi_{IV}/kT)$ factor in I-V characteristics.

On the other hand, the density of bright (high SBH) spots is much higher in Fig. 4(a) than that in Fig. 4(b). With this fact, it is possible to explain qualitatively why ϕ_{CV} is larger for the interface annealed at 500 °C than that annealed at 550 °C. In the C-V measurements, SBH is determined by the dependence of depletion-layer width on applied reverse bias. The

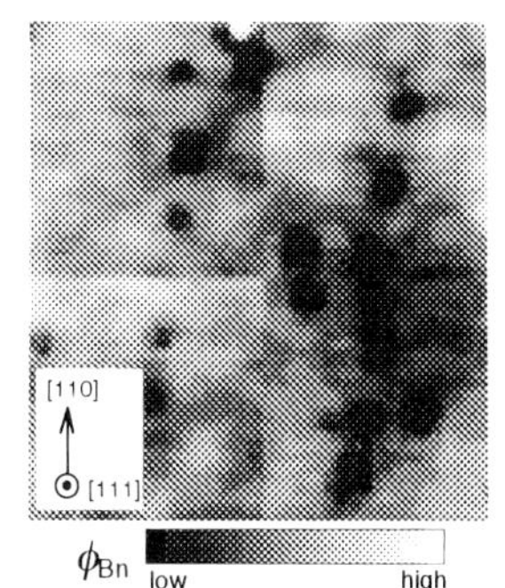

(a) photoyield map (reverse contrast). 75 × 75 μm².

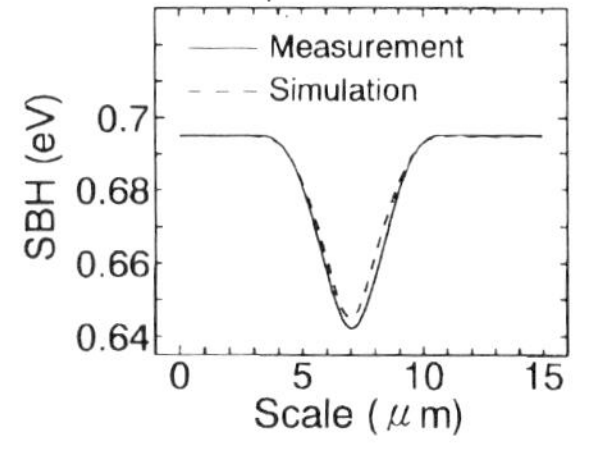

(b) SBH profile around a dark spot

Fig. 3. Al/n-Si(111) depoisted at RT and then annealed at 400 °C.

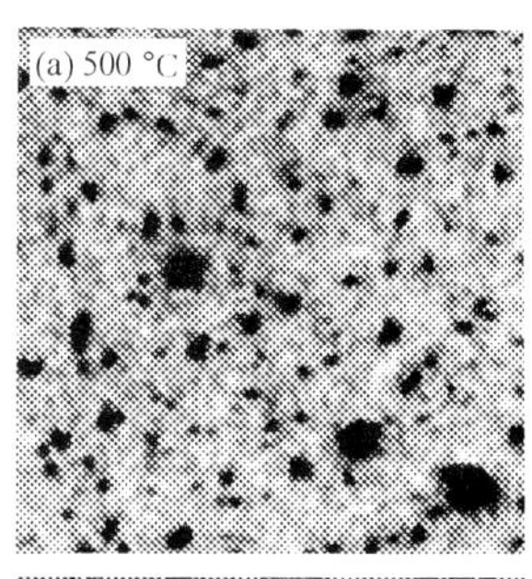

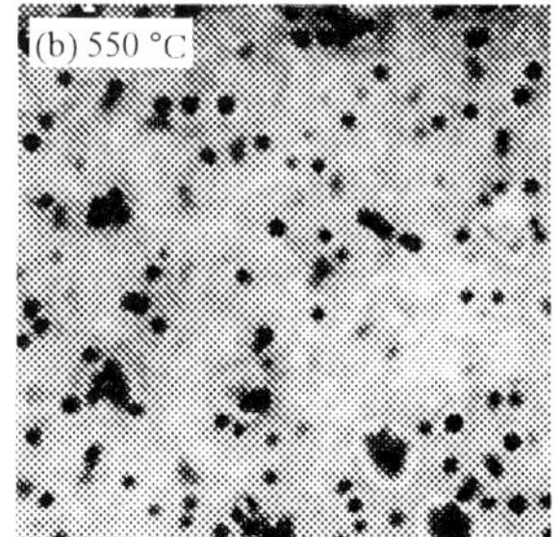

Fig. 4. Photoyield maps for Al/n-Si(111) depoisted at 220 °C and then annealed at (a) 500 and (b) 550 °C. 75 × 75 μm².

depletion layer expands not only vertically but also laterally. When the density of high SBH spots increases, the depletion regions can overlap with one another laterally; for the present substrate the depletion layer width is approximately 1.5 μm, which is comparable to the distance between high SBH spots. Therefore, the effective area of the high SBH region defined by the measured capacitance becomes larger than the real one.

It is interesting that the average distance between the adjacent bright spots is in the same order as the average size of type-A grains observed in the as-deposited state. We speculate that the high SBH region was attributed to a p^+-n junction; recrystallized Al-doped Si was locally formed in the vicinity of grain boundary upon annealing. Since the higher the annealing temperature, the more rapidly the grain boundaries disappear, the amount of Si diffusion into Al is reduced resulting in lower density of high SBH spots (Miura et al. 1992).

5. Conclusions

I have demonstrated that scanning internal-photoemission microscopy (SIPM) is a powerful technique to reveal microscopic inhomogeneity in SBH's at buried metal-semiconductor interfaces. Although the developed SIPM has a spatial resolution of several mm, sophisticated image processing techniques, if combined, will improve the spatial resolution as high as a submicron region.

Acknowledgments

The author would like to thank Kenji Shiojima, Shuji Miyazaki, Yoshinao Miura and Kazuyuki Hirose for their enthusiastic collaboration in these studies.

References

Beam III E and Chung D D L 1985 *Thin Solid Films* **128** 321
DiStefano T H 1971 *Appl. Phys. Lett.* **19** 280
Kaiser W J and Bell L D 1988 *Phys. Rev. Lett.* **60** 1406
Miura Y, Hirose K, Aizawa K, Ikarashi N and Okabayashi H 1992 *Appl. Phys. Lett.* **61** 1057
Miyazaki S, Okumura T, Miura Y, Aizawa K and Hirose K 1994a *Inst. Phys. Conf. Ser.* No. 135 (Bristol: IOP) p. 361
Miyazaki S, Okumura T, Miura Y and Hirose K 1994b *Control of Semiconductor Interfaces* (Amsterdam: Elsevier) p. 255
Miyazaki S, Okumura T, Miura Y and Hirose K 1995 this volume
Ohdomari I and Tu K N 1980 *J. Appl. Phys.* **54** 3735
Okumura T and Tu K N 1987 *J. Appl. Phys.* **61** 2955
Okumura T and Shiojima K 1989 *Jpn. J. Appl. Phys.* **28** L1108
Poate J M, Tu K N and Mayer J W 1978 *Thin Films-Interdiffusion and Reaction* (New York: Wiley)
Shiojima K and Okumura T 1991a *Jpn. J. Appl. Phys.* **30** 2127
Shiojima K and Okumura T 1991b *Proc. IEEE International Reliability Phys. Symp. , Las Vegas*, p. 234
Sze S M 1981 *Physics of Semiconductor Devices 2nd ed.*. (New York: Wiley) p. 279
Tung R T 1992 *Phys. Rev.* **B45** 13509

Inst. Phys. Conf. Ser. No 149
Paper presented at DRIP 95

The role of defects in electroless metal deposition on silicon (100)

P. Gorostiza*, J. Servat, F. Sanz+ and J. R. Morante

+ Departament de Química Física and Departament de Fisica Aplicada i Electrònica, Universitat de Barcelona. Diagonal 645, Barcelona 08028, Spain.

Abstract. The first stages of platinum electroless deposition on p-Si (100) from hydrogenfluoride solutions are studied using several techinques. Platinum nuclei are deposited on a silicon substrate by immersion in an aqueous HF solution containing a platinum salt. Tapping Mode AFM and TEM have been used to characterize the samples morphologically, and EDS and XPS to identify the chemical nature of the main features. Platinum silicide is formed in the interphase between nuclei and sustrate. The influence of crystal defects is analyzed from a qualitative point of view.

1. Introduction

A variety of Chromium-based HF etchants has been widely used in the past 40 years to reveal dislocations and lattice defects in silicon [1-7] or to delineate p-n junctions [8]. More recently, other HF etchants [9-11] have been developed for those purposes but still little attention is paid to the detailled study of the etching/metal deposition process.

Etching pits and metal deposits form the slip lines and haze that reveal the defects in wafers subjected to these etchants. HF is known as a SiO_2 etchant as well as fluoride anions as a Si etchant at sufficiently high pH. Aqueous HF etching of silicon surface leaves an H-terminated surface, forming a *passivated* surface which takes several hours to grow a native oxide at atmospheric conditions. In a solution containing HF and metal ions, the following redox reactions may occur [12]:

$$Si^0(s) + 6F^-(aq) \rightarrow SiF_6^{2-}(aq) + 4e^- \qquad -E^0_{SiF6/Si} = +1.2\ V$$
$$Me^{m+}(aq) + ne^- \rightarrow Me^{(m-n)+}(aq/s) \qquad E^0_{red}$$

where Me^{ox} and Me^{red} represent the metal in its oxidized and reduced form, respectively. If the energy balance is favorable, the process takes place at the wafer surface, involving the simultaneous etching of the silicon substrate and the reduction of the metal ions, with electron exchanging through the substrate-metal interface [13]. The preferential sites for these reactions to occur are the defect sites, which may correspond to inhomogeneities in the H-passivating layer.

(*) e-mail: pgorosti@mafalda.qui.ub.es

In this way, dislocations and lattice defects are revealed either by the metal nuclei deposited on them or by metal-induced etching pits. The first case corresponds to metal ions that reduce just to deposit on the surface as metal (e.g. $Cu^{+2} \rightarrow Cu^0$, $Ag^{+1} \rightarrow Ag^0$) and the second case corresponds to metal ions that still remain in solution after reduction (e.g. Chromium reduction is $Cr^{+6} \rightarrow Cr^{+3}$ in Secco etch). Metal deposition from HF-containing solutions is known as *electroless* metal deposition since it does not require the external application of a potential, it does not proceed as an electrochemical process but it involves an electrochemical mechanism.

In this work the electroless deposition of Pt on Si (100) from an HF aqueous solution has been studied by Tapping Mode Atomic Force Microscopy (TMAFM), Transmission Electron Microscopy (TEM) and X-ray Photoelectron Spectroscopy (XPS). We have focused on the early stages of the process and the understanding of their relationship with the presence of defects in the substrate.

2. Experimental

All solutions were prepared with p.a. grade reagents and triply-deionized water (18 MΩ resistivity: Milli-Q, Millipore), which was also used for all water rinses. Etching solutions were freshly-prepared from HF 40% "Suprapur" grade (Merck) and Milli-Q water.

The substrates used in this study were silicon (100) CZ and FZ wafers both p-doped with Boron and with resistivities ranging between 12.0 and 18.5 Ω cm for CZ wafers and between 0.8 and 1.2 Ω cm for FZ wafers. Some FZ samples were implanted with an argon ion gun to induce lattice surface defects. Implantation energy was 30 KeV and argon doses were 10^{13} cm^{-2} and 10^{14} cm^{-2}. Previously to each deposition experiment the native oxide layer was etched in an aqueous HF solution (40% HF solution was diluted ten times) during a few seconds, until they show an hydrophobic behaviour.

Plating solutions were obtained by dissolving small amounts of a metal salt in the aqueous HF etching solution. Sodium hexachloroplatinate (IV) powder, $Na_2PtCl_6 \cdot 6H_2O$ and palladium (II) chloride, $PdCl_2$ (both supplied by Johnson Matthey GmbH) were used to yield a metal concentration in solution of 1.0 mM in every case. Immediately after removal of the protective oxide layer, samples were immersed in a 10 cm^3 beaker of the corresponding solution and held vertically during deposition. Typical deposition times ranged between 1 min and 30 min and no agitation was applied to the system during deposition. All depositions were carried out at controlled ambient conditions. After deposition, samples to be imaged by AFM were withdrawn from the solution and rinsed thoroughly in flowing ultrapure water, blown dry with pure argon, attached to a holder and immediately imaged.

Cross-sections of the samples for TEM imaging were prepared by the conventional procedure. Samples to be analyzed by XPS were withdrawn from the solution, rinsed and dried following the same procedure as for AFM and immediately introduced in the XPS UHV chamber. Cross sections of the samples were imaged in a Philips CM30 (300 kV) TEM equipped with an Oxford Link EDS. XPS measurements were performed in a Physical Electronics PHI 5500 system.

In order to reliably characterize the samples by TMAFM, a good survey of images was taken in different regions of each sample. The experiments were repeated several times and showed good reproducibility. A Nanoscope III electronics with a Multimode AFM stand was employed to characterize both the substrate and the deposited surfaces. Contact AFM images of silicon and silicon oxide are difficult to obtain due to the interaction with the Si_3N_4 tip but in Tapping Mode AFM we found that lateral forces are minimized and such problems are not encountered on silicon. Nevertheless, platinum-deposited samples often yield artifacts and tip effects as multiple tips even when using new tips. These effects are attributable to the high tendency of noble metals to adhere to on silicon surfaces and they have been reported before [12].

3. Results and discussion

Two main changes occur on the surface of a silicon sample when it is immersed in a metal plating solution during a short period of time: metal nuclei deposit and grow on the silicon surface and the roughness measured on the silicon surface between these nuclei increases. Nuclei have submicrometric size and follow a progressive nucleation process because at any time different sized nuclei are found.

Figure 1 shows a TMAFM image of a p-doped CZ silicon sample after 30 minutes platinum deposition in a solution of pH=1 (HF:H_2O 1:10). Nuclei of hemispherical shape are observed on the silicon substrate, which really correspond to platinum nuclei. Figure 2 is a TMAFM image of the substrate surface between the nuclei. The average roughness is 27 Å (RMS), while silicon samples etched in a blank solution (no Platinum salt added) have a surface roughness less than 5 Å. TMAFM images also revealed some deeper pits scattered on the sample. As an example, the pit shown in the middle of figure 2 has a depth of about 500 Å. It is not completely clear the origin of these pits, whether they can be correlated with structural defects or with nuclei that are torn off either by the cleaning process or during the AFM scanning.

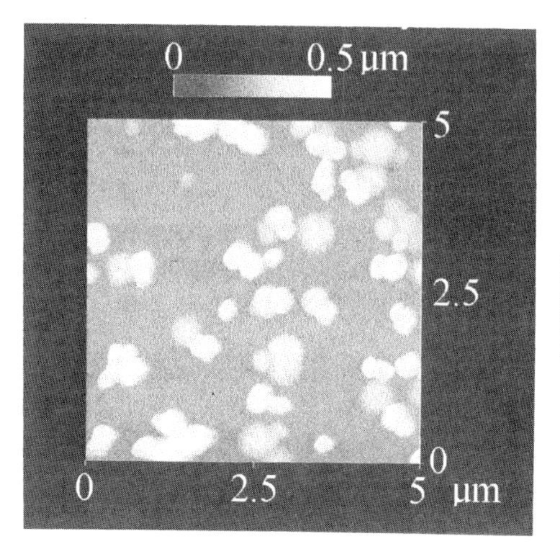

Fig. 1: TMAFM image of Pt deposited on CZ Si (100) from a solution of pH=1 for 30 min.

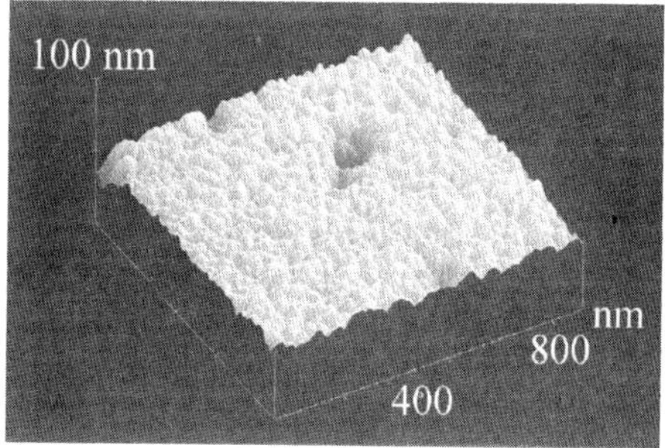

Fig.2: TMAFM image of the Si substrate between the nuclei in fig. 1.

The deposited nuclei have a polycrystalline structure with domains of about 50 Å, where atomic resolution was achieved by HRTEM [12]. Figure 3 shows a cross-sectional bright field TEM image of a sample deposited under the same conditions of those used in figures 1 and 2. Silicon surface around the nucleus has been more etched than the rest of the surface, indicating a strong local reaction. Of course electrons must be supplied from the substrate to reduce platinum on the nucleus external peel. The convolution of the nuclei sides with TMAFM tips prevents to observe the complete profile of the nuclei in this kind of deposited samples. In this way TEM images are complementary to the AFM images to define the morphology of the deposit.

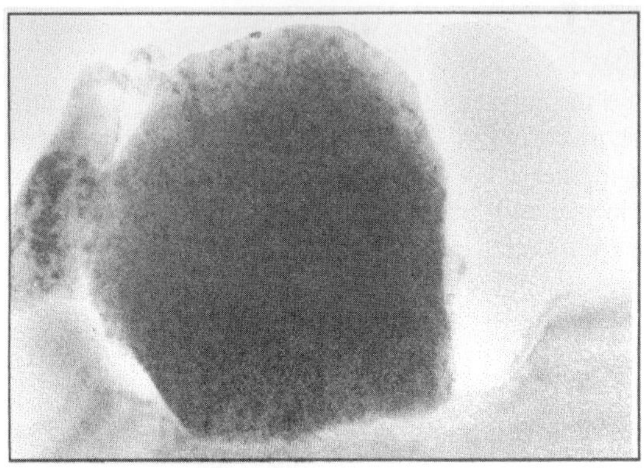

Fig. 3: TEM cross section of a Pt nucleus deposited on Si (100). Conditions as fig.1. Width: 1μm.

XPS results [14] make evident the existence of a thin layer of platinum silicide at the interphase between the nuclei and the substrate. Most probably the region between the nucleus base level and the original surface level may contain this silicide, originated by platinum diffusion. The corresponding XPS analysis show that Pt_2Si is formed, as indicated by the silicide bond contribution (99.6 eV) [15] in the silicon region of the spectrum and the Pt_2Si bonds (72.5 eV and 75.8 eV) in the platinum region.

From the above results it seems that platinum deposits on silicon in a process that is a competition between two mechanisms: the electrochemical deposition of pure platinum and the local formation of platinum-silicon bonds. Fluoride anions (or the corresponding hydrogen difluoride anions HF_2^-) etch silicon atoms from the surface at high pH [16] and promote the formation of charge carriers necessary for the electrochemical mechanism. Before proceeding to study the influence of the crystal defects on the deposition process, we have analyzed the behavior of the system at pH=3.

Figure 4 shows a TMAFM image of a p-doped CZ silicon sample with platinum deposited on during 30 minutes from a solution in which pH=3. In this sample the nuclei are *nailed* on the substrate and in some cases can be found underneath the surface level. This can be best appreciated in the TMAFM nucleus profile. The corresponding TEM cross-section at pH=3 is shown in figure 5, where the silicon etching has resulted in the penetration of the nuclei into the substrate. The degree of penetration varies between nuclei but their wedge shape is very reproducible.

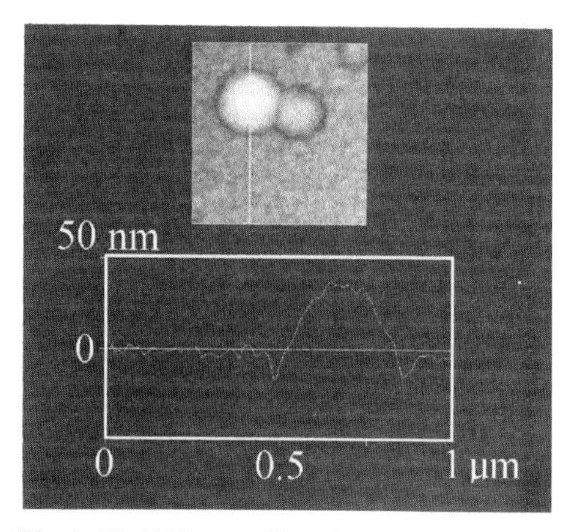

Fig.4: TMAFM profile of a *nailed* nucleus deposited at pH=3 .

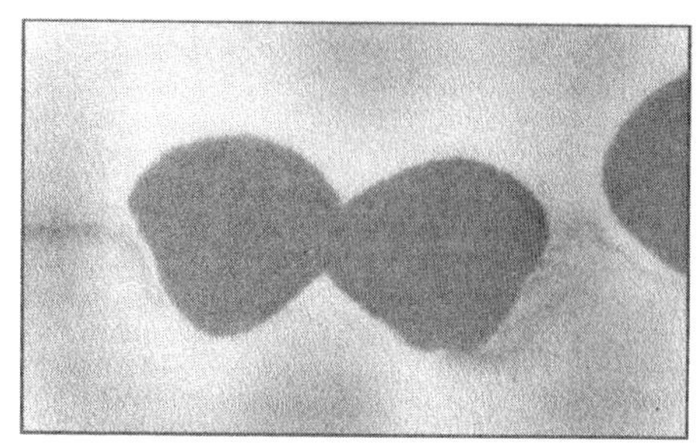

Fig.5: TEM cross section of Pt nuclei deposited on Si (100). Conditions as fig. 4. Width: 1μm

Atomic concentration profiles calculated from XPS results for samples deposited at pH=1 and pH=3 are compared in figure 6. Platinum penetrates twice as deep in pH=3 samples than in pH=1 samples, as could be expected from the surface morphology. Moreover, analyzing the XPS spectra along the depth profile for the case of pH=1, metallic platinum has the larger contribution in the first spectra and decreases untill a value that remains constant (fig. 7). Platinum silicide increases up to a constant value. In the case of pH=3, along the depth profile the silicide and the metallic platinum mantein the relative percentage [17]. It is to be expected that in the case of pH=3 the silicide is located mainly in a thin layer in the lateral walls of the nailed nuclei.

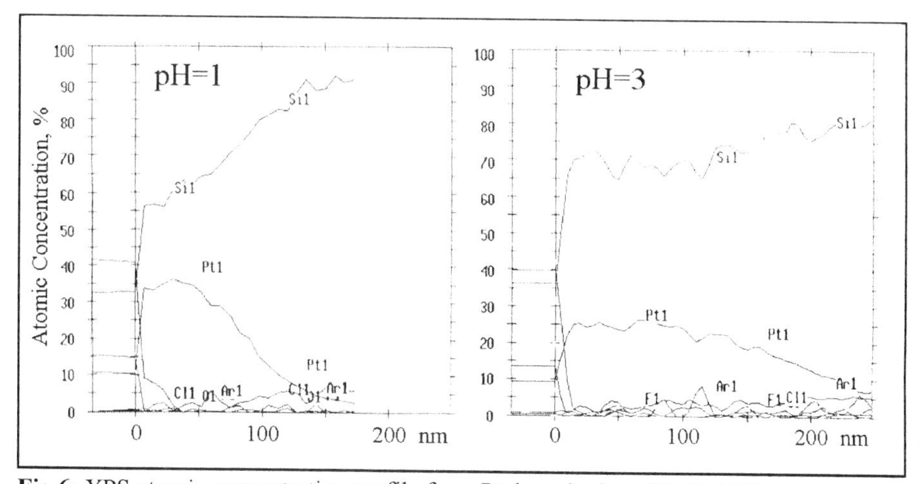

Fig.6: XPS atomic concentration profile from Pt deposited on CZ Si (100).

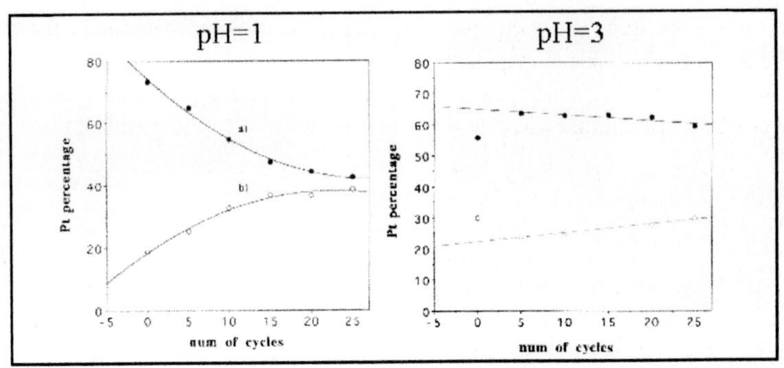

Fig.7: XPS atomic concentration percentage of Pt deposited on Si (100). a) metallic Pt, b) Pt silicide.

The electrochemical mechanism favours revealing the initiation (defect) sites at the nanometer scale and then, a series of experiments were performed depositing platinum from a pH=3 solution using samples with a different nominal quantity of defects. The samples used in this case were FZ Si (100) *as-grown*, which in some cases are argon ion implanted samples in order to induce additional defects on the surface. An increase by a factor 10 on the deposition rate with respect to CZ silicon samples was really obtained in these FZ samples when deposition was made from the same departure solution. Thus, the deposition time has been decreased to 5 min against the 30 min used to prepare the CZ silicon samples The FZ series includes the blank silicon wafer and two different implanted samples and the TMAFM images after depositing platinum are presented in Fig. 8.

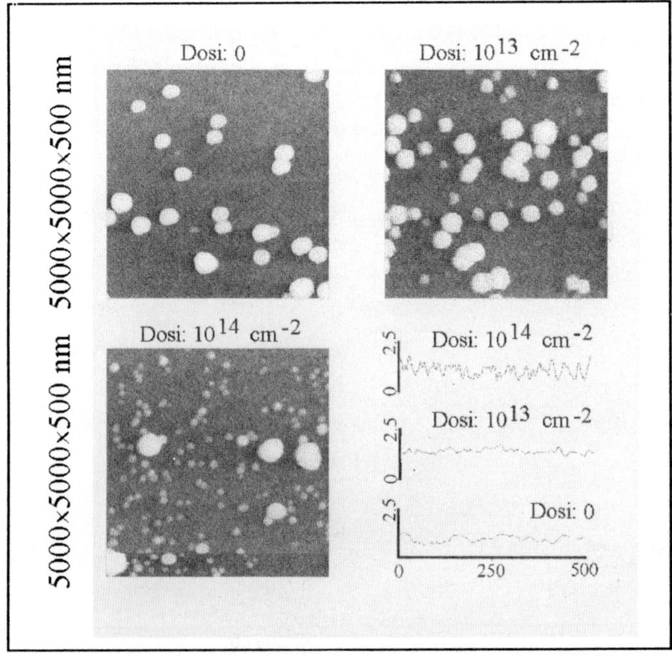

Fig.8: TMAFM images and profiles of Pt deposited on FZ Si (100) from a solution of pH=3 during 5 min. for different defect dosage.

An increase in the number of deposited nuclei is observed which corresponds to the increasing defect concentration. An ncrease in the resulting substrate roughness between the nuclei is associated to the amount of planinum deposited as indicated in the same figure. It is interesting to denote that an increase of the etching rate due to change the solution pH promotes an increase of platinum deposition but an increase of platinum deposition also increases the final surface etching.

4. Conclusions

Tapping Mode AFM is a powerful tool to analyse the silicon surface and to follow defect sites., in the case that the adequate metal solution was chosen.
Electroless platinum deposition on silicon (100) surface follows a progressive nucleation process and has been found to depend on doping type and level as well as on pH. Metal nuclei grow following an electrochemical mechanism mainly at the expenses of the surrounding silicon material, although the entire surface undergoes etching.
The interphase between the nuclei and the substrate is formed by a thin layer of platinum silicide whose thickness and shape depend on the pH of the depositing solution.
There is a direct relationship between the amount of metal nuclei deposited on the substrate and the concentration of the defects induced by Ar Ion Implantation. On the other hand, Ion Implantation processes enhance the electroless suitability.

References

1. W. C. Dash, *J. Appl. Phys.*, **27**, 1193 (1956)
2. E. Sirtl and A. Adler, *Z. Metallk.*, **52**, 529 (1961)
3. F. Secco, *J. Electrochem. Soc.*, **119**, 948 (1972)
4. M. W. Jenkins, *J. Electrochem. Soc.* **124** , 757 (1974)
5. D. G. Schimmel, *J. Electrochem. Soc.* **123** , 734 (1976)
6. H. Seiter, in *Semiconductor Silicon 1977*, H. R. Huff and E. Sirtl, Editors, PV77-2, p. 187, The Electrochemical Society Proceedings Series, Princeton, NJ (1977)
7. K. H. Yang, in *Semiconductors Processing*, ASTM STP 850, D. C. Gupta, Editor, p. 309, American Society for Testing and Materials, Philadelphia (1984)
8. J. C. Carter and A. G. R. Evans, *Electronics Lett.*, **27**, 2135 (1991)
9. Y. Saito, Eur. Pat. 0281115 A2 (1988)
10. T. C. Chandler, *J. Electrochem. Soc.*, **137**, 944 (1990)
11. K. Graff and P. Heim, *J. Electrochem. Soc.*, **141**, 2821 (1994)
12. L.A. Nagahara, T. Ohmori, K. Hashimoto and A. Fujishima, *J. Vac. Sci. Technol. A*, **11**, 763 (1993)
13. P. Gorostiza, J. Servat, J. R. Morante and F. Sanz, *Thin Solid Films*, in press.
14. P. Gorostiza, J. Servat, F. Sanz and J. R. Morante, Proceedings of the J. Electrochem. Soc. Meeting, fall 1995. SOTAPOCS XXIII ; PV 95-21, The Electrochemical Society, Pennington NJ, 1995, p.379
15. J.F. Moulder, W.F. Stickle, P.E. Sobol, K.D. Bomben, *Handbook of X-ray Photoelectron Spectroscopy*, (1992).
16. S. Verhaverbeke, I. Teerlinck, C. Vinckier, G. Stevens, R. Cartuyvels and M. M. Heyns, *J. Electrochem. Soc.*, **141**, 2852 (1994)
17. P. Gorostiza, J. Servat, J. R. Morante and F. Sanz, MRS 1995 Fall Meeting, Symposium H, Boston.

Inst. Phys. Conf. Ser. No 149
Paper presented at DRIP 95

Applications of Atomic Force Microscopy for Silicon Wafer Characterization

M. Suhren, D. Gräf, R. Schmolke, H. Piontek and P. Wagner

Wacker Siltronic GmbH, P.O. Box 1140, 84479 Burghausen, Germany

Abstract. AFM (Atomic Force Microscopy) is a highly sensitive tool for the analysis of the microroughness of Si wafers and for the investigation of crystal defects. AFM images of atomic steps were used for verification of the vertical AFM calibration on slightly misoriented Si(111) wafers after chemical etching and epitaxial deposition. The roughness analysis of etched, polished and epitaxial Si(100) wafers shows a reduction of the surface roughness by chemomechanical polishing by more than two orders of magnitude compared to an etched surface. The morphology of crystal originated particles on Si(100) appears as smooth surface depression after polishing and sharply defined pit after SC1 treatment.

1. Introduction

Atomic Force Microscopy (AFM) has become a highly sensitive tool for the characterization of Si wafer surfaces [1]. The method can be applied in air, allows direct imaging of surface morphology with vertical resolution on the atomic scale and detection of roughness components in the lateral nm range, extending the range accessible by conventional light scattering or profiling techniques [2]. AFM can be compared with other methods determining the surface roughness by analyzing the Power Spectral Densities (PSD), as demonstrated in [3].

In the present paper we want to focus on two aspects of Si wafer characterization by AFM, roughness analysis and characterization of localized surface irregularities and defects, getting increasingly important due to shrinking dimensions in ULSI (Ultra Large Scale Integration) device design.

The AFM measurements were performed in air with a commercial large sample stage equipment for wafers up to 200 mm diameter from Digital Instrument (AFM Nanoscope III, Dimension 5000). All data were collected in the tapping mode to avoid lateral forces (friction or drag) on the sample damaging the surface.

2. Stepped wafers for AFM calibration

The calibration of AFMs and other scanning probe microscopes (SPMs) with vertical atomic resolution requires the preparation of samples providing structures with defined dimensions on the Å scale. The structural dimensions are determined by the lattice constant of the material. Samples with well-resolved step heights of 0.31 nm, defined by the atomic structure of the Si(111) surface, were prepared by depositing an epitaxial layer and by chemical etching.

2.1 Chemically etched wafers

Si (111) surfaces with a misorientation of less than 0.1° in $[11\bar{2}]$ direction were used for atomic step preparation. A thermal oxide of 100 nm was grown by oxidizing the wafers in dry oxygen at 1000 °C. The wafers were cleaned in SC1 solution at room temperature for 10 min, rinsed with DI water and spin dried. The wafers were then etched in buffered HF at pH 5 for 2 min, rinsed with DI water for 10 sec, etched in NH_4F at pH 7.8 for 30 sec, rinsed with DI water for 10 sec and spin dried, resulting in an hydrophobic surface [4]. A stable, hydrophilic surface state was generated by treating the wafers for 10 min with SC1 solution at room temperature, followed by DI water rinse for 10 sec and spin drying.

AFM images were taken of the hydrophobic (fig. 1a) as well as of the hydrophilic (fig. 1b) surface. The AFM images display pronounced terrace structures with triangular etch pits [5]. The width of the terraces varies between 250 and 450 nm. A value of 416 nm was obtained from Fourier analysis of a line scan (fig. 2a), equivalent to a misorientation of 0.043°.

On the hydrophobic, hydrogen terminated surface the image was taken directly after NH_4F/HF etching, DI water rinse and spin drying (fig. 1a). Small particles, appearing as white dots in the AFM image are detected on the surface. Additional treatment of the wafer with SC1 solution generating a hydrophilic surface covered by a native oxide does not change the appearance of the surface structure, but results in a significantly reduced particle level (fig. 1b). This observation confirms the efficiency of SC1 solution in particle removal.

A quantitative analysis of the step height was done by section analysis of the hydrophilic surface (fig. 2a). An average terrace height of 0.31 nm was detected, in excellent agreement with the nominal step height of 0.314 nm of atomic steps on Si(111). The root-mean-square (rms) roughness Rq obtained from a 5 μm line scan perpendicular to the steps is 0.109 nm. This result is in good agreement with the theoretical Rq of 0.09 nm calculated for a stepped Si(111) surface with 0.314 nm step height and ideally flat terraces. This value is estimated to be decreased by less than 10 % due to the finite radius of the AFM tip for the present terrace width. Rq obtained from the 5*5 μm^2 area scan of the hydrophobic surface is 0.122 nm which is only slightly enhanced as compared to the line scan since the surface is almost atomically flat in the direction parallel to the steps.

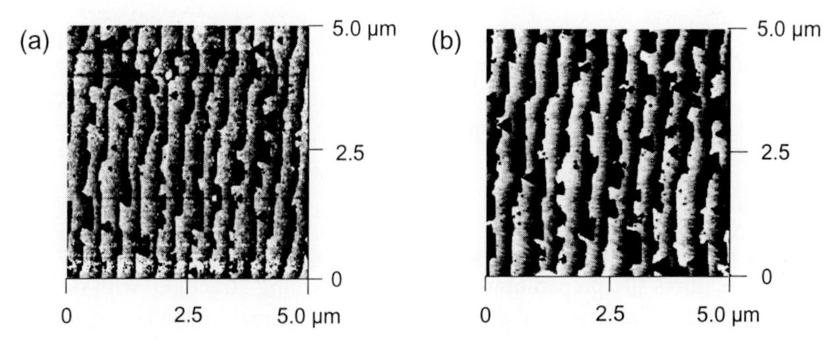

Fig. 1 AFM images of an etched Si(111) wafer, misoriented by 0.043°, hydrophobic (a) and hydrophilic surface (b)

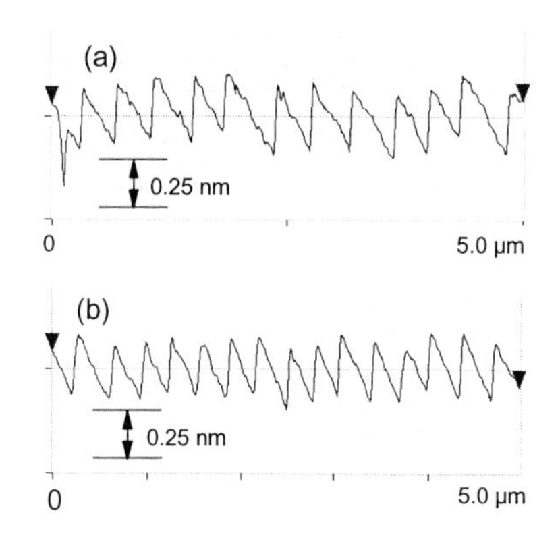

Fig. 2 Section analysis of an etched Si(111) wafer, misorientation: 0.043° (a) and an epitaxial Si(111) wafer, misorientation: 0.057 ° (b)

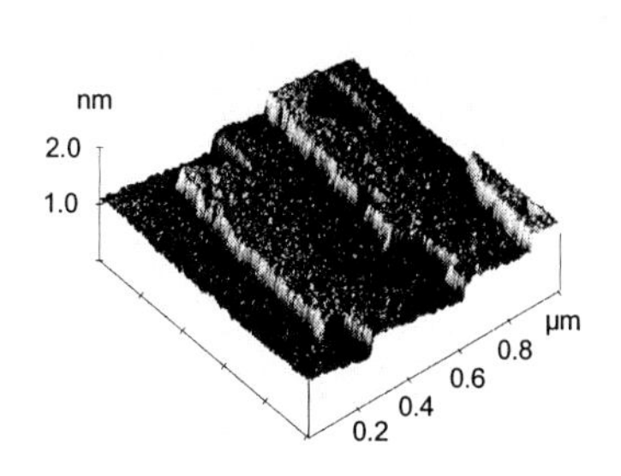

Fig. 3 AFM image of an etched Si(111) wafer, misoriented by 0.043°, hydrophilic surface

The (111) terraces between two step edges are atomically flat, provided no triangular pits are present. The Rq on such terraces should therefore be very close to the noise level of the AFM and can be used to check the stability of the instrument. A Rq of 0.032 nm was calculated from a 250*750 nm^2 area on a center terrace (fig. 3). This value is very close to the noise level of 0.030 nm obtained from a "zero-scan", i.e. data acquisition of 512*512 data points on a fixed spot on the sample.

2.2 Epi wafers

An homoepitaxial layer is deposited on Si wafers for the most advanced applications in device manufacturing. Epi layers provide very high crystalline perfection with low grown-in defect densities and low oxygen content. The roughness of epi wafers depends on the parameters chosen for epi deposition and can be comparable to polished wafers (see section 4).

Si(111) and Si(100) wafers were chosen with defined misorientation <0.1° in $[11\bar{2}]$ and $[01\bar{1}]$ direction, respectively, for preparing atomic steps. A 4 µm thick epitaxial layer was deposited from $SiHCl_3$ at 1100°C. AFM images of the Si(111) surface were taken of the hydrophobic surface immediately after epi deposition and of the hydrophilic surface after subsequent SC1 treatment, DI water rinse and spin drying (fig. 4a). The morphology is similar on both hydrophilic and hydrophobic surfaces, respectively.

The steps on epi wafers are not absolutely straight, due to the growth of "epi fingers". By section analysis an average step width of 311 nm and a step height of 0.31 nm was obtained in excellent agreement with the step height observed on the etched wafer and the nominal step height of 0.314 nm (fig. 2b). The Rq calculated from a line scan is 0.101 nm, which is lower as compared to the etched wafer with etch pits (fig. 1b) and closer to the theoretical value of 0.09 nm. A Rq of 0.108 nm was obtained from the 5*5 μm^2 area scan.

On a Si(100) surface a much higher step density is observed than on a Si(111) surface for the same misorientation angle (fig. 4b), in accordance with the small nominal step height of 0.135 nm on Si(100) surfaces. The steps appear less pronounced than on the Si(111)

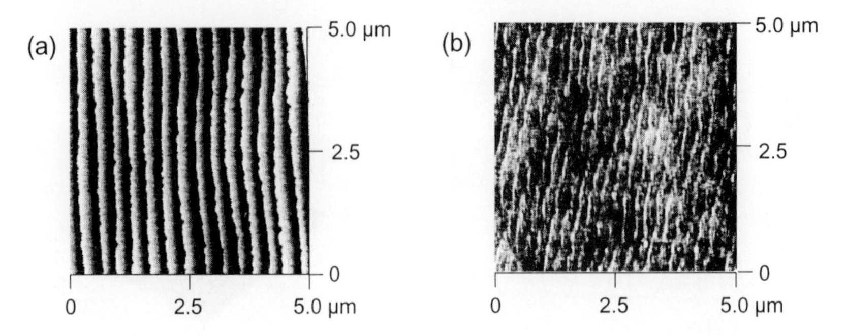

Fig. 4 AFM images of an Si(111) epi wafer, misoriented by 0.057° (a) and of an Si(100) epi wafer, misoriented by 0.038° (b)

surface. Due to the small step height, an overall smoother surface is observed on the Si(100) as compared to the Si(111) surface, with a Rq of 0.086 nm.

3. Roughness analysis of etched, polished and epi wafers

Chemomechanical polishing (CMP) is the technology applied since more than two decades to produce ultraflat Si wafer surfaces. An alkaline polishing slurry is used for Si stock removal by combining the effects of chemical etching and mechanical action. The surface morphology is a result of the polishing conditions applied (pH, size distribution and concentration of the slurry particles, pressure, temperature, polishing pad, polishing time) and material parameters of the Si wafers (crystal orientation, doping type, doping level and density and distribution of crystal defects) [6].

The last processing step before CMP is etching to remove the damaged surface layer introduced in the wafering process. The Si(100) surface after an acid etching step (HNO_3/HF) is characterized by an irregular structure, apparent in the AFM image (fig. 5a) and a large microroughness with a Rq of 130 nm.

Polishing at high pH and with a high polishing pressure (stock removal polishing) resulting in large silicon removal rates, establishes wafer geometry and reduces surface roughness by about two orders of magnitude. A Rq of 0.22 nm was found (fig. 5b) and the surface structure is still somewhat irregular. Note the different scan sizes and vertical scales for the etched surface (fig. 5a, scan size 100*100 μm^2, vertical scale 500 nm per division) in contrast to the polished and epitaxial surfaces (fig. 5b,c,d, scan size 1*1 μm^2, vertical scale 1 nm per division).

The Si removal rate is reduced in the final polishing sequence as compared to stock removal polishing by operating at lower pH and lower polishing pressure, producing a microscopically smooth surface (fig. 5c). A Rq as low as 0.068 nm was measured.

The surface of a Si(100) wafer with an epi layer is extremely smooth, the Rq calculated from the image is as low as 0.050 nm (fig. 5d). The thickness of the epi layer is 17 µm, deposited at 1100°C.

4. Characterization of localized defects: COPs

The density, radial distribution and size distribution of crystal defects strongly varies as a function of the pulling conditions of CZ silicon crystals [7]. In the wafer polishing process, depending on the polishing conditions applied, such crystal related defects may be

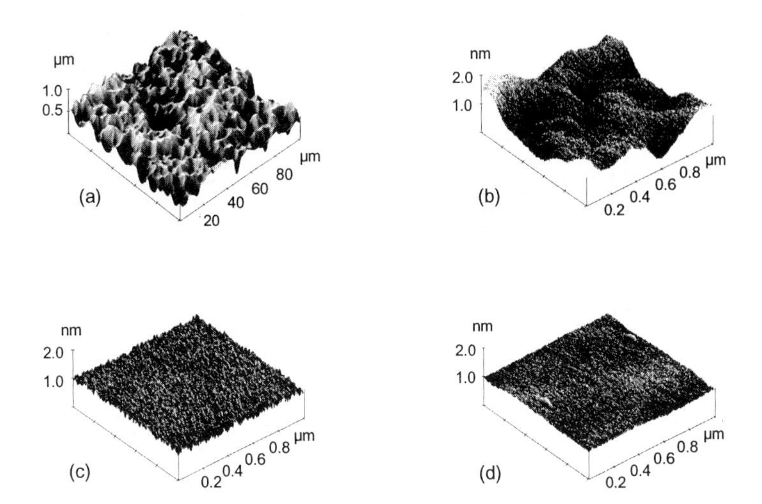

Fig. 5 Comparison of AFM maps of Si(100) surfaces after the successive processing steps etching (a), stock removal polishing (b), final polishing (c) and epi deposition (d). Note the different scales for the lateral and vertical scan widths

selectively etched to a certain extend by the alkaline slurries [5]. Treating wafers in hot SC1 ($NH_3/H_2O_2/H_2O$) solution is a well established method to delineate crystal related defects. LPDs (Light Point Defects) originating from crystal related defects and detected at the wafer surface after SC1 treatment are commonly referred to as COPs (Crystal Originated Particles) [8].

The coordinates of LPDs after polishing as well as of COPs after additional SC1 treatment were determined on Si(100) wafers by a SSIS (Surface Scanning Inspection System) along with other surface irregularities and particles, since this technique allows rapid mapping and defect localization on the entire wafer (< 1 min for a 200 mm wafer) [9]. A high accuracy of the coordinates is required as input for AFM investigation of localized defects, due to the limited AFM scan size (100*100 μm^2 for our instrument) and the long measuring time (about 10 min per map).

The AFM map of a LPD originating from a crystal defect after polishing reveals a shallow depression with a smooth curvature without sharp edges (fig. 6a). The depth of this LPD is about 65 nm, the width about 1.3 μm. COPs with sharp edges and an almost square shape on Si(100) surfaces depicting the fourfold symmetry are developed after 4h SC1 treatment at 85°C (NH_3:H_2O_2:H_2O=1:1:5, Si removal about 150 nm per side on Si(100) wafers). The depth of the COP presented in fig. 6b is about 140 nm, the width about 300 nm.

The different shapes of the LPDs and COPs can be understood in terms of different conditions for Si removal in the CMP process and by SC1 treatment. The Si removal is almost homogeneous due to the fluid dynamics achieved by the mechanical action of slurry particles and polishing pad, although preferential etching by the alkaline slurries used in CMP should be expected. Therefore the LPDs observed after CMP are shallow, with a smooth curvature. In contrast the square shape on Si(100) and the sharp edges of COPs after SC1 treatment are determined by the preferential etching action of the alkaline solution.

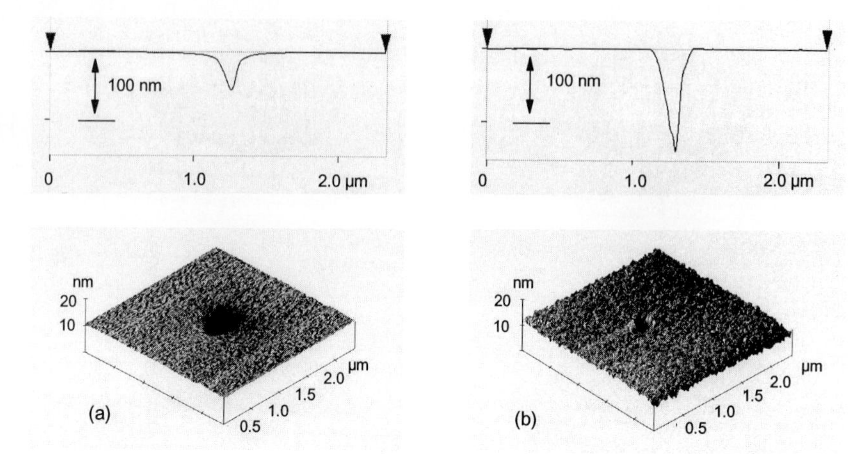

Fig. 6 AFM images and AFM profiles of a LPD after polishing (a) and a COP after SC1 treatment (b)

5. Summary

The feasibility of using atomic steps on Si(111) surfaces for the verification of the vertical atomic resolution of SPMs was demonstrated. AFM was applied for roughness analysis and defect characterization on Si wafers. A reduction of the surface roughness by CMP by more than two orders of magnitude was measured. Rq as low as 0.068 nm and 0.050 nm were observed on polished and epitaxial wafers. COPs were localized by SSIS and the morphology determined by AFM imaging. A square structure of the COPs is obtained after 4h SC1 etching on Si(100) surfaces, depicting the crystal structure.

Acknowledgments: The assistance of E.-P. Mayer in preparing the epi wafers is highly appreciated. This work has been supported by the Federal Department of Research and Technology of Germany under contract number M 2793 F. The authors alone are responsible for the content.

References

[1] F. Chollet, M. Cayamax, W. Vandervorst and E. André 1995 *Proceedings of the Satellite Symposium to ESSDERC* **95-30** (The Hague; B.O. Kolbesen, C. Claeys, P. Stallhofer, eds.) 263

[2] P.O. Hahn, M. Grundner, A. Schnegg and H. Jacob 1988 *The Physics and Chemistry of SiO2 and the Si-SiO2 interface* (Plenum Publishing Corporation, C.R. Helms and B.E. Deal, eds.) 401
J.M. Bennett, L. Mattson 1989 *Introduction to Surface Roughness and Scattering* (Washington D.C., Optical Society of America)

[3] P. Wagner, H.A. Gerber 1995 *Results of a Round Robin of SEMI Europe with Respect to Surface Roughness of Silicon Wafers; Technical Conference: Particles, Haze and Microroughness on Si Wafers* (Semicon/Europe, Geneva, Switzerland, April 4-6)

[4] P. Jacob and Y.J. Chabal, 1991 *J. Chem. Phys.* **95** 2897

[5] H.E. Hessel, A. Feltz, M. Reiter, U. Memmert and R.J. Behm, 1991 *Chem. Phys. Lett.* **186** 275

[6] H. Fusstetter, A. Schnegg, D. Gräf, H. Kirschner, M. Brohl and P. Wagner 1995 *Mat. Res. Soc. Symp. Proc.* (to be published), MRS San Francisco, April 17-21 1994

[7] M. Brohl, D. Gräf, P. Wagner, H.A. Gerber and H. Piontek 1994 *ECS Fall Meeting* vol 94-2 (Miami Beach, Oct. 9-14) 619
Y. Yamagishi, I. Fusegawa, N. Fujimaki and M. Katayama 1992 Semicond. Sci. Technology **7** A135
T. Abe and H. Takeno 1992 *Mat. Res. Soc. Symp. Proc.* **262** 3

[8] J. Ryuta, E. Morita, T. Tanaka and Y. Shimanuki 1990 *Jpn. J. Appl. Phys.* **29** L1947
J. Ryuta, E. Morita, T. Tanaka and Y. Shimanuki 1992 *Jpn. J. Appl. Phys.* **31** L293

[9] P. Wagner 1995 *Proceedings of the Satellite Symposium to ESSDERC* **95-30** (The Hague; B.O. Kolbesen, C. Claeys, P. Stallhofer, eds.) 236

Inst. Phys. Conf. Ser. No 149
Paper presented at DRIP 95

Microscopic correspondence between Schottky-barrier height and interface morphology at thermally degraded Al/(111)Si contacts

S. Miyazaki, T. Okumura, Y. Miura* and K. Hirose*a)

Department of Electronics & Information Eng. Faculty of Eng., Tokyo Metropolitan University
1.1, Minami-ohsawa, Hachiohji, Tokyo 192-03, Japan
*Microelectronics Research Laboratories, NEC Co.,
34, Miyukigaoka, Tsukuba, Ibaraki 305, Japan

Abstract. We have measured microscopic distribution of Schottky barrier heights (SBHs) at epitaxial-Al/n-Si(111) interfaces by using scanning internal-photoemission microscopy. The SBH distribution at the interface was homogeneous in the as-deposited state. After annealing, the inhomogeneity of SBH arose and it caused the discrepancy between two SBHs determined by *I-V* and *C-V* methods. The thickness distribution of recrystallized p^+-Si layer, formed during annealing process, was responsible for the distribution of SBH.

1. Introduction

The electrical and thermal stability of metal-semiconductor interface is of prime importance in semiconductor devices. The interfacial-defect formation, electrode interdiffusion, and chemical reaction produce inhomogeneity in interfacial structures and electrical properties during device fabrication processes. Aluminum and its alloy are widely used in silicon integrated circuit technology and satisfy the requirements of low resistance. However, the Al-Si contact produces problems such as spiking after annealing (Sze 1985). Chino (1973) has shown that the Schottky barrier height (SBH) of Al-Si contacts increases from 0.71 eV to 0.81 eV accompanying with the formation of the alloying pits at the Al-Si interface. Miura et al. (1992) have shown that the SBHs ($q\phi_B^{I\text{-}V}$) of the epitaxial Al-Si contacts before and after annealing, determined by current-voltage (*I-V*) method, agree within 10 meV. However, the discrepancy between $q\phi_B^{I\text{-}V}$ and the SBH ($q\phi_B^{C\text{-}V}$) determined by capacitance-voltage (*C-V*) method increased up to 70 meV after annealing at 500 °C.

In this paper, we show that the discrepancy between $q\phi_B^{I\text{-}V}$ and $q\phi_B^{C\text{-}V}$ after annealing is attributed to the inhomogeneity at the Al-Si interface. Furthermore, the spatial distribution of the SBH is caused by the distribution of the thickness of the recrystallized p^+-layer formed between Al and Si. The spatial distributions of the

a) present address: Institute of Space and Astronautical Science, 3.1.1, Yoshinodai, Sagamihara, Kanagawa 229, Japan

SBH were investigated by using scanning internal-photoemission microscopy (Shiojima et al 1991, Miyazaki et al 1994).

2. Experimental

Epitaxial Al films were grown on n-Si (111) substrates with resistivity of 10 - 20 Ωcm. The substrates were chemically cleaned and dipped in dilute HF (H_2O : HF = 30 : 1). After N_2 blowing, they were quickly loaded into the vacuum chamber (base pressure < 10^{-7} Torr). It is thought that the Si surfaces treated in this way remain terminated by atomic hydrogen at least up to 300 °C. Next, Al was deposited by electron beam evaporation at the growth rate of 3 nm/s at the substrate temperature of 220 °C. The thickness of Al film was about 100 nm. Then, one set of samples were annealed at 500 °C for 1 hour in a N_2 atmosphere. After annealing, the samples were quenched in water. Schottky diodes, 500 μm in diameter, were fabricated with the as-deposited and the annealed samples as follows. Au-Ti was deposited on Al through a stainless-steel shadow mask, then the samples were etched in H_2PO_4 to define the diodes. The as-deposited sample revealed by x-ray diffraction measurement that the Al film was (111) oriented epitaxially and consisted mainly of "type-A" and partially of "type-B". The spacings between grain boundaries were in the range of 0.1 - 5 μm. After annealing at temperatures above 450 °C, it was found that the Al/Si interface changed to type A and the grain boundaries disappeared completely (Miura et al 1992).

Schottky barrier heights (SBHs) were determined by means of current-voltage (*I-V*) and capacitance-voltage (*C-V*) methods. The SBH values, determined by *I-V*, were corrected with the imaging-force lowering of the SBH.

Scanning internal-photoemission microscopy (SIPM) is a technique, which enables us to visualize inhomogeneity of the Schottky-barrier height of a buried metal-semiconductor interface (Shiojima et al 1991, Miyazaki et al 1994). The outline of the technique is as follows: A light from a pigtail-type 1.3 μm laser diode with a maximum power of 2 mW is transmitted through the single-mode fiber that has a core diameter of 5 μm, collimated by an infrared objective lens (NA = 0.75, magnification = 80) and then focused at the metal-semiconductor interface through the semiconductor substrate. Accompanying with the photon absorption, photocurrent (Y) due to the so-called internal-photoemission effect is generated, and its magnitude of dependence on the light wavelength reflects the SBH. Thus, imaging of the electrical properties at the interface is realized by raster scanning of the laser beam. A beam diameter of less than 5 μm was achieved in free space. The wavelength between 1.29 μm and 1.32 μm of the laser beam is varied by controlling the operating temperature of laser diode between 0 °C and 60 °C and the local SBH is determined with a Fowler's formula ($Y^{1/2} = B\,(h\nu\ -q\phi_B)$, Y is the internal photoemission current, B is a constant).

The morphologies of the Si surface were characterized with Normarski-type optical microscope and AFM after removing Au-Ti-Al films with aqua.

3. Results and Discussions

Figure 1 shows a summary of the Schottky barrier heights (SBHs) for Al-Si interfaces before and after annealing at 500 °C for 1 hour. The $q\phi_B^{I-V}$ and $q\phi_B^{C-V}$ agree well within 5 meV in the as-deposited state. However, the discrepancy between these values increases to 70 meV after annealing. After annealing, $q\phi_B^{I-V}$ decreases by 0.01 eV to 0.71 eV and $q\phi_B^{C-V}$ increases by 0.06 eV to 0.78 eV. We suppose that the discrepancy between those values is associated with the inhomogeneities of the SBH (Tung 1992), which is formed by the nonuniform reaction between Al and Si upon annealing. The decrease of $q\phi_B^{I-V}$ is inconsistent with Chino's results (1973). On the other hand, Basterfield et al. (1975) have shown that the interface between Al and Si of the quenched sample is smooth compared with that of the slow-cooled sample. Furthermore, they have explained the increase of SBH as follows. Si dissolves into Al during annealing. Then the recrystallized Si forms between Al and Si, resulting in an Al doped p-type layer on cooling. As the recrystallized p^+-layer on an n-type substrate increases the potential barrier for electrons, and thus $q\phi_B^{I-V}$ apparently increases (Basterfield et al 1975). Chino (1973) has shown that $q\phi_B^{I-V}$ of Al-Si contacts decreases by less than 0.01 eV with the annealed temperature until up to 450 °C. In the case of the quenched sample, as well as the sample annealed at below 450 °C, the recrystallization of p^+-layer on an n-type substrate is little. The lower recrystallization of p^+-layer on an n-type substrate is probably associated with the SBH decrease.

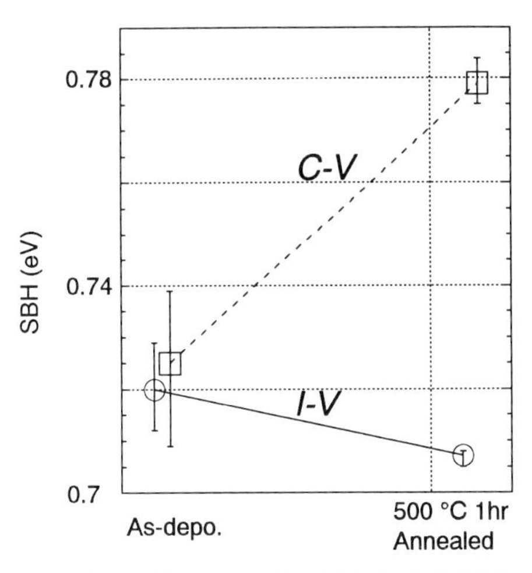

Fig. 1. Summary of the SBHs for Al/(111)Si determined by I-V and C-V methods.

Figure 2 shows the typical images of the internal photoemission current (Y) for the Al-Si interfaces (a) in the as-deposited state and (b) annealed at 500 °C. The measured area is 75 × 75 μm² and one pixel area is 0.5 × 0.5 μm² in both cases. The contrast uses a scale such that the average value through the entire region becomes

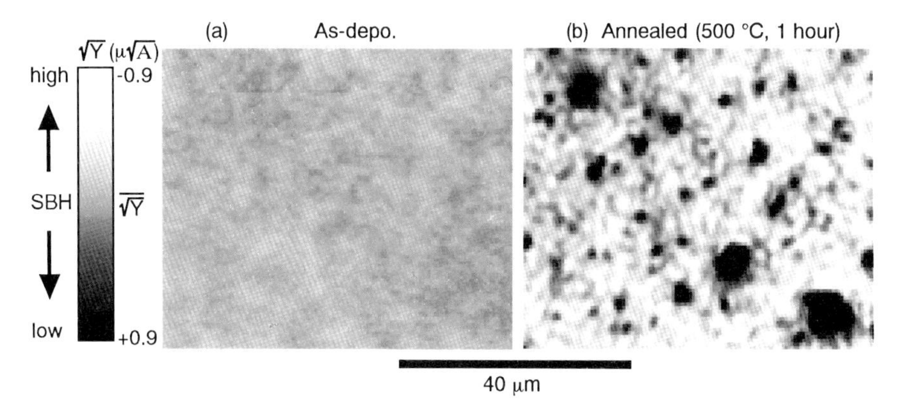

Fig. 2. The typical images of the internal-photoemission current for (a) the as-deposited and (b) the annealed samples at 500 °C. $\overline{\sqrt{Y}}$ is the average value through the entire region for each sample.

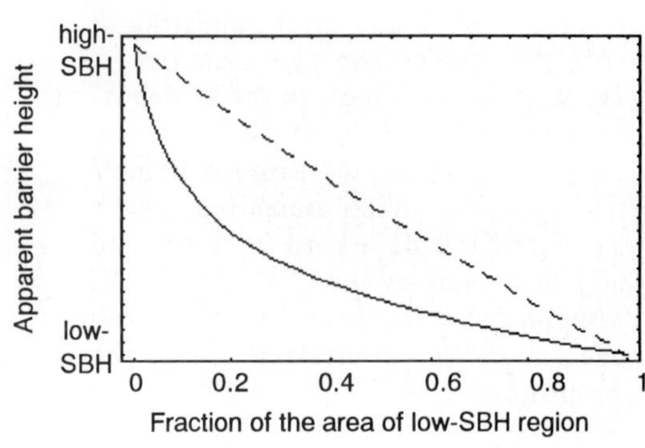

Fig. 3. The computed curves of apparent SBHs for *I-V* and *C-V* as a function of the fraction of the contact area occupied by low-SBH region.

the mid-range for each sample. The full range of the scale (square root of photocurrent) corresponds to the photocurrent variation of 1.8 $\mu A^{1/2}$. The bright and the dark in the scale correspond to the highest-SBH and the lowest-SBH, respectively. The interface for the as-deposited sample is more homogeneous than that for the annealed sample. Since the interface for the as-deposited sample is uniform, $q\phi_B{}^{I\text{-}V}$ and $q\phi_B{}^{C\text{-}V}$ agree well. On the other hand, the several inhomogeneities are observed in the image for the annealed sample. A large number of the low-SBH (dark) and the high-SBH (bright) spots are observed in the entire region of Fig. 2(b). The diameters of the low-SBH and high-SBH spots are in the range of 1 - 6 μm and 1 - 2 μm, respectively. The nonuniform reaction between Al and Si during annealing causes the parallel contact, that is the distribution of the different SBH in the interface. Figure 3 shows the relationship between the apparent SBHs, ($q\phi_B{}^{I\text{-}V}$ and $q\phi_B{}^{C\text{-}V}$), and the fraction of the area of low-SBH region, in case of coexistence of the two regions with the different SBHs (Ohdomari et al 1980). It reveals that $q\phi_B{}^{I\text{-}V}$ is smaller than $q\phi_B{}^{C\text{-}V}$ in the parallel contact, because $q\phi_B{}^{I\text{-}V}$ is much affected by the low-SBH region compared with $q\phi_B{}^{C\text{-}V}$. From those results, we believe that the discrepancy of the SBHs determined by the different methods is caused by the spatial distribution of the SBH.

Figure 4 shows the relation between (a) the photocurrent image and (b) the optical micrograph for the Al/Si annealed at 500 °C. Figure 4(b) shows the morphologies of the Si surface, for the same area observed in Fig. 4(a), after removing Al film and observed by using a Normarski-type optical microscope. The low-SBH spots (dark) in Fig. 4(a) correspond exactly to the pits in Fig. 4(b). On the other hand, we did not observe the region in Fig. 4(b), to which the high-SBH spots (bright) correspond. Fig. 5 shows the more detailed surface morphology of the Si

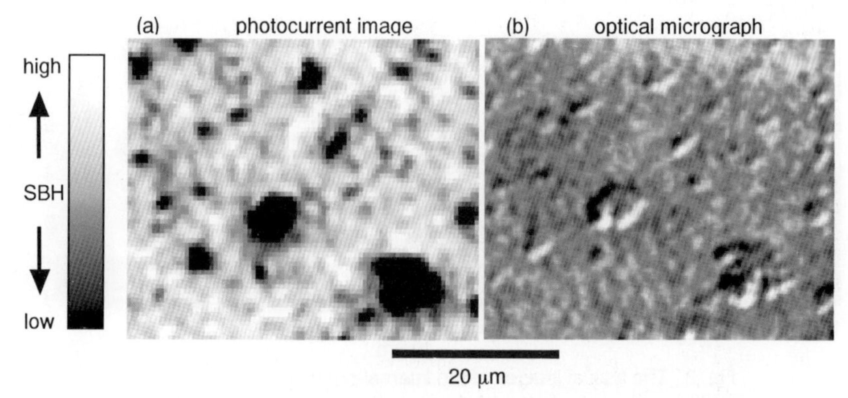

Fig. 4. (a) The typical image of photoemission current at Al/(111)Si interface.
(b) The optical micrograph of Si surface in the same area after removing Al.

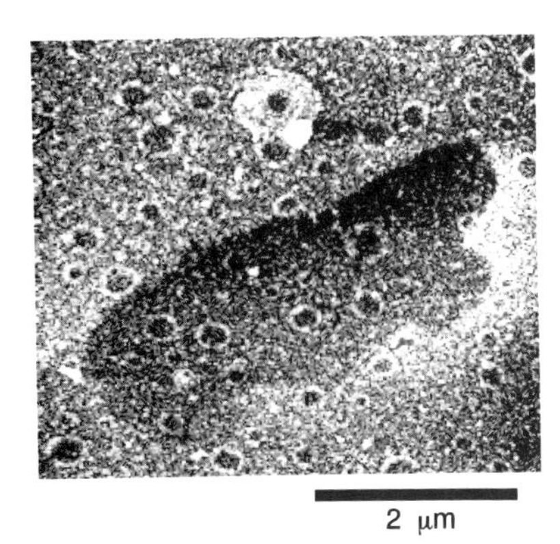

Fig. 5. The AFM image of Si surface after removing Al.

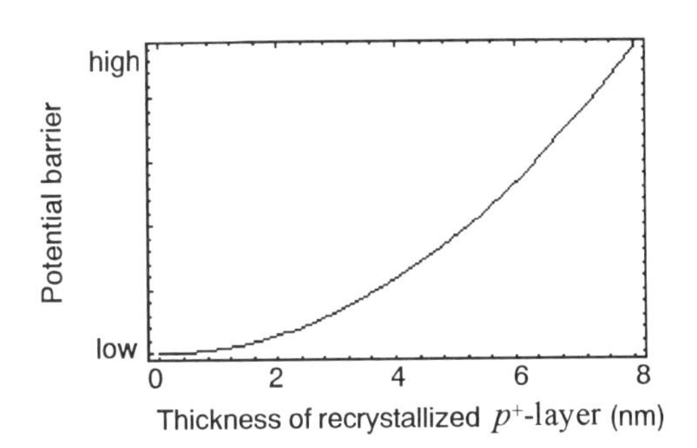

Fig. 6. The computed curves of the potential barrier for electrons as a function of the fraction of thickness for recrystallized p^+-layer.

surface obtained by AFM. The observation area is 5 × 5 μm². A number of pits are observed in the AFM image after removing Al film. Two types of the pits are observed. One pit (area 2 × 4.5 μm²) is the same type as the low-SBH spots, as observed in Fig. 4(a). The bottom of the pit is 3 - 5 nm deep compared with the surrounding region. SIPM images corresponding to the other pits (diameter 0.2 - 0.3 μm) in AFM image are not observed in Fig. 4(a), since the diameter of the pits is much smaller than the spatial resolution of SIPM (~ 5 μm). The surface morphology, which corresponds to the high-SBH regions, is not observed in Fig. 5 similar to that in Fig. 4(b). We believe that the surface morphology, which corresponds to the high-SBH, should exist in the Si surface, but not detected. The reason that the high-SBH regions have not been observed by optical micrograph and AFM techniques is speculated as follows. The change of the edge of the high-SBH region might be spatially gradual, while that of the low-SBH region is sharp. We expect that optical micrograph and AFM techniques are not sensitive in gradual change in height. Figure 6 shows the relationship between SBH and thickness of the p^+-layer. It is clear that the change of SBH is strongly sensitive to the thickness of the p^+-layer in the high-SBH region. We believe that the difference in thickness between the high-SBH region and the surrounding region is too small to be detected by structural analysis, but enough to be detected electrically in an SBH change. As a result, the high-SBH region is observed in the photocurrent image measured by using SIPM, while it is not observed in the optical micrograph and AFM images.

Finally, we discuss the inhomogeneous formation of the recrystallized p^+-layer. Miura et al. (1992) have shown that the type A and type B grains of Al exist on Si and the spacings between them are in the range of 0.1 - 5 μm before annealing. Furthermore, they have proposed that the Si dissolves into Al mainly through the grain boundaries and recrystallizes to form the p^+-layer in their vicinity. If we assume that a recrystallized p^+-Si layer is mainly formed in the vicinity of grain boundaries, the p^+-layer is thin under the grain and thick near the grain boundary. As a result, the sizes of the low-SBH spots, where the p^+-layer is thin, are influence by the various sizes of the grains. In fact, the range of sizes of the low-SBH spots (1 - 6 μm) and the those of the grains (0.1 - 5 μm) are in the same order. On the other hand, the sizes of the high-SBH spots , where the p^+-layer is thick around the grain boundary, only vary between 1 - 2 μm. The our results are consistent with Miura's assumption.

4. Conclusions

We have measured microscopic distribution of Schottky barrier heights (SBHs) at epitaxial-Al/n-Si(111) interfaces by using scanning internal-photoemission microscopy. The Si surface after removing Al has been investigated by using optical micrograph and AFM. In the as-deposited sample, the distribution of SBH at the interface was homogeneous, and hence the SBHs determined by conventional *I-V* and *C-V* methods agreed well. After annealing, the inhomogeneity of SBH was observed at the interface, and it could cause the discrepancy between $q\phi_B^{I\text{-}V}$ and $q\phi_B^{C\text{-}V}$. A recrystallized p^+-Si layer was mainly forms in the vicinity of grain boundaries during annealing process, and its thickness variation causes local fluctuation of SBH.

Acknowledgments

The authors are grateful to H. Okabayashi and K. Aizawa for helping to fabricate the present samples and express hers gratitude to C. Kaneshiro for helping to measure the AFM image. This work was partly supported by the JSPS Japan.

Refernces

Basterfield J, Shannon J M, Gill A 1975 Solid-st. Electron 18, 290

Chino K 1973 Solid-St. Electron. 16, 119

Miura Y, Hirose K, Aizawa K, Ikarashi N, Okabayashi H 1992 Appl. Phys. Lett. 61, 1057

Miyazaki S, Okumura T, Miura Y, Aizawa K, Hirose K 1994 Inst. Phys. Conf. Ser. No. 135, p.361 (IOP, Briston), presented at DRIP 5

Ohdomari I, Tu K N 1980 J. Appl. Phys. 51, 3735

Shiojima K, Okumura T 1991 Jpn. J. Appl. Phys. 30, 2127

Sze S M 1985 Semiconductor devices Physics and Technology (Wiley, New York)

Tung R T 1992 Phys. Rev. B 45, 13509

Inst. Phys. Conf. Ser. No 149
Paper presented at DRIP 95

Fine structure observed in the thermal emission process for defects in semiconductors

L. Dobaczewski and M. Surma

Institute of Physics, Polish Academy of Sciences
al. Lotników 32/46, 02-668 Warsaw, POLAND

Abstract. High-resolution Laplace-transform DLTS technique has been used to study the influence of small disturbances on the carrier emission process for transition metal- and thermal donors-related defects in silicon. For the iron-boron pair in the p-type silicon, two different configurations of the defect have been observed: stable and metastable. For both of them the influence of the magnetic field on the hole emission is demonstrated and a possible microscopic structure of the metastable defect configuration is discussed. Due to the resolution of the method it was also possible to demonstrate a complex character of the electron emission for two charge states of thermal donors in the n-type silicon and observe the influence of the magnetic field on the process.

1. Introduction

Although the standard Deep Level Transient Spectroscopy (DLTS) gained an unquestioned position as a tool for semiconducting material characterisation, it still has limited abilities for the identification of the defect. A price to pay for the very high sensitivity offered by this technique is its rather poor resolution. Local defect environment sensitive methods have to be able to discriminate interactions between a defect and its environment of the order of milielectronvolts or less in the energy scale. The defect energy is modified in this energy range by, e.g., magnetic field, uniaxial stress, or crystal disorder. Only the electric field can change the defect emission rate so much that it is easily observed in the standard DLTS experiments as the so-called Poole-Frenkel or phonon-assisted tunnelling effects. The other interactions mentioned above are almost undetectable by the standard DLTS.

It has been already demonstrated that the resolution of DLTS can be substantially increased when the so-called Laplace transform DLTS is used [1]. This method allows the observation of the influence of small disturbances on the process of carrier thermal emission. As a result, in some cases, it provides a new insight into defect microscopy or, alternatively, makes possible to distinguish defects with slightly different emission characteristics. A good example of the alloy disorder influence on the electron emission are DX centres in III-V semiconductor alloys. In this case, it was possible to compare how the alloy disorder affects the emission process when the central defect atom sits in a different crystal sublattice, and thus, to determine the defect microscopic structure [2]. It was found that the electron emission from the well-known EL2 centre in GaAs is also modified by the defect environment, similarly to the case of the DX centres. However, in this case it was not the crystal itself which disturbs the emission but other defect centres which presumably accompany EL2. As a result, the electron emission observed by the standard DLTS, and commonly associated with this defect, is in reality a convolution of processes having different spatial and capture characteristics [3].

Transition metal (TM) impurities are among the most common contaminants in silicon and they have received a growing interest from a point of view of high-quality Si-crystal production [4]. Besides an initial residual contamination of as-grown crystals, these metals influence the device performance if introduced unintentionally during the silicon crystallisation process. DLTS is generally used for the TM detection and TM-related defect characterisation. However, in silicon there is a number of other defects which may have very similar emission characteristics as the TM-related defects. For example, in p-type boron-doped silicon there are many defects which have an activation energy between 0.18 and 0.25eV for the hole emission. Among them the most typical are: divacancy, boron-vacancy, and boron-oxygen complexes. These defects produce DLTS peaks which usually overlap with the peaks related to some transition metal impurities, e.g., nickel or cobalt. For the peaks where such a coincidence may occur, the Laplace DLTS, indeed, reveals a fine structure in the hole emission process, in contrast to the standard DLTS peaks where this coincidence is not expected, the Laplace method showed no structure. This coincidence between the TM- and other defect-related DLTS peaks can be the reason for some scatter in the transition metal defect parameters reported in the literature [5].

The problem of the low resolution of the DLTS technique is even more severe for the case of thermal donors (TD) in n-type silicon. It has already been well established that these defects always form a family [6]. In our previous study, we have demonstrated that when the thermal donor family is at early stage of formation (low concentration of defects) the Laplace DLTS technique can resolve its particular components [7]. For higher concentrations, a large number of defect configurations within the family makes it impossible for the Laplace method to deconvolute the DLTS signal, and show broad featureless peaks in the spectra similarly to the standard method.

In this work, the Laplace DLTS measurements of the iron-boron pair and thermal donors related defects when the Si samples are placed in a magnetic field are presented. The results demonstrate that the method is able to observe small changes in the carrier emission process produced by the field. In general, the magnetic field affects both the initial and the final states of the transition. In the p-type silicon, where the holes are heavier and less mobile than the electrons in n-type silicon, formation of the Landau subbands is much less effective. Consequently, the Zeeman splitting effect for the initial state contributes to the overall changes in the process in a similar magnitude as the modification of the valence band density of states. However, in the n-type silicon the changes in the emission process come predominantly from the changes in the conduction band (CB), thus, in this case the results for the thermal donors are less conclusive. When in the same sample in a magnetic field two different ionisation processes are observed one can minimise a participation of the unknown changes in the conduction or valence bands, by taking into account only the difference between the energies of these two defects. This approach allowed us to speculate about the effective g-factor for the metastable configuration of the iron-boron pair.

2. Experimental details, results and discussion

The samples used in our study were grown by the vapour-phase epitaxy on p^+- or p-type (boron doped) substrates. The active layer was lightly doped with boron and covered by the n^+-type layer to form the p-n junction. Prior to the growth process, the back surface of the substrate was intentionally covered by the 4N purity TM Fe-foil. For the studies of the thermal donor fine structure, P-doped floating zone (FZ) silicon crystals have been used. The silicon for TD studies was not subjected to any heat treatment, thus only residual content of the thermal donors generated during crystal cooling process was observed (concentration less than $10^{12}cm^{-3}$). In this case, Au-Schottky diodes have been fabricated on top of the layer.

It has already been well-established that iron forms numerous complexes and the iron atom itself forms multiple energy states associated with different charge states in the silicon band gap. The most well-known defect complexes iron forms with shallow acceptors [8-10]. The main feature of these complexes is that they are easily formed and dissociated [9-12]. A stable configuration of iron in the p-type silicon is the iron-acceptor pair. Its formation is driven by an electrostatic attraction of the two constituents of the defect. In the boron-doped silicon the iron-boron pair forms the very characteristic 0.1eV energy level (above the valence band). The Electron Spin Resonance (ESR) measurements revealed that the symmetry of this pair is trigonal with the interstitial position of iron next to the substitutional boron [13]. This pair, and consequently, this energy state, can be easily annihilated by a short period of annealing at approximately 200°C [11]. Alternatively, this pair also dissociates at low temperatures in regions where an intensive electron-hole recombination process occurs [12]. After the dissociation process a much deeper (0.43eV above the valence band) energy state appears in the band gap. This state is associated with the isolated interstitial iron atom.

For the case of the iron-aluminium pair it was evidenced that there is an intermediate state in the dissociation process. The iron atom before it leaves the immobile aluminium acceptor takes the position of the second-nearest neighbour (along the <100> direction). In this interstitial position it forms another pair with the orthorhombic symmetry [10]. This metastable configuration forms in the band gap an energy level shallower than that for the stable configuration.

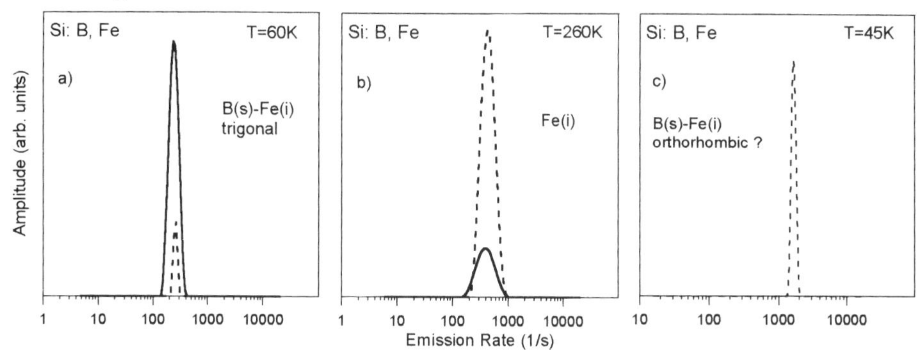

Fig. 1 Laplace DLTS spectra of three Fe-related defects in B-doped silicon. Solid lines show the spectra when the defects are in their inital (stable) configurations, dashed lines - after the electron injection to the space charge region. a) Fe_i-B_s pair (trigonal), b) isolated Fe_i, and c) Fe_i-B_s pair (orthorhombic).

Fig.1a shows the Laplace DLTS spectrum of the Fe_i-B_s pair in the stable configuration: only one peak is present in the spectrum. This peak decreased in size (dashed line) when the spectrum was taken after electron injection into the p-n junction. The electron-hole recombination process occurring in the junction space charge region causes the pair to dissociate, and, as a result, a new energy levels appear in the band gap. At much higher temperatures one observes a small peak which after electron injection increases (Fig. 1b). This new peak we associate with the energy level of the isolated iron atom in the interstitial position. At temperatures lower than 55K for moderate electron injection, an additional peak appears (Fig.

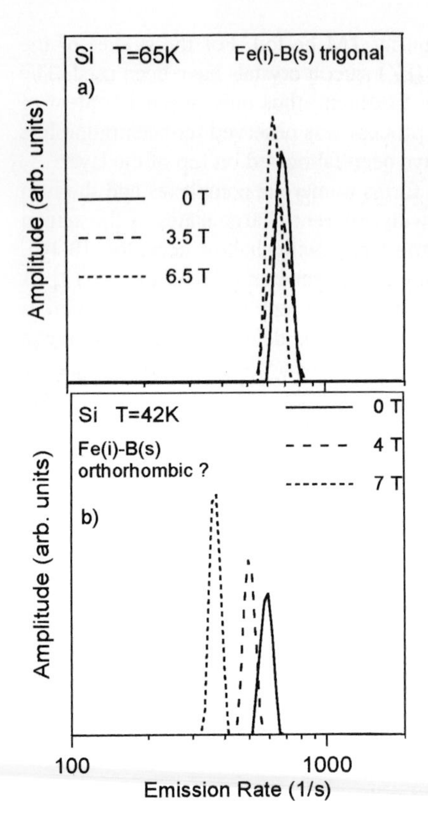

Fig. 2 Laplace DLTS spectra of the stable a) and metastable b) configurations of the Fe_i-B_s pair in different magnetic fields.

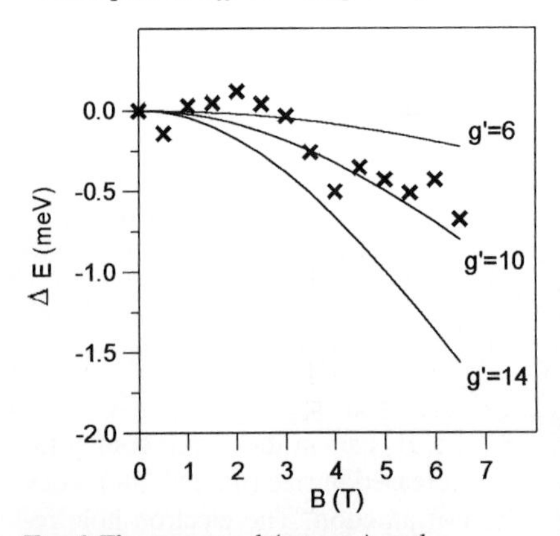

Fig. 3 The measured (crosses) and calculated (lines) relative shifts of the trigonal and orthorhombic iron-boron pair energy levels. Different effective g-factors (g') for the spin equal to 1/2 were assumed.

1c). This new centre was not observed prior to injection (even in a residual concentration) and it can be completely annihilated by a long period of electron injection. The hole emission activation energy for this new defect is 0.074eV. Room temperature annealing of the sample for a few hours restores the initial amplitudes for all three peaks. The metastable appearance of the 0.074eV level and its obvious correlation with the presence of iron in the crystal can be, by an analogy to the case of the iron-aluminium pair, considered as a creation of the second-nearest neighbour configuration of the iron-boron pair (with the orthorhombic local symmetry). This pair configuration has not been reported in the electrical measurements yet, however, recently it has been observed in the ESR experiment [14].

Despite the fact that at least one of the peaks seen in Fig. 1 is related to a complex, the spectra show no structure. All of these centres behave as ideal point defects producing in the Laplace DLTS spectra sharp lines. These sharp lines and that in these two cases they are observed at relatively low temperatures provide very advantageous experimental conditions for investigation of the influence of the magnetic field on the emission process. From the ESR measurements performed on the iron-boron pair it is known that the spin of the defect ground state is 3/2. This should make the Zeeman effect relatively large and easily observed. Fig. 2 shows the effect of the magnetic field on the emission rate for both configurations of the iron-boron pair. For the stable configuration of the defect (Fig. 2a) the change of the emission rate with the magnetic field is less pronounced then it is for the metastable defect configuration (Fig. 2b).

In the magnetic field the initial state of the emission can split into sublevels according to the Zeeman effect. The occupancy between the sublevels is governed by the Boltzman distribution. As the temperatures, at which the DLTS experiments are carried out, are relatively high in DLTS spectra one sees only some average energy level shifting with the increasing field. Knowing the spin of the ground state, the zero-field splitting of the sublevels (equal to 2.7cm^{-1} [13]) and the temperature one can evaluate the position of this average energy level for a given splitting. The final state for both processes is a free hole in the valence band.

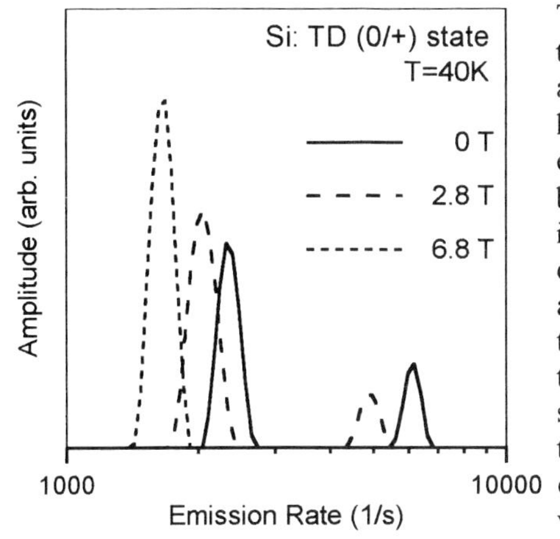

Fig. 4 Laplace DLTS spectra of the TD (0/+) energy state in different magnetic fields.

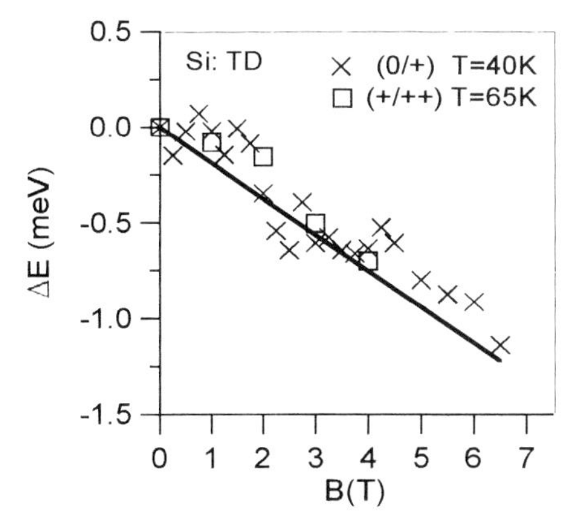

Fig. 5. The shift of the dominant peaks in the spectra for two charge states of TD is shown. The line demonstrates the defect energy shift which one could expect if all changes in the emission process were solely due to the shift of the first Landau subband in CB.

The band density of states can be modified by the field forming the Landau subbands. This is a very difficult process to evaluate. The free hole mobility (μ_h) in silicon is very low, so even for the highest magnetic fields, which can be achieved in our system (7T), a product $\mu_h B$ is less than 0.1. This would mean that the Landau level formation is very ineffective. In our approach, in order to disregard the influence of the field on the valence band, we have assumed that both states of the iron-boron pair interact similarly with the heavy and light hole bands. In this case one can subtract the shifts of both energy levels and analyse their relative motion with the field. Crosses in Fig. 3 show the result of such a subtraction.

The lines in Fig. 3 represent the difference between the average energy level of the trigonal iron-boron pair split by the magnetic field and a hypothetical level corresponding to the orthorhombic configuration of the pair. For the calculations different effective g-factors (g') and the effective defect spin equal to 1/2 were assumed. The experimental results are close to the calculated line if g' for the orthorhombic iron-boron pair approaches 10. It is obvious that the accuracy of this result is rather low and cannot be compared to the direct ESR measurements. However, one has to keep in mind that this particular experiment attempts for the first time to link the DLTS with magnetic measurements and to identify the defects successfully measured by the standard DLTS with those observed in the ESR experiments.

The metastability of the Fe_i-B_s pair may have some resemblance to the case of the DX centres in III-V materials. In DX the metastable configuration of the defect is formed by a bond-breaking process. Presumably, for the Fe_i-B_s pair the atom rearrangement is much more dramatic, i.e., it is associated with the pair constituents separation. This process may result in much higher energy barriers for this configuration formation than it was found for DX. The detailed analysis of the creation of the metastable configuration of the Fe_i-B_s pair will be presented elsewhere.

The Laplace DLTS spectra for the two charge states of the thermal donors in n-type silicon show two or three peaks [7]. Fig. 4 demonstrates the shift of the spectrum for the (0/+) level in the magnetic field and Fig. 5 shows the effect of the magnetic field on the dominant peaks in both spectra. In Fig. 5 the influence of the field on the emission is already recalculated onto the energy scale. Although both energy levels differ in the number of electrons bound by the centre, and consequently, in their spins, they move in a similar manner with the magnetic field. The line in Fig. 5 demonstrates what shift of the defect energy one could expect if all changes in the emission process were solely due to the shift of the first Landau subband in the conduction band. If the Zeeman effect considerably contributes to the total effect then this shift should be even more pronounced. From Fig. 5, one sees that there is no systematic deviation of the experimental points from the line. Consequently, if there is any participation of the Zeeman effect in the overall shift of the levels it must be hidden in the experimental error. It seems that the conduction band of silicon contributes to strongly to the total changes in the emission rate making the observations of the magnetic field effect on the defect ground state much more difficult than it is in the p-type crystals of silicon.

3. Summary

It is demonstrated that the high-resolution Laplace-transform DLTS technique offers sufficient resolution for studying the influence of small disturbances on the carrier emission process for the TM- and TD-related defects in silicon. For the iron-boron pair in p-type silicon two different configurations of the defect have been observed: stable and metastable and for both of them the influence of the magnetic field on the hole emission has been demonstrated. This metastable defect appears only when the stable configuration of the iron-boron pair dissociates and before the isolated interstitial iron centres are formed. This fact and a very close resemblance to the case of the metastable configuration of the iron-aluminium pair allows us to conclude that the metastable defect observed in this work can be identified with the iron-boron pair with the orthorhombic symmetry.

For the measurements of the carrier emission process in the magnetic field a crucial factor is how the modification of the band density of states participates in this process. It was found that for n-type silicon the Landau subbands formation is very strong and it obscures the effect of the field on the defect ground state. However, in p-type silicon the influence of the field on the defects ground state can be observed by Laplace DLTS.

The authors would like to thank Drs P. Kamiński and R. Kozłowski for providing the samples. This work has been financially supported by the State Committee for Scientific Research grant No 8 P302 117 05 and the EC grant No CIPA-CT94-0172.

References

[1] Dobaczewski L, Kaczor P, Hawkins I D, and Peaker A R 1994 *J. Appl. Phys*, **76**, 194
[2] Dobaczewski L, Kaczor P, Missous M, Peaker A R, and Żytkiewicz Z R 1992 *Phys. Rev. Lett.* **68**, 2508; 1995 *J. Appl. Phys.* **78**, 2468
[3] Dobaczewski L, Kaczor P, and Peaker A R 1993 *Mat. Science Forum* Vols 143-147, p. 1001
[4] A review of the TM defects in silicon if given in: Weber E R 1983 *Appl. Phys. A*, **30**, 1
[5] see, e.g., in *Landolt-Börnstein* New Series III/22b, ed. Fumino Shimura, (Academic Press Inc., Boston, 1994, Chapt. 4.2.3, p. 270
[6] Michel J and Kimerling L C, in *Semiconductors and Semimetals*, ed. O Madelung and M Shulz (Springer-Verlag, Berlin, 1989), Vol. 42, p. 257
[7] Dobaczewski L, Kamiński P, Kozłowski R, and Surma M 1995, *Proceedings of the 18th Conference on the Defects in Semiconductors*, Sendai, (in print)
[8] Wünstel K, Froehner K -H, and Wagner P 1983, *Physica*, **116B**, 301
[9] Kimerling L C and Benton J L1983, *Physica*, **116B**, 297
[10] Chantre A and Bois D1985, *Phys. Rev.* B, **31**, 7979
[11] Gao X, Mollenkopf H, and Yee S1991, *Appl. Phys. Lett.*, **59**, 2133
[12] Nakashima H, Sadoh T, and Tsurushima T1993, *J. Appl. Phys.*, **73**, 2803
[13] Gehlhoff W and Segsa K S, 1983 *phys. stat. sol* (b), **115**, 443
[14] Sakauchi S, Suezawa M, and Sumino K 1995 *Proceedings of the 18th Conference on the Defects in Semiconductors*, Sendai, (in print)

Inst. Phys. Conf. Ser. No 149
Paper presented at DRIP 95

Recombination activity of oxygen precipitation related defects in Si

W Seifert[1], M Kittler[1], J Vanhellemont[2], E Simoen[2], C Claeys[2] and F G Kirscht[3]

[1] Institut für Halbleiterphysik, W.-Korsing-Str. 2, D-15230 Frankfurt (Oder), Germany
[2] IMEC, Kapeldreeef 75, B-3001 Leuven, Belgium
[3] Siltec Silicon, POB 7748, Salem, Oregon 97303-0139, USA

Abstract. The recombination activity of oxygen precipitation related defects in n- and p-type Si is studied using electron beam induced current (EBIC) as method for both imaging of recombination sites and quantification of the defect recombination activity. A clear correlation is established between minority carrier diffusion length and defect density/recombination activity. The defects in as-processed samples exhibit low activity at 300 K and become strongly active upon cooling to 80 K, thus indicating that shallow levels control defect activity at temperatures below 300 K. Defects in n- and p-type material are found to behave in the same way. The EBIC results are correlated with those of complementary techniques.
Furthermore, it is shown that contamination by iron can dramatically increase the defect activity at 300 K.

1. Introduction

The impact of the amount and the state of oxygen on the electrical properties of Cz-Si has been the subject of extensive investigations performed by a variety of different characterization methods for more than thirty years already [1]. The electrical activity of precipitation related extended crystal defects is receiving a renewed interest recently due to the increased use of internal gettering in advanced silicon technology. Knowledge about the behaviour of such defects may be important also for bulk devices like solar cells as precipitation of oxygen can reduce the diffusion length and degrade the cell performance. Despite the large research efforts in the past the electrical activity of oxygen precipitation related defects in silicon is still far from being really understood. In particular, it is not clear whether the oxide precipitate, i.e. mainly the oxide/silicon interface (see [2]), or dislocations and stacking faults often surrounding the precipitate are causing the lifetime reduction. Moreover, many of the early experiments were probably not performed under the cleanest conditions. This complicates the picture because of the likely unintentional introduction of metals leading to decoration of the defects.
Results on the recombination activity of oxygen precipitation related defects formed under today´s processing standards are reported in the present paper. The presented EBIC results are part of a more extended study involving also structural analysis, photoluminescence, lifetime measurements and diode characterization. The influence of metal contamination on recombination activity is demonstrated as well.

2. Experimental

n^+p diodes were fabricated on p-type Cz-grown wafers with interstitial oxygen contents between $7 \cdot 10^{17}$ cm^{-3} and $11 \cdot 10^{17}$ cm^{-3}. Before the diode process different thermal treatments were performed. A set of n-type wafers of similar oxygen content received the same treatment as in the diode process but no diodes were fabricated on these substrates. A detailed description of the diode process can be found elsewhere [3].

Prior to EBIC investigations, the device structures were removed by etching (no mechanical polishing as this can produce artefacts, removed layer about 10 μm) and blocking Schottky contacts were prepared by evaporation. A Cambridge 360 SEM equipped with an Oxford cold stage and a Matelect Induced Signal Monitor was used to visualize recombination active defects and to measure their recombination activity between 80 and 300 K. The beam energy was varied between 15 and 40 keV and the beam current was kept below 50 pA for low injection level. Measurements of EBIC contrasts were carried out on stored images. The minority carrier diffusion length was determined from the decay of the charge collection efficiency between 30 and 40 keV [4] averaging over an area of typically 0.1 ... 0.2 mm^2.

Data on material recombination properties were also obtained by complementary methods. So, the effective recombination lifetime in the surface near layer was obtained from an analysis of the I-V characteristics of the diodes fabricated on p-type wafers [3]. The minority carrier lifetime in the bulk of both n- and p-type wafers at moderate injection level was determined by 10 GHz microwave absorption (MWA) measurements of carrier decay after a 1.06 μm wavelength pulse of a Nd:YAG laser. About 50 μm of Si was removed from both sides so that the lifetime in the bulk of the wafer was addressed.

TEM was applied to characterize the bulk defects and to estimate their density and distribution. The TEM analyses were done on the Jeol 1250 microscope of the University of Antwerp with an accelerating voltage of 1000 kV.

Other characterization methods were photoluminescence spectroscopy at 4 K used to monitor the dislocation D bands, deep level transient spectroscopy to determine deep levels in the diodes and Fourier transform IR spectroscopy at room temperature to monitor the interstitial oxygen content (for more details of these techniques see [3, 5]).

Fe contamination experiments were carried out on another set of n-type samples. The samples received a precipitation anneal simulating a device process and were contaminated with Fe by the following procedure: Fe in-diffusion from the backside at 950°C for 15 min, fast quenching to room temperature and subsequent annealing at 300°C for 20 min in a N_2 ambient [6] .

3. Results and discussion

TEM reveals typical oxygen precipitation related defects after the diode process: precipitate/dislocation complexes, isolated plate-like SiO_x precipitates and also small bulk stacking faults. In all cases the density of precipitate/dislocation complexes is at least one order of magnitude higher than that of isolated precipitates and at least two orders of magnitude higher than the stacking fault density. The density of the defects as established by TEM varies between $1 \cdot 10^9$ cm^{-3} and a few 10^{10} cm^{-3}, depending on the annealing procedure and the initial oxygen content. The density of defects estimated from EBIC images at 20 keV is usually much lower (see Table 1). The reason of this systematic difference is not yet

Table 1: Main data of the samples investigated: interstitial oxygen content before and after processing, defect densities determined by TEM and EBIC, and lifetime data at T = 300 K as obtained from MWA, I-V characteristics of the fabricated diodes and EBIC diffusion length measurements. In the latter case, minority carrier diffusivities of $D_n = 35\ cm^2/s$ and $D_p = 12\ cm^2/s$ were taken for p- and n-type samples, respectively, to calculate the lifetime from the diffusion length.
(- value not determined, * diffusion length at the limit of EBIC technique)

sample, conduct. type	initial O_i ($10^{17}\ cm^{-3}$)	final O_i ($10^{17}\ cm^{-3}$)	N_{TEM} ($10^9\ cm^{-3}$)	N_{EBIC} ($10^9\ cm^{-3}$)	carrier lifetime (μs) MWA	I-V	EBIC
T1, p	9.3	9.0	1 - 2	0.5	4.2	1.6	0.7 - 1.8
T2, p	9.2	-	1 - 2	0.4	1.6	0.66	0.4 - 0.7
T3, p	9.2	8.7	1	0.1	3.2	7.2	≥ 3*
T26, p	11.1	3.0	50	6	0.62	0.062	0.2
T28, p	10.8	3.1	10	0.8	0.75	0.75	0.7
T11, n	10.2	6.7	10	2.5	1	-	> 5*
T12, n	10.0	6.8	10	1.7	1.2	-	≥ 8*
T13, n	9.9	5.8	3	0.5	1.1	-	≥ 8*

clear. Inhomogeneous defect distribution and differences in electrical activity are possible explanations.

Table 1 shows also room temperature lifetime data obtained by 3 different methods. The agreement is remarkably good taking into account the differences in the injection conditions and in the volume probed. Fig.1 compares diffusion lengths of p-type samples obtained from diode characteristics and EBIC measurements. Apart from sample T3 which has a diffusion length too large to be accurately determined by EBIC the data fit fairly well. The reason of this agreement is that the underlying processes are similar for both methods. In case of diode measurements the diffusion component of the diode current, J_d, is first determined by a careful analysis, correcting for both perimeter effects and diode ideality [3]. Recombination lifetime τ_r and diffusion length L can then be estimated using the relation

$$J_d \approx q\sqrt{\frac{D_n}{\tau_r}}\frac{n_i^2}{N_A} = q\frac{D_n}{L}\frac{n_i^2}{N_A} \qquad (1),$$

q being the electron charge, D_n the minority carrier (electron) diffusivity, n_i the intrinsic carrier concentration and N_A the dopant(acceptor) concentration. L and τ_r are effective values

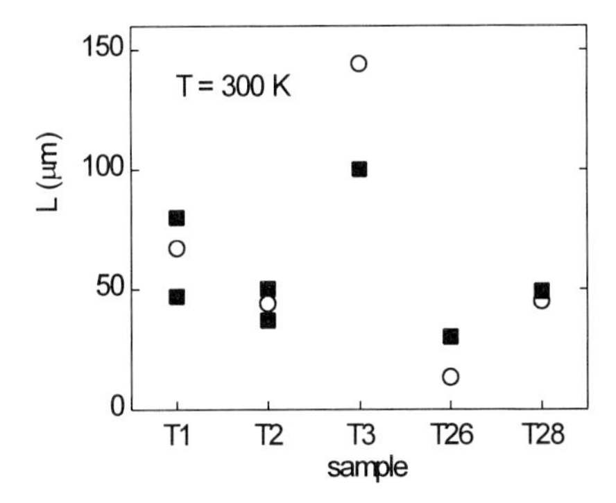

Fig. 1: Comparison of diffusion lengths at T = 300 K obtained from diode I-V measurements (o) and EBIC (■)

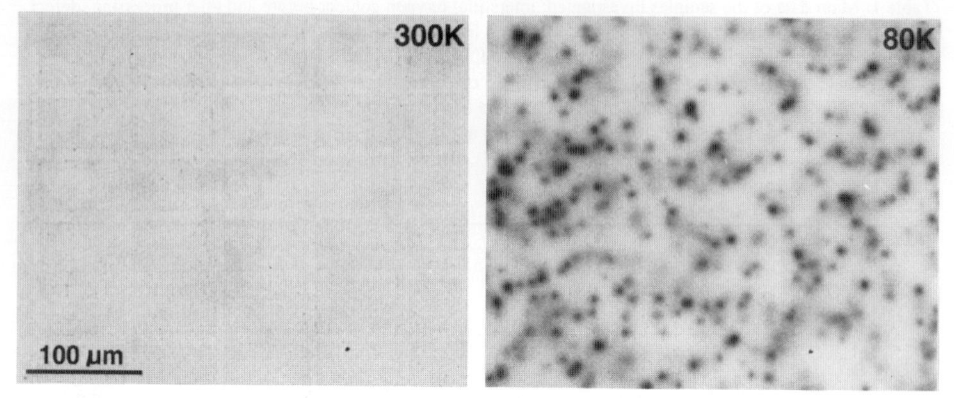

Fig. 2: EBIC micrographs (E_o = 30 keV) of a sample with oxygen precipitation related defects at 300 K and 80 K. The defects are nearly inactive at room temperature. The density of defects seen in the EBIC image is approximately $2 \cdot 10^5$ cm^{-2}. The information depth for defect imaging is about 6 µm.

of diffusion length and lifetime which take account of the depth profile of recombination properties in the material near to the junction. The EBIC current depends, essentially, on the same effective diffusion length. This similarity even allows, at least for high densities of defects, to judge about the diode leakage on the basis of EBIC measurements [7].

Fig. 2 shows EBIC micrographs of sample T1. At room temperature the defects are invisible in EBIC while pronounced recombination activity is evident at 80 K. Similar behaviour was observed for all samples investigated, irrespective of defect density and conductivity type. In agreement with the temperature dependence of EBIC contrast, the diffusion length drops rapidly upon cooling. The diffusion length of the material L can be considered to consist of two components - a component related to extended defects, L_{def}, and a background component, L_b, due to point defects or extended defects of low activity/high density which cannot be spatially resolved by EBIC:

$$L^{-2} \approx L_{def}^{-2} + L_b^{-2} \qquad (2).$$

L_{def}^{-2} is defined by the density of extended defects and by their specific activity (EBIC contrast). Fig. 3 demonstrates that EBIC contrast and diffusion length have an identical

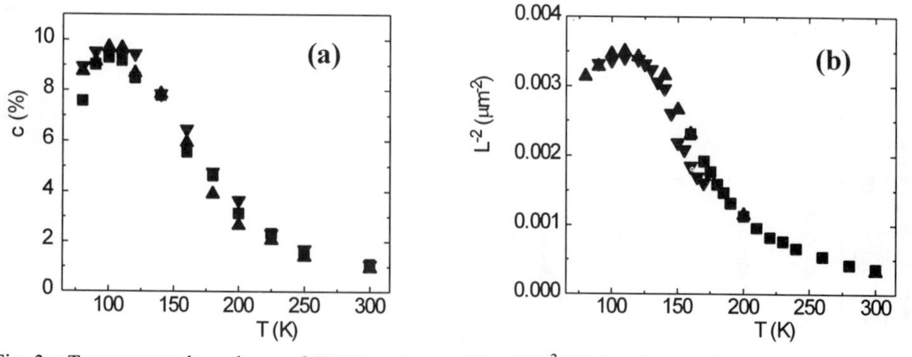

Fig. 3: Temperature dependence of EBIC contrast c (a) and L^{-2} (b) for the sample shown in Fig. 1. The strong similarity of the two curves proves that bulk recombination is controlled by the extended lattice defects.

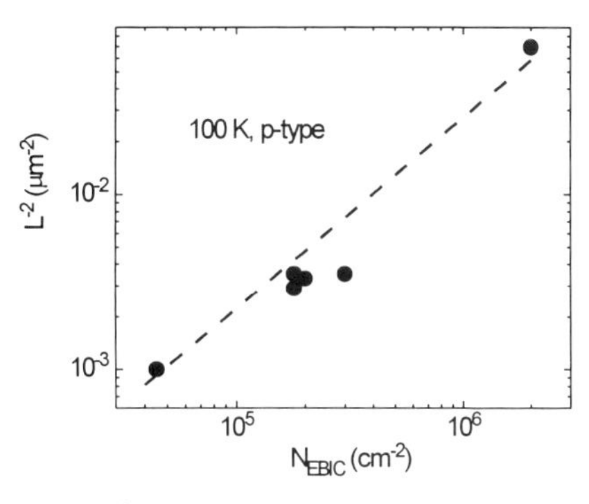

Fig. 4: L^{-2} vs defect density seen by low temperature EBIC. The dashed lines stands for proportionality between L^{-2} and N_{EBIC}

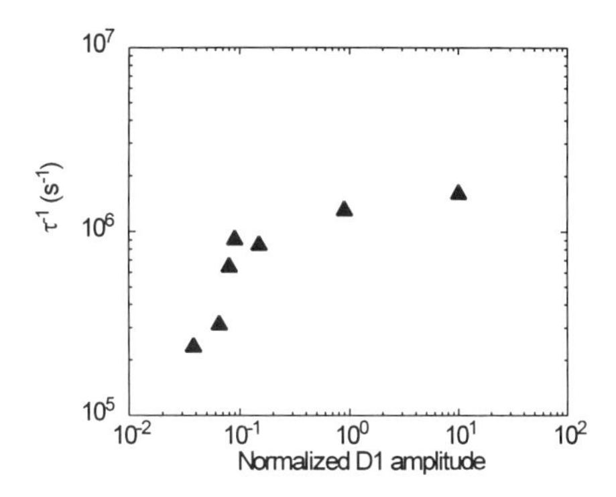

Fig. 5: Correlation between amplitude of D1 line and inverse lifetime at 300 K measured by MWA

temperature dependence, indicating that the extended defects dominate bulk recombination for $T < 300$ K. Another argument for the extended defects is the dependence of the diffusion length on the density of recombination active defects shown in Fig. 4. A nearly linear relationship is observed. Also at 300 K, where the recombination activity of the individual defects is low , a clear correlation between material recombination properties and density of oxygen precipitation related defects is established [5,6].

The type of contrast behaviour shown in Figs. 2 and 3 is characteristic of uncontaminated samples and points to clean processing conditions. The contrast increase at low temperatures indicates that, between 300 and 80 K, the recombination activity of oxygen precipitation related defects is controlled by rather shallow centres [8]. A determination of the exact position in the band gap is difficult, unfortunately, because a quantitative physical model is not available so far. The conclusion about shallow traps is in certain contradiction to DLTS investigations which revealed minority carrier traps at E_C - 0.43 eV and E_C - 0.23 eV [3]. One can, however, imagine that these traps may provide a substantial recombination path near room temperature while shallower centres not detected by DLTS take over at low temperatures.

Photoluminescence investigations of the samples reveal as most prominent features the well known dislocation related D1 and D2 lines at 807 and 874 meV, respectively [9]. Fig. 5 illustrates that there exists a correlation between the amplitude of the D lines and the inverse

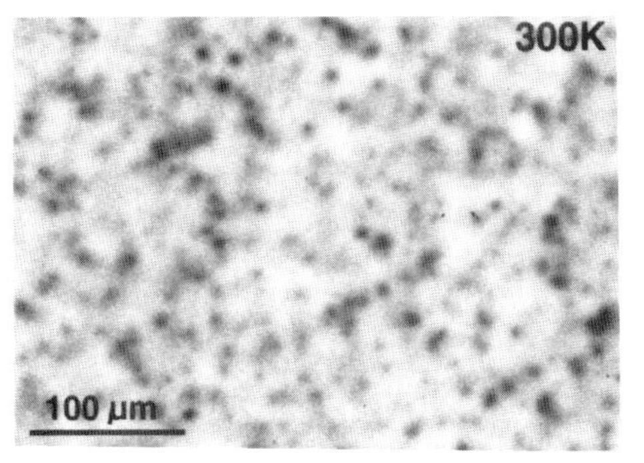

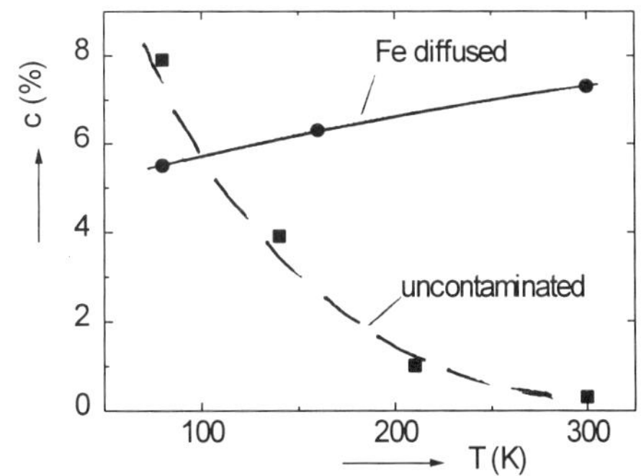

Fig.6 : EBIC micrograph (E_o = 30 keV) of oxygen precipitation related defects in a Fe-diffused sample and typical temperature dependence of defect contrast. For comparison the temperature dependence of the contrast in an uncontaminated sample is given, too.

of the lifetime. This suggests that the dislocations surrounding the oxide particles may be responsible for the lifetime reduction.

This conclusion is indirectly obtained also from the contrast behaviour of the oxygen precipitation related defects. Low activity at 300 K and strong activity at 80 K is found for the as-processed, ′clean′ samples. A completely different picture indicating deep centres is obtained after Fe-contamination (Fig. 6), with low diffusion lengths and defects active already at room temperature. The temperature dependence of recombination activity and the influence of contamination strongly resemble the behaviour of dislocations. According to a recent detailed study [10], ′clean′ dislocations are inactive at room temperature and exhibit large activity at 80 K while contaminated dislocations are active at 300 K and show a slight decrease of activity upon cooling. This is just what is found for the defects studied here.

4. Conclusions

It is shown that temperature dependent EBIC is a useful tool to study oxygen precipitation related defects. A clear correlation is established between minority carrier diffusion length and defect density/recombination activity.

It is found that the precipitation related defects have only low recombination activity at 300 K and become strongly active upon cooling to 80 K, thus indicating that shallow levels control defect activity at temperatures below 300 K. Defects in n- and p-type material are found to behave in the same way.

Furthermore, it is shown that contamination can strongly increase the impact of the precipitation related defects on lifetime/diffusion length.

Because of the similarity to the recombination behaviour of dislocations it is suggested that dislocations around the oxygen precipitates play the main part in recombination. This is in agreement with PL results. More detailed investigations on samples with different stages of dislocation formation, including precipitates without secondary defects, and with different contamination levels are required to clearly differentiate between the possible lifetime factors.

Acknowledgements

Two of the authors (W. S. and M. K.) acknowledge the support of parts of the work by the German Federal Ministry of Research under contract No. 0329536H.

References

[1] Bender H and Vanhellemont J 1994 *′Oxygen in Silicon′* in *Handbook on Semiconductors Vol. 3* ed. T.S. Moss and S. Mahajan (New York: Elsevier) 1637-1753
[2] Hwang J M and Schroder D K 1986 *J. Appl. Phys.* **59** 2476-2487
[3] Vanhellemont J, Simoen E, Kaniava A, Libezny M, and Claeys C 1995 *J. Appl. Phys.* **77** 5669-5676
[4] Kittler M, Lärz J, Morgenstern G and Seifert W 1991 *J. Phys. IV* **1** C6-173-179
[5] Vanhellemont J, Kaniava A, Libezny M, Simoen E, Kissinger G, Gaubas E, Claeys C and Clauws P 1995 *MRS Symp. Proc. Vol.* **378** 35-40
[6] Seifert W, Kittler M, Seibt M, Buczkowski A 1995 *Proc. Gadest′95, Solid-State Phenomena* **47-48** 365-370
[7] Kittler M and Seifert W 1989 *phys. stat. sol. (a)* **99** 559-572
[8] Kittler M and Seifert W 1994 *Mater. Sci. Eng. B* **24** 78-81
[9] Weronek K, Weber J and Queisser H J 1993 phys. stat. sol. (a) **137** 543-548
[10] Kittler M, Ulhaq-Bouillet C and Higgs V 1995 *J. Appl. Phys.* **78** 4573-4583

Inst. Phys. Conf. Ser. No 149
Paper presented at DRIP 95

Noncontact separate measurements of bulk lifetime and surface recombination velocities in silicon wafers with oxidized mirror, etched, sliced, and sandblasted surfaces

Y. Ogita, M. Minegishi, H. Higuma, Y. Shigeto* and K. Yakushiji*

Kanagawa Institute of Technology, Atsugi, Kanagawa 243–02 Japan
*Showa Denko K. K., Chichibu, Saitama 369–18 Japan

Abstract. The true pthotoconductivity decay curves as expected theoretically have been measured for Si wafers with various surfaces by UHF wave under the bias light illumination. The bulk lifetime, the front and back surface recombination velocities have been separately determined for the wafers with 620 μm thick for a practical use and with sliced, oxidized sliced, oxidized mirror, oxidized etched, and oxidized sandblasted surfaces. The effect of the bias light illumination has been considered experimentally from the illumination intensity dependencies.

1. Introduction

The minority carrier recombination lifetime is known to be very sensitive to the defects, heavy metal contaminations and surface properties in silicon crystals [1–5] . The recombination lifetime measurement is one of the most powerful technique to evaluate the crystal quality in silicon wafers and wafer processes for the device fabrication. Ogita has proposed a separation measurement method of the bulk lifetime and surface recombination velocities in the wafer with different surface recombination velocities at the front and back surfaces of the wafer, named as bi–surface photoconductivity decay(BSPCD)[6–7]. The method have separated successfully them for the sliced wafers with 0.7–2 mm thick[6] and poly–silicon extrinsic gettering Si wafers[8]. However, for the wafer with a thickness less than 0.7mm, the bulk lifetimes has been determined to be smaller. For the wafer with the oxidized surface, the bulk lifetime has not been determined exactly because of the decay curve with a round shape unexpected theoretically.

In this paper, we have examined the separation between the bulk lifetime and surface recombination velocities from the PCD curves measured under the bias light illumination employed newly for the wafer with 620μm for a practical use or 1.08mm. We have also examined it for the wafer having the combination the oxidized sliced–surface with the sliced–surface or the oxidized–sliced surface, and the combination the oxidized mirror–surface with oxidized sandblasted–surface or the oxidized etched–surface. Further, the effect of the bias light illumination has been considered experimentally

2. *Measurement method and samples*

The sequence for the separation by BSPCD method is illustrated in Fig. 2. The wafer model for the BSPCD is shown in top of the figure. Where, the surface recombination velocity in the front surface (S_0–surface) and the back surface (S_w–surface) of the Si wafer is denoted by S_0 and S_w, respectively. The bulk lifetime is denoted by τ_b. The respective PCD curve is measured for the pulse photoexcitation on respective surfaces. The inverse of the gradient of the linear tail decay gives an apparent (or effective) lifetime τ_a shown in medium in the figure. The normalized intercept $\sigma_{01}(0)/\sigma_0 0$ and $\sigma_{w1}(0)/\sigma_w(0)$ are given by the extrapolation to t=0 of the respective asymptote of the tail decay. S_0 and S_w are determined by solving the simultaneous equations obtained from the theoretical analysis[7]. τ_b is determined through the theoretical relation between S_0 , S_w and τ_a[7].

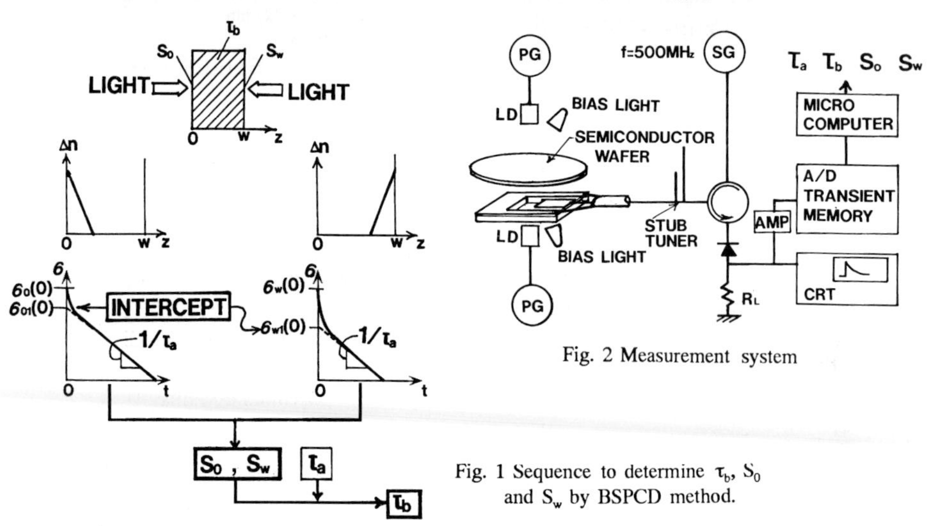

Fig. 2 Measurement system

Fig. 1 Sequence to determine τ_b, S_0 and S_w by BSPCD method.

The experimental scheme to measure photoconductivity decay curves is shown in Fig. 2. The wafer sample placed apart by about 1mm from the coplanar strip line to couple with the 500 MHz UHF wave supplied through the stub tuner and the circulator from the oscillator with a output power of 100 mW. Here, the lower frequency was used to prevent the skin effect. The sample was photoexcited to generate the excess carriers by the LD (LD–65, Laser Diode Lab.) with a optical power of several watts, a wave length of 904 nm and a pulse width of 60 ns from both sides of the sample. The sample was also illuminated by the lamps (Manabeam, Matshushita Dennsi. K.K.) with 60 Watts as the bias light from both sides. The lamp much included the wavelength longer 700Å. The change of reflective UHF wave induced by the photoconductivity change was received by the same slot line and detected by the diode, then amplified by 10–100 times by the amplifier(EX–31, NF Co., Ltd.) and A/D converted and eventually displayed on the computer monitor.

Two kinds of 5 inch P–Si–CZ wafers with 10 Ωcm, (100) plane, different thickness and various surfaces, sliced from the same ingot were employed in this experiment, as shown in Fig. 3 and 4. The wafer labeled as sample A owns a wafer thickness of 1.08mm and the S_0–surface with the oxidized sliced–surface and the S_w–surface whose half is the oxidized sliced–surface and the residual half is the sliced–surface obtained by removing the oxidized layer by the wet etching, as shown in Fig. 3. The oxide was produced at 1000 °c to be the thickness of 800Å. The another wafer labeled as sample B owns a thinner wafer thickness of 620 μm for a practical use and the S_0–surface with the oxidized mirror–surface and the S_w–surface whose half is the oxidized etched–surface and residual half is the oxidized sandblasted–surface, as shown in Fig. 4. The oxide was produced at 1100°c to be the thickness of 1000Å.

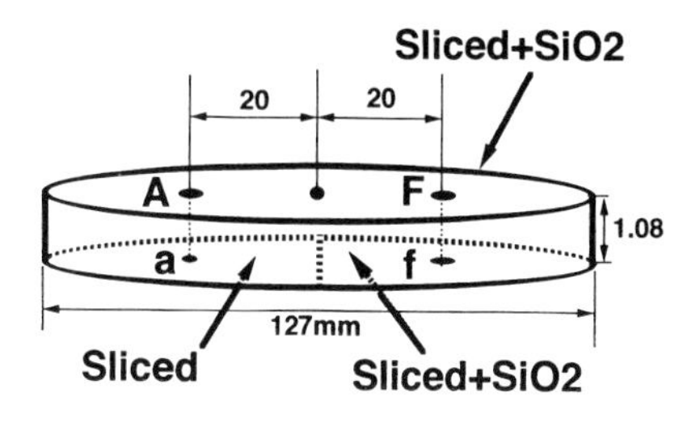

Fig. 3 Si wafer of sample A.

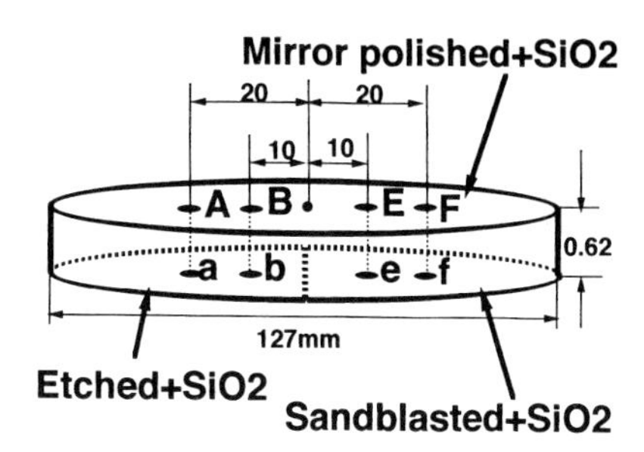

Fig. 4 Si wafer of sample B.

3. Experimental results and discussion

Fig. 5 and 6 show PCD curves measured at the points as A–a and F–f for sample A, respectively. The notation in the figures indicates the photoexcited surface; e.g., the curve notated as "Sliced+SiO_2" was measured for the photoexcitation of the oxidized sliced–surface. But, for only FIg. 6, F or f indicated in the figure denotes the surface photoexcited. Fig. 7 shows the typical example of S_0 dependencies of PCD calculated analytically, in the

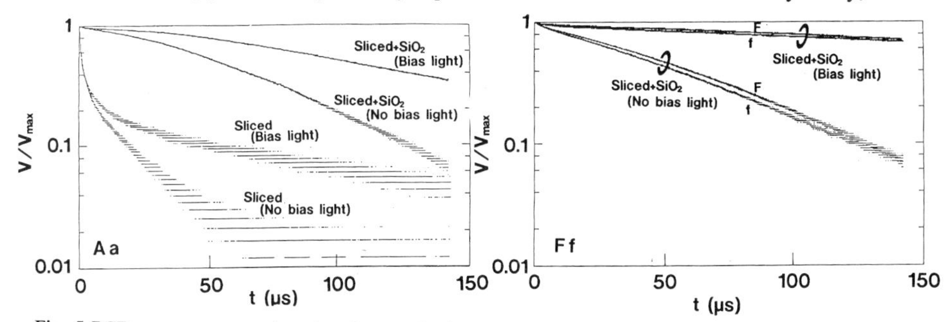

Fig. 5 PCD curves measured at A–a in sample A.

Fig. 6 PCD curves measured at F–f in sample A.

case that the S_0–surface is photoexcited[6]. The upper "sliced+SiO_2" curves in Fig. 5 seem to correspond to the curve for S_0 =1 cm/s. The lower "sliced" curves seem to correspond to that for the larger S_0 in Fig. 7. This means that S_0 in S_0–surface will be much smaller than S_w in S_w–surface. Because S_0 will decrease due to the existence of SiO_2. From Figs. 5 and 6, we can see that the bias light illumination make extremely slow the decay. In addition, it also make the round–shape–decay as seen in Fig. 5 and 6 linear shape as fit to the expectation derived theoretically shown in Fig. 7. Using the PCD curves under bias light, the separation result by BSPCD is shown in Table 1. Small S_0 at point A or F and small S_w at f are valid due to the existence of SiO_2. Large S_w at a is due to the slice surface. τ_b at A–a is very close to that at the point F–f. Further, it closes to the bulk lifetimes of 650μs determined for the wafer thickness of 0.7–2mm sliced from the same ingot[7].

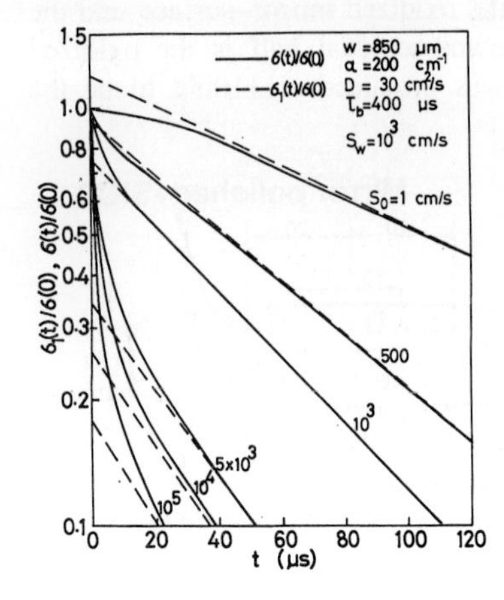

Fig. 7 PCD curves calculated analytically for various S_0.

	τ_{ao} (μs)	τ_{aw} (μs)	τ_a (μs)	τ_b (μs)	S_0 (cm/s)	S_w (cm/s)
Aa	111	115	113	624	18.4	6285
Ff	420	424	422	659	30.7	66.08

Table 1 τ_b, S_0 and S_w determined by BSPCD method for sample A.(τ_{a0} and τ_{aw} obtained for photoexcitaion of S_0–surface and S_w–surface, respectively.)

	τ_{ao} (μs)	τ_{aw} (μs)	τ_a (μs)	τ_b (μs)	S_0 (cm/s)	S_w (cm/s)
Aa	411	413	412	412	<1	<1
Bb	422	422	422	422	<1	<1
Ee	83.2	83.5	83.3	565	<1	1042
Ff	81.0	81.3	81.2	626	<1	1112

Table 2 τ_b, S_0 and S_w determined by BSPCD method for sample A.(τ_{a0} and τ_{aw} obtained for photoexcitaion of S_0–surface and S_w–surface, respectively.)

Fig. 8 shows the PCD curves measured at the points B–b for the sample B. Both decays under no bias light illumination seem to be a slightly round shape and to bend at the midpoint. Such curve does not estimated theoretically as shown in Fig. 7. On the other hand, the curves under the bias light illumination seem to be linear. The linear curves become consistent with the theoretical result. Thus, the separation can be made, the result determined from the curves is shown together the result at the another point A–a. S_0 and S_w are very small. Because the velocities are decreased due to the oxidation of the mirror or etched surface. The determined τ_b is smaller compared with the ingot lifetime of 650 μs. We think that it comes from the introduction of the contaminants in the oxidation process.

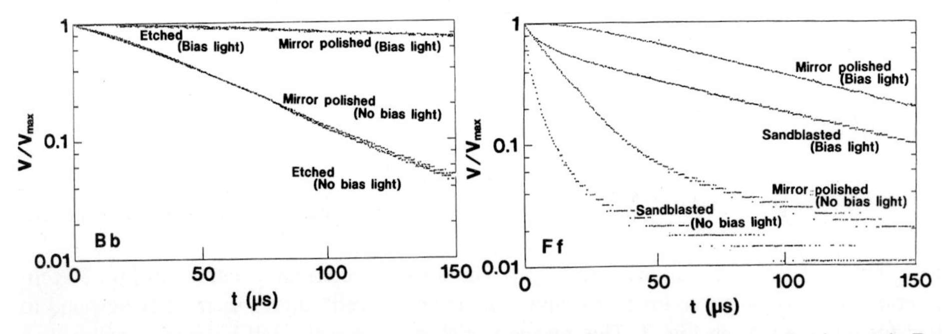

Fig. 8 PCD curves measured at B–b in sample B. Fig. 9 PCD curves measured at F–f in sample B.

Fig. 9 shows the PCD curves measured at the point F–f for the sample B. The curves under no bias light illumination decay fast in spite of the expectation of decreased S_0 due to the oxidation. However, under the bias light illumination, the curves decay as expected theoretically. Thus, we can determine them separately using the curves. The result is shown together the result at the another point E–e in Table 2. S_0s are small but S_ws are not small, in spite of the oxidized surface. This is because the S_w–surface is the sandblasted surface. Obtained τ_bs of 565 and 626μs close to the ingot lifetime of 650 μs. This is considered due to recovery of τ_b by gettering effect of the sandblasted surface through the oxidation process. τ_b at A–a and B–b in sample B is small. This can be explained by the reason that τ_b does not recovered because the oxidized mirror or etched surface scarcely does not yield the gettering action.

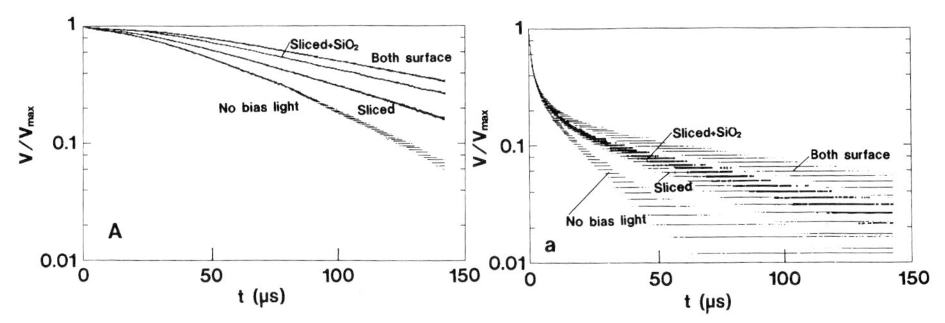

Fig. 10 PCD curves measured at A in sample A for different illumination surface of bias light. (Oxidized sliced–surface photoexcited)

Fig. 11 PCD curves measured at a in sample A for different illumination surface of bias light. (Sliced–surface photoexcited)

5. Discussion of bias light illumination effect

In order to clear the effect of the bias light illumination, some experiments were carried out. Fig. 10 and 11 show PCD curves measured for different illumination surfaces of the bias light to see the influence of the bias light on the surface. Fig. 10 shows PCD curves for

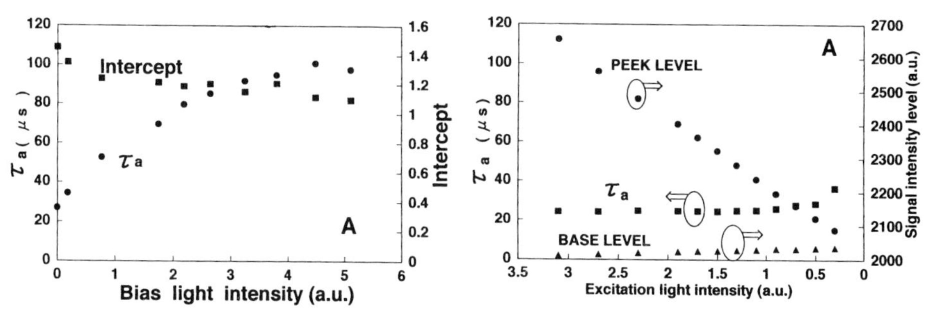

Fig. 12 τ_a and intercept as a function of bias light intensity.(At A in sample A)

Fig. 13 τ_a and injection level as a function of photoexcitaion intensity.

photoexcitation on the oxidized sliced–surface. Fig. 11 shows them for photoexcitation on the sliced–surface. As seen from the comparison between both figures, the effect of the bias light illumination on the oxidized sliced–surface is stronger than that in the illumination on the sliced–surface. This will indicates that the oxidized sliced–surface has more recombination centers compared with the sliced–surface. However, as seen from the both figure, the illumination on both surface is most effective.

On the other hand, it has been observed that stronger bias light intensity can bring the true PCD cures. Fig. 12 shows τ_a as a function of bias light intensity for photoexcited oxidized sliced–surface. We can see that τ_a is strongly dependent on the bias light intensity. However, the intercept does not show the pronounced variation with increase of the intensity. As seen in Fig. 7, the intercept is changed largely with increase of S_0. Thus, we can say that the bias light does not change S_0 or S_w dominantly. Saitoh et. al. showed that S_0 is decreased with a bias light intensity for a oxidized mirror–surface in solar devices[9]. Our result is not agreement with their result. This will come from the difference of the interface between their and our samples.

The intensity change of the bias light effectively changes the injection carrier level. Thus, the change of τ_a in Fig. 12 may be caused by a carrier injection level. So, τ_a was measured as a function of photoexcitation intensity taken to be the same intensity as the bias light intensity. The result is shown together the injection level(the peak level minus the base level) in Fig. 13. We see that τ_a scarcely does not change with carrier injection level. It has been confirmed that temperature rise(40°c) of the sample by the bias light illumination does not influenced τ_a and the surface recombination velocities[10].

Conclusions

τ_b, S_0 and S_w for the wafers with various surface has been determined using BSPCD from the PCD curves measured under the bias light illumination. The bias light illumination has been very effective to obtain true PCD curves and to separate them for the wafer having such as oxidized sliced, sliced oxidized mirror, oxidized etched, oxidized sandblasted surface. The BSPCD method with bias light illumination has determined τ_b, S_0 and S_w for the thinner wafer with thickness of 620 μm for a practical use. The effect of the bias light illumination has been experimentally considered. It has led that the bias light scarcely does not control the surface recombination velocity. τ_a depends on the bias light intensity but the intercept concerned with surface recombination velocities does not depend on it. The injection change induced by the bias light illumination scarcely does not cause changes of τ_a, S_0 and S_w. The temperature rise by the bias light illumination also does not influence τ_a, S_0 and S_w.

References

[1] A Y Liang and C J Varker 1980 Lifetime Factors in Silicon ASTM STP712 (Philadelphia, PA: ASTM)

[2] Y Ogita and H Takai 1982 Semiconductor World **1** 39–44

K Katayama and F Shimura 1992 Proc. Diagnostic Techniques for Semiconductor

[3] Materials and Devices **92–2** (Pennington NJ: The Electrochemical Society) 184

[4] M Hourai T Naridomi Y Oka K Murakami S Sumita N Fujino and T Shiraiwa 1988 Jpn. J. Appl. Phys. **27** L2361

[5] Y Hayamizu T Hamaguti S Ushio T Abe and F Shimura 1991 J. Appl. Phys. **69** 3077

[6] Y Ogita 1990 177th Electrochemical Society Meeting **90–1** (Pennington NJ: The electrochemical Society) 792

[7] Y Ogita J. Appl. Phys. 1995 (to be submitted)

[8] H Daio Y Uematsu and Y Ogita 1995 Extended Abstracts of the 42th Spring Meeting, The Japan Society of Applied Physics and Related Society **No. 2**(Tokyo: The Japan Society of Applied Physics)834(31a–H–7)

[9] H Nomura T Iga T Saitoh and T Uematsu 1993 Extended Abstracts of the 40th Spring Meeting, The Japan Society of Applied Physics and Related Society **No. 2**(Tokyo: The Japan Society of Applied Physics) 786(1a–ZC–1)

[10]Y Ogita S Uehara M Shigeto and K Yakushiji 1995 Extended Abstracts of the 42th Spring Meeting, The Japan Society of Applied Physics and Related Society **No.2**(Tokyo: The Japan Society of Applied Physics)600(26a–ZP–7)

Inst. Phys. Conf. Ser. No 149
Paper presented at DRIP 95

LIGHT SCATTERING TOMOGRAPHY STUDY OF LATTICE DEFECTS IN HIGH QUALITY AS-GROWN CZ SILICON WAFERS AND THEIR EVOLUTION DURING GATE OXIDATION

J. Vanhellemont[1], G. Kissinger[2], D. Gräf[3], K. Kenis[1], M. Depas[1], P. Mertens[1], U. Lambert[3], M. Heyns[1], C. Claeys[1], H. Richter[2] and P. Wagner[3]

[1]IMEC, Kapeldreef 75, B-3001 Leuven, Belgium
[2]IHP, P.O. Box 409, D-15204 Frankfurt (Oder), Germany
[3]Wacker Siltronic GmbH, P.O. Box 1140, D-84479 Burghausen, Germany

Abstract. Wafers from silicon ingots grown in the vacancy rich regime with different crystal cooling rates have been used for investigations by infrared light scattering tomography (IR-LST), preferential defect etching and gate oxide integrity (GOI) tests. GOI evaluation was done for 6.4 nm and 15 nm gate oxides. A clear correlation is obtained between substrate defects observed by LST and Secco etching after gate oxidation and GOI. In a second experiment the effect of intentional contamination with 10^{12} Fe cm^{-2} before gate oxidation is evaluated.

1. Introduction

Lattice defects in as-grown silicon can have an important impact on the gate oxide integrity in advanced submicron technologies [1-5]. Due to their low density and small size, few tools are available to study the nature of these defects which is therefore not yet fully understood. The grown-in defect distribution and type is known however to depend strongly on silicon crystal pulling conditions and has a strong impact on GOI as is shown in the present study.

2. Experimental

150 and 200 mm diameter p-type Cz wafers from crystals grown with different cooling ramps are used in the present study. In all cases however the ratio of pulling speed over thermal gradient was below the critical value ($\approx 1.3\text{x}10^{-3}$ cm^{2}min^{-1}K^{-1}) for stacking fault ring formation, i.e. in the "vacancy-rich regime" [6]. Wafers are labelled A,B,C and D with decreasing cooling rate [7]. The crystal defect density is studied with IR-LST using a MILSA IRHQ-2 instrument [8] and using non-agitated Secco etching for 30 min. This procedure reveals "flow pattern defects" (FPD's) which in literature are often associated with vacancy related clusters as their density decreases strongly for crystals grown in the "interstitial-rich" regime.

After cleaning, the 150 mm wafers are ramped-up in a 5% O_2/95% N_2 atmosphere in which also the 6.5 nm oxides are grown while the 15 nm gate oxides are grown in O_2, both at 900°C. In a second experiment 10^{12} cm^{-2} Fe contamination is introduced on the wafer surface by dipping in an iron spiked solution before the 15 nm gate oxidation. 1, 4 and 16 mm^2 capacitors are made by deposition of 500 nm in situ doped polycrystalline silicon followed by wet etching. Backside contacts are made by Al/Si sputtering and a 30 min sintering at 435°C.

3. Observations and Discussion

Infrared light scattering tomography

The density of grown-in defects (N_{LST}) larger than the size detection limit of LST is very low and a large volume has thus to be scanned in order to detect a significant number of them. In the present study cross-section samples are studied therefore using the low defect density mode whereby the 5 µm wide probing beam is scanned over a length of 2.3 mm while observing the whole wafer thickness. By this approach a volume of about 5.75×10^{-6} cm^3 is probed yielding a lowest defect density detection limit of 2×10^5 cm^{-3} for a single scan. Under these conditions several tens of defects are detected per scan in the as-grown wafers. The LST defect densities of "equivalent" as-grown wafers are larger for 150 mm wafers as compared to 200 mm (Fig. 1). In general there is a trend to have a decreasing LST defect density towards the centre of the wafers due to the decreasing cooling rate. The apparent different behaviour of the A1 wafer is probably due to the fact that the size of the inclusions decreases away from the centre and most of them are no longer observed by LST.

It is interesting to note that the effective scattering size (proportional to the scattering intensity to the power 1/6) and the number of LST defects increases after the 6.4 and 15 nm gate oxidation (Figures 2 and 3). The equivalent SiO_2 sizes in Fig. 2 and Table I are average values obtained on ten defects and are calculated from the effective scattering size using a calibration factor obtained from silicon substrates with known silicon oxide precipitate sizes. The figure illustrates also that before gate oxidation a correlation exists between N_{LST}, LST defect size and cooling rate: the lower the cooling rate, the lower the defect density and the larger their size. A similar behaviour is observed for the "crystal originated particles" (COP's) observed on the wafer surface with visible light scattering tools [7]. Both the data points for the as-grown wafers and after 15 nm gate oxidation can be fitted well with an exponential relation.

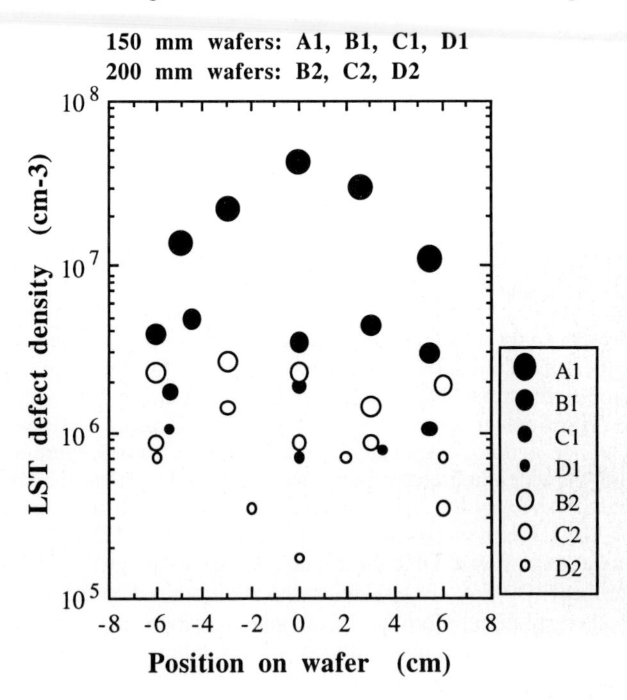

Fig. 1: LST defect density for 150 and 200 mm wafers.

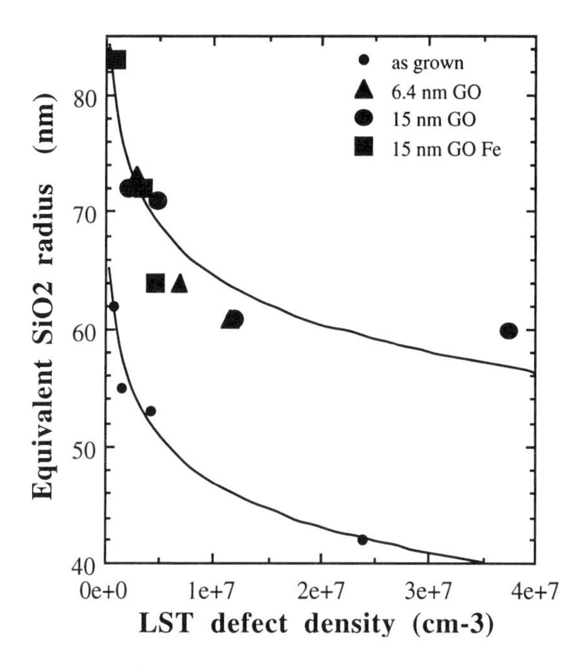

Fig. 2: LST defect size vs. LST defect density before and after gate oxidation.

The observed growth by the gate oxidation suggests that at least part of the observed LST defects are silicon oxide precipitates and that due to the size detection limit (estimated 20 nm equivalent SiO_2 radius) one observes only the tail of the precipitate size distribution. By the gate oxidation step this size distribution shifts to larger values leading to the apparent larger (observable) LST defect density. Further work is going on to study the size increase of the LST defects during prolonged thermal treatments in order to confirm (or not) that the growth kinetics is determined by interstitial oxygen diffusion. This approach will also allow to estimate the grown-in size distribution as a function of the pulling conditions.

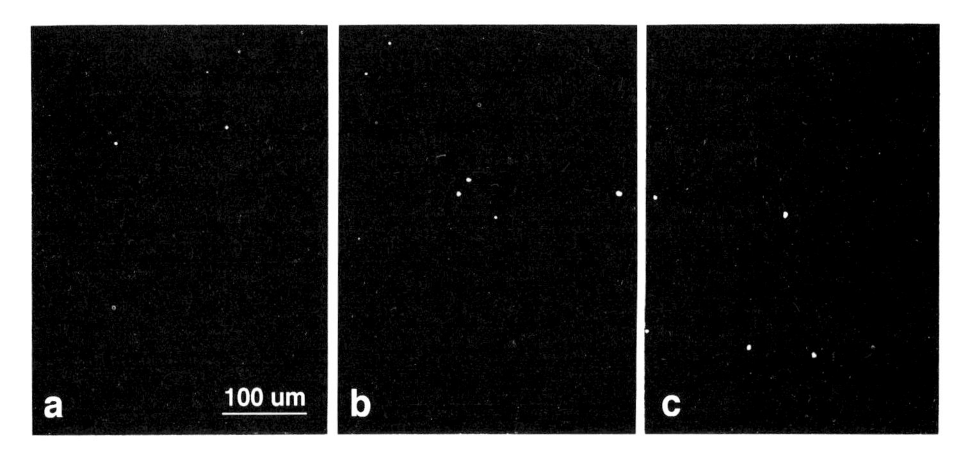

Fig. 3: Typical LST images of a B-type 150 mm wafer: (a) as grown and after 15 nm gate oxidation without (b) and with Fe contamination (c).

Table I. IR laser scattering intensity I_{LST} (arbitrary units) averaged for ten defects. The equivalent SiO_2 precipitate radius r_{LST} (nm). (- = no sample; no = not observed)

Wafer	A1	B1	C1	D1
I_{LST} as-grown	263	552	631	875
I_{LST} 6.4 nm	-	850	978	1460
I_{LST} 15 nm	822	866	1380	1436
I_{LST} 15 nm Fe	no	1003	1409	2146
r_{LST} as-grown	42	53	55	62
r_{LST} 6.4 nm	-	61	64	73
r_{LST} 15 nm	60	61	71	72
r_{LST} 15 nm Fe	no	64	72	83

Defect etching

FPD's have been studied both for the as-grown wafers and after gate oxidation using non-agitated Secco etching. Results for the as-grown wafers are summarised in Fig. 4. As illustrated in Fig. 5, both before and after gate oxidation the FPD density (N_{FPD}) correlates well with N_{LST} following approximately a linear relation. To reveal possible surface effects in some samples 20 μm was polished off before FPD etching. After gate oxidation N_{FPD} remains quasi constant in agreement with their reported high anneal out temperature (> 1100°C). The apparent low N_{FPD} for the highest N_{LST} values (A wafers) is probably again related with the small size of the clusters not yielding a FPD. Further experiments are ongoing to study the evolution of FPD defects after different gate oxidations and thermal pre-treatments.

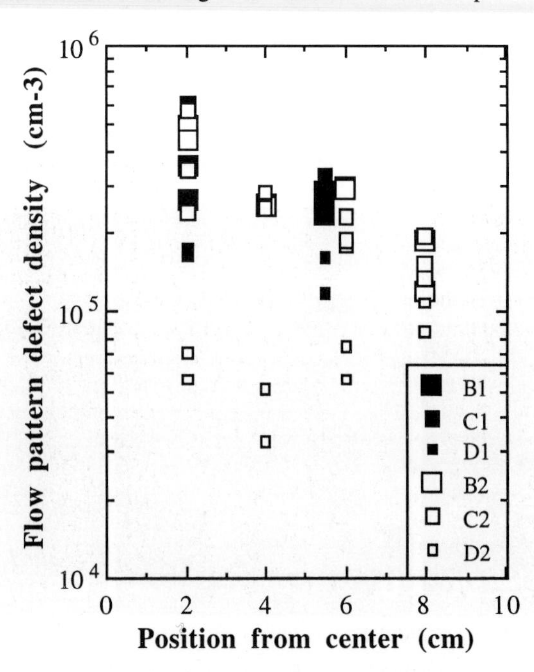

Fig. 4: FPD density before gate oxidation.

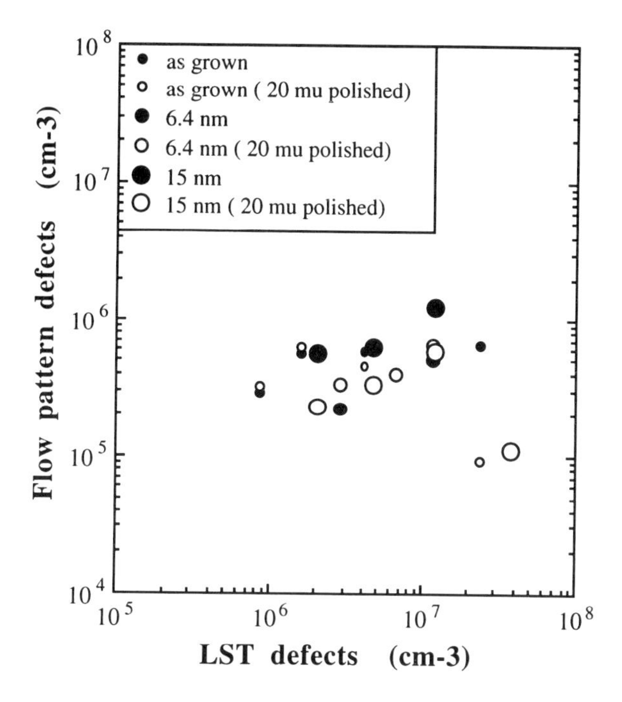

Fig. 5: FPD density before and after gate oxidation versus the LST defect density.

Gate oxide integrity

The gate oxide defect density (N_{GOD}) in the different substrates is given in Fig. 6, revealing a clear correlation with N_{LST} after gate oxidation both for the 6.4 and 15 nm gate oxides [9]. The gate oxide defect density increases thus with increasing cooling rate.

The breakdown field decrease with increasing oxide thickness is related with the presence of silicon oxide precipitates and D-defects in the substrate [4,10]. Up to about 25 nm oxide thickness, the decrease can be explained for Cz silicon mainly by the presence of silicon oxide precipitates which are incorporated in the gate oxide. The thicker the oxide grows, the more precipitates are encountered and the higher the gate oxide defect density becomes. Above a critical gate oxide thickness however, which will depend on the average silicon oxide precipitate size, the harmful effect of the incorporated precipitates decreases.

As the near surface defects have been removed by the polishing process leading to the shallow LPD's (≈ 3-4 nm) which are observed before cleaning, the oxide starts to encounter only the bulk defects after consuming a few nm of silicon, one of the reasons why the substrate effect is not yet very pronounced for the 6.5 nm oxide. For thicker gate oxides, up to 100 nm gate oxide thickness, a decrease of gate oxide yield is still observed which according to Park [4] is mainly due to the presence of D-defects which are probably large vacancy clusters.

Fig. 6 shows that the introduction of 10^{12} Fe cm^{-2} has a relatively small effect on the gate oxide defect density for the 15 nm gate oxides. For the A-type wafer the yield was however too low to extract reliable gate oxide defect densities which however should be considerably higher than the value for the uncontaminated wafers which are the average of two wafers of each type. The bulk lifetime of the wafers is on the contrary strongly affected by the iron contamination and drops after gate oxidation to a value of less than 1 µs. For the 15 nm gate oxidation, the substrate effect dominates thus over the one of iron contamination for the 10^{12} Fe cm^{-2} level used in the present experiment.

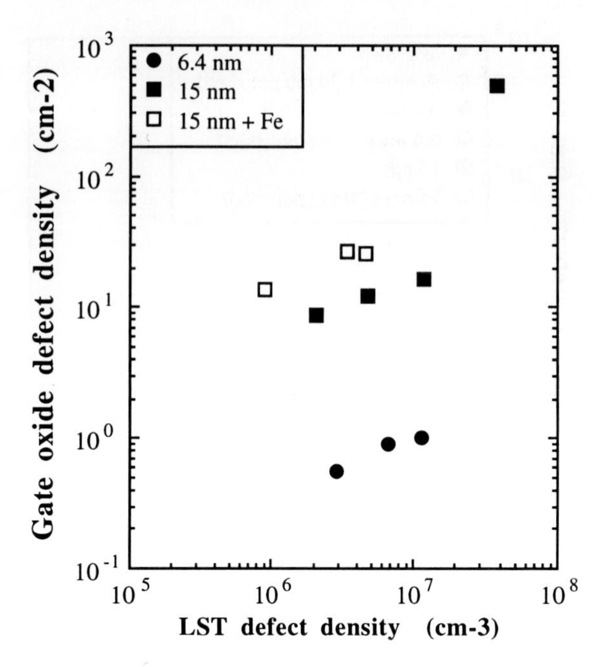

Fig. 6: N_{GOD} versus N_{LST} after oxidation.

Summary

A clear effect of the pulling conditions of CZ crystals on the gate oxide integrity was observed. A strong correlation exists between the density of silicon oxide precipitates observed by light scattering tomography and the gate oxide defect density. Growth of the LST defects as well as an apparent increase of density is observed after gate oxidation. Further work is needed to study the correlation between LST and GO defects and to establish the real nature of the LST defects: voids and/or silicon oxide precipitates? 10^{12} cm^{-2} Fe does not have a strong effect on 15 nm gate oxide integrity while the bulk carrier lifetime is reduced to about 1 µs.

References

[1] H. Yamagishi, I. Fusegawa, N. Fujimaki and M. Katayama, Semicond. Sci. Technol. **7**, A135 (1992).
[2] Y. Satoh, Y. Murakami, H. Furuya and T. Shingyouji, Appl. Phys. Lett. **64**, 303 (1992).
[3] S. Sadamitsu, S. Umeno, Y. Koike, M. Hourai, S. Sumita and T. Shigematsu, Jpn. J. Appl. Phys. **32**, 3675 (1993).
[4] J.-G. Park, S. Ushio, H. Takeno, K.-C. Cho, J.-K. Kim and G.A. Rozgonyi, The Electrochemical Society Proceedings Vol. **94-33**, 53 (1994).
[5] D. Zemke, P. Gerlach, W. Zulehner and K. Jacobs, J. Cryst. Growth **139**, 37 (1994).
[6] W. von Ammon, E. Dornberger, H. Oelkrug and H. Weidner, J. Cryst. Growth **151**, 273 (1995).
[7] M. Brohl, D. Gräf, P. Wagner, U. Lambert, H.A. Gerber and H. Piontek, The Electrochem. Soc. Extended Abstracts **94-2**, 619 (1994).
[8] G. Kissinger, J. Vanhellemont, D. Gräf, C. Claeys and H. Richter, these proceedings.
[9] J. Vanhellemont, G. Kissinger, D. Gräf, K. Kenis, M. Depas, P. Mertens, U. Lambert, M. Heyns, C. Claeys, H. Richter and P. Wagner, Proc. 18th Int. Conf. on Defects in Semiconductors (ICDS18), July 23-28, 1995, Sendai, Japan, in press.
[10] Y. Satoh, Y. Murakami, H. Furuya and T. Shingyouji, Appl.Phys.Lett. 64, 303 (1994).

Inst. Phys. Conf. Ser. No 149
Paper presented at DRIP 95

Detection of Interstitial Oxygen in CZ Silicon Wafers by Light Scattering

Nobuhito NANGO and Tomoya OGAWA*

Ratoc System Engineering Co.
The 1st Kirin Buldg.
540, Waseda-Tsurumakicho, Shinjuku-ku,
Tokyo, 162, Japan, FAX +81-3-3208-8201

* Department of Physics, Gakushuin University
Mejiro, Tokyo, 171, Japan, FAX +81-3-3590-2602

Abstract Residual and interstitial oxygen atoms in Cz silicon wafers are one of the most important factors requiring in silicon technology.
Light scattering technique was tentatively applied to the problem mentioned above.

1. ANALYTICAL INTRODUCTION

An optical beam passing through a material is determined by the total sum of radiation emitted from every dipole which is generated by the light illuminating the material. Moment of the dipole is given by displacement of electrons in the material, which is caused by the incident light.
The displacement x of a bound electron is obtained by the following equation of motion:

$$(1) \qquad d^2x/dt^2 + \gamma dx/dt + \Omega^2 x = (q/m)E_o e^{i\omega t},$$

where q, m, Ω, γ and $m\Omega^2 x$ are, respectively, the charge, mass, resonant angular frequency, damping factor and restoring force of the electron, and $E_o e^{i\omega t}$ is the electric field of the incident beam. Thus, the polarization due to the displaced electron is given by

$$(2) \qquad p = qx = (q^2/m)E_o e^{i\omega t}/(\Omega^2 - \omega^2 + i\gamma\omega).$$

From eq.(2) the electrons displaced by the electric field of an infrared beam will, therefore, be outershell electrons and/or electrons trapped in deep levels, because the energy of infrared radiations is estimated as about one electron volt by

hc/λ, where h is Planck constant, c the light velocity and λ is the wavelength. The outershell electrons mainly contribute to chemical bonds in a material whereas almost all electrons included in the materials will be polarized by x-rays because these photon energies are a few ten keV.

Interference of the radiation emitted from the two dipoles at positions A_1 and A_2 is given by

$$(3) \qquad F = \{ 1 + e^{i(2\pi/\lambda)(S\ r)\}},$$

where r is the vector indicating the position A_1 to A_2 and its magnitude is the distance between them. S is the unit vector to indicate the difference between unit vectors indicating the incident and scattered light, respectively. $(2\pi/\lambda)(S\ r)$ is, therefore, the phase difference between the radiations scattered at A_1 and A_2, respectively.

When more than two dipoles are generated by the incident light within a unit volume, amplitude caused by the dipole-radiations from all those dipoles is proportional to the factor:

$$(4) \qquad F = \sum e^{i(2\pi/\lambda)(S\ r)\}},$$

if all the dipolemoments induced by the light are equal. This is called a scattering factor.

When density of the electrons is given by $\rho(r)$, the scattering factor is given by

$$(5) \qquad F = \int_V \rho(r) e^{i(2\pi/\lambda)(S\ r)\}} dV,$$

where dV is a volume element located at r and V is volume of the specimen illuminated by the light.

If the electron density $\rho(r)$ is completely uniform throughout the inside of the material and thus given by ρ_o, the scattering factor is given by

$$(6) \qquad F = \rho_o \int_V e^{i(2\pi/\lambda)(S\ r)\}} dV$$

$$= (4\pi w^3 \rho_o/3)\Upsilon.$$

Here, w is the beam radius inside the illuminated material and Υ is given by

$$(7) \qquad \Upsilon = 3\{\sin z - z \cos z\}/z^3$$

where $z = 4\pi w \sin\theta/\lambda$ and θ is the angle between the incident and scattered light.

Since w is usually large enough compared with λ, w is substituted to infinity and then F has a finite value only when $\theta = 0$. This indicates that the incident light beam is propagating straight away without any scattering at all.

If the electron density deviates from ρ_o, it is then given by

$$(8) \qquad \rho(\mathbf{r}) = \rho_o + \Delta\rho(\mathbf{r}),$$

and the scattering factor is, therefore, given by

$$(9) \qquad F = \int_V \{\rho_o + \Delta\rho(\mathbf{r})\} e^{i(2\pi/\lambda)(S\ \mathbf{r})\}}\, dV$$
$$= \rho_o \int_V e^{i(2\pi/\lambda)(S\ \mathbf{r})\}}\, dV + \int_V \Delta\rho(\mathbf{r}) e^{i(2\pi/\lambda)(S\ \mathbf{r})\}}\, dV.$$

Here, the second term will be finite at $\theta \neq 0$ if w is large enough compared to λ.

Equation (9) indicates that non-uniform distribution of the electrons is the source of light scattering. Since distribution of interstitial atoms and vacancies in a Si wafer must be inhomogeneous, they act as the light scatterers and the light scattered by the atoms and vacancies will, therefore, be observed.

2. EXPERIMENTAL RESULTS AND DISCUSSION

To observe the light scattered by interstitial atoms, especially oxygen atoms in Si wafers, an IR beam of TEM_{00} mode from a Nd YAG laser with 1 watt was focused into a few microns inside the wafer and the scattered light was then clearly observed as shown in figs. 1 and 2 which were respectively obtained in a slow and normal pulling crystal. Similar pictures were obtained by denuded zones, which would relate to the quality of these zones. Measurements concerned with the interstitial oxygen atoms are still being refined as related with the other physical properties such as the absorbance of the atoms by FT-IR spectroscopy.

When light with a wavelength shorter than the absorption edge was used here, a sort of multiple reflection of an incident laser beam was observed in both the slowly and normally pulled crystals as well as in the denuded zones generated by the intrinsic gettering treatments, because of the scattered light is proportional to the fourth power of the light frequency. The multiple scattering is sometimes very effective to estimate density of the scatterers but is not simple to be handled analytically.

3. CONCLUSION

It was analytically and experimentally confirmed that interstitial atoms act as light scattering centers while the experimental results were not qualitatively discussed.

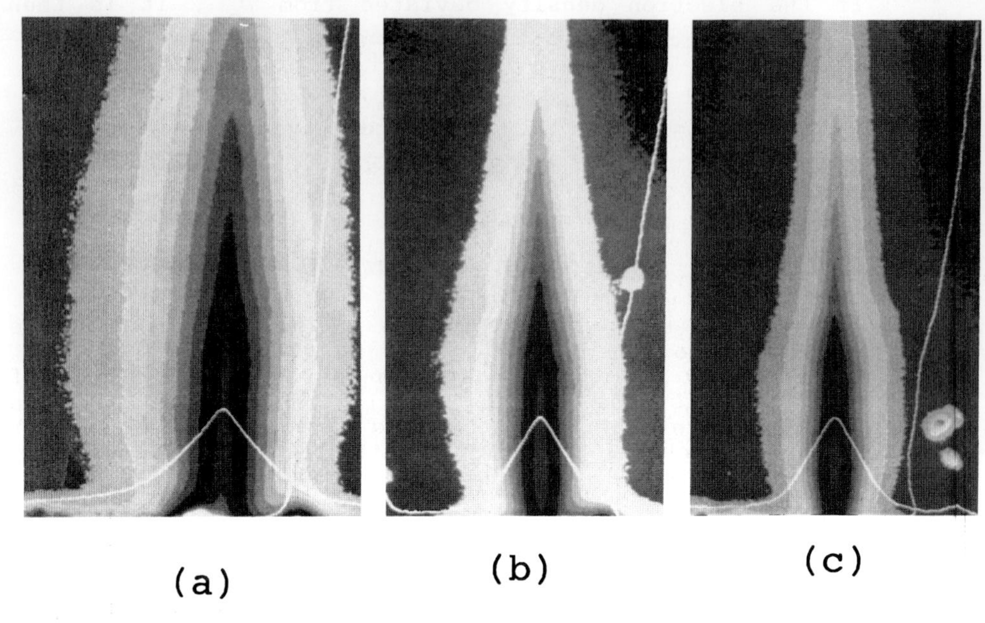

Fig.1
Scattered laser beam images in Cz-Si wafers which were illuminated by a TEM_{oo} mode of 1.06 μm radiation from a YAG laser where the beam was focused into 5 μm in diameter in the wafer.

(a) The image obtained from a highly concentrated wafer, where the interstitial oxygen atom concentration is estimated as 29.20 ppma by an FT-IR measurement. A half width of the scattered beam is measured as 102 pixels on the vidicon, which is a parameter to indicate magnitude of the scattering.

(b) The image obtained from a wafer with 24.17 ppma concentration of interstitial oxygen atoms. The width is measured as 76 pixels.

(c) The image obtained from a wafer with 14.14 ppma concentration, where the width is 64 pixels.

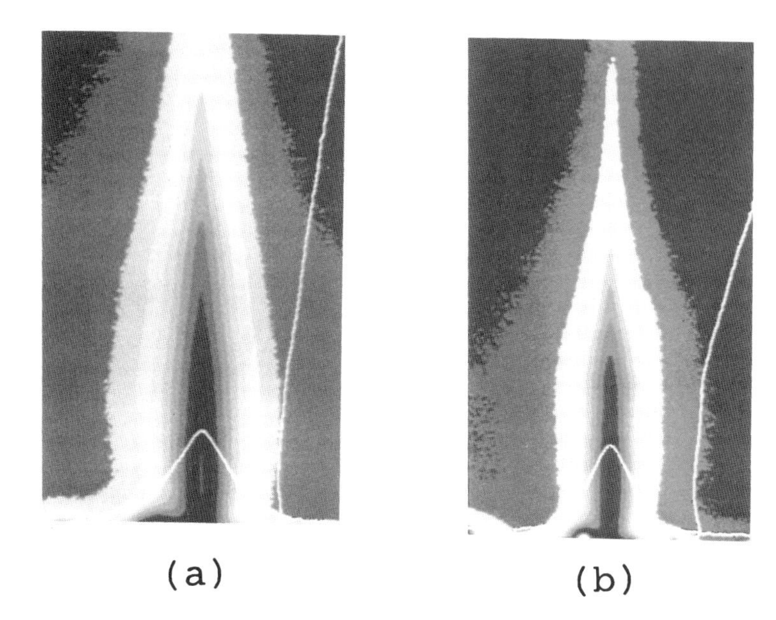

Fig.2
Scattered laser beam images in Cz-Si wafers, where the optical condition was very similar to the condition of fig.1

(a) The beam image obtained from a slowly pulled crystal.

(b) The beam image obtained from a crystal grown by a usually used condition.

Inst. Phys. Conf. Ser. No 149
Paper presented at DRIP 95

Possibilities of application of elastic mid-IR light scattering for inspection of internal gettering operations

O V Astafiev, A N Buzynin, V P Kalinushkin[1], D I Murin and V A Yuryev[2]

General Physics Institute of the Russian Academy of Sciences, 38, Vavilov Street, Moscow, GSP-1, 117942, Russia

Abstract. A method of low-angle mid-IR light scattering is shown to be applicable for the contactless and non-destructive inspection of the internal gettering process in CZ Si crystals. A classification of scattering inhomogeneities in initial crystals and crystals subjected to the getting process is presented.

1. Introduction

Recently, the internal gettering process became one of the main operations for manufacturing of semiconductor devices of CZ Si. However, methods for the direct inspection of the internal gettering efficiency and stability have been practically absent thus far. The purpose of this paper is to present such a method developed on the basis of law-angle IR-light scattering technique (LALS) [1], which has been successfully applied thus far for the investigation of large-scale electrically active defect accumulations (LSDAs) in semiconductor crystals (see e.g. [2] and references therein).

A method of LALS was applied at the first time to the investigation of the influence of both the internal and external gettering processes on large-scale impurity accumulations (LSIAs[3]) in crystals of the industrial CZ Si:B in Ref. [3]. The conclusions were made in Ref. [3] that (i) the external gettering process resulted in a considerable decrease of the impurity concentration in LSIAs and (ii) new defects arose in the crystal bulk as a result of the internal gettering process which became a predominating type of defects.

The current work presents an application of LALS with the non-equilibrium carrier photoexcitation [1, 4] to the studies of the internal gettering process in addition to the conventional LALS measurements.

[1] E-mail: VKALIN@LDPM.GPI.RU.
[2] E-mail: VYURYEV@LDPM.GPI.RU.
[3] LSIAs are a type of LSDAs which contain mainly impurities rather than intrinsic defects.

The LALS temperature dependances are also presented and the activation energies of the centers constituting the LSIAs are estimated in this work.

2. Experimental

A continuous 10.6-μm emission of a CO_2-laser was used as the source of the probe radiation in LALS. All the details of this technique are described in Refs. [1, 2]. We would like to remind only that such parameters of LSDAs as their effective sizes and the product of the LSDA concentration by the square of the deviation of the free carrier concentration (or the square of the dielectric constant variation) in LSDAs $C\Delta n_{ac}^2$ (or $C\Delta\varepsilon_{ac}^2$) can be calculated from the light-scattering diagrams.

The investigation of the influence of a sample temperature on its light scattering enables the estimation of the thermal activation energies (ΔE) of impurities and defects composing the LSDAs — $I_{sc} \sim \Delta n_{ac}^2$ [5]. During the low-temperature measurements, the sample temperature varied from 80 to 300K.

The influence of non-equilibrium carrier photoexcitation on light scattering was studied as well. The essence of the experiments consist in the following. If a crystal contains large-scale centers of recombination (e.g. precipitates and their colonies, stacking faults, swirls, *etc.*), regions with decreased concentration of non-equilibrium carriers are formed around these centers during the process of the non-equilibrium carrier generation. These regions scatter the light like usual nonuniformities and when pulse generation of carrier is used, the scattered light pulses are observed in LALS. Selecting this pulsed component, it is possible to register the light scattering by recombination defects (RDs). Then the usual procedure of the light-scattering diagram measurement and treatment is applied to estimate the dimensions of the depleted regions around RDs [1, 4]. In this work, the non-equilibrium carrier was generated by 40-ns pulses of $YAG:Nd^{3+}$-laser at the wavelength of 1.06 μm, frequency of 1 kHz and mean power of 1 W. The photoexcitation at this wavelength pumps whole the crystal bulk practically uniformly, as the absorption for this wavelength is not too high but sufficient to produce the efficient enough electron-hole pair generation. The scheme of the used instrument is described in detail in Refs. [1, 4].

Electron beam induced current (EBIC) and selective etching (SE) were used to reveal the defects as well. During the sample preparation for EBIC, a special technique was used which included the plasma etching of the sample surface in special regime before the Schottky barrier was created. This technique considerably increases EBIC sensitivity to RDs in bulk Si [6].

About 40 wafers of dislocation-free Si were studied. The crystals were produced by Czochralski method and doped with boron (CZ Si:B) up to the specific resistivity from 1 to 40 Ω cm. The crystals were produced at three different establishments and subjected to the internal gettering process at five different establishments. In this paper, the effect of different gettering regimes on defects is summarized. In the experiments on LALS, the two following schemes of experiment were used. In one scheme, a preliminary study of the as-grown substrates, which then were subjected to the internal gettering process and examined by LALS, was carried out. In the other scheme the substrates were cut into several parts. Some of these parts were subjected to the gettering process and the other parts were used as the reference samples.

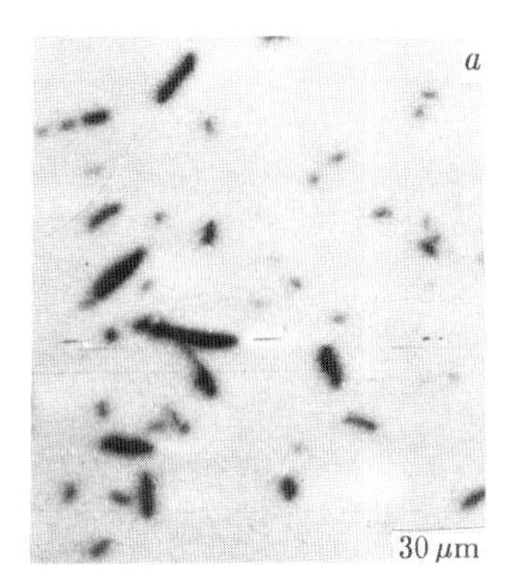

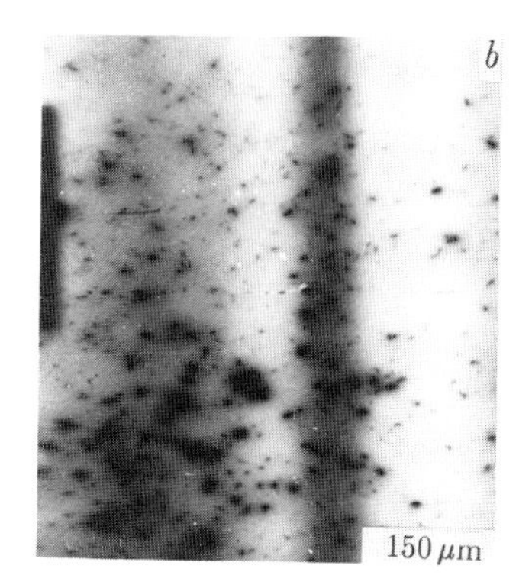

Figure 1. Typical EBIC pictures of impurity accumulations in the as-grown sample (*a*) and after the internal gettering process (*b*).

Experiments on EBIC and SE were carried out only in accordance with the second scheme.

3. Results

3.1. As-grown samples

The studied here initial wafers contained a standard for CZ Si:B set of LSIAs [7]. In these samples, so-called *cylindrical defects* (CDs) with the lengths from 15 to 40 μm and diameters from 5 to 10 μm were observed in the EBIC microphotographs (Fig. 1 (*a*)). The concentration of CDs was estimated as 10^6–$10^7\,\mathrm{cm}^{-3}$. In the scattering diagram (Fig. 2, curve 1) these defects correspond to the sections at $\theta < 7°$. So-called *spherical defects* (SDs), the concentration of which in initial samples did not exceed $10^5\,\mathrm{cm}^{-3}$, were also observed (Fig. 1 (*a*)). The dimensions of these defects were from 5 to 15 μm. These defects correspond to the sections at $\theta > 7°$ in the scattering diagram (Fig. 2, curve 1).

The depleted regions around RDs in initial samples mainly had the dimensions less than 4 μm and looked as a "plateau" in the diagrams (Fig. 3, curve 1, $\theta > 5°$). Sometimes the SDs were also observed (Fig. 3, curve 1, $\theta < 5°$).

3.2. After internal gettering

The main result for the crystals subjected to the internal gettering process are as follows:

1. As a result of the internal gettering, SDs with the dimensions from 10 to 30 μm (Fig. 1(*b*); Fig. 2, curve 2), became the predominating type of defects, and I_{sc} for these defects became by two orders of magnitude greater in comparison with that for the initial material. I_{sc} for CDs changed rather weakly and this led to prevalence of scattering by SDs in the scattering diagrams of the substrates subjected to the internal gettering. We can conclude that the concentration of SDs considerably increased as a consequence of the internal gettering (Fig. 1 (*b*)). The increase of I_{sc} and the SD concentration after the internal gettering was the common phenomenon for all the studied samples. This phenomenon did not depend on the gettering regime. The increase of SD-related I_{sc}

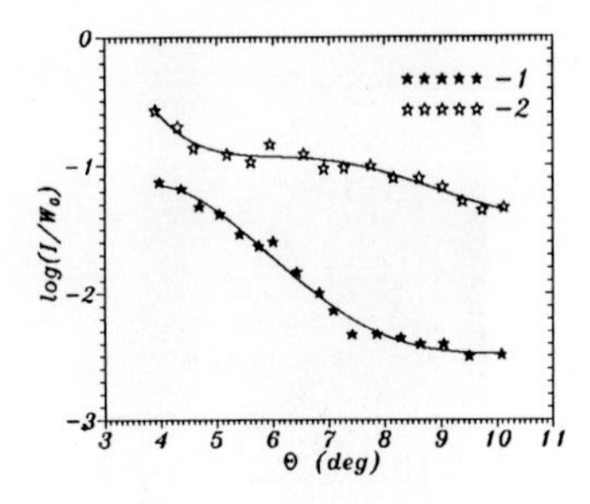

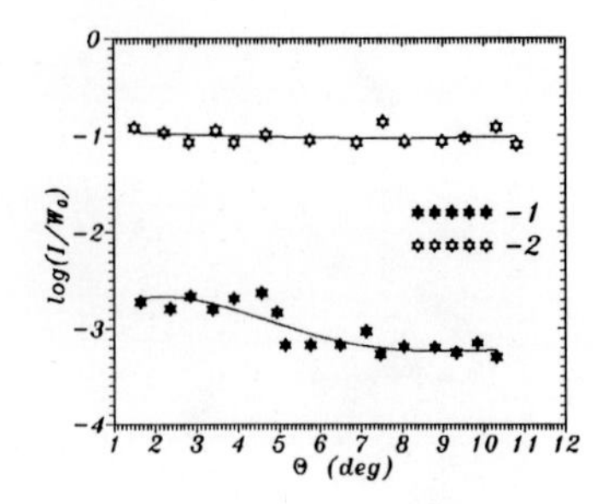

Figure 2. Characteristic light-scattering diagrams for CZ Si:B: (1) as-grown, (2) after internal gettering.

Figure 3. Characteristic light-scattering diagrams for RDs in CZ Si:B measured with photoexcitation: (1) as-grown, (2) after internal gettering.

correlated with the appearance of the gettering defects revealed by SE. We did not obtain a proportional dependance of SD-related I_{sc} on epd, however .

2. It was found that I_{sc} significantly increased after the internal gettering and a good correlation of I_{sc} with epd was observed (Fig. 3, curve 2).

3. The values of the activation energies (ΔE) of the centers predominating in SDs, perhaps, are defined by the growth conditions and the thermal prehistory of a sample. For example in Fig. 4, SD-related I_{sc} temperature dependances are shown for two samples grown at different establishments and subjected to the internal gettering. It is seen that these dependances and the values of ΔE are absolutely different for these samples— $\Delta E = 130$–170 meV for sample 1 and $\Delta E = 60$–90 meV for sample 2. So, different point centers constituted SDs in these samples after the internal gettering. Nowadays we have not got enough data to be sure what of the following factors defines the SD composition: the initial material parameters or the gettering process peculiarities. We think that the first factor is more important.

4. Discussion

On the basis of the above we assume that RDs in the substrates subjected to the internal gettering are defects of structure (most likely, they are precipitates and their colonies) which are formed in the wafer bulk during the gettering process, and are the gettering defects. As for SDs, they are the impurity atmospheres around these defects. These atmospheres are formed when impurities flow to the gettering defects [4]. These conclusions are confirmed by the correlation of RD-related I_{sc} with epd and the increase of SD-related I_{sc} after the appearance of a great number of defects revealed by SE.

For any inhomogeneity, $I_{sc} \sim C\Delta\varepsilon_{ac}^2$ where C is the defect concentration and $\Delta\varepsilon_{ac}$ is the deviation of the dielectric constant inside them. In the case of RDs, $\Delta\varepsilon_{ac}$ is

[4] It is important to note that impurity atmospheres consist at least of two components: the dissolved impurities and the impurity precipitates. In the conventional LALS experiments—without photoexcitation—only the first component is observed.

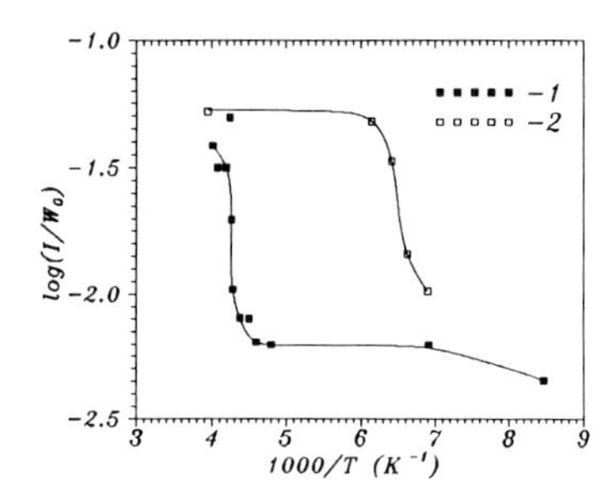

Figure 4. Temperature dependances of the light-scattering intensity for two different CZ Si:B samples after the internal gettering process.

determined first of all by the generated excess carrier concentration. At equal levels of photoexcitation RD-related $I_{sc} \sim C$ for the crystals with different C. So I_{sc} in this case is a straightforward measure of the RD concentration in crystals. In the case of impurity atmospheres, the situation is much more complicated. $\Delta\varepsilon_{ac}$ in them is controlled by many parameters: the average concentration of impurities in SDs, the ratio of the dissolved and precipitated impurity concentrations, the compensation degree, *etc.* The gettering process may change both C and any of these parameters which will result in the violation of the I_{sc} proportionality to C in experiments. Hence in the case of SDs, we are forced to rely only upon a qualitative correlation which is observed in the experiment [5].

On the basis of the above, we suggest the following method for the examination of the gettering process in silicon substrates.

The RD-related light scattering measurements (LALS with photoexcitation) enable the inspection for the presence and stability of the gettering defects in the substrate bulk.

The investigation of SD-related light scattering (conventional LALS measurements) enables the inspection the presence and dimensions of the impurity atmospheres around the gettering defects, i.e. the efficiency of the gettering operations.

Note that the main advantage of this technique is its applicability for the input and technological step inspection of substrates during the whole technological cycle. It enables the examination of stability and efficiency of the gettering process after any high-temperature operations. The equipment may be easily adapted for the technological process; it allows one to carry out the express (for 1 or 2 minutes) testing and mapping of substrates of any diameter.

The studies of I_{sc} temperature dependances enable the analysis of the impurity atmospheres composition, i.e. they allows one to determine what impurities are gathered by the gettering defects from a free zone. In this case it is reasonable to speak of a random inspection or laboratory research.

5. Conclusion

In conclusion, we would like to pay your attention upon a very promising potentiality.

[5] Now new techniques of scanning LALS (SLALS and OLALS) have been developed which enable the direct visualization of defects and accurate estimation of their parameters [8]. These techniques are very promising for the internal gettering investigation [9].

It is possible to carry out the nondestructive input and technological step inspection for the presence of RDs not only in a wafer bulk but directly in the boundary zone. The proposed technique is analogous to the method of RD revealing in a substrate bulk, but instead of the "bulk" excitation of the electron-hole pairs it is proposed to use the "surface" photoexcitation (e.g. using short pulses of the second harmonic of YAG:Nd^{3+}-laser, $\lambda = 0.53\,\mu m$). In this case, the non-equilibrium carriers will penetrate into the depth of 10–20-μm subsurface layer, and just this layer will be analyzed for the content of recombination-active defects of structure and precipitates. The main difficulty here is the small I_{sc} but the preliminary experiments demonstrated the possibility of registration of light scattering from layers with the RD concentration down to 10^4–$10^5\,cm^{-3}$. This technique is undoubtedly very promising since it will allow one to directly determine a degree of purification of the free zone rather than processes developing in the wafer bulk. For instance, it would enable the inspection of the effect of precipitates "germinating" in the free zone during the technological cycle. Note also that this technique is suitable not only for solving the problem of gettering but also for the examination of any epitaxial and boundary layers.

References

[1] Kalinushkin V P 1988 *Proc. Inst. Gen. Phys Acad Sci USSR* vol 4 *Laser Methods of Defect Investigations in Semiconductors and Dielectrics* (New York: Nova) pp 1–79

[2] Kalinushkin V P, Yuryev V A and Astafiev O V 1995 *Proc. 1st Int. Conf. on Materials for Microelectronics, Barcelona, 17–19 October 1994* (*Mater. Sci. Technol.*) in press

[3] Gulidov E N, Kalinushkin V P, Murin D I *et al* 1985 *Sov. Phys.–Microelectronics* **14** (2) 130–3

[4] Kalinushkin V P, Murin D I, Murina T M *et al* 1986 *Sov. Phys.–Microelectronics* **15** (6) 523–7

[5] Kalinushkin V P, Yuryev V A, Murin D I and Ploppa M G 1992 *Semicond. Sci. Technol.* **7** A255–62
Kalinushkin V P, Yuryev V A and Murin D I 1991 *Sov. Phys.–Semicond.* **25** (5) 798–806
Voronkov V V, Voronkova G I, Kalinushkin V P *et al* 1984 *Sov. Phys.–Semicond.* **18** (5) 938–40
Zabolotskiy S E, Kalinushkin V P, Murin D I *et al* 1987 *Sov. Phys.–Semicond.* **21** (8) 1364–8

[6] Buzynin A N, Butylkina N A, Lukyanov A E *et al* 1988 *Bul. Acad. Sci. USSR. Phys. Ser.* **52** (7) 1387–90

[7] Buzynin A N, Zabolotskiy S E, Kalinushkin V P *et al* 1990 *Sov. Phys.–Semicond.* **24** (2) 264–70
Astafiev O V, Buzunin A N, Buvaltsev A I *et al* 1994 *Sov. Phys.–Semicond.* **28** (3) 407–15

[8] Astafiev O V, Kalinushkin V P and Yuryev V A 1995 *Mater. Sci. Eng.* B **33** in press; Astafiev O V, Kalinushkin V P and Yuryev V A 1996 *Proc. 6th Int. Conf. on Defect Recognition and Image Processing in Semiconductors, Estes Park, Colorado, 3–6 December 1995*

[9] Astafiev O V, Kalinushkin V P and Yuryev V A 1995 *Proc. 1995 MRS Spring Meeting* vol 378 ed S Ashok, J Chevallier *et al* (Pittsburgh: Material Research Society) in press

Inst. Phys. Conf. Ser. No 149
Paper presented at DRIP 95

Growth studies at bulk III-Vs by image processing

J Donecker[1], G Hempel, J Kluge, M Seifert and B Lux

Institut für Kristallzüchtung,
Rudower Chaussee 6, D - 12489 Berlin, Germany

Abstract. The patterns of inhomogeneities in GaAs and InP are studied by scattering and diffraction of light. An adapted version of laser scattering tomography is used for observations with short exposure times and large fields. The information about the three-dimensional distribution of the scatterers in GaAs are evaluated by video travels through the crystal and images of intensities added in interesting directions. Near-infrared transmission and striation distance mapping act like special data compression techniques due to their optical principles.
In general, columnar extension of cellular patterns and striations could not be detected in s.i. GaAs. Long-range correlations exist for lineages and slip lines. The comparison with the behaviour of striations in doped InP cannot confirm the idea that cellular patterns in GaAs originate from constitutional supercooling during solidification.

1. Introduction

The inhomogeneities in III-V wafers are often remains due to the growth and annealing processes. Therefore any information about the inhomogeneities is welcome to evaluate and optimize growth and annealing technology. The paper deals with the two-and three-dimensional spatial distributions of inhomogeneities in regions with extensions of some millimeters. The enormous quantities of information in such volumes require appropriate representations and compressions of the data to discuss the results. This is done by video travels through representative regions of the crystal, processing of image sequences and appropriate optical methods.

Since 1976 the old technique of ultramicroscopy has been revived by introducing modern optical components and data processing. Meanwhile, a lot of insights about the capability of this so-called laser scattering tomography (LST) for the characterization of near-infrared transmissive semiconductors has been obtained [1, 2] and complete LST equipment is available. Recently the studies concentrate to high resolutions and sensitivities to investigate electronic circuits, too.

[1] E-mail: ur@ikz.FTA-Berlin.de

The capability of LST to give information about the three-dimensional distribution of scatterers [3] is a nearly exceptional feature in relation to other characterization methods which inform mostly about the properties of the existing surfaces or a thin adjacent region of the sample. The results of surface measurements concerning the bulk suffer from uncertainties about the third dimension. Especially in the case of growth studies the third dimension is imperative.

2. Experimentals

The s.i. GaAs crystals are grown in [001] direction in our institute by the LEC technique and cut into thick wafers with polished (001) surfaces. The wafers used here were bisected by cleaving. In such way wafers with at least two perpendicular optical surfaces, (001) and (110), are obtained. Wafers of commercial suppliers are measured for comparison.

Several sophisticated equipments are described for LST [4]. We abandon the scanning of a plane of the sample by a moving irradiated line. To get high repetition rates of images and large fields we used an intersecting plane of light for irradiation and for observation a silicon CCD target with good signal-to-noise ratio at the wavelength $1.064\mu m$ of the Nd:YAG laser (fig. 1). The laser beam was shaped planarly by an anamorphotic lens system. The thickness of the light plane usually amounts to $20\mu m$ inside the crystal. The dark current problem often solved by linear scanning is overcome by real-time subtraction of the dark image from the signal image. The intensity dependency of the signal was approximately linear. The images could be stored, accumulated, recursive filtered, recorded or printed in subsequent equipment.

In such way integration times below one second are achieved. This makes it convenient to scan the volume of the crystal by the light plane and to take video movies of this motion. Natural impressions of such travels through the crystal are obtained by recursive filtering.

Travelling distances of a few cm are possible before absorption requires integration times above one second. The microscope has to be translated simultaneously in the ratio $\frac{(n-1)}{n}$ to the path of the sample to conserve focusing at the position of the light

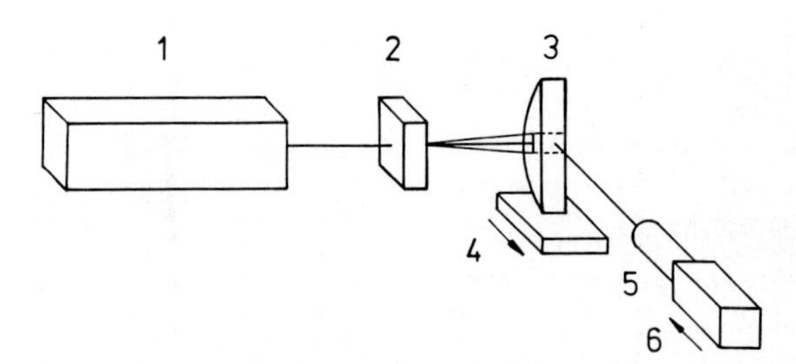

Figure 1. Experimental arrangement. 1 – Nd:YAG laser, 2 – anamorphotic beam shaper, 3 – sample (in "wafer view" position) at a translation stage – 4, translated zoom microscope (5) with CCD camera – 6.

plane in the GaAs sample with the refraction index of $n = 3.4$. The quality of the images decreases at great depths of the irradiated plane, because the objective of the microscope is not corrected for immersion in a thick high-refractive medium. A compromise has to be made between exposure times and optical resolution by reducing the numerical aperture of the objective.

Movements of the sample along certain crystallographic orientations could be used by appropriate translations of the sample relative to the zoom microscope used for imaging at the CCD target. Mainly, two directions of observation are used. Firstly, view at the (001) wafer surface with cleaved edge oriented vertically - called "wafer view" here. Fig. 1 shows this sample orientation. Further, view at the (110) longitudinal surface, obtained by cleaving the wafer diametrically - called "longitudinal view".

3. Experimental results and discussion

LST images like published ones [1, 2, 3, 4, 5] are observed in as-grown and in annealed samples showing precipitates threaded at dislocations arranged in cells. Fig. 2 shows precipitation filled and empty areas. The scattered intensities of filled and empty regions differ by more than two orders of magnitude in general. Therefore, the empty regions contain only particles with volumes below one tenth of that in filled regions, if we suppose scattering intensities to be proportional to the sixth power of the diameter. The well-introduced term "cells" seems to be a little bit incorrect, because the empty regions are not surrounded by skinlike layers in general. Sometimes, the cells look more like holes or cavities in an otherwise filled field.
Fig. 2 shows a couple of LST images taken in a distance of $250\mu m$. This distance corresponds to the mean radius of larger cells in "wafer view". The pattern differences in the image pair implies, the cells do not extent columnarly in growth direction.
An extensive possibility for the recognition of pattern development in growth direction is to look at quick-motion pictures of travels through the crystal. Sequences of "wafer view" images recorded in growth direction show convincingly the lack of columnar shapes

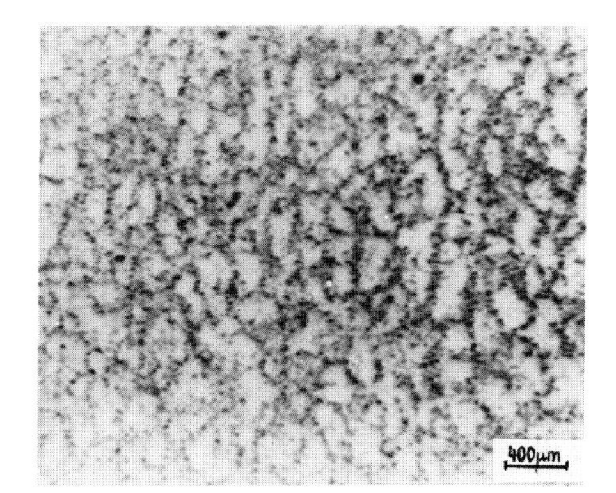

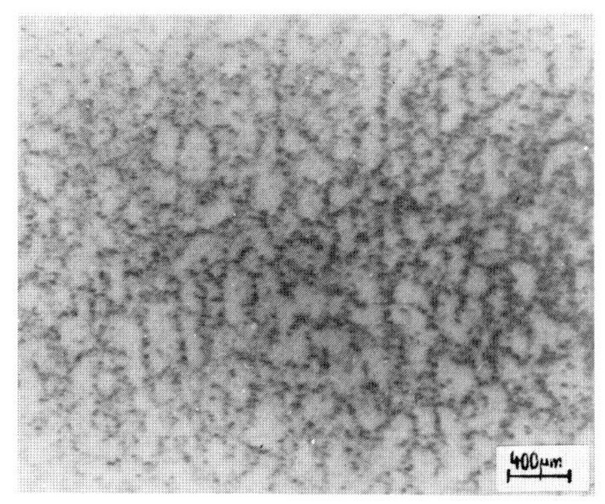

Figure 2. LST images at half radius of a s.i. GaAs wafer taken in a distance of $250\mu m$ in growth direction in "wafer view" arrangement near lineages.

of the cells. The cells change their positions in an arbitrary way without detectable correlation to the former position in the videos, aside the exceptions discussed below. A cell influences its next and next-nearest neighbours, but long range effects in preferred directions could not be seen.

Also "longitudinal views" do not confirm the existence of preferred orientations of cells. Especially, the direction of growth is not preferred. In "longitudinal view" images two phenomena are striking: slip lines and planes filled with precipitates. The latter correspond to lineages. They are extended in growth and in radial direction parallel to <110>. This direction is preferred crystallographically, too. Therefore, these observations are not a proof for grown-in columnar structures.

As fig. 2 shows, the cell patterns change by translation of the light plane. Already for shifts above $20\mu m$ significant changes of the patterns appear (measured with smaller thicknesses of the light plane then mentioned above). Doubts about the relevance of measurements of the etch pit density (e.p.d.) measured at the wafer surface to characterize the surrounding crystal are dissipated by the fact that a dislocation cannot terminate inside a crystal. Therefore, the mean e.p.d. seems to be conserved in subsequent planes in spite of different images of the cell patterns, if the surface intersects slip lines and dislocation loops are distributed randomly or irrelevant for the e.p.d. values.

Further, we used images of added intensities for each pixel during the travel through the crystal to reveal correlated structural elements. Fig. 3a shows a result for addition along 1.5 mm in the direction of growth ("wafer view"). Very faint structures are obtained similar to the structures in the LST images in fig. 2. The number of elongated cells is strongly increased, whereas small details are omitted due to the averaging effect. The operation of addition of the intensities should be improved by a quantitative treatment as used in [5] for single images to obtain images related to the volume of precipitates. In the present state the images demonstrate the existence of long-range correlations in parts of the crystal near <110> directions. Investigating the videos for "longitudinal view" we assured, the elongated objects in fig. 3a are precipitate filled planes with the

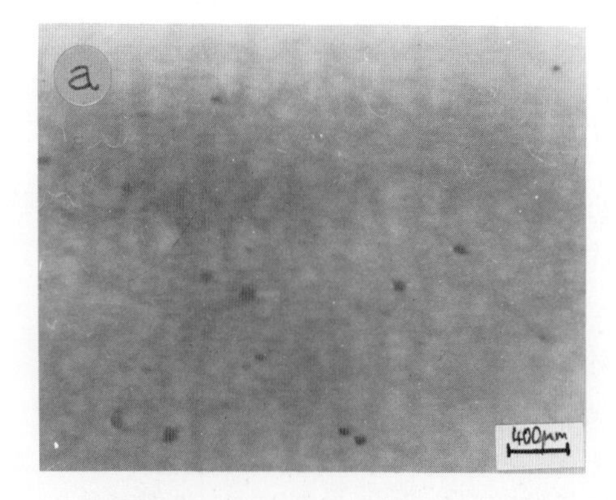

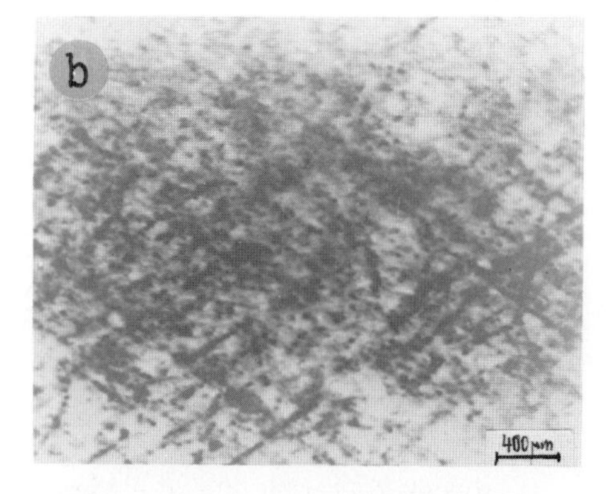

Figure 3. a: Image of added intensities of LST images in "wafer view" for a depth of 1.5 mm (at the position of the observed area of the images in fig. 2). **b:** Image of added intensities of LST images in "longitudinal view".

behaviour of the lineages at the wafer surface. The objects leading to the elongated cell images do not consist of inclined columnar cells. The added "longitudional view" images are determined by slip lines (fig. 3b).

Larger depths of integration could not be obtained because of the resolution of our digitizer, and we transferred our studies to near-infrared transmission. The objects responsible for NIR transmission differ from the precipitates, but coincidences are shown sometimes up to the same wavelength for maximal scattering and absorption [6].

It is well-known, that transmission images of thick samples do not display the real cellular structure [2]. We want to treat the transmission like a special type of image processing. To simplify the discussion, we approximate the cells by absorbing hollow spheres and assume linear superposition of transmission losses, because of the small absorption coefficient. If equal spheres are randomly stacked, the transmission image depends obviously on the transmitted thickness, or number of transmitted cells, respectively. The real cell diameter is displayed for transmitted thicknesses below the diameter of the smallest cell. For very large thicknesses, the image does not display the objects. It consists of noise. However, if the cells show some kind of order in the direction of light, then the transmission patterns reflect coincident walls of spheres in the direction of light. Cells with large diameters are preferred. The transmission image of a thick sample is low pass filtered for the spatial frequencies as by the modulation transfer function of an overstrained objective.

The critical point for observations in thick samples is to distinct between correlation and noise induced parts in the images, because noise from randomly superposed isotropic objects can create closed pattern elements like cells. In the NIR transmission shown in fig. 4ä (taken at the same position as figs. 2 and 3a) we see elongated patterns, besides the isotropic cells of doubtful origin. The elongated patterns correspond to the lineages

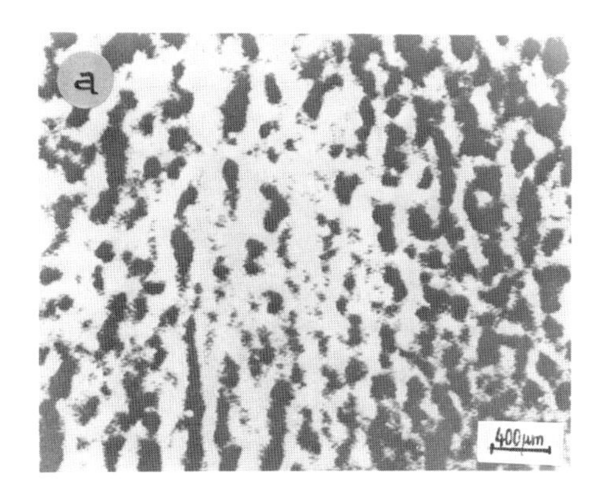

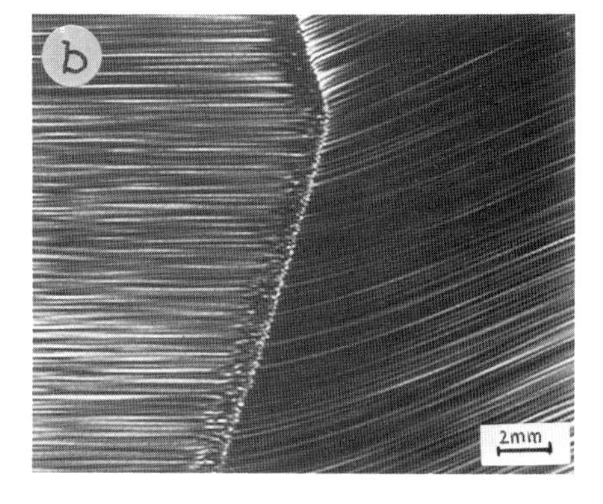

Figure 4. a: Near-infrared transmission at 1μm through the sample of a thickness of 4 mm at the same position of the observed area as in fig. 2.
b: S doped InP crystal imaged with the light of the first order of diffraction (striation distance mapping). Longitudional cut of the end of the crystal. Left edge of the picture corresponds to the axis of the crystal grown downward.(Original picture coloured).

treated above. We think, one cannot draw the conclusion of the existence of columnar structure from the appearance of cellular patterns in transmission without additional experiments.

Using diffraction of light, it is possible to obtain the information about the striation distances in parallel from the whole sample [7]. This corresponds to the compression of an enormous amount of data in one colour-coded picture. Fig. 4b shows a longitudinal cut of the end of an InP:S crystal in the reflected diffracted light as an example. The surface was etched by a DSL-type solution to get a doping related surface profile modulated by the striations.

The obvious lines in fig. 4b in horizontal direction represent groups of rotational striations with equal distances. They appear in the core of the crystal and outside. Although at the surface of GaAs striations are detected by etching, we did not observe recurring structures in the direction of striations in the "longitudional views" in LST, neither in the distances calculated for rotational striations nor in the much larger distances of possible striation groups.

The temperature and concentration conditions at the growth interface are reflected in the striations. Cellular structures due to constitutional supercooling are generated by the temperature and concentration conditions, too. Therefore, the grown-in cells have to be associated with the striations. We think, the lack of influences of striations on the cells has to be explained by post-growth development of the main parts of cellular structures. This agrees with the above observations of the isotropic shape of cells and the exclusion of columnar extension of cell patterns, in general. In the case of lineages we observed extensions in growth direction, but growth is not the only explanation of it. From the fact that the lineages in our crystals do not show any relicts of striations, we conclude, they are of post-growth origin, too.

Acknowledgments

The authors are indebted to Prof. J. Bohm, Prof. P. Rudolph and Dr. K. Böttcher for helpful discussions.

References

[1] Ogawa T 1988 *J. Crystal Growth* **88** 332–40

[2] Fillard J P 1988 *Rev. Phys. Appl.* **23** 765–77

[3] Todoroki S, Sakai K and Ogawa T 1990 *J. Crystal Growth* **103** 116–19

[4] Ogawa T and Nango N 1986 *Rev. Sci. Instrum.* **57** 1135–39

[5] Fawcett T J and Brozel M R 1990 *J. Crystal Growth* **103** 78–84

[6] Ogawa T and Kojima T 1987 *DRIP II (Elsevier) MS Monograph* **44** 207-14

[7] Donecker J 1994 *Inst. of Physics, Conf. Ser.*, **135** 311–14
Donecker J 1995 *J. Crystal Growth*, submitted

Inst. Phys. Conf. Ser. No 149
Paper presented at DRIP 95

Dislocation Lines and Walls in Vapor Phase Grown ZnSe Crystals Studied by Light Scattering Tomography

Minya MA and Tomoya OGAWA

Department of physics, Gakushuin University
Mejiro, Tokyo, 171, Japan
FAX +81-3-3590-2602

Abstract ZnSe crystals are very promising host materials of blue laser diodes (LD). Here, dislocations are studied by light scattering tomography (LST). Dislocation walls where dislocation lines were randomly piled in thin and flat regions of about 10 μm thick on some {111} and {110} planes in the crystals were found by LST.

1. Introduction

Defects in ZnSe crystals grown by a vapor phase were studied by light scattering tomography (LST) (Moriya and Ogawa, 1980, 1981), cathodoluminescence mode of scanning electron microscopy (CL-SEM) and molten KOH etching (Ma and Ogawa, 1995). The crystals purchased from Russia did not include any stacking faults or twinning planes, as confirmed by Laue spots of x-ray diffraction, while most ZnSe crystals have usually the twins and stacking faults. Although the reason for the piling up of dislocation lines within flat regions is not yet known, morphology of the walls was studied by layer-by-layer tomography (Todoroki et al., 1990).

2. Observation of dislocation lines and walls

Dislocation lines and walls observed at 200 μm intervals by layer-by-layer LST are respectively shown in fig.1 (a), (b), (c) and (d) where the two bold dotted lines in each photograph are images of the walls on {$\bar{1}$11} and the others are images of the lines. The walls appear to be constructed by many dislocation lines piled up within thin and flat regions since the LST images of the walls are composed of many dots along the lines. The dots are believed to correspond with dislocations belonging to the wall. Many similar walls were observed on {110} planes.

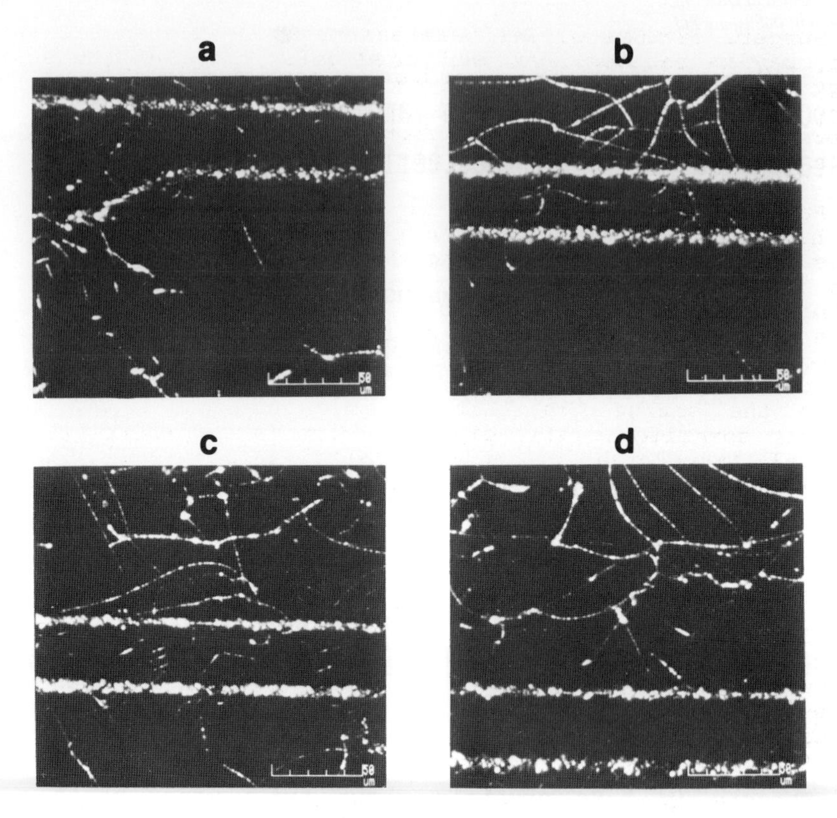

Fig.1
Light scattering tomographs successively taken by layer-by-layer method with 200 μm intervals along (a), (b) (c) and (d)

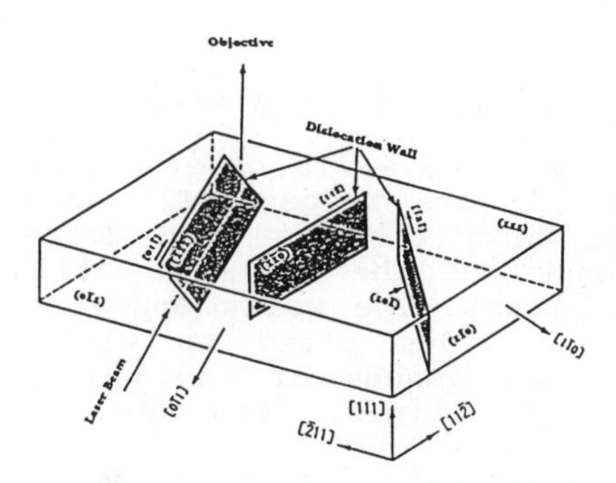

Fig.2
Schematic drawing of crystallographic orientations between dislocation walls, shape of a specimen and direction of incident laser beam for observation of those defects

Burgers vectors of dislocations on those planes are very short, may be shortest, as mentioned below, so that the self-energy of dislocations is minimum on the planes because the energy is proportional to the square of the magnitude of Burgers vector. This is the reason why the dislocation walls locate only on {111} and {110} planes.

Packing density of atoms on {111} and {110} planes is very high, may be highest, from the following reasons: {111} planes have the highest packing density of the same kind of atoms in a ZnSe crystal with the zincblende structure; the packing density on {110} planes is nearly double of {111} planes because {110} planes include equal amounts of Zn and Se atoms, which makes {110} planes electrically neutral. Therefore, a ZnSe crystal will be cleaved along {110} planes because the charge neutrality on the planes is the most important condition. {111} planes, however, are positively or negatively charged according to the planes composed of only Zn or only Se atoms, and dislocations slip along these {111} planes because the magnitude of Burgers vector is minimum on {111} planes. Here atomic species of dangling bonds of the slipping dislocations must coincide with those of the {111} planes.

Since some dislocation lines reacted among themselves, they constructed many Y shape nodes and pseudo-hexagonal nets as shown in fig.3. The Y nodes were not deformed nor did they move even if the specimen was fractured by an extremely large compressive force, which was confirmed by fragments of the fractured specimen.

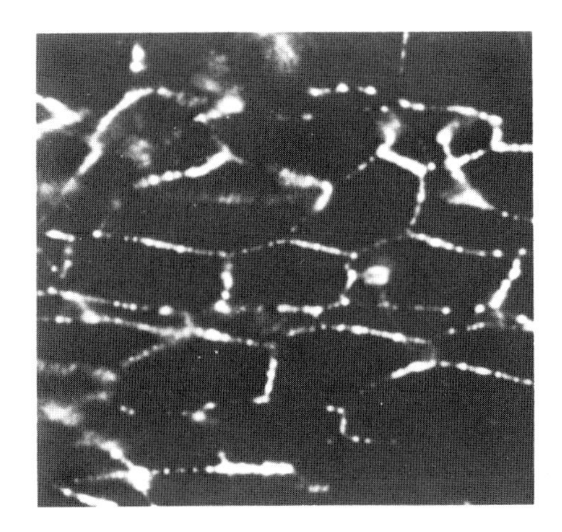

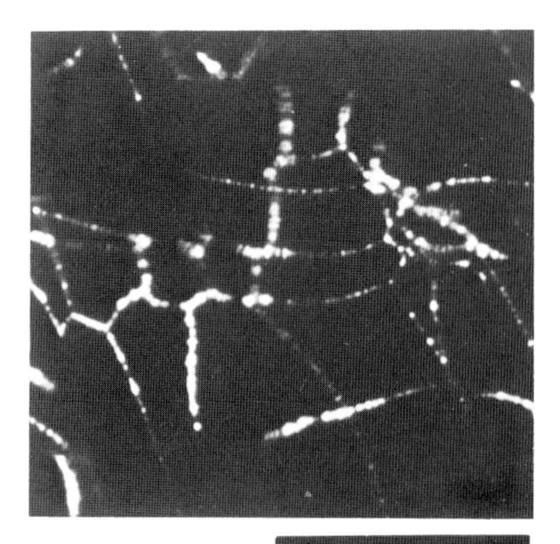

Fig.3
Dislocation nets and Y nodes observed by LST

CL-SEM images indicate clearly undecorated dislocations as well as scratch flaws which act as killer centers (fig.4). CL-SEM is a complemental method since most images observed optically here were caused by decorated defects. Of course, undecorated defects will be optically detected when the scattering vector is precisely adjusted to a characteristic plane of the fault to be studied, such as the plane determined by line and Burgers vector of an edge dislocation (Moriya and Ogawa, 1980, 1981), while the intensity scattered by the undecorated faults is fairly weak and also sensitive to the matching condition of the scattering vector for the characteristic direction.

Fig.4
CL-SEM image of a dislocation wall and slip lines
The bold straight line with many half arcs is the image of a dislocation wall; other straight lines clearly observed are, unfortunately, scratch lines due to the polishing; many parallel and vague lines are slip lines.

Since etching is one of the most orthodox methods to characterize dislocations, molten KOH etching was developed here to identify the nature of the lines inside the walls, where etching temperature and duration period were, respectively, $300^{o}C$ and about 3 sec.

The straight black line in fig.5 was generated due to cleavage of the specimen caused by thermal stresses, and the etch pits by the dislocation wall were an array of acute triangles, which were are generated because of inclination of

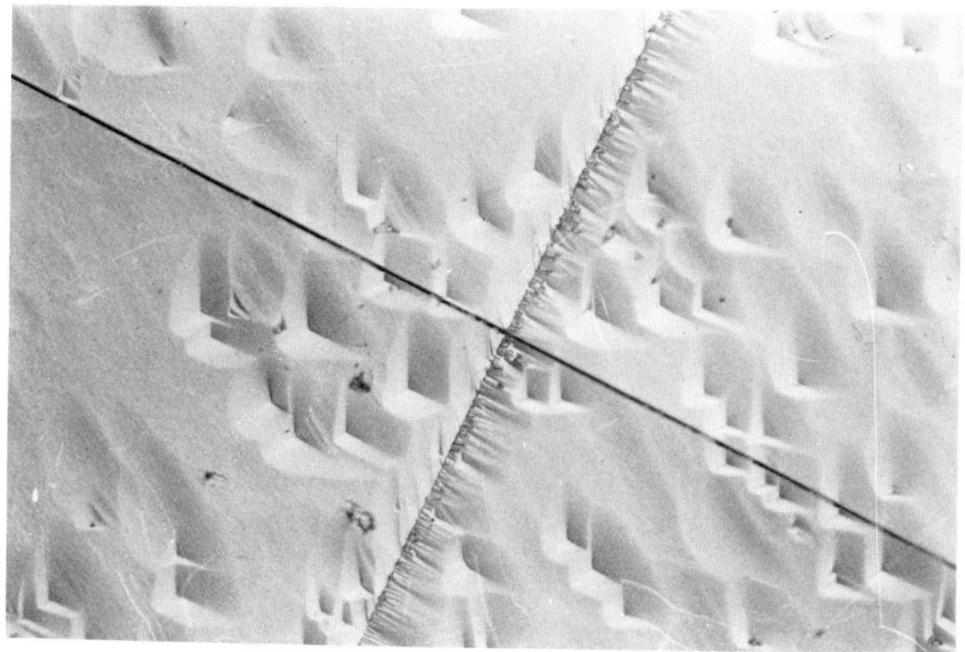

Fig.5
Pits caused by molten KOH etching at $300^{o}C$ for about 3 sec
Array of many acute triangles is caused by a dislocation wall and the sharp straight line is made by cleavage due to thermal stresses of high temperature etching

the dislocations belonging to the wall, since the wall locates on ($\bar{1}11$) while the surface of the specimen was (111) as shown in fig.2.

3. Conclusion

LST is a very useful method with which to study the morphology of defects in crystals.

The dislocation walls in ZnSe crystals observed here were a new type of dislocation morphology for the first in II-VI compound semiconductors.

4. References

Ma M and Ogawa T 1995 Phil. Mag. A. **72** 113.

MORIYA K and OGAWA T 1980 Phil. Mag. A **41** 191, 1981 ibid A 1085.

Todoroki S, Sakai K and Ogawa T 1990 J Crystal Growth **103** 116.

Inst. Phys. Conf. Ser. No 149
Paper presented at DRIP 95

Optical beam-induced scattering mode of mid-IR laser microscopy: a method for defect investigation in near-surface and near-interface regions of bulk semiconductors

O V Astafiev[1], V P Kalinushkin and V A Yuryev[2]

General Physics Institute of the Russian Academy of Sciences, 38, Vavilov Street, Moscow, GSP-1, 117942, Russia

Abstract. This paper presents a new technique of optical beam-induced scattering of mid-IR-laser radiation, which is a special mode of the recently developed scanning mid-IR-laser microscopy. The technique in its present form is designed for investigation of large-scale recombination-active defects in near-surface and near-interface regions of semiconductor wafers. However, it can be easily modified for the defect investigations in the crystal bulk. Being in many respects analogous to EBIC, the present technique has some indisputable advantages, which enable its application for both non-destructive laboratory investigations and quality monitoring in the industry.

1. Introduction

A method of low-angle mid-IR-light scattering (LALS) in combination with current carrier photoexcitation in the studied sample has been used for the investigation of large-scale centers of recombination and gluing in the bulk of semiconducting materials for many years [1]. This method has not been applied to the investigation of these defects in near-surface layers of semiconductors until recently, however, although it is evidently promising for both the investigation of near-surface and near-interface regions of semiconducting materials and the inspection of "working" layers of technological semiconductor wafers as well as the studies of large-scale recombination-active defects (LSRDs) directly in "working" layers of semiconductor substrates: in contrast to e.g. SEM in the EBIC mode, it requires practically no special preparation of substrate surfaces and consequently affects the physical properties of neither the studied near-surface layers nor the investigated defects[3]; the method allows one to investigate the interfaces of semiconductors as

[1] E-mail: ASTF@KAPELLA.GPI.RU.
[2] E-mail: VYURYEV@LDPM.GPI.RU.
[3] The only requirement imposed by this method upon a substrate — polishing on both sides, while a considerable part of technological substrates is polished on one side — seem to be removed by use of the methods of the laser heterodyne microscopy [2].

well as the surfaces covered with dielectric coatings (without coating removing, until the substrate is metallized) that, as far as we know, impossible to do using any of presently existing methods of material investigation and diagnostics. Nevertheless the first works demonstrating the applicability of LALS with surface photoexcitation to investigation of LSRDs in near-surface regions of semiconductors using Ge single crystals as an example were made by us very recently [3]. A method for visualization of LSRDs presented in the current paper — the optical beam-induced scanning low-angle mid-IR-light scattering technique (OLALS) — has become a direct logical development of these works [4]. A scanning dark-field mid-IR-laser microscope (scanning LALS or SLALS) recently proposed for the investigation of the large-scale electrically-active defect accumulations (LSDAs) in semiconductors [5–8] was applied in this technique as a basic instrument.

2. Experimental details

2.1. Optical diagram

An ideal optical diagram of the SLALS microscope in the OLALS mode (Fig. 1) gives a sufficiently clear idea of the instrument used in the present work. Note only the 55-mW He-Ne laser oscillating at the wavelength of 633 nm was applied as a light source for the "surface" photoexcitation of the studied samples, the modulated emission of which was focussed on the sample surface in the back focus of the lens *L1* in the spot with the dimensions of around 50 μm. The imperfection of the optical and mechanical parts of the laboratory prototype we used impeded us to reach the optimal focusing of the exciting beam [4]. Nonetheless even the available rather imperfect instrument allowed us to obtain the images of LSRDs in near-surface layers of silicon wafers and demonstrate the serviceability of the proposed technique.

10.6-μm CO_2-laser emission was used as a probe beam to produce a scattered wave. A liquid nitrogen cooled MCT photoresistor was used as an IR detector.

To obtain images of LSRDs we used the lock-in detection at the modulation frequency of the Ne-Ne laser emission (the probe CO_2-laser radiation was not modulated in the OLALS mode), while the lock-in detection at the frequency of the CO_2-laser light modulation was applied to obtain images of the same regions of the wafers without photoexcitation (in the basic mode of the SLALS microscope).

2.2. Samples

76-mm 381$\pm$15 μm thick wafers of single crystalline FZ n-Si:P grown in the <100> direction were used as experimental samples. Their specific resistivity ranged from 16 to 24 μm. The samples were chemical-mechanically polished on one side by the manufacturer ("Wacker"), the other side of the samples was mechanically polished up to the

[4] The microscope resolution in the OLALS mode is controlled by the effective dimensions of the domain, in which the non-equilibrium carrier exist, i.e. at the optimal focusing of the exciting beam the sizes of the light spot must be as small as possible and not exceed several μm. The effective dimensions of the non-equilibrium carrier domain — and the resolution — must be controlled only by the carrier lifetime, diffusion coefficient and the surface recombination velocity (we mean materials with μs lifetimes). In addition, the SLALS microscope itself imposes some restrictions on the sizes of the scattering domain — the domain of non-equilibrium carriers. Due to spacial filtering applied it depresses the images of too large objects [7, 8].

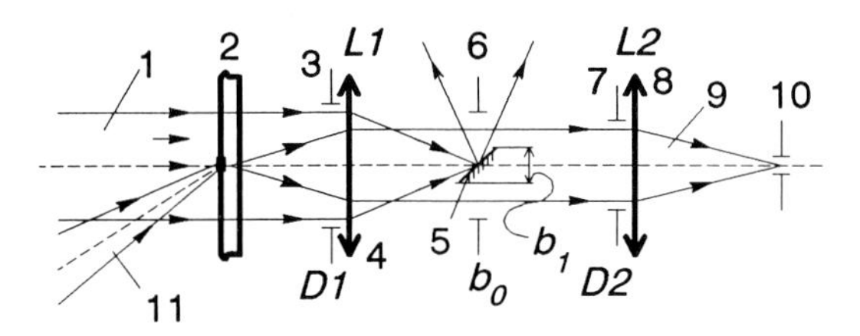

Figure 1. Optical diagram of the SLALS microscope: (1) probe mid-IR-laser beam, (2) sample, (3,6,7) diaphragms, (4,8) lenses, (9) scattered wave, (10) IR photodetector, (11) exciting laser beam (used in OLALS mode).

optical precision grade at GPI of RAS. In addition, we studied a sample of 76-mm CZ n-Si:P grown in the <100> direction (type KEF–4.5) on one side of which 1200 Å thick SiO_2 layer was produced by oxidation process (this sample was taken from the CCD matrix production cycle). The opposite side of this sample was polished mechanically up to the optical precision grade.

3. OLALS images

3.1. LSRDs in subsurface region

Fig. 2 (*a*) presents an OLALS image of the subsurface layer of chemical-mechanically polished side of one of the wafers. The same region scanned without photoexcitation (SLALS image) is given in Fig. 2 (*b*). The average signal level in Fig. 2 (*a*) is 5–7 times greater than that in Fig. 2 (*b*). The image contrast in Fig. 2 (*b*) — white spots — is determined by LSDAs in the crystal bulk (these defects were studied by LALS e.g. in [1, 9]), while the contrast in Fig. 2 (*a*) — dark objects — is caused by domains with the enhanced recombination rate in the near-surface layer of the wafer, i.e. LSRDs. It is seen that most of these LSRD-rich domains look like traces of scratches but no scratches were observed on this side of the wafer by conventional methods.

Fig. 2 (*c*) shows an OLALS image of the opposite — mechanically polished side — of the same wafer, and Fig. 2 (*d*) presents an image of the same region as Fig. 2 (*c*) in the SLALS mode. The average signal levels in Figs. 2 (*b*) and 2 (*d*) practically equal, while that in Fig. 2 (*c*) is 10–15 times lower than in Figs. 2 (*b*) and 2 (*d*). A chaotic conglomeration of small LSDAs is seen in Fig. 2 (*c*) in the damaged by mechanical polishing subsurface layer.

As it should be expected, absolutely different LSRD-rich domains were observed on the wafer sides subjected to different polish processes.

Note that similar inferences were previously made by us from the experiments on LALS with surface photoexcitation of Ge samples [3]. The difference of the data presented in Ref. [3] from the results of this work consists in the following: the big stripe-like LSRDs observed in the present work on the sample side subjected to chemical-mechanical polish cannot be revealed from the light-scattering diagrams in LALS. That is why only defects similar to impurity clouds [1, 9] were observed in Ref. [3] (as it was established in Ref. [3], the latter are also LSRDs). On the mechanically polished side of

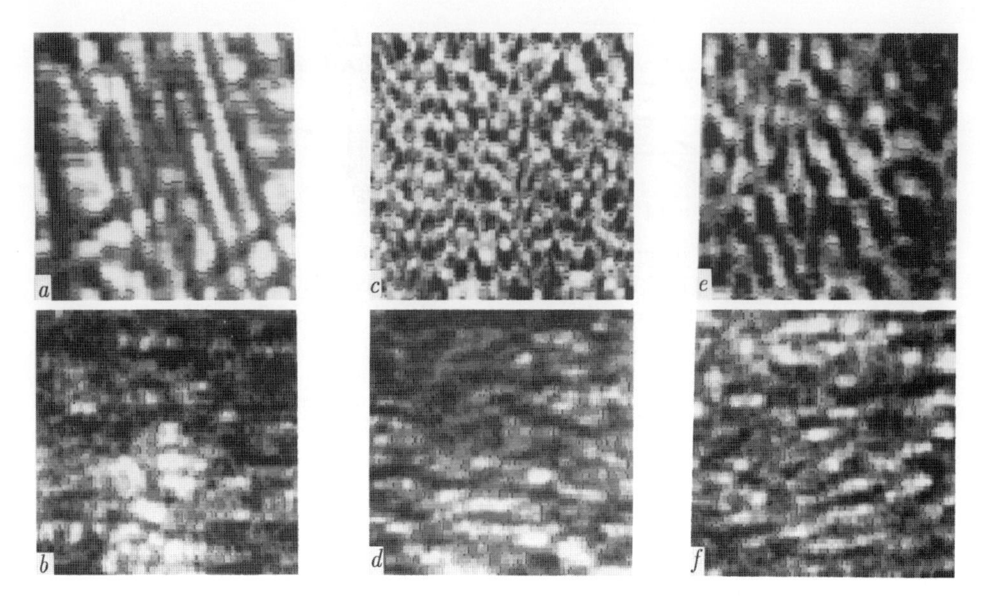

Figure 2. OLALS (*a,c,e*) and SLALS (*b,d,f*) images of defects in FZ Si:P (*a–d*) and CZ Si:P (*e,f*): chemical-mechanical polish (*a,b*), mechanical polish (*c,d*), under 1200 Å thick SiO_2 layer (*e,f*); $1\times1\,mm^2$.

the samples, like in the present work, rather intense light scatter by small LSRDs was registered in the near-surface layer [3].

Remark also that different schemes of elecron–hole pairs were used in the present work and in the work [3]. In the current work, as mentioned above, the non-equilibrium carriers in the near-surface layer were excited by the focussed light beam in a quasi-continuous regime, and a whole domain of the non-equilibrium carriers scattered light, while in [3], the photoexcitation was produced with a wide light beam in a pulse regime and a surface inhomogeneity in the distribution of non-equlibrium carrier concentration inside the sight spot caused by subsurface LSRDs was a source of the mid-IR-light scatter. In both cases, the light scatter intensity must be proportional to the square of the non-equilibrium carrier concentration and, in case of linear recombination, the square of the photoexcitation power. The real excitation power dependances will be discussed below.

3.2. LSRDs under SiO_2 layer

Fig. 2 (*e*) presents an OLALS image of the silicon wafer surface under the oxide layer. The dark spots in the image represent the domains with low non-equilibrium carrier lifetime [5]. As far as we know, such defects of a near-interface semiconductor layer directly under the oxide layer have never been observed thus far. It is clear that such a strong nonuniformity of recombination properties of the subinterface layer must disastrously affect the quality of devices, in particular CCD matrices, from the production cycle of which the studied sample was taken. It is obvious also that these defects arose due to wafer processing before the oxidation.

[5] The closely spaced parallel stripes are the noises brought in the picture by the scanner mechanics.

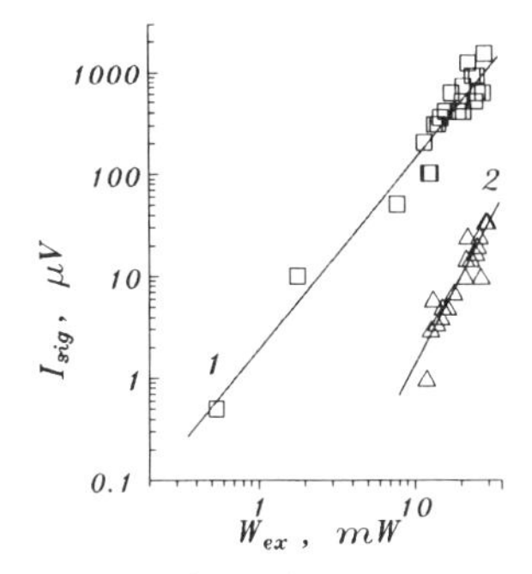

Figure 3. Dependances of IR-photodetector signal on the absorbed power of the exciting radiation: chemical-mechanical (1) and mechanical (2) polish, FZ Si:P wafer depicted in Fig. 2.

Fig. 2 (*f*) shows a SLALS image of the same region on the sample (the white spots are the images of LSDAs in the wafer bulk). The pattern in this picture is absolutely different from that in Fig. 2 (*e*) because different imperfections were revealed in each case.

3.3. OLALS signal dependence on photoexcitation power

Fig. 3 demonstrates the dependencies of the scattered CO_2-laser light intensity registered by the detector on the power of He-Ne laser radiation absorbed by the crystal for the FZ Si:P wafer sides subjected to chemical-mechanical (1) and mechanical polish processes, OLALS images of which are shown in Fig. 2. The square dependence obtained for chemical-mechanically polished side verifies the fact that this surface is imaged in the rays scattered by the non-equilibrium carrier domain and the recombination is linear [6]. The situation is much more sophisticated for the mechanically polished side, for which the cubic dependence was obtained. The similar dependence was previously obtained by us for the mechanically lapped surface of the Ge sample in Ref. [3]. As of now, we have got no perspicuous explanation for these dependances. Perhaps the cubic dependence is conditioned by the recombination processes in the vicinity of a huge amount of potential barriers in the damaged by mechanical polish subsurface layer of the samples [10]. It is obvious, though, that in this case the surface image is also formed in the rays scattered by the non-equilibrium carrier domain.

4. Conclusion

Summarizing the above we can conclude that this work presents an optical non-destructive technique which can be a powerful tool for the investigation of near-surface and near-interface layers of semiconducting materials. The described method is in some respects analogous in its physics to EBIC or OBIC but as distinct to the latter ones, it requires neither special preparation of a sample surface nor Schottky barrier or *p–n*

[6] It is generally known that the light scattering intensity is proportional to the square of the dielectric constant deviation in the scatterer.

junction. It enables the investigation of semiconductor interfaces and surfaces coated with dielectric layers and does not require to remove the coatings.

To demonstrate the method potentialities, the images of LSRDs situated in the vicinity of the chemical-mechanically and mechanically polished surfaces of single crystalline silicon were obtained without preliminary preparation of the surfaces. The images of LSRDs located near the surface, on which the SiO_2 layer was created by the industrial process of oxidation, were also obtained without removing the coating and surface processing. As far as we know, such images can be obtained by means of no presently used methods of microscopy. The defects observed on the Si crystal side subjected to chemical-mechanical polish and under the oxide layer, caused by the technological treatments of the wafers, undoubtedly must disastrously affect the serviceability of devices fabricated of such wafers.

We should remark in the conclusion that the technique described is easily adaptable for work directly in the technological line of semiconductor devices production and can serve as a promising tool e.g. for monitoring of the "working" layer quality. Some potential industrial applications of the technique are disused in Refs. [6, 8].

References

[1] Zubov B V, Kalinushkin V P, Krynetsky B B *et al* 1974 *JETP Letters* **20** (3) 167–71
Voronkov V V, Murina T M, Voronkova G I *et al* 1978 *Sov. Phys.–Solid State* **20** (5) 1365–8
Kalinushkin V P, Murin D I, Murina T M *et al* 1986 *Sov. Phys.–Microelectronics* **15** (6) 523–7
Kalinushkin V P 1988 *Proc. Inst. Gen. Phys Acad Sci USSR* vol 4 *Laser Methods of Defect Investigations in Semiconductors and Dielectrics* (New York: Nova) pp 1–79

[2] Protopopov V V and Ustinov N D 1985 *Laser Heterodyning* (Moscow: Nauka)

[3] Kalinushkin V P, Murin D I, Yuryev V A, Astafiev O V and Buvaltsev A I 1994 *Proc. SPIE* **2332** 146–53
Astafiev O V, Buvaltsev A I, Kalinushkin V P, Murin D I and Yuryev V A 1995 *Phys. Chem. Mech. Surf.* (4) 79–83

[4] Astafiev O V, Kalinushkin V P and Yuryev V A 1995 *Journ. Tech. Phys. Letters* **21** (11) 52–60

[5] Astafiev O V, Kalinushkin V P and Yuryev V A 1994 *Proc. SPIE* **2332** 138–45; 1995 *Mater. Sci. Eng.* B **33** in press; 1995 *Proc. 9th Int. Conf. on Microscopy of Semiconducting Materials, Oxford, 20–23 March 1995* (*IOP Conf. Ser.*) in press
Yuryev V A, Kalinushkin V P and Astafiev O V 1995 *Semiconductors* **29** (3) 455-8

[6] Astafiev O V, Kalinushkin V P, Yuryev V A *et al* 1995 *Proc. 1995 MRS Spring Meeting* vol 378 ed S Ashok, J Chevallier *et al* (Pittsburgh: Material Research Soc.) in press

[7] Astafiev O V, Kalinushkin V P and Yuryev V A 1996 *Microelectronics* **25** in press

[8] Kalinuhkin V P, Yuryev V A and Astafiev O V 1995 *Proc. 1st Int. Conf. on Materials for Microelectronics, Barcelona, 17–19 October 1994* (*Mater. Sci. Technol.* **11**) in press

[9] Voronkov V V, Voronkova G I, Golovina V N *et al* 1981 *J. Cryst. Growth* **52** 939–42
Zabolotsky S E, Kalinushkin V P, Murina T M, Ploppa M G and Tempelhoff K 1985 *Phys. Status Solidi* (a) **88** 539–42

[10] Zegrya G G and Kharchenko V A 1992 *JETP* **101** (1) 327–43

Index